1,001 Calculus
Practice Problems

FOR

DUMMIES

A Wiley Brand

1,001 Calculus Practice Problems

FOR DUMMIES®

A Wiley Brand

by PatrickJMT

1,001 Calculus Practice Problems For Dummies®

Published by: **John Wiley & Sons, Inc.,** 111 River St., Hoboken, NJ 07030-5774, www.wiley.com

For general information on our other products and services, please contact our Customer Care Department within the U.S. at 877-762-2974, outside the U.S. at 317-572-3993, or fax 317-572-4002. For technical support, please visit www.wiley.com/techsupport.

Wiley publishes in a variety of print and electronic formats and by print-on-demand. Some material included with standard print versions of this book may not be included in e-books or in print-on-demand. If this book refers to media such as a CD or DVD that is not included in the version you purchased, you may download this material at http://booksupport.wiley.com. For more information about Wiley products, visit www.wiley.com.

Library of Congress Control Number: 2013954232

ISBN 978-1-118-49671-8 (pbk); ISBN 978-1-118-49670-1 (ebk); ISBN 978-1-118-49673-2 (ebk)

Manufactured in the United States of America

V10013167_081919

Contents at a Glance

Table of Contents

Introduction

T his book is intended for a variety of calculus students. Perhaps you want a supplement to your current calculus class or you're looking to brush up on a course you took long ago. Or maybe you're teaching yourself and need a comprehensive book of extra practice problems.

The 1,001 questions in this book cover calculus concepts that a high school student would encounter in a calculus course in preparation for the AP exam. It also covers most of the concepts that a calculus student could expect to see in the first two semesters of a three-semester calculus course. The types of questions are questions that I regularly assigned when teaching both as homework questions or are questions that a student could've expected to see on a quiz or test.

Jump around the book as you like. You can find a robust algebra and trigonometry review at the beginning of the book to make sure that you're prepared for calculus. The number-one reason students have difficulty in calculus is not calculus itself but having a weak background in algebra and trigonometry. If you're rusty on the fundamentals, spend time on those first two chapters before jumping into the rest of the text!

As with many things worth doing in life, there's no shortcut to becoming proficient in mathematics. However, by practicing the problems in this book, you'll be on your way to becoming a much stronger calculus student.

What You'll Find

The 1,001 calculus practice problems in the book are divided into 15 chapters, with each chapter providing practice of the mechanical side of calculus or of applications of calculus. Some of the questions have a diagram or graph that you need in order to answer the question.

The end of the book provides thorough and detailed solutions to all the problems. If you get an answer wrong, try again before reading the solution! Knowing what *not* to do is often a great starting point in discovering the correct approach, so don't worry if you don't immediately solve each question; some problems can be quite challenging.

Beyond the Book

This book provides a lot of calculus practice. If you'd also like to track your progress online, you're in luck! Your book purchase comes with a free one-year subscription to all 1,001 practice questions online. You can access the content whenever you want. Create your own question sets and view personalized reports that show what you need to study most.

What you'll find online

The online practice that comes free with the book contains the same 1,001 questions and answers that are available in the text. You can customize your online practice to focus on specific areas, or you can select a broad variety of topics to work on — it's up to you. The online program keeps track of the questions you get right and wrong so you can easily monitor your progress.

This product also comes with an online Cheat Sheet that helps you increase your odds of performing well in calculus. Check out the free Cheat Sheet at www.dummies.com/cheatsheet/1001calculus. (No PIN required. You can access this info before you even register.)

How to register

To gain access to the online version of all 1,001 practice questions in this book, all you have to do is register. Just follow these simple steps:

1. **Find your PIN access code.**

   • **Print book users:** If you purchased a hard copy of this book, turn to the inside of the front cover of this book to find your access code.

   • **E-book users:** If you purchased this book as an e-book, you can get your access code by registering your e-book at dummies.com/go/getaccess. Go to this website, find your book and click it, and answer the security question to verify your purchase. Then you'll receive an e-mail with your access code.

2. **Go to learn.dummies.com and click Already have an Access Code?**

3. **Enter your access code and click Next.**

4. **Follow the instructions to create an account and establish your personal login information.**

That's all there is to it! You can come back to the online program again and again — simply log in with the username and password you chose during your initial login. No need to use the access code a second time.

If you have trouble with the access code or can't find it, please contact Wiley Product Technical Support at 877-762-2974 or http://support.wiley.com.

Your registration is good for one year from the day you activate your access code. After that time frame has passed, you can renew your registration for a fee. The website gives you all the important details about how to do so.

Where to Go for Additional Help

Calculus is hard, so don't become overwhelmed if a particular topic isn't immediately easy to you. This book has many practice problems of varying difficulty, so you can focus on those problems that are most appropriate for you.

Practice Problems For Dummies

In addition to getting help from your friends, teachers, or coworkers, you can find a variety of great materials online. If you have internet access, a simple search often turns up a treasure trove of information. You can also head to www.dummies.com to see the many articles and books that can help you in your studies.

1,001 Calculus Practice Problems For Dummies gives you just that — 1,001 practice questions and answers in order for you to practice your calculus skills. If you need more in-depth study and direction for your calculus courses, you may want to try out the following *For Dummies* products (or their companion workbooks):

- ✔ *Calculus For Dummies:* This book provides instruction parallel to the 1,001 calculus practice problems found here.

- ✔ *Calculus II For Dummies:* This book provides content similar to what you may encounter in a second-semester college calculus course.

- ✔ *Pre-Calculus For Dummies:* Use this book to brush up on the foundational skills and concepts you need for calculus — solving polynomials, graphing functions, using trig identities, and the like.

- ✔ *Trigonometry For Dummies:* Try this book if you need a refresher on trigonometry.

Part I
The Questions

 Visit www.dummies.com for free access to great *For Dummies* content online.

In this part . . .

*T*he only way to become proficient in math is through a lot of practice. Fortunately, you have now 1,001 practice opportunities right in front of you. These questions cover a variety of calculus-related concepts and range in difficulty from easy to hard. Master these problems, and you'll be well on your way to a very solid calculus foundation.

Here are the types of problems that you can expect to see:

- Algebra review (Chapter 1)
- Trigonometry review (Chapter 2)
- Limits and continuity (Chapter 3)
- Derivative fundamentals (Chapters 4 through 7)
- Applications of derivatives (Chapter 8)
- Antiderivative basics (Chapters 9 and 10)
- Applications of antiderivatives (Chapter 11)
- Antiderivatives of other common functions and L'Hôpital's rule (Chapter 12)
- More integration techniques (Chapters 13 and 14)
- Improper integrals, the trapezoid rule, and Simpson's rule (Chapter 15)

Chapter 1

Algebra Review

· ·

Performing well in calculus is impossible without a solid algebra foundation. Many calculus problems that you encounter involve a calculus concept but then require many, many steps of algebraic simplification. Having a strong algebra background will allow you to focus on the calculus concepts and not get lost in the mechanical manipulation that's required to solve the problem.

The Problems You'll Work On

In this chapter, you see a variety of algebra problems:

- ✔ Simplifying exponents and radicals
- ✔ Finding the inverse of a function
- ✔ Understanding and transforming graphs of common functions
- ✔ Finding the domain and range of a function using a graph
- ✔ Combining and simplifying polynomial expressions

What to Watch Out For

Don't let common mistakes trip you up. Some of the following suggestions may be helpful:

- ✔ Be careful when using properties of exponents. For example, when multiplying like bases, you add the exponents, and when dividing like bases, you subtract the exponents.
- ✔ Factor thoroughly in order to simplify expressions.
- ✔ Check your solutions for equations and inequalities if you're unsure of your answer. Some solutions may be extraneous!
- ✔ It's easy to forget some algebra techniques, so don't worry if you don't remember everything! Review, review, review.

Simplifying Fractions

1–13 Simplify the given fractions by adding, subtracting, multiplying, and/or dividing.

1. $\dfrac{1}{2} + \dfrac{3}{4} - \dfrac{5}{6}$

2. $\dfrac{11}{3} + \dfrac{3}{5} + \dfrac{17}{20}$

3. $\left(\dfrac{2}{3}\right)(21)\left(\dfrac{11}{14}\right)$

4. $\dfrac{5/6}{15/22}$

5. $\dfrac{5}{yx} + \dfrac{7}{x^2} + \dfrac{10x}{y}$

6. $\dfrac{x}{x-1} - \dfrac{x-4}{x+1}$

7. $\left(\dfrac{x^2-1}{xy^2}\right)\left(\dfrac{y^3}{x+1}\right)$

8. $\dfrac{\dfrac{x^2-5x+6}{6xy^3}}{\dfrac{x^2+3x-10}{10x^3y^2}}$

9. $\dfrac{x^3 y^4 z^5}{y^2 z^{-8}}$

10. $\dfrac{(x+3)^2 x^4 (y+5)^{14}}{x^6 (x+3)(y+5)^{17}}$

11. $\left(\dfrac{4x^2 y^{100} z^{-3}}{18x^{15} y^4 z^8}\right)^0$

12. $\dfrac{x^4 y^3 z^2 + x^2 y z^4}{x^2 y z}$

13. $\dfrac{\left(x^2 y^3\right)^4 \left(y^4\right)^0 z^{-2}}{\left(xz^2\right)^3 y^{-5}}$

Simplifying Radicals

14–18 Simplify the given radicals. Assume all variables are positive.

14. $\sqrt{50}$

15. $\dfrac{\sqrt{8}\sqrt{20}}{\sqrt{50}\sqrt{12}}$

16. $\sqrt{20x^4 y^6 z^{11}} \sqrt{5xy^2 z^7}$

17. $\sqrt[3]{x^4 y^8 z^5}\, \sqrt[3]{x^7 y^4 z^{10}}$

18. $\dfrac{\sqrt[3]{8x^3 y^6}\,\sqrt[5]{x^{10} y^{14}}}{\sqrt[7]{x^{14} y^{14}}}$

Writing Exponents Using Radical Notation

19–20 *Convert between exponential and radical notation.*

19. Convert $4^{1/3} x^{3/8} y^{1/4} z^{5/12}$ to radical notation. (**Note:** The final answer can have more than one radical sign.)

20. Convert $\sqrt[3]{4x^2 y}\,\sqrt[5]{z^4}$ to exponential notation.

The Horizontal Line Test

21–23 *Use the horizontal line test to identify one-to-one functions.*

21. Use the horizontal line test to determine which of the following functions is a one-to-one function and therefore has an inverse.

(A) $y = x^2 + 4x + 6$

(B) $y = |2x| - 1$

(C) $y = \dfrac{1}{x^2}$

(D) $y = 3x + 8$

(E) $y = \sqrt{25 - x^2}$

22. Use the horizontal line test to determine which of the following functions is a one-to-one function and therefore has an inverse.

(A) $y = x^2 - 4$

(B) $y = x^2 - 4, x \geq 0$

(C) $y = x^2 - 4, -2 \leq x \leq 8$

(D) $y = x^2 - 4, -12 \leq x \leq 6$

(E) $y = x^2 - 4, -5.3 \leq x \leq 0.1$

23. Use the horizontal line test to determine which of the following functions is a one-to-one function and therefore has an inverse.

(A) $y = x^4 + 3x^2 - 7$

(B) $y = 4|x| + 3$

(C) $y = \cos x$

(D) $y = \sin x$

(E) $y = \tan^{-1} x$

Find Inverses Algebraically

24–29 *Find the inverse of the one-to-one function algebraically.*

24. $f(x) = 4 - 5x$

25. $f(x) = x^2 - 4x, x \geq 2$

26. $f(x) = \sqrt{8 - 5x}$

27. $f(x) = 3x^5 + 7$

28. $f(x) = \dfrac{2 - \sqrt{x}}{2 + \sqrt{x}}$

29. $f(x) = \dfrac{2x - 1}{x + 4}$

The Domain and Range of a Function and Its Inverse

30–32 Solve the given question related to a function and its inverse.

30. The set of points $\{(0, 1), (3, 4), (5, -6)\}$ is on the graph of $f(x)$, which is a one-to-one function. Which points belong to the graph of $f^{-1}(x)$?

31. $f(x)$ is a one-to-one function with domain $[-2, 4)$ and range $(-1, 2)$. What are the domain and range of $f^{-1}(x)$?

32. Suppose that $f(x)$ is a one-to-one function. What is an expression for the inverse of $g(x) = f(x + c)$?

Linear Equations

33–37 Solve the given linear equation.

33. $3x + 7 = 13$

34. $2(x + 1) = 3(x + 2)$

35. $-4(x + 1) - 2x = 7x + 3(x - 8)$

36. $\dfrac{5}{3}x + 5 = \dfrac{1}{3}x + 10$

37. $\sqrt{2}\left(x + 3\right) = \sqrt{5}\left(x + \sqrt{20}\right)$

Quadratic Equations

38–43 Solve the quadratic equation.

38. Solve $x^2 - 4x - 21 = 0$.

39. Solve $x^2 + 8x - 17 = 0$ by completing the square.

40. Solve $2x^2 + 3x - 4 = 0$ by completing the square.

41. Solve $6x^2 + 5x - 4 = 0$.

42. Solve $3x^2 + 4x - 2 = 0$.

43. Solve $x^{10} + 7x^5 + 10 = 0$.

Solving Polynomial Equations by Factoring

44–47 Solve the polynomial equation by factoring.

44. $3x^4 + 2x^3 - 5x^2 = 0$

45. $x^8 + 12x^4 + 35 = 0$

46. $x^4 + 3x^2 - 4 = 0$

47. $x^4 - 81 = 0$

Absolute Value Equations

48–51 Solve the given absolute value equation.

48. $|5x - 7| = 2$

49. $|4x - 5| + 18 = 13$

50. $|x^2 - 6x| = 27$

51. $|15x - 5| = |35 - 5x|$

Solving Rational Equations

52–55 Solve the given rational equation.

52. $\dfrac{x+1}{x-4} = 0$

53. $\dfrac{1}{x+2} + \dfrac{1}{x} = 1$

54. $\dfrac{x+5}{x+2} = \dfrac{x-4}{x-10}$

55. $\dfrac{1}{x-2} - \dfrac{2}{x-3} = \dfrac{-1}{x^2 - 5x + 6}$

Polynomial and Rational Inequalities

56–59 Solve the given polynomial or rational inequality.

56. $x^2 - 4x - 32 < 0$

57. $2x^4 + 2x^3 \geq 12x^2$

58. $\dfrac{(x+1)(x-2)}{(x+3)} < 0$

59. $\dfrac{1}{x-1} + \dfrac{1}{x+1} > \dfrac{3}{4}$

Absolute Value Inequalities

60–62 Solve the absolute value inequality.

60. $|2x - 1| < 4$

61. $|5x - 7| > 2$

62. $|-3x + 1| \leq 5$

Graphing Common Functions

63–77 Solve the given question related to graphing common functions.

63. What is the slope of the line that goes through the points (1, 2) and (5, 9)?

64. What is the equation of the line that has a slope of 4 and goes through the point (0, 5)?

65. What is the equation of the line that goes through the points (–2, 3) and (4, 8)?

66. Find the equation of the line that goes through the point (1, 5) and is parallel to the line $y = \dfrac{3}{4}x + 8$.

67. Find the equation of the line that goes through the point (3, –4) and is perpendicular to the line that goes through the points (3, –4) and (–6, 2).

68. What is the equation of the graph of $y = \sqrt{x}$ after you stretch it vertically by a factor of 2, shift the graph 3 units to the right, and then shift it 4 units upward?

69. Find the vertex form of the parabola that passes through the point (0, 2) and has a vertex at (–2, –4).

70. Find the vertex form of the parabola that passes through the point (1, 2) and has a vertex at (–1, 6).

71. A parabola has the vertex form $y = 3(x + 1)^2 + 4$. What is the vertex form of this parabola if it's shifted 6 units to the right and 2 units down?

72. What is the equation of the graph of $y = e^x$ after you compress the graph horizontally by a factor of 2, reflect it across the y-axis, and shift it down 5 units?

73. What is the equation of the graph of $y = |x|$ after you stretch the graph horizontally by a factor of 5, reflect it across the x-axis, and shift it up 3 units?

74. Find the equation of the third-degree polynomial that goes through the points (–4, 0), (–2, 0), (0, 3), and (1, 0).

75. Find the equation of the fourth-degree polynomial that goes through the point (1, 4) and has the roots –1, 2, and 3, where 3 is a repeated root.

76. A parabola crosses the x-axis at the points (–4, 0) and (6, 0). If the point (0, 8) is on the parabola, what is the equation of the parabola?

77. A parabola crosses the x-axis at the points (–8, 0) and (–2, 0), and the point (–4, –12) is on the parabola. What is the equation of the parabola?

Domain and Range from a Graph

78–80 Find the domain and range of the function with the given graph.

78.

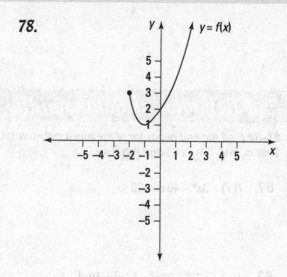

79.

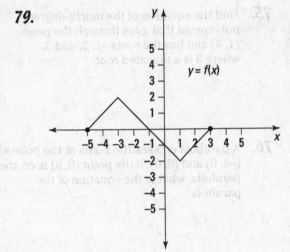

80.

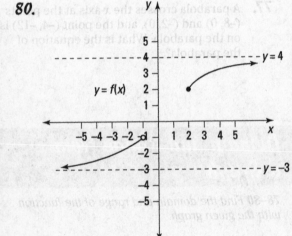

End Behavior of Polynomials

81–82 Find the end behavior of the given polynomial. That is, find $\lim\limits_{x \to -\infty} f(x)$ and $\lim\limits_{x \to \infty} f(x)$.

81. $f(x) = 3x^6 - 40x^5 + 33$

82. $f(x) = -7x^9 + 33x^8 - 51x^7 + 19x^4 - 1$

Adding Polynomials

83–87 Add the given polynomials.

83. $(5x + 6) + (-2x + 6)$

84. $(2x^2 - x + 7) + (-2x^2 + 4x - 9)$

85. $(x^3 - 5x^2 + 6) + (4x^2 + 2x + 8)$

86. $(3x + x^4 + 2) + (-3x^4 + 6)$

87. $(x^4 - 6x^2 + 3) + (5x^3 + 3x^2 - 3)$

Subtracting Polynomials

88–92 Subtract the given polynomials.

88. $(5x - 3) - (2x + 4)$

89. $(x^2 - 3x + 1) - (-5x^2 + 2x - 4)$

90. $(8x^3 + 5x^2 - 3x + 2) - (4x^3 + 5x - 12)$

91. $(x + 3) - (x^2 + 3x - 4) - (-3x^2 - 5x + 6)$

92. $(10x^4 - 6x^3 + x^2 + 6) - (x^3 + 10x^2 + 8x - 4)$

Multiplying Polynomials

93–97 Multiply the given polynomials.

93. $5x^2(x - 3)$

94. $(x + 4)(3x - 5)$

95. $(x - y + 6)(xy)$

96. $(2x - 1)(x^2 - x + 4)$

97. $-x(x^4 + 3x^2 + 2)(x + 3)$

Long Division of Polynomials

98–102 Use polynomial long division to divide.

98. $\dfrac{x^2 + 4x + 6}{x - 2}$

99. $\dfrac{2x^2 - 3x + 8}{x + 4}$

100. $\dfrac{x^3 - 2x + 6}{x - 3}$

101. $\dfrac{3x^5 + 4x^4 - x^2 + 1}{x^2 + 5}$

102. $\dfrac{3x^6 - 2x^5 - x^4 + x^3 + 2}{x^3 + 2x^2 + 4}$

Chapter 2
Trigonometry Review

• •

In addition to having a strong algebra background, you need a strong trigonometry skill set for calculus. You want to know the graphs of the trigonometric functions and to be able to evaluate trigonometric functions quickly. Many calculus problems require one or more trigonometric identities, so make sure you have more than a few of them memorized or at least can derive them quickly.

The Problems You'll Work On

In this chapter, you solve a variety of fundamental trigonometric problems that cover topics such as the following:

- ✔ Understanding the trigonometric functions in relation to right triangles
- ✔ Finding degree and radian measure
- ✔ Finding angles on the unit circle
- ✔ Proving identities
- ✔ Finding the amplitude, period, and phase shift of a periodic function
- ✔ Working with inverse trigonometric functions
- ✔ Solving trigonometric equations with and without using inverses

What to Watch Out For

Remember the following when working on the trigonometry review questions:

- ✔ Being able to evaluate the trigonometric functions at common angles is very important since they appear often in problems. Having them memorized will be extremely useful!

- ✔ Watch out when solving equations using inverse trigonometric functions. Calculators give only a single solution to the equation, but the equation may have many more (sometimes infinitely many solutions), depending on the given interval. Thinking about solutions on the unit circle is often a good way to visualize the other solutions.

- ✔ Although you may be most familiar with using degrees to measure angles, radians are used almost exclusively in calculus, so learn to love radian measure.

- ✔ Memorizing many trigonometric identities is a good idea because they appear often in calculus problems.

Basic Trigonometry

103–104 *Evaluate* sin θ, cos θ, *and* tan θ *for the given right triangle. Remember to rationalize denominators that contain radicals.*

103.

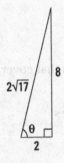

$\sqrt{65}$ 7

θ

4

104.

8

$2\sqrt{17}$

θ

2

105–108 *Evaluate the trig function. Remember to rationalize denominators that contain radicals.*

105. Given $\sin\theta = \dfrac{3}{7}$, where $\dfrac{\pi}{2} < \theta < \pi$, find $\cot\theta$.

106. Given $\cos\theta = \dfrac{3}{4}$, where $\dfrac{3\pi}{2} < \theta < 2\pi$, find $\csc\theta$.

107. Given $\tan\theta = -\dfrac{8}{5}$, where $\sin\theta > 0$ and $\cos\theta < 0$, find $\sin(2\theta)$.

108. Given $\cot\theta = \dfrac{-9}{2}$, where $\sin\theta < 0$, find $\cos(2\theta)$.

Converting Degree Measure to Radian Measure

109–112 *Convert the given degree measure to radian measure.*

109. 135°

110. –280°

111. 36°

112. –315°

Converting Radian Measure to Degree Measure

113–116 Convert the given radian measure to degree measure.

113. $\frac{7\pi}{6}$ rad

114. $\frac{11\pi}{12}$ rad

115. $\frac{-3\pi}{5}$ rad

116. $\frac{-7\pi}{2}$ rad

Finding Angles in the Coordinate Plane

117–119 Choose the angle that most closely resembles the angle in the given diagram.

117. Using the diagram, find the angle measure that most closely resembles the angle θ.

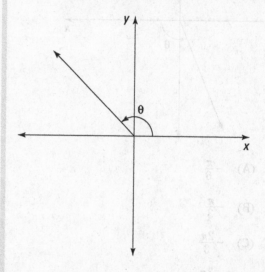

(A) $\frac{\pi}{3}$

(B) $\frac{3\pi}{4}$

(C) π

(D) $\frac{7\pi}{6}$

(E) $\frac{5\pi}{3}$

118. Using the diagram, find the angle measure that most closely resembles the angle θ.

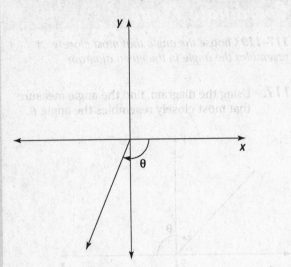

(A) $-\dfrac{\pi}{6}$

(B) $-\dfrac{\pi}{3}$

(C) $-\dfrac{2\pi}{3}$

(D) $-\dfrac{3\pi}{2}$

(E) $-\dfrac{11\pi}{6}$

119. Using the diagram, find the angle measure that most closely resembles the angle θ.

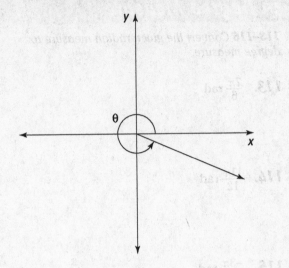

(A) $\dfrac{5\pi}{6}$

(B) $\dfrac{7\pi}{6}$

(C) $\dfrac{4\pi}{3}$

(D) $\dfrac{3\pi}{2}$

(E) $\dfrac{11\pi}{6}$

Finding Common Trigonometric Values

120–124 *Find* $\sin\theta$, $\cos\theta$, *and* $\tan\theta$ *for the given angle measure. Remember to rationalize denominators that contain radicals.*

120. $\theta = \dfrac{\pi}{4}$

121. $\theta = \dfrac{5\pi}{6}$

122. $\theta = \dfrac{-2\pi}{3}$

123. $\theta = -135°$

124. $\theta = 180°$

Simplifying Trigonometric Expressions

125–132 *Determine which expression is equivalent to the given one.*

125. $\sin\theta\cot\theta$

   (A) $\cos\theta$

   (B) $\sin\theta$

   (C) $\sec\theta$

   (D) $\csc\theta$

   (E) $\tan\theta$

126. $\sec x - \cos x$

   (A) 1

   (B) $\sin x$

   (C) $\tan x$

   (D) $\cos x \cot x$

   (E) $\sin x \tan x$

127. $(\sin x + \cos x)^2$

   (A) $2 + \sin 2x$

   (B) $2 + \cos 2x$

   (C) $1 + \sec 2x$

   (D) $1 + \sin 2x$

   (E) $1 + \cos 2x$

128. $\sin(\pi - x)$

   (A) $\cos x$

   (B) $\sin x$

   (C) $\csc x$

   (D) $\sec x$

   (E) $\tan x$

129. $\sin x \sin 2x + \cos x \cos 2x$

   (A) $\cos x$

   (B) $\sin x$

   (C) $\csc x$

   (D) $\sec x$

   (E) $\tan x$

130. $\dfrac{1}{1-\cos\theta} + \dfrac{1}{1+\cos\theta}$

(A) $2\sin^2\theta$

(B) $2\tan^2\theta$

(C) $2\sec^2\theta$

(D) $2\csc^2\theta$

(E) $2\cot^2\theta$

131. $\dfrac{\sin x}{1-\cos x}$

(A) $\csc x + \cot x$

(B) $\sec x + \cot x$

(C) $\csc x - \cot x$

(D) $\sec x - \tan x$

(E) $\csc x - \tan x$

132. $\cos(3\theta)$

(A) $5\cos^3\theta - 3\cos\theta$

(B) $2\cos^3\theta - 3\cos\theta$

(C) $4\cos^3\theta - 3\cos\theta$

(D) $4\cos^3\theta + 3\cos\theta$

(E) $2\cos^3\theta + 5\cos\theta$

Solving Trigonometric Equations

133–144 Solve the given trigonometric equations. Find all solutions in the interval $[0, 2\pi]$.

133. $2\sin x - 1 = 0$

134. $\sin x = \tan x$

135. $2\cos^2 x + \cos x - 1 = 0$

136. $|\tan x| = 1$

137. $2\sin^2 x - 5\sin x - 3 = 0$

138. $\cos x = \cot x$

139. $\sin(2x) = \dfrac{1}{2}$

140. $\sin 2x = \cos x$

141. $2\cos x + \sin 2x = 0$

142. $2 + \cos 2x = -3 \cos x$

143. $\tan(3x) = -1$

144. $\cos(2x) = \cot(2x)$

Amplitude, Period, Phase Shift, and Midline

145–148 Determine the amplitude, the period, the phase shift, and the midline of the function.

145. $f(x) = \frac{1}{2} \sin\left(x + \frac{\pi}{2}\right)$

146. $f(x) = -\frac{1}{4} \cos(\pi x - 4)$

147. $f(x) = 2 - 3 \cos(\pi x - 6)$

148. $f(x) = \frac{1}{2} - \sin\left(\frac{1}{2}x + \frac{\pi}{2}\right)$

Equations of Periodic Functions

149–154 Choose the equation that describes the given periodic function.

149.

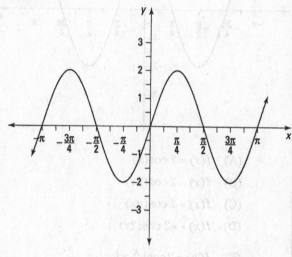

(A) $f(x) = 2 \sin(2x)$

(B) $f(x) = -2 \sin(2x)$

(C) $f(x) = 2 \sin(x)$

(D) $f(x) = 2 \sin(\pi x)$

(E) $f(x) = 2 \sin\left(\frac{\pi}{2} x\right)$

150.

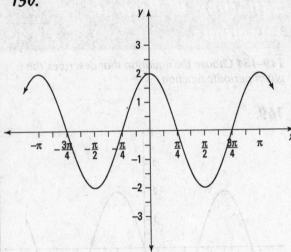

(A) $f(x) = 2\cos(x)$

(B) $f(x) = 2\cos(2x)$

(C) $f(x) = 2\cos(\pi x)$

(D) $f(x) = -2\cos(2x)$

(E) $f(x) = 2\cos\left(\dfrac{\pi}{2}x\right)$

151.

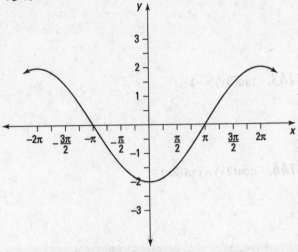

(A) $f(x) = 2\cos(2x) + 1$

(B) $f(x) = -2\cos(2x) + 2$

(C) $f(x) = 2\cos(2x)$

(D) $f(x) = -2\cos\left(\dfrac{1}{2}x\right)$

(E) $f(x) = 2\cos(\pi x)$

152.

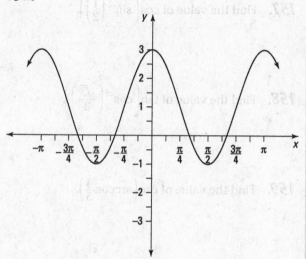

(A) $f(x) = -2\cos(2x)$

(B) $f(x) = -2\cos(2x) + 2$

(C) $f(x) = 2\cos(2x) + 1$

(D) $f(x) = 2\cos(\pi x) + 1$

(E) $f(x) = 2\cos\left(\dfrac{\pi}{2}x\right) + 1$

153.

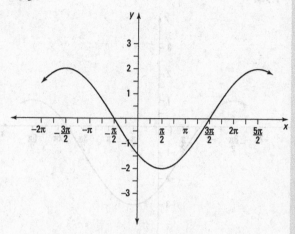

(A) $f(x) = 2\cos\left(\dfrac{1}{2}x - \dfrac{\pi}{4}\right)$

(B) $f(x) = 2\cos\left(\dfrac{1}{2}x - \dfrac{\pi}{2}\right)$

(C) $f(x) = -2\cos\left(\dfrac{1}{2}x - \dfrac{\pi}{4}\right)$

(D) $f(x) = -2\cos\left(x - \dfrac{\pi}{4}\right)$

(E) $f(x) = -2\cos\left(\dfrac{1}{2}x - \dfrac{\pi}{2}\right)$

154.

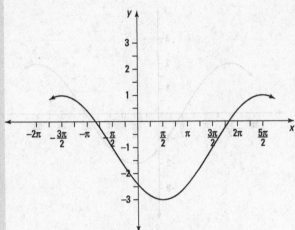

(A) $f(x) = 2\cos\left(\frac{1}{2}x - \frac{\pi}{4}\right) - 1$

(B) $f(x) = -2\cos\left(\frac{1}{2}x - \frac{\pi}{4}\right) + 1$

(C) $f(x) = -2\cos\left(x - \frac{\pi}{4}\right) - 1$

(D) $f(x) = -2\cos\left(\frac{1}{2}x - \frac{\pi}{4}\right) - 1$

(E) $f(x) = -2\cos\left(4x - \frac{\pi}{4}\right) - 1$

Inverse Trigonometric Function Basics

155–160 Evaluate the inverse trigonometric function for the given value.

155. Find the value of $\sin^{-1}\left(\frac{\sqrt{3}}{2}\right)$.

156. Find the value of $\arctan(-1)$.

157. Find the value of $\cos\left(\sin^{-1}\left(\frac{1}{2}\right)\right)$.

158. Find the value of $\tan\left(\cos^{-1}\left(\frac{\sqrt{3}}{2}\right)\right)$.

159. Find the value of $\csc\left(\arccos\frac{4}{5}\right)$.

160. Find the value of $\sin\left(\tan^{-1}(2) + \tan^{-1}(3)\right)$.

Solving Trigonometric Equations Using Inverses

161–166 Solve the given trigonometric equation using inverses. Find all solutions in the interval $[0, 2\pi]$.

161. $\sin x = 0.4$

162. $\cos x = -0.78$

163. $5 \sin(2x) + 1 = 4$

164. $7 \cos(3x) - 1 = 3$

165. $2 \sin^2 x + 8 \sin x + 5 = 0$

166. $3 \sec^2 x + 4 \tan x = 2$

Chapter 3

Limits and Rates of Change

. . . .

Limits are the foundation of calculus. Being able to work with limits and to understand them conceptually is crucial, because key ideas and definitions in calculus make use of limits. This chapter examines a variety of limit problems and makes the intuitive idea of continuity formal by using limits. Many later problems also involve the use of limits, so although limits may go away for a while during your calculus studies, they'll return!

The Problems You'll Work On

In this chapter, you encounter a variety of problems involving limits:

- ✔ Using graphs to find limits
- ✔ Finding left-hand and right-hand limits
- ✔ Determining infinite limits and limits at infinity
- ✔ Practicing many algebraic techniques to evaluate limits of the form 0/0
- ✔ Determining where a function is continuous

What to Watch Out For

You can use a variety of techniques to evaluate limits, and you want to be familiar with them all! Remember the following tips:

- ✔ When substituting in the limiting value, a value of zero in the denominator of a fraction doesn't automatically mean that the limit does not exist! For example, if the function has a removable discontinuity, the limit still exists!

- ✔ Be careful with signs, as you may have to include a negative when evaluating limits at infinity involving radicals (especially when the variable approaches negative infinity). It's easy to make a limit positive when it should have been negative!

- ✔ Know and understand the definition of *continuity*, which says the following: A function $f(x)$ is continuous at a if $\lim_{x \to a} f(x) = f(a)$.

Finding Limits from Graphs

167–172 Use the graph to find the indicated limit.

167.

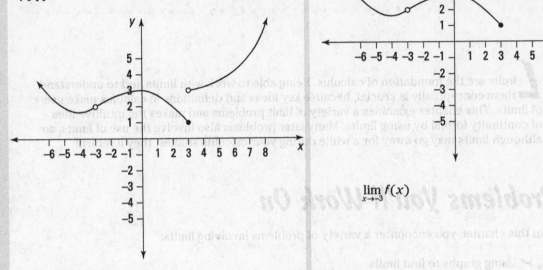

$$\lim_{x \to 3^-} f(x)$$

168.

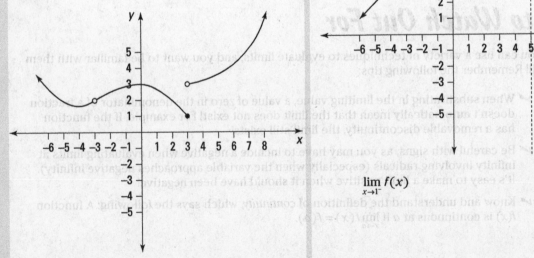

$$\lim_{x \to 3^+} f(x)$$

169.

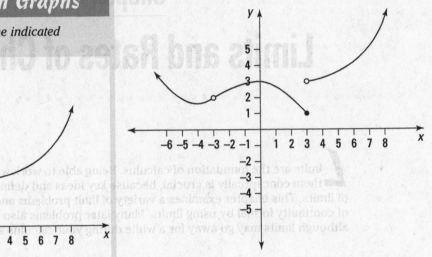

$$\lim_{x \to -3} f(x)$$

170.

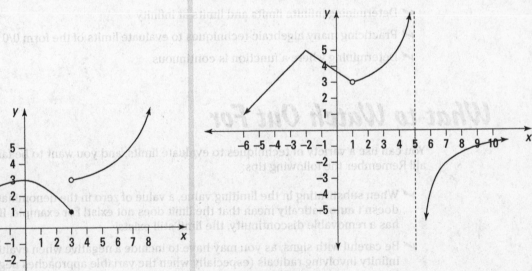

$$\lim_{x \to 1^-} f(x)$$

171.

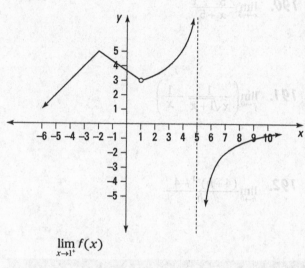

$$\lim_{x \to 1^+} f(x)$$

172.

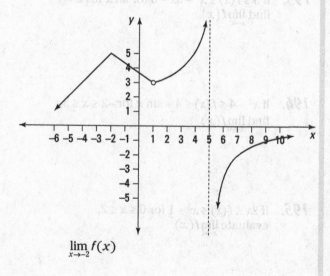

$$\lim_{x \to -2} f(x)$$

173–192 Evaluate the given limit.

173. $\lim\limits_{x \to 3} \dfrac{x^2 - 2x - 3}{x - 3}$

174. $\lim\limits_{x \to 2} \dfrac{x^2 + 3x - 10}{x^2 - 8x + 12}$

175. $\lim\limits_{x \to -5} \dfrac{x^2 + 5x}{x^2 - 25}$

176. $\lim\limits_{x \to 4} \dfrac{4 - x}{2 - \sqrt{x}}$

177. $\lim\limits_{x \to 0} \left(\dfrac{1}{x} - \dfrac{1}{x^2 + x} \right)$

178. $\lim\limits_{x \to 4} |x - 4|$

179. $\lim\limits_{x \to -1} \dfrac{x^3 + 1}{x + 1}$

180. $\lim\limits_{h \to 0} \dfrac{\sqrt{4 + h} - 2}{h}$

181. $\lim\limits_{x\to 3}\dfrac{x^4-81}{x-3}$

182. $\lim\limits_{x\to 0^+}\left(\dfrac{2}{x^2}+\dfrac{2}{|x|}\right)$

183. $\lim\limits_{x\to 0^-}\left(\dfrac{2}{x^2}-\dfrac{2}{|x|}\right)$

184. $\lim\limits_{x\to \frac{4}{3}}\dfrac{3x^2-4x}{|3x-4|}$

185. $\lim\limits_{x\to 5}\dfrac{x^2-25}{3x^2-16x+5}$

186. $\lim\limits_{x\to 3}\dfrac{\sqrt{x+3}-\sqrt{2x}}{x^2-3x}$

187. $\lim\limits_{h\to 0}\dfrac{(2+h)^3-8}{h}$

188. $\lim\limits_{x\to 4^+}\dfrac{x-4}{|x-4|}$

189. $\lim\limits_{x\to 5^-}\dfrac{x-5}{|x-5|}$

190. $\lim\limits_{x\to -5}\dfrac{\frac{1}{5}+\frac{1}{x}}{x+5}$

191. $\lim\limits_{x\to 0}\left(\dfrac{1}{x\sqrt{1+x}}-\dfrac{1}{x}\right)$

192. $\lim\limits_{h\to 0}\dfrac{(4+h)^{-1}-4^{-1}}{h}$

Applying the Squeeze Theorem

193–198 Use the squeeze theorem to evaluate the given limit.

193. If $5\le f(x)\le x^2+3x-5$ for all x in $[2,\infty)$, find $\lim\limits_{x\to 2^+}f(x)$.

194. If $x^2+4\le f(x)\le 4+\sin x$ for $-2\le x\le 5$, find $\lim\limits_{x\to 0}f(x)$.

195. If $2x\le f(x)\le x^3+1$ for $0\le x\le 2$, evaluate $\lim\limits_{x\to 1}f(x)$.

196. Find the limit: $\lim\limits_{x\to 0}x^4\cos\left(\dfrac{2}{x^2}\right)$.

197. Find the limit: $\lim\limits_{x\to0^+} x^2 \sin\left(\dfrac{2}{\sqrt{x}}\right)$.

205. $\lim\limits_{x\to2} \dfrac{\sin(x-2)}{x^2+x-6}$

198. Find the limit: $\lim\limits_{x\to0^+} \sqrt[3]{x}\left(3-\sin^2\left(\dfrac{\pi}{x}\right)\right)$.

206. $\lim\limits_{x\to0} \dfrac{\sin x}{x+\tan x}$

Evaluating Trigonometric Limits

199–206 Evaluate the given trigonometric limit. Recall that $\lim\limits_{x\to0} \dfrac{\sin x}{x}=1$ and that $\lim\limits_{x\to0} \dfrac{\cos x-1}{x}=0$.

199. $\lim\limits_{x\to0} \dfrac{\sin(5x)}{x}$

200. $\lim\limits_{x\to0} \dfrac{2\cos x-2}{\sin x}$

201. $\lim\limits_{x\to\frac{\pi}{4}} \dfrac{\cos 2x}{\sin x-\cos x}$

202. $\lim\limits_{x\to0} \dfrac{\sin(5x)}{\sin(9x)}$

203. $\lim\limits_{x\to0} \dfrac{\tan(7x)}{\sin(3x)}$

204. $\lim\limits_{x\to0} \dfrac{\sin^3(2x)}{x^3}$

Infinite Limits

207–211 Find the indicated limit using the given graph.

207.

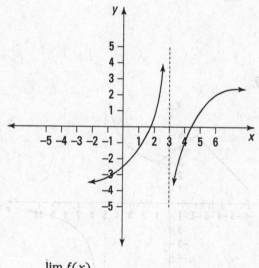

$$\lim\limits_{x\to3^-} f(x)$$

208.

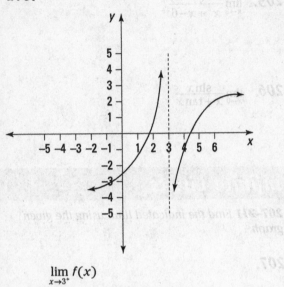

$$\lim_{x \to 3^+} f(x)$$

210.

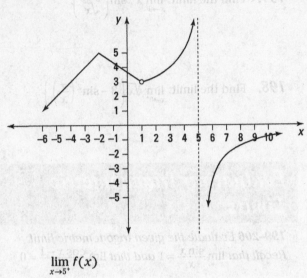

$$\lim_{x \to 5^+} f(x)$$

209.

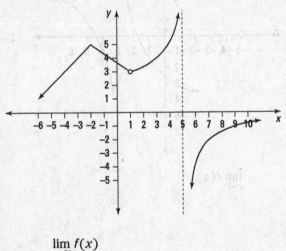

$$\lim_{x \to 5^-} f(x)$$

211.

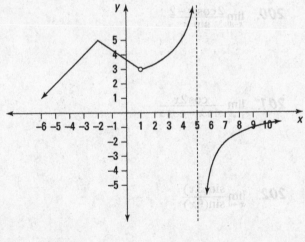

$$\lim_{x \to 5} f(x)$$

212–231 *Find the indicated limit.*

212. $\displaystyle\lim_{x\to 1^+}\frac{3}{x-1}$

213. $\displaystyle\lim_{x\to 1^-}\frac{3}{x-1}$

214. $\displaystyle\lim_{x\to \frac{\pi}{2}^+}(\tan x)$

215. $\displaystyle\lim_{x\to \pi^-}\frac{x^2}{\sin x}$

216. $\displaystyle\lim_{x\to 5^-}\frac{x+3}{x-5}$

217. $\displaystyle\lim_{x\to 0^-}\frac{1-x}{e^x-1}$

218. $\displaystyle\lim_{x\to 0^-}\cot x$

219. $\displaystyle\lim_{x\to 2}\frac{4e^x}{|2-x|}$

220. $\displaystyle\lim_{x\to \frac{1}{2}^+}\frac{x^2+1}{x\cos(\pi x)}$

221. $\displaystyle\lim_{x\to 5}\frac{\sin x}{(5-x)^4}$

222. $\displaystyle\lim_{x\to 0}\frac{x+5}{x^4(x-6)}$

223. $\displaystyle\lim_{x\to 1}\frac{3x}{e^x-e}$

224. $\displaystyle\lim_{x\to 0^+}\frac{x-1}{x^2(x+2)}$

225. $\displaystyle\lim_{x\to e^-}\frac{x^3}{\ln x-1}$

226. $\displaystyle\lim_{x\to e^2}\frac{-x}{\ln x-2}$

227. $\displaystyle\lim_{x\to 2}\frac{x+2}{x^2-4}$

228. $\displaystyle\lim_{x\to 25}\frac{5+\sqrt{x}}{x-25}$

229. $\displaystyle\lim_{x\to 0}\frac{x^2+4}{x^2(x-1)}$

230. $\lim\limits_{x \to 0^-} \dfrac{x-1}{x^2(x+2)}$

231. $\lim\limits_{x \to 3} \dfrac{3+x}{|3-x|}$

Limits from Graphs

232–235 Find the indicated limit using the given graph.

232.

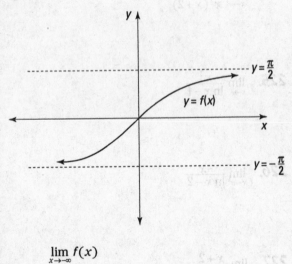

$\lim\limits_{x \to -\infty} f(x)$

233.

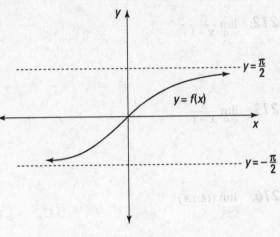

$\lim\limits_{x \to \infty} f(x)$

234.

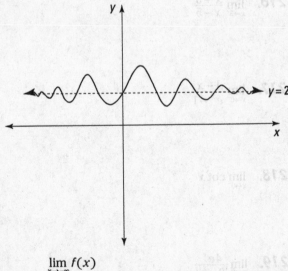

$\lim\limits_{x \to -\infty} f(x)$

235.

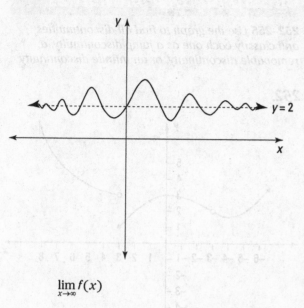

$y = 2$

$$\lim_{x \to \infty} f(x)$$

Limits at Infinity

236–247 Find the indicated limit.

236. $\lim\limits_{x \to \infty} \dfrac{1}{3x+4}$

237. $\lim\limits_{x \to -\infty} \left[x^2(x+1)(3-x) \right]$

238. $\lim\limits_{x \to \infty} \dfrac{3x+4}{x-7}$

239. $\lim\limits_{x \to \infty} \cos x$

240. $\lim\limits_{x \to \infty} \dfrac{5x^4+5}{\left(x^2-1\right)\left(2x^3+3\right)}$

241. $\lim\limits_{x \to -\infty} \dfrac{x}{\sqrt{x^2+1}}$

242. $\lim\limits_{x \to -\infty} \dfrac{8x^4+3x-5}{x^2+1}$

243. $\lim\limits_{x \to -\infty} \dfrac{\sqrt{9x^{10}-x}}{x^5+1}$

244. $\lim\limits_{x \to \infty} \dfrac{\sqrt{9x^{10}-x}}{x^5+1}$

245. $\lim\limits_{x \to \infty} \left(\sqrt{x} - x \right)$

246. $\lim\limits_{x \to \infty} \left(\sqrt{x^4+3x^2} - x^2 \right)$

247. $\lim\limits_{x \to -\infty} \left(x + \sqrt{x^2+5x} \right)$

Horizontal Asymptotes

248–251 Find any horizontal asymptotes of the given function.

248. $y = \dfrac{1+3x^4}{x+5x^4}$

249. $y = \dfrac{5-x^2}{5+x^2}$

250. $y = \dfrac{\sqrt{x^4+x}}{3x^2}$

251. $y = \dfrac{x}{\sqrt{x^2+2}}$

Classifying Discontinuities

252–255 Use the graph to find all discontinuities and classify each one as a jump discontinuity, a removable discontinuity, or an infinite discontinuity.

252.

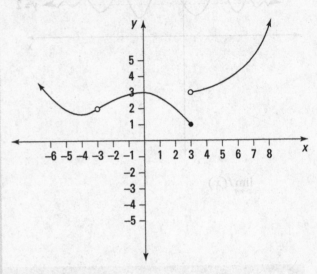

253.

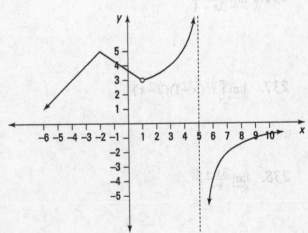

254.

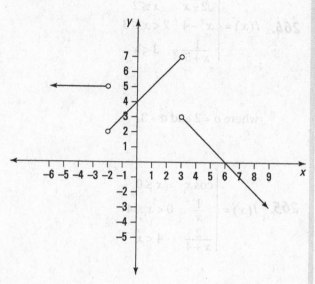

255.

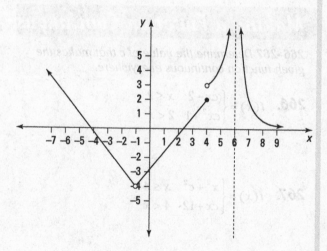

256–261 Determine whether the function is continuous at the given value of a. If it's continuous, state the value at $f(a)$. If it isn't continuous, classify the discontinuity as a jump, removable, or infinite discontinuity.

256. $f(x) = \begin{cases} \dfrac{1}{x-2} & x \neq 2 \\ 3 & x = 2 \end{cases}$

where $a = 2$

257. $f(x) = \begin{cases} 1+x^2 & x \leq 1 \\ 4\sqrt{x}-2 & x > 1 \end{cases}$

where $a = 1$

258. $f(x) = \begin{cases} \dfrac{x^2-x-6}{x-3} & x \neq 3 \\ 5 & x = 3 \end{cases}$

where $a = 3$

259. $f(x) = \begin{cases} \dfrac{4-\sqrt{x}}{16-x} & x \neq 16 \\ \dfrac{1}{8} & x = 16 \end{cases}$

where $a = 16$

260. $f(x) = \begin{cases} \dfrac{x+6}{|x+6|} & x \neq -6 \\ -1 & x = -6 \end{cases}$

where $a = -6$

261. $f(x) = \begin{cases} \dfrac{x^3+1}{x+1} & x \neq -1 \\ 2 & x = -1 \end{cases}$

where $a = -1$

262–265 *Determine whether the function is continuous at the given values of a. If it isn't continuous, classify each discontinuity as a jump, removable, or infinite discontinuity.*

262. $f(x) = \begin{cases} 2+x^2 & x \leq 0 \\ 2\cos x & 0 < x \leq \pi \\ \sin x - 2 & \pi < x \end{cases}$

where $a = 0$ and $a = \pi$

263. $f(x) = \begin{cases} x+2 & x \leq 1 \\ 2x^2 & 1 < x \leq 3 \\ x^3 & 3 < x \end{cases}$

where $a = 1$ and $a = 3$

264. $f(x) = \begin{cases} \sqrt{2-x} & x \leq 2 \\ x^2 - 4 & 2 < x \leq 3 \\ \dfrac{1}{x+5} & 3 < x \end{cases}$

where $a = 2$ and $a = 3$

265. $f(x) = \begin{cases} \cos x & x \leq 0 \\ \dfrac{1}{x} & 0 < x \leq 4 \\ \dfrac{2}{x+4} & 4 < x \end{cases}$

where $a = 0$ and $a = 4$

Making a Function Continuous

266–267 *Determine the value of c that makes the given function continuous everywhere.*

266. $f(x) = \begin{cases} cx - 2 & x \leq 2 \\ cx^2 + 1 & 2 < x \end{cases}$

267. $f(x) = \begin{cases} x^2 + c^2 & x \leq 4 \\ cx + 12 & 4 < x \end{cases}$

The Intermediate Value Theorem

268–271 *Determine which of the given intervals is guaranteed to contain a root of the function by the intermediate value theorem.*

268. By checking only the endpoints of each interval, determine which interval contains a root of the function $f(x) = x^2 - \frac{3}{2}$ by the intermediate value theorem:

(A) $[-5, -4]$

(B) $[-4, -3]$

(C) $[0, 1]$

(D) $[1, 2]$

(E) $[5, 12]$

269. By checking only the endpoints of each interval, determine which interval contains a root of the function $f(x) = 3\sqrt{x} - 4x + 5$ by the intermediate value theorem:

(A) $[0, 1]$

(B) $[1, 4]$

(C) $[4, 9]$

(D) $[9, 16]$

(E) $[16, 25]$

270. By checking only the endpoints of each interval, determine which interval contains a solution to the equation $2(3^x) + x^2 - 4 = 32$ according to the intermediate value theorem:

(A) $[0, 1]$

(B) $[1, 2]$

(C) $[2, 3]$

(D) $[3, 4]$

(E) $[4, 5]$

271. By checking only the endpoints of each interval, determine which interval contains a solution to the equation $4|2x - 3| + 5 = 22$ according to the intermediate value theorem:

(A) $[0, 1]$

(B) $[1, 2]$

(C) $[2, 3]$

(D) $[3, 4]$

(E) $[4, 5]$

(C) $[0, 1]$

(C) $[4, 8]$

(C) $[2, 3]$

(C) $[2, 3]$

(D) $[3, 4]$

(E) $[4, 5]$

Chapter 4

Derivative Basics

· ·

The derivative is one of the great ideas in calculus. In this chapter, you see the formal definition of a derivative. Understanding the formal definition is crucial, because it tells you what a derivative actually is. Unfortunately, computing the derivative using the definition can be quite cumbersome and is often very difficult. After finding derivatives using the definition, you see problems that use the power rule, which is the start of some techniques that make finding the derivative much easier — although still challenging in many cases.

The Problems You'll Work On

In this chapter, you see the definition of a derivative and one of the first shortcut formulas, the power rule. Here's what the problems cover:

✔ Using a variety of algebraic techniques to find the derivative using the definition of a derivative

✔ Evaluating the derivative at a point using a graph and slopes of tangent lines

✔ Encountering a variety of derivative questions that you can solve using the power rule

What to Watch Out For

Using the definition of a derivative to evaluate derivatives can involve quite a bit of algebra, so be prepared. Having all the shortcut techniques is very nice, but you'll be asked to find derivatives for complicated functions, so the problems will still be challenging! Keep some of the following points in mind:

✔ Remember your algebra techniques: factoring, multiplying by conjugates, working with fractions, and more. Many students get tripped up on one part and then can't finish the problem, so know that many problems require multiple steps.

✔ When interpreting the value of a derivative from a graph, think about the slope of the tangent line on the graph at a given point; you'll be well on your way to finding the correct solution.

✔ Simplifying functions using algebra and trigonometric identities before finding the derivative makes many problems much easier. Simplifying is one of the very first things you should consider when encountering a "find the derivative" question of any type.

Determining Differentiability from a Graph

272–276 Use the graph to determine for which values of x the function is not differentiable.

272.

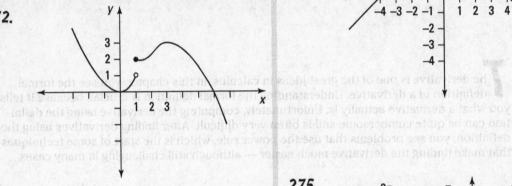

273.

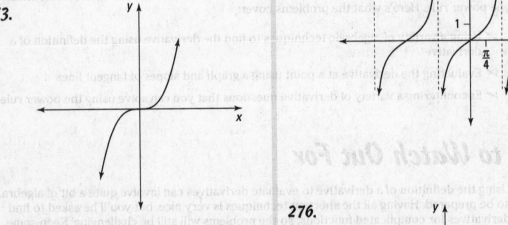

274.

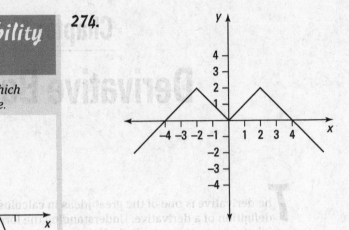

275.

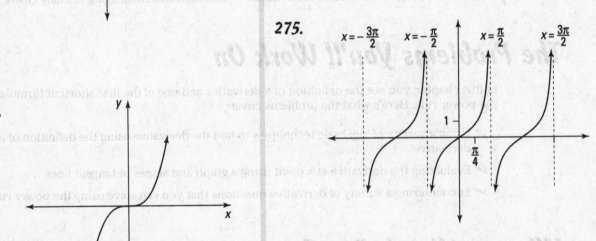

276.

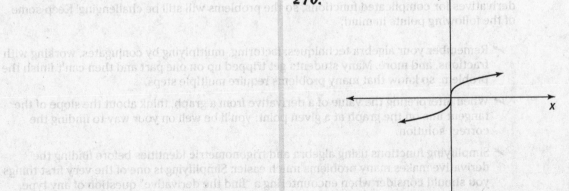

Finding the Derivative by Using the Definition

277–290 Find the derivative by using the definition $f'(x) = \lim\limits_{h \to 0} \dfrac{f(x+h)-f(x)}{h}$.

277. $f(x) = 2x - 1$

278. $f(x) = x^2$

279. $f(x) = 2 + x$

280. $f(x) = \dfrac{1}{x}$

281. $f(x) = x^3 - x^2$

282. $f(x) = 3x^2 + 4x$

283. $f(x) = \sqrt{x}$

284. $f(x) = \sqrt{2 - 5x}$

285. $f(x) = \dfrac{1}{\sqrt{x-1}}$

286. $f(x) = x^3 + 3x$

287. $f(x) = \dfrac{1}{x^2 + 5}$

288. $f(x) = \dfrac{2x+1}{x+4}$

289. $f(x) = \dfrac{x^2+1}{x+3}$

290. $f(x) = \sqrt[3]{2x+1}$

Finding the Value of the Derivative Using a Graph

291–296 *Use the graph to determine the solution.*

291.

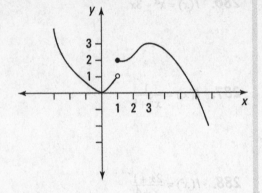

Estimate the value of $f'(3)$ using the graph.

292.

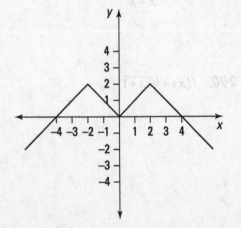

Estimate the value of $f'(-1)$ using the graph.

293.

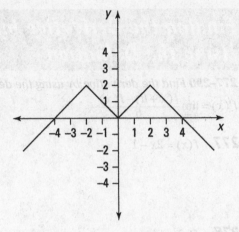

Estimate the value of $f'(-3)$ using the graph.

294.

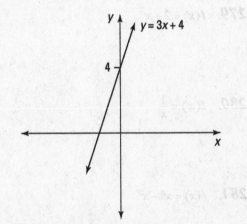

Based on the graph of $y = 3x + 4$, what does $f'(-22\pi^3)$ equal?

295.

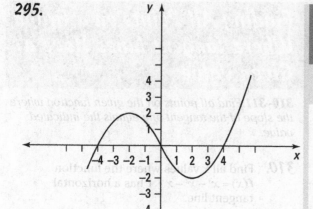

Based on the graph, arrange the following from smallest to largest: $f'(-3)$, $f'(-2)$, and $f'(1)$.

296.

Based on the graph, arrange the following from smallest to largest: $f'(1)$, $f'(2)$, $f'(5)$, and 0.1.

Using the Power Rule to Find Derivatives

297–309 *Use the power rule to find the derivative of the given function.*

297. $f(x) = 5x + 4$

298. $f(x) = x^2 + 3x + 6$

299. $f(x) = (x + 4)(2x - 1)$

300. $f(x) = \pi^3$

301. $f(x) = \sqrt{5x}$

302. $f(x) = \dfrac{3x^2 + x - 4}{x^3}$

303. $f(x) = \sqrt{x}\left(x^3 + 1\right)$

304. $f(x) = \dfrac{\sqrt{3}}{x^4} + 2x^{1.1}$

305. $f(x) = 4x^{-3/7} + 8x + \sqrt{5}$

306. $f(x) = \left(\dfrac{1}{x^3} - \dfrac{2}{x}\right)\left(x^2 + x\right)$

307. $f(x) = (x^{-3} + 4)(x^{-2} - 5x)$

308. $f(x) = 4x^4 - x^2 + 8x + \pi^2$

309. $f(x) = \sqrt{x} - \dfrac{1}{\sqrt[4]{x}}$

Finding All Points on a Graph Where Tangent Lines Have a Given Value

310–311 Find all points on the given function where the slope of the tangent line equals the indicated value.

310. Find all x values where the function $f(x) = x^3 - x^2 - x + 1$ has a horizontal tangent line.

311. Find all x values where the function $f(x) = 6x^3 + 5x - 2$ has a tangent line with a slope equal to 6.

Chapter 5

The Product, Quotient, and Chain Rules

• •

This chapter focuses on some of the major techniques needed to find the derivative: the product rule, the quotient rule, and the chain rule. By using these rules along with the power rule and some basic formulas (see Chapter 4), you can find the derivatives of most of the single-variable functions you encounter in calculus. However, after using the derivative rules, you often need many algebra steps to simplify the function so that it's in a nice final form, especially on problems involving the product rule or quotient rule.

The Problems You'll Work On

Here you practice using most of the techniques needed to find derivatives (besides the power rule):

✔ The product rule

✔ The quotient rule

✔ The chain rule

✔ Derivatives involving trigonometric functions

What to Watch Out For

Many of these problems require one calculus step and then many steps of algebraic simplification to get to the final answer. Remember the following tips as you work through the problems:

✔ Considering simplifying a function before taking the derivative. Simplifying before taking the derivative is almost always easier than finding the derivative and then simplifying.

✔ Some problems have functions without specified formulas in the questions; don't be thrown off! Simply proceed as you normally would on a similar example.

✔ Many people make the mistake of using the product rule when they should be using the chain rule. Stop and examine the function before jumping in and taking the derivative. Make sure you recognize whether the question involves a product or a composition (in which case you must use the chain rule).

✔ Rewriting the function by adding parentheses or brackets may be helpful, especially on problems that involve using the chain rule multiple times.

Using the Product Rule to Find Derivatives

312–331 Use the product rule to find the derivative of the given function.

312. $f(x) = (2x^3 + 1)(x^5 - x)$

313. $f(x) = x^2 \sin x$

314. $f(x) = \sec x \tan x$

315. $f(x) = \dfrac{x}{\cos x}$

316. $f(x) = 4x \csc x$

317. Find $(fg)'(4)$ if $f(4) = 3$, $f'(4) = 2$, $g(4) = -6$, and $g'(4) = 8$.

318. $f(x) = \dfrac{g(x)}{x^3}$

319. $f(x) = (\sec x)(x + \tan x)$

320. $f(x) = \left(x^2 + x\right) \csc x$

321. $f(x) = 4x^3 \sec x$

322. $f(x) = \dfrac{\cot x}{x^{1/2}}$

323. Assuming that g is a differentiable function, find an expression for the derivative of $f(x) = x^2 g(x)$.

324. Assuming that g is a differentiable function, find an expression for the derivative of $f(x) = \dfrac{1 + x^2 g(x)}{x}$.

325. Find $(fg)'(3)$ if $f(3) = -2$, $f'(3) = 4$, $g(3) = -8$, and $g'(3) = 7$.

326. $f(x) = x^2 \cos x \sin x$

327. Assuming that g is a differentiable function, find an expression for the derivative of

$$f(x) = \frac{2 + xg(x)}{\sqrt{x}}.$$

328. $f(x) = \left(\frac{1}{x^2} - \frac{2}{x}\right)(\tan x)$

329. $f(x) = \frac{x - 2x\sqrt[3]{x}\cos x}{x^2}$

330. Assuming that g is a differentiable function, find an expression for the derivative of

$$f(x) = \frac{4g(x)}{x^5}.$$

331. Assuming that g and h are differentiable functions, find an expression for the derivative of $f(x) = [g(x)h(x)]\sin x$.

Using the Quotient Rule to Find Derivatives

332–351 Use the quotient rule to find the derivative.

332. $f(x) = \frac{2x + 1}{3x + 4}$

333. $f(x) = \frac{2x}{5 + x^2}$

334. $f(x) = \frac{\sin x - 1}{\cos x + 1}$

335. $f(x) = \frac{x}{x^5 + x^3 + 1}$

336. Assuming that f and g are differentiable functions, find the value of $\left(\dfrac{f}{g}\right)'(4)$ if $f(4) = 5$, $f'(4) = -7$, $g(4) = 8$, and $g'(4) = 4$.

337. $f(x) = \frac{x + 2}{3x + 5}$

338. $f(x) = \frac{x^2}{3x^2 - 2x + 1}$

339. $f(x) = \frac{\sin x}{x^3}$

340. $f(x) = \frac{(x - 1)(x + 2)}{(x - 3)(x + 1)}$

341. $f(x) = \frac{\sec x}{1 + \sec x}$

342. $f(x) = \dfrac{4x^2}{\sin x + \cos x}$

343. Assuming that f and g are differentiable functions, find the value of $\left(\dfrac{f}{g}\right)'(5)$ if $f(5) = -4$, $f'(5) = 2$, $g(5) = -7$, and $g'(5) = -6$.

344. $f(x) = \dfrac{x^2}{1 + \sqrt{x}}$

345. $f(x) = \dfrac{\sin x}{\cos x + \sin x}$

346. $f(x) = \dfrac{x^2}{1 + \sqrt[3]{x}}$

347. $f(x) = \dfrac{\sqrt{x} + 2}{\sqrt{x} + 1}$

348. $f(x) = \dfrac{\tan x - 1}{\sec x + 1}$

349. $f(x) = \dfrac{x}{x + \dfrac{4}{x}}$

350. Assuming that g is a differentiable function, find an expression for the derivative of $f(x) = \dfrac{g(x)}{x^3}$.

351. Assuming that g is a differentiable function, find an expression for the derivative of $f(x) = \dfrac{\cos x}{g(x)}$.

Using the Chain Rule to Find Derivatives

352–370 Use the chain rule to find the derivative.

352. $f(x) = \left(x^2 + 3x\right)^{100}$

353. $f(x) = \sin(4x)$

354. $f(x) = \sqrt[3]{1 + \sec x}$

355. $f(x) = \dfrac{1}{\left(x^2 - x\right)^5}$

356. $f(x) = \csc\left(\dfrac{1}{x^2}\right)$

357. $f(x) = \left(x + \cos^2 x\right)^5$

358. $f(x) = \dfrac{\sin(\pi x)}{\cos(\pi x) + \sin(\pi x)}$

359. $f(x) = \cos(x \sin x)$

360. $f(x) = x \sin \sqrt[3]{x}$

361. $f(x) = \dfrac{x}{\sqrt{3 - 2x}}$

362. $f(x) = \sec^2 x + \tan^2 x$

363. $f(x) = \dfrac{x}{\sqrt{1 + x^2}}$

364. $f(x) = \left(x^2 + 1\right)\sqrt[4]{x^3 + 5}$

365. $f(x) = \left(x^4 - 1\right)^3 \left(x^5 + 1\right)^6$

366. $f(x) = \dfrac{(x - 1)^3}{\left(x^2 + x\right)^5}$

367. $f(x) = \sqrt{\dfrac{x + 1}{x - 1}}$, where $x > 1$

368. $f(x) = \sin\left(\sin\left(\sin\left(x^2\right)\right)\right)$

369. $f(x) = \sqrt[3]{x + \sqrt{x}}$

370. $f(x) = (1 + 5x)^4 \left(2 + x - x^2\right)^7$

More Challenging Chain Rule Problems

371–376 Solve the problem related to the chain rule.

371. Find all x values in the interval $[0, 2\pi]$ where the function $f(x) = 2\cos x + \sin^2 x$ has a horizontal tangent line.

372. Suppose that H is a function such that $H'(x) = \dfrac{2}{x}$ for $x > 0$. Find an expression for the derivative of $F(x) = [H(x)]^3$.

373. Let $F(x) = f(g(x))$, $g(2) = -2$, $g'(2) = 4$, $f'(2) = 5$, and $f'(-2) = 7$. Find the value of $F'(2)$.

374. Let $F(x) = f(f(x))$, $f(2) = -2$, $f'(2) = -5$, and $f'(-2) = 8$. Find the value of $F'(2)$.

375. Suppose that H is a function such that $H'(x) = \dfrac{2}{x}$ for $x > 0$. Find an expression for the derivative of $f(x) = H(x^3)$.

376. Let $F(x) = f(g(x))$, $g(4) = 6$, $g'(4) = 8$, $f'(4) = 2$, and $f'(6) = 10$. Find the value of $F'(4)$.

Chapter 6

Exponential and Logarithmic Functions and Tangent Lines

- -

After becoming familiar with the derivative techniques of the power, product, quotient, and chain rules, you simply need to know basic formulas for different functions. In this chapter, you see the derivative formulas for exponential and logarithmic functions. Knowing the derivative formulas for logarithmic functions also makes it possible to use logarithmic differentiation to find derivatives.

In many examples and applications, finding either the tangent line or the normal line to a function at a point is desirable. This chapter arms you with all the derivative techniques, so you'll be in a position to find tangent lines and normal lines for many functions.

The Problems You'll Work On

In this chapter, you do the following types of problems:

- ✔ Finding derivatives of exponential and logarithmic functions with a variety of bases
- ✔ Using logarithmic differentiation to find a derivative
- ✔ Finding the tangent line or normal line at a point

What to Watch Out For

Although you're practicing basic formulas for exponential and logarithmic functions, you still use the product rule, quotient rule, and chain rule as before. Here are some tips for solving these problems:

- ✔ Using logarithmic differentiation requires being familiar with the properties of logarithms, so make sure you can expand expressions containing logarithms.
- ✔ If you see an exponent involving something other than just the variable x, you likely need to use the chain rule to find the derivative.
- ✔ The tangent line and normal line are perpendicular to each other, so the slopes of these lines are opposite reciprocals.

Derivatives Involving Logarithmic Functions

377–385 Find the derivative of the given function.

377. $f(x) = \ln(x)^2$

378. $f(x) = (\ln x)^4$

379. $f(x) = \ln\sqrt{x^2 + 5}$

380. $f(x) = \log_{10}\left(x + \sqrt{6 + x^2}\right)$

381. $f(x) = \log_6\left|\dfrac{-1 + 2\sin x}{4 + 3\sin x}\right|$

382. $f(x) = \log_7\left(\log_8 x^5\right)$

383. $f(x) = \ln|\sec x + \tan x|$

384. $f(x) = -\dfrac{\sqrt{x^2 + 1}}{x} + \ln\left(x + \sqrt{x^2 + 1}\right)$

385. $f(x) = \dfrac{\log_5\left(x^2\right)}{(x+1)^3}$

Logarithmic Differentiation to Find the Derivative

386–389 Use logarithmic differentiation to find the derivative.

386. $f(x) = x^{\tan x}$

387. $f(x) = (\ln x)^{\cos x}$

388. $f(x) = \sqrt{\dfrac{x^2 - 4}{x^2 + 4}}$

389. $f(x) = \dfrac{(x-1)\sin x}{\sqrt{x+2}(x+4)^{3/2}}$

Finding Derivatives of Functions Involving Exponential Functions

390–401 Find the derivative of the given function.

390. $f(x) = e^{5x}$

391. $f(x) = e^{\sin x + x^4}$

392. $f(x) = (x^3 + 1)2^x$

393. $f(x) = \log_5 5^{x^2+3}$

394. $f(x) = e^{x^3}(\sin x + \cos x)$

395. $f(x) = 5^{\sqrt{x}}\sin x$

396. $f(x) = (4^{-x} + 4^x)^3$

397. $f(x) = \ln\left(\dfrac{1+e^x}{1-e^x}\right)$

398. $f(x) = \dfrac{6^x + x}{x^2 + 1}$

399. $f(x) = \dfrac{4}{e^x + e^{-x}}$

400. $f(x) = \dfrac{8^{x^2+1}}{\cos x}$

401. $f(x) = x6^{-5x}$

Finding Equations of Tangent Lines

402–404 Find the equation of the tangent line at the given value.

402. $f(x) = 3\cos x + \pi x$ at $x = 0$

403. $f(x) = x^2 - x + 2$ at $(1, 2)$

404. $f(x) = \dfrac{e^{x^2}}{x}$ at $x = 2$

Finding Equations of Normal Lines

405–407 Find the equation of the normal line at the indicated point.

405. $f(x) = 3x^2 + x - 2$ at $(3, 28)$

406. $f(x) = \sin^2 x$ at $x = \dfrac{\pi}{4}$

407. $f(x) = 4 \ln x + 2$ at $x = e^2$

Chapter 7

Implicit Differentiation

· ·

*W*hen you know the techniques of implicit differentiation (this chapter) and logarithmic differentiation (covered in Chapter 6), you're in a position to find the derivative of just about any function you encounter in a single-variable calculus course. Of course, you'll still use the power, product, quotient, and chain rules (Chapters 4 and 5) when finding derivatives.

The Problems You'll Work On

In this chapter, you use implicit differentiation to

- ✔ Find the first derivative and second derivative of an implicit function
- ✔ Find slopes of tangent lines at given points
- ✔ Find equations of tangent lines at given points

What to Watch Out For

Lots of numbers and variables are floating around in these examples, so don't lose your way:

- ✔ Don't forget to multiply by *dy/dx* at the appropriate moment! If you aren't getting the correct solution, look for this mistake.
- ✔ After finding the second derivative of an implicitly defined function, substitute in the first derivative in order to write the second derivative in terms of *x* and *y*.
- ✔ When you substitute the first derivative into the second derivative, be prepared to further simplify.

Using Implicit Differentiation to Find a Derivative

408–413 Use implicit differentiation to find $\dfrac{dy}{dx}$.

408. $x^2 + y^2 = 9$

409. $y^5 + x^2y^3 = 2 + x^2y$

410. $x^3y^3 + x\cos(y) = 7$

411. $\sqrt{x+y} = \cos\left(y^2\right)$

412. $\cot\left(\dfrac{y}{x}\right) = x^2 + y$

413. $\sec(xy) = \dfrac{y}{1+x^2}$

Using Implicit Differentiation to Find a Second Derivative

414–417 Use implicit differentiation to find $\dfrac{d^2y}{dx^2}$.

414. $8x^2 + y^2 = 8$

415. $x^5 + y^5 = 1$

416. $x^3 + y^3 = 5$

417. $\sqrt{x} + \sqrt{y} = 1$

Finding Equations of Tangent Lines Using Implicit Differentiation

418–422 Find the equation of the tangent line at the indicated point.

418. $x^2 + xy + y^2 = 3$ at $(1, 1)$

419. $3\left(x^2 + y^2\right)^2 = 25\left(x^2 - y^2\right)$ at $(2, 1)$

420. $x^2 + 2xy + y^2 = 1$ at $(0, 1)$

421. $\cos(xy) + x^2 = \sin y$ at $\left(1, \dfrac{\pi}{2}\right)$

422. $y^2(y^2 - 1) = x^2 \tan y$ at $(0, 1)$

Chapter 8

Applications of Derivatives

· ·

*W*hat good are derivatives if you can't do anything useful with them? Well, don't worry! There are tons and tons of useful applications involving derivatives. This chapter illustrates how calculus can help solve a variety of practical problems, including finding maximum and minimum values of functions, approximating roots of equations, and finding the velocity and acceleration of an object, just to name a few. Without calculus, many of these problems would be very difficult indeed!

The Problems You'll Work On

This chapter has a variety of applications of derivatives, including

- ✔ Approximating values of a function using linearization
- ✔ Approximating roots of equations using Newton's method
- ✔ Finding the optimal solution to a problem by finding a maximum or minimum value
- ✔ Determining how quantities vary in relation to each other
- ✔ Locating absolute and local maxima and minima
- ✔ Finding the instantaneous velocity and acceleration of an object
- ✔ Using Rolle's theorem and the mean value theorem

What to Watch Out For

This chapter presents a variety of applications and word problems, and you may have to be a bit creative when setting up some of the problems. Here are some tips:

- ✔ Think about what your variables represent in the optimization and related-rates problems; if you can't explain what they represent, start over!
- ✔ You'll have to produce equations in the related-rates and optimization problems. Getting started is often the most difficult part, so just dive in and try different things.
- ✔ Remember that *linearization* is just a fancy way of saying "tangent line."
- ✔ Although things should be set up nicely in most of the problems, note that Newton's method doesn't always work; its success depends on your starting value.

Finding and Evaluating Differentials

423–425 Find the differential dy and then evaluate dy for the given values of x and dx.

423. $y = x^2 - 4x$, $x = 3$, $dx = \frac{1}{6}$

424. $y = \frac{1}{x^2 + 1}$, $x = 1$, $dx = -0.1$

425. $y = \cos^2 x$, $x = \frac{\pi}{3}$, $dx = 0.02$

Using Linearizations to Estimate Values

429–431 Estimate the value of the given number using a linearization.

429. Estimate $7.96^{2/3}$ to the thousandths place.

430. Estimate $\sqrt{102}$ to the tenths place.

431. Estimate $\tan 46°$ to the thousandths place.

Finding Linearizations

426–428 Find the linearization L(x) of the function at the given value of a.

426. $f(x) = 3x^2$, $a = 1$

427. $f(x) = \cos x + \sin x$, $a = \frac{\pi}{2}$

428. $f(x) = \sqrt[3]{x^2 + x}$, $a = 2$

Understanding Related Rates

432–445 Solve the related-rates problem. Give an exact answer unless otherwise stated.

432. If V is the volume of a sphere of radius r and the sphere expands as time passes, find $\frac{dV}{dt}$ in terms of $\frac{dr}{dt}$.

433. A pebble is thrown into a pond, and the ripples spread in a circular pattern. If the radius of the circle increases at a constant rate of 1 meter per second, how fast is the area of the circle increasing when the radius is 4 meters?

434. If $y = x^4 + 3x^2 + x$ and $\frac{dx}{dt} = 4$, find $\frac{dy}{dt}$ when $x = 3$.

435. If $z^3 = x^2 - y^2$, $\frac{dx}{dt} = 3$, and $\frac{dy}{dt} = 2$, find $\frac{dz}{dt}$ when $x = 4$ and $y = 1$.

436. Two sides of a triangle are 6 meters and 8 meters in length, and the angle between them is increasing at a rate of 0.12 radians per second. Find the rate at which the area of the triangle is increasing when the angle between the sides is $\frac{\pi}{6}$. Round your answer to the nearest hundredth.

437. A ladder 8 feet long rests against a vertical wall. If the bottom of the ladder slides away from the wall at a rate of 3 feet per second, how fast is the angle between the top of the ladder and the wall changing when the angle is $\frac{\pi}{3}$ radians?

438. The base of a triangle is increasing at a rate of 2 centimeters per minute, and the height is increasing at a rate of 4 centimeters per minute. At what rate is the area changing when $b = 20$ centimeters and $h = 32$ centimeters?

439. At noon, Ship A is 150 kilometers east of Ship B. Ship A is sailing west at 20 kilometers per hour, and Ship B is sailing north at 35 kilometers per hour. How quickly is the distance between them changing at 3 p.m.? Round your answer to the nearest hundredth.

440. A particle moves along the curve $y = \sqrt[3]{x} + 1$. As the particle passes through the point $(8, 3)$, the x coordinate is increasing at a rate of 5 centimeters per second. How quickly is the distance from the particle to the origin changing at this point? Round your answer to the nearest hundredth.

441. Two people start walking from the same point. One person walks west at 2 miles per hour, and the other walks southwest (at an angle 45° south of west) at 4 miles per hour. How quickly is the distance between them changing after 40 minutes? Round your answer to the nearest hundredth.

442. A trough is 20 feet long, and its ends are isosceles triangles that are 5 feet across the top and have a height of 2 feet. If the trough is being filled with water at a rate of 8 cubic feet per minute, how quickly is the water level rising when the water is 1 foot deep?

443. An experimental jet is flying with a constant speed of 700 kilometers per hour. It passes over a radar station at an altitude of 2 kilometers and climbs at an angle of 45°. At what rate is the distance from the plane to the radar station increasing 2 minutes later? Round your answer to the hundredths place.

444. A lighthouse is located on an island 5 kilometers away from the nearest point P on a straight shoreline, and the light makes 6 revolutions per minute. How fast is the beam of light moving along the shore when it's 2 kilometers from P?

445. Gravel is being dumped into a pile that forms the shape of a cone whose base diameter is twice the height. If the gravel is being dumped at a rate of 20 cubic feet per minute, how fast is the height of the pile increasing when the pile is 12 feet high?

Finding Maxima and Minima from Graphs

446–450 Use the graph to find the absolute maximum, absolute minimum, local maxima, and local minima, if any. Note that endpoints will not be considered local maxima or local minima.

446.

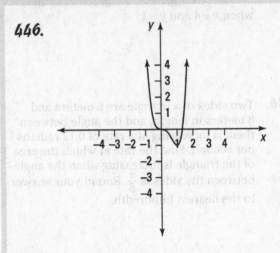

447.

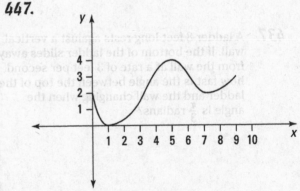

448.

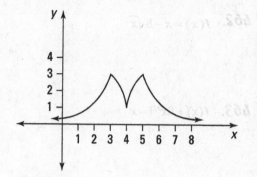

Using the Closed Interval Method

451–455 Find the absolute maximum and absolute minimum of the given function using the closed interval method.

451. $f(x) = 3x^2 - 12x + 5$ on $[0, 3]$

449.

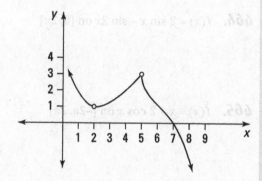

452. $f(x) = x^4 - 2x^2 + 4$ on $[-2, 3]$

453. $f(x) = \dfrac{x}{x^2 + 1}$ on $[0, 3]$

450.

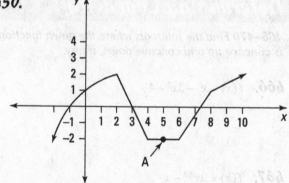

454. $f(t) = t\sqrt{4 - t^2}$ on $[-1, 2]$

455. $f(x) = x - 2\cos x$ on $[-\pi, \pi]$

Point A corresponds to which of the following?

   I.   local maximum

   II.   local minimum

   III. absolute maximum

   IV. absolute minimum

Finding Intervals of Increase and Decrease

456–460 Find the intervals of increase and decrease, if any, for the given function.

456. $f(x) = 2x^3 - 24x + 1$

457. $f(x) = x\sqrt{x+3}, x \geq -3$

458. $f(x) = \cos^2 x - \sin x$ on $[0, 2\pi]$

459. $f(x) = 2\cos x - \cos 2x$ on $0 \leq x \leq 2\pi$

460. $f(x) = 4\ln x - 2x^2$

Using the First Derivative Test to Find Local Maxima and Minima

461–465 Use the first derivative to find any local maxima and any local minima.

461. $f(x) = 2x^3 - 3x^2 - 12x$

462. $f(x) = x - 8\sqrt{x}$

463. $f(x) = 6x^{2/3} - x$

464. $f(x) = 2\sin x - \sin 2x$ on $[0, 2\pi]$

465. $f(x) = x + 2\cos x$ on $[-2\pi, 2\pi]$

Determining Concavity

466–470 Find the intervals where the given function is concave up and concave down, if any.

466. $f(x) = x^3 - 3x^2 + 4$

467. $f(x) = 9x^{2/3} - x$

468. $f(x) = x^{1/3}(x + 1)$

469. $f(x) = (x^2 - 4)^3$

470. $f(x) = 2 \cos x - \sin(2x)$ on $[0, 2\pi]$

477. $f(x) = x^4 - 4x^2 + 1$

Identifying Inflection Points

471–475 Find the inflection points of the given function, if any.

471. $f(x) = \dfrac{1}{x^2 - 9} = \left(x^2 - 9\right)^{-1}$

478. $f(x) = 2x^2(1 - x^2)$

479. $f(x) = \dfrac{x}{x^2 + 4}$

472. $f(x) = 2x^3 + x^2$

480. $f(x) = 2 \sin x - x$ on $[0, 2\pi]$

473. $f(x) = \dfrac{\sin x}{1 + \cos x}$ on $[0, 2\pi]$

Applying Rolle's Theorem

481–483 Verify that the function satisfies the hypotheses of Rolle's theorem. Then find all values c in the given interval that satisfy the conclusion of Rolle's theorem.

474. $f(x) = 3 \sin x - \sin^3 x$ on $[0, 2\pi]$

481. $f(x) = x^2 - 6x + 1$, $[0, 6]$

475. $f(x) = x^{5/3} - 5x^{2/3}$

482. $f(x) = x\sqrt{x + 8}$, $[-8, 0]$

Using the Second Derivative Test to Find Local Maxima and Minima

483. $f(x) = \cos(2\pi x)$, $[-1, 1]$

476–480 Use the second derivative test to find the local maxima and local minima of the given function.

476. $f(x) = \sqrt[3]{\left(x^2 + 1\right)^2}$

Using the Mean Value Theorem

484–486 Verify that the given function satisfies the hypotheses of the mean value theorem. Then find all numbers c that satisfy the conclusion of the mean value theorem.

484. $f(x) = x^3 + 3x - 1$, [0, 2]

485. $f(x) = 2\sqrt[3]{x}$, [0, 1]

486. $f(x) = \dfrac{x}{x+2}$, [1, 4]

Applying the Mean Value Theorem to Solve Problems

487–489 Solve the problem related to the mean value theorem.

487. If $f(1) = 12$ and $f'(x) \geq 3$ for $1 \leq x \leq 5$, what is the smallest possible value of $f(5)$? Assume that *f* satisfies the hypothesis of the mean value theorem.

488. Suppose that $2 \leq f'(x) \leq 6$ for all values of *x*. What are the strictest bounds you can put on the value of $f(8) - f(4)$? Assume that *f* is differentiable for all *x*.

489. Apply the mean value theorem to the function $f(x) = x^{1/3}$ on the interval [8, 9] to find bounds for the value of $\sqrt[3]{9}$.

Relating Velocity and Position

490–492 Use the position function s(t) to find the velocity and acceleration at the given value of t. Recall that velocity is the change in position with respect to time and acceleration is the change in velocity with respect to time.

490. $s(t) = t^2 - 8t + 4$ at $t = 5$

491. $s(t) = 2 \sin t - \cos t$ at $t = \dfrac{\pi}{2}$

492. $s(t) = \dfrac{2t}{t^2 + 1}$ at $t = 1$

Finding Velocity and Speed

493–497 Solve the given question related to speed or velocity. Recall that velocity is the change in position with respect to time.

493. A mass on a spring vibrates horizontally with an equation of motion given by $x(t) = 8 \sin(2t)$, where *x* is measured in feet and *t* is measured in seconds. Is the spring stretching or compressing at $t = \dfrac{\pi}{3}$? What is the speed of the spring at that time?

494. A stone is thrown straight up with the height given by the function $s = 40t - 16t^2$, where s is measured in feet and t is measured in seconds. What is the maximum height of the stone? What is the velocity of the stone when it's 20 feet above the ground on its way up? And what is its velocity at that height on the way down? Give exact answers.

495. A stone is thrown vertically upward with the height given by $s = 20t - 16t^2$, where s is measured in feet and t is measured in seconds. What is the maximum height of the stone? What is the velocity of the stone when it hits the ground?

496. A particle moves on a vertical line so that its coordinate at time t is given by $y = t^3 - 4t + 5$ for $t \geq 0$. When is the particle moving upward, and when is it moving downward? Give an exact answer in interval notation.

497. A particle moves on a vertical line so that its coordinate at time t is given by $y = 4t^2 - 6t - 2$ for $t \geq 0$. When is the particle moving upward, and when it is it moving downward? Give your answer in interval notation.

Solving Optimization Problems

498–512 Solve the given optimization problem. Recall that a maximum or minimum value occurs where the derivative is equal to zero, where the derivative is undefined, or at an endpoint (if the function is defined on a closed interval). Give an exact answer, unless otherwise stated.

498. Find two numbers whose difference is 50 and whose product is a minimum.

499. Find two positive numbers whose product is 400 and whose sum is a minimum.

500. Find the dimensions of a rectangle that has a perimeter of 60 meters and whose area is as large as possible.

501. Suppose a farmer with 1,500 feet of fencing encloses a rectangular area and divides it into four pens with fencing parallel to one side. What is the largest possible total area of the four pens?

502. A box with an open top is formed from a square piece of cardboard that is 6 feet wide. Find the largest volume of the box that can be made from the cardboard.

503. A box with an open top and a square base must have a volume of 16,000 cubic centimeters. Find the dimensions of the box that minimize the amount of material used.

504. Find the point(s) on the ellipse $8x^2 + y^2 = 8$ farthest from $(1, 0)$.

505. Find the point on the line $y = 4x + 6$ that is closest to the origin.

506. A rectangular poster is to have an area of 90 square inches with 1-inch margins at the bottom and sides and a 3-inch margin at the top. What dimensions give you the largest printed area?

507. At which x values on the curve $f(x) = 2 + 20x^3 - 4x^5$ does the tangent line have the largest slope?

508. A rectangular storage container with an open top is to have a volume of 20 cubic meters. The length of the base is twice the width. The material for the base costs $20 per square meter. The material for the sides costs $12 per square meter. Find the cost of the materials for the cheapest such container. Round your answer to the nearest cent.

509. A piece of wire that is 20 meters long is cut into two pieces. One is shaped into a square, and the other is shaped into an equilateral triangle. How much wire should you use for the square so that the total area is at a maximum?

510. A piece of wire that is 20 meters long is cut into two pieces. One is bent into a square, and the other is bent into an equilateral triangle. How much wire should you use for the square so that the total area is at a minimum?

511. The illumination of a light source is directly proportional to the strength of the light source and inversely proportional to the square of the distance from the source. Two light sources, one five times as strong as the other, are placed 20 feet apart, and an object is placed on the line between them. How far from the bright light source should the object be placed so that the object receives the least illumination?

512. Find the area of the largest rectangle that can be inscribed in the ellipse $\frac{x^2}{4} + \frac{y^2}{9} = 1$.

Doing Approximations Using Newton's Method

513–515 Find the fifth approximation of the root of the equation using the given first approximation.

513. $x^3 + 4x - 4 = 0$ using $x_1 = 1$. Round the solution to the fifth decimal place.

514. $x^4 - 18 = 0$ using $x_1 = 2$. Round the solution to the seventh decimal place.

515. $x^5 + 3 = 0$ using $x_1 = -1$. Round the answer to the seventh decimal place.

Approximating Roots Using Newton's Method

516–518 Find the root using Newton's method.

516. Use Newton's method to find the root of $\cos x = x$ correct to five decimal places.

517. Use Newton's method to find the root of $x^3 - x^2 - 2$ in the interval $[1, 2]$ correct to five decimal places.

518. Use Newton's method to find the positive root of $\sqrt{x+1} = x^2$ correct to five decimal places.

Chapter 9

Areas and Riemann Sums

• •

This chapter provides some of the groundwork and motivation for antiderivatives. Finding the area underneath a curve has real-world applications; however, for many curves, finding the area is difficult if not impossible to do using simple geometry. Here, you approximate the area under a curve by using rectangles and then turn to Riemann sums. The problems involving Riemann sums can be quite long and involved, especially because shortcuts to finding the solution do exist; however, the approach used in Riemann sums is the same approach you use when tackling definite integrals. It's worth understanding the idea behind Riemann sums so you can apply that approach to other problems!

The Problems You'll Work On

This chapter presents the following types of problems:

✔ Using left endpoints, right endpoints, and midpoints to estimate the area underneath a curve

✔ Finding an expression for the definite integral using Riemann sums

✔ Expressing a given Riemann sum as a definite integral

✔ Evaluating definite integrals using Riemann sums

What to Watch Out For

Here are some things to keep in mind as you do the problems in this chapter:

✔ Estimating the area under a curve typically involves quite a bit of arithmetic but shouldn't be too difficult conceptually. The process should be straightforward after you do a few problems.

✔ The problems on expressing a given Riemann sum as a definite integral don't always have unique solutions.

✔ To evaluate the problems involving Riemann sums, you need to know a few summation formulas. You can find them in any standard calculus text if you don't remember them — or you can derive them!

Calculating Riemann Sums Using Left Endpoints

519–522 Find the Riemann sum for the given function with the specified number of intervals using left endpoints.

519. $f(x) = 2 + x^2, 0 \leq x \leq 2, n = 4$

520. $f(x) = \sqrt[3]{x} + x, 1 \leq x \leq 4, n = 5$. Round your answer to two decimal places.

521. $f(x) = 4 \ln x + 2x, 1 \leq x \leq 4, n = 7$. Round your answer to two decimal places.

522. $f(x) = e^{3x} + 4, 1 \leq x \leq 9, n = 8$. Give your answer in scientific notation, rounded to three decimal places.

Calculating Riemann Sums Using Right Endpoints

523–526 Find the Riemann sum for the given function with the specified number of intervals using right endpoints.

523. $f(x) = 1 + 2x, 0 \leq x \leq 4, n = 4$

524. $f(x) = x \sin x, 2 \leq x \leq 6, n = 5$. Round your answer to two decimal places.

525. $f(x) = \sqrt{x} - 1, 0 \leq x \leq 5, n = 6$. Round your answer to two decimal places.

526. $f(x) = \dfrac{x}{x+1}, 1 \leq x \leq 3, n = 8$. Round your answer to two decimal places.

Calculating Riemann Sums Using Midpoints

527–530 Find the Riemann sum for the given function with the specified number of intervals using midpoints.

527. $f(x) = 2\cos x$, $0 \le x \le 3$, $n = 4$. Round your answer to two decimal places.

528. $f(x) = \dfrac{\sin x}{x+1}$, $1 \le x \le 5$, $n = 5$. Round your answer to two decimal places.

529. $f(x) = 3e^x + 2$, $1 \le x \le 4$, $n = 6$. Round your answer to two decimal places.

530. $f(x) = \sqrt{x} + x$, $1 \le x \le 5$, $n = 8$. Round your answer to two decimal places.

Using Limits and Riemann Sums to Find Expressions for Definite Integrals

531–535 Find an expression for the definite integral using the definition. Do not evaluate.

531. $\displaystyle\int_1^4 \sqrt{x}\,dx$

532. $\displaystyle\int_0^\pi \sin^2 x\,dx$

533. $\displaystyle\int_1^5 \left(x^2 + x\right)dx$

534. $\displaystyle\int_0^{\pi/4} \left(\tan x + \sec x\right)dx$

535. $\displaystyle\int_4^6 \left(3x^3 + x^2 - x + 5\right)dx$

Finding a Definite Integral from the Limit and Riemann Sum Form

536–540 Express the limit as a definite integral. Note that the solution is not necessarily unique.

536. $\displaystyle\lim_{n\to\infty}\sum_{i=1}^{n}\frac{3}{n}\left(4+\frac{3i}{n}\right)^{6}$

537. $\displaystyle\lim_{n\to\infty}\sum_{i=1}^{n}\frac{\pi}{3n}\sec\left(\frac{i\pi}{3n}\right)$

538. $\displaystyle\lim_{n\to\infty}\sum_{i=1}^{n}\frac{1}{n}\left(6+\frac{i}{n}\right)$

539. $\displaystyle\lim_{n\to\infty}\sum_{i=1}^{n}\frac{\pi}{2n}\left(\cos\left(\frac{i\pi}{2n}\right)+\sin\left(\frac{i\pi}{2n}\right)\right)$

540. $\displaystyle\lim_{n\to\infty}\sum_{i=1}^{n}\frac{5}{n}\sqrt{\frac{5i}{n}+\frac{125i^{3}}{n^{3}}}$

Using Limits and Riemann Sums to Evaluate Definite Integrals

541–545 Use the limit form of the definition of the integral to evaluate the integral.

541. $\displaystyle\int_{0}^{2}(1+2x)\,dx$

542. $\displaystyle\int_{0}^{4}\left(1+3x^{3}\right)dx$

543. $\displaystyle\int_{1}^{4}(4-x)\,dx$

544. $\displaystyle\int_{0}^{3}\left(2x^{2}-x-4\right)dx$

545. $\displaystyle\int_{1}^{3}\left(x^{2}+x-5\right)dx$

Chapter 10

The Fundamental Theorem of Calculus and the Net Change Theorem

• •

U sing Riemann sums to evaluate definite integrals (see Chapter 9) can be a cumbersome process. Fortunately, the fundamental theorem of calculus gives you a much easier way to evaluate definite integrals. In addition to evaluating definite integrals in this chapter, you start finding *antiderivatives,* or indefinite integrals. The net change theorem problems at the end of this chapter offer some insight into the use of definite integrals.

Although the antiderivative problems you encounter in this chapter aren't too complex, finding antiderivatives is in general a much more difficult process than finding derivatives, so consider yourself warned! You encounter many challenging antiderivative problems in later chapters.

The Problems You'll Work On

In this chapter, you see a variety of antiderivative problems:

✔ Finding derivatives of integrals

✔ Evaluating definite integrals

✔ Computing indefinite integrals

✔ Using the net change theorem to interpret definite integrals and to find the distance and displacement of a particle

What to Watch Out For

Although many of the problems in the chapter are easier antiderivative problems, you still need to be careful. Here are some tips:

✔ Simplify before computing the antiderivative. Don't forget to use trigonometric identities when simplifying the integrand.

✔ You don't often see problems that ask you to find derivatives of integrals, but make sure you practice them. They usually aren't that difficult, so they make for easier points on a quiz or test.

✔ Note the difference between distance and displacement; distance is always greater than or equal to zero, whereas displacement may be positive, negative, or zero! Finding the distance traveled typically involves more work than simply finding the displacement.

Using the Fundamental Theorem of Calculus to Find Derivatives

546–557 Find the derivative of the given function.

546. $f(x) = \int_0^x \sqrt{1+4t}\, dt$

547. $f(x) = \int_3^x \left(2+t^6\right)^4 dt$

548. $f(x) = \int_0^x t^3 \cos(t)\, dt$

549. $f(x) = \int_x^4 e^{t^2} dt$

550. $f(x) = \int_{\sin x}^0 \left(1-t^2\right) dt$

551. $f(x) = \int_{\ln(x^2+1)}^4 e^t \, dt$

552. $f(x) = \int_{\cos x}^1 \left(t^2 + \sin t\right) dt$

553. $f(x) = \int_{1/x}^1 \sin^2(t)\, dt$

554. $f(x) = \int_1^{x^3} \frac{1}{t+t^4}\, dt$

555. $f(x) = \int_{\tan x}^{x^3} \frac{1}{\sqrt{3+t}}\, dt$

556. $f(x) = \int_{2x}^{6x} \frac{t^2-1}{t^4+1}\, dt$

557. $f(x) = \int_{\log_5 x}^{x^2} 5^t \, dt$

Working with Basic Examples of Definite Integrals

558–570 Evaluate the definite integral using basic antiderivative rules.

558. $\int_1^3 5\, dx$

559. $\int_0^{\pi/4} \cos x \, dx$

560. $\int_0^{\pi/4} \sec^2 t \, dt$

561. $\int_0^{\pi/3} \sec x \tan x \, dx$

562. $\int_1^2 4^x \, dx$

563. $\int_0^{\pi} (x - \sin x) \, dx$

564. $\int_1^4 (x + x^3) \, dx$

565. $\int_1^8 2\sqrt[3]{x} \, dx$

566. $\int_{\pi/4}^{\pi/2} (\csc^2 x - 1) \, dx$

567. $\int_0^2 (1 + 2y - 4y^3) \, dy$

568. $\int_0^1 (\sqrt{x} + x) \, dx$

569. $\int_{-3}^{\pi/2} f(x) \, dx$, where

$$f(x) = \begin{cases} x, & -3 \le x \le 0 \\ \cos x, & 0 < x \le \dfrac{\pi}{2} \end{cases}$$

570. $\int_{-3}^4 |x - 2| \, dx$

Understanding Basic Indefinite Integrals

571–610 Find the indicated antiderivative.

571. $\int x^{3/4} \, dx$

572. $\int \dfrac{5}{x^7} \, dx$

573. $\int (x^3 + 3x^2 - 1) \, dx$

574. $\int (3\cos x - 4\sin x)\,dx$

575. $\int (4\cos^2 x + 4\sin^2 x)\,dx$

576. $\int \tan^2 x\,dx$

577. $\int (3x^2 + 2x + 1)\,dx$

578. $\int \left(\dfrac{2}{3}x^{4/3} + 5x^4\right)dx$

579. $\int (5 + x + \tan^2 x)\,dx$

580. $\int 6\sqrt{x}\,\sqrt[3]{x}\,dx$

581. $\int \sqrt{5x}\,dx$

582. $\int \sqrt[6]{\dfrac{4}{x}}\,dx$

583. $\int \dfrac{1 - \sin^2 x}{3\cos^2 x}\,dx$

584. $\int x^2 (x^2 + 3x)\,dx$

585. $\int \dfrac{x^2 + x + 1}{x^4}\,dx$

586. $\int (1 + x^2)^2\,dx$

587. $\int (2 - \cot^2 x)\,dx$

588. $\int \dfrac{\sqrt{x} - 4x^2}{x}\,dx$

589. $\int (\sqrt[3]{x} + 1)^2\,dx$

590. $\int \sqrt{x}\left(1+\dfrac{1}{\sqrt[3]{x}}\right)dx$

591. $\int \dfrac{\cos x}{\sin^2 x}\,dx$

592. $\int \dfrac{x+5x^6}{x^3}\,dx$

593. $\int \dfrac{x+1}{\sqrt[7]{x}}\,dx$

594. $\int \dfrac{1+\sin^2 x}{\sin^2 x}\,dx$

595. $\int (1-x)(2+x)\,dx$

596. $\int \sec x\left(\sec x-\cos x\right)dx$

597. $\int \dfrac{x^2-5x+6}{x-2}\,dx$

598. $\int \dfrac{x^3+1}{x+1}\,dx$

599. $\int \left(4x^{1.6}-x^{2.8}\right)dx$

600. $\int \dfrac{5+x}{x^{2.1}}\,dx$

601. $\int \left(x^{7/2}-x^{1/2}+\dfrac{1}{x^{1/5}}\right)dx$

602. $\int \dfrac{\sin 2x}{4\cos x}\,dx$

603. $\int \sin^2 x\left(1+\cot^2 x\right)dx$

604. $\int \left(\dfrac{4x^3+5x-2}{x^3}\right)dx$

605. $\int \left(3\sin^2 x+3\cos^2 x+4\right)dx$

606. $\int \dfrac{x^3 - 25x}{x+5} dx$

607. $\int \dfrac{\tan^2 x}{\sin^2 x} dx$

608. $\int \dfrac{\cos 2x}{\cos x + \sin x} dx$

609. $\int \dfrac{\cos 2x}{\sin^2 x} dx$

610. $\int \dfrac{\cos x + \cos x \tan^2 x}{\sec^2 x} dx$

Understanding the Net Change Theorem

611–619 Use the net change theorem to interpret the given definite integral.

611. If $w'(t)$ is the rate of a baby's growth in pounds per week, what does $\int_0^2 w'(t) dt$ represent?

612. If $r(t)$ represents the rate at which oil leaks from a tanker in gallons per minute, what does $\int_0^{180} r(t) dt$ represent?

613. A new bird population that is introduced into a refuge starts with 100 birds and increases at a rate of $p'(t)$ birds per month. What does $100 + \int_0^6 p'(t) dt$ represent?

614. If $v(t)$ is the velocity of a particle in meters per second, what does $\int_0^{10} v(t) dt$ represent?

615. If $a(t)$ is the acceleration of a car in meters per second squared, what does $\int_3^5 a(t) dt$ represent?

616. If $P'(t)$ represents the rate of production of solar panels, where t is measured in weeks, what does $\int_2^4 P'(t) dt$ represent?

617. The current in a wire $I(t)$ is defined as the derivative of the charge, $Q'(t) = I(t)$. What does $\int_{t_1}^{t_2} I(t) dt$ represent if t is measured in hours?

618. $I'(t)$ represents the rate of change in your income in dollars from a new job, where t is measured in years. What does $\int_0^{10} I'(t) dt$ represent?

619. Water is flowing into a pool at a rate of $w'(t)$, where t is measured in minutes and $w'(t)$ is measured in gallons per minute. What does $\int_{60}^{120} w'(t)\,dt$ represent?

Finding the Distance Traveled by a Particle Given the Velocity

625–629 A particle moves according to the given velocity function over the given interval. Find the total distance traveled by the particle. **Remember:** Velocity is the rate of change in position with respect to time.

Finding the Displacement of a Particle Given the Velocity

620–624 A particle moves according to the given velocity function over the given interval. Find the displacement of the particle. **Remember:** Displacement is the change in position, and velocity is the rate of change in position with respect to time.

620. $v(t) = 2t - 4,\ 0 \le t \le 5$

621. $v(t) = t^2 - t - 6,\ 2 \le t \le 4$

622. $v(t) = 2\cos t,\ 0 \le t \le \pi$

623. $v(t) = \sin t - \cos t,\ \dfrac{-\pi}{6} \le t \le \dfrac{\pi}{2}$

624. $v(t) = \sqrt{t} - 4,\ 1 \le t \le 25$

625. $v(t) = 2t - 4,\ 0 \le t \le 5$

626. $v(t) = t^2 - t - 6,\ 2 \le t \le 4$

627. $v(t) = 2\cos t,\ 0 \le t \le \pi$

628. $v(t) = \sin t - \cos t,\ \dfrac{-\pi}{6} \le t \le \dfrac{\pi}{2}$

629. $v(t) = \sqrt{t} - 4,\ 1 \le t \le 25$

Finding the Displacement of a Particle Given Acceleration

*630–632 A particle moves according to the given acceleration function over the given interval. First, find the velocity function. Then find the displacement of the particle. **Remember:** Displacement is the change in position, velocity is the rate of change in position with respect to time, and acceleration is the rate of change in velocity with respect to time.*

630. $a(t) = t + 2, v(0) = -6, 0 \leq t \leq 8$

631. $a(t) = 2t + 1, v(0) = -12, 0 \leq t \leq 5$

632. $a(t) = \sin t + \cos t, v\left(\frac{\pi}{4}\right) = 0, \frac{\pi}{6} \leq t \leq \pi$

Finding the Distance Traveled by a Particle Given Acceleration

*633–635 A particle moves according to the given acceleration function over the given interval. Find the total distance traveled by the particle. **Remember:** Displacement is the change in position, velocity is the rate of change in position with respect to time, and acceleration is the rate of change in velocity with respect to time.*

633. $a(t) = t + 2, v(0) = -6, 0 \leq t \leq 8$

634. $a(t) = 2t + 1, v(0) = -12, 0 \leq t \leq 5$

635. $a(t) = \sin t + \cos t, v\left(\frac{\pi}{4}\right) = 0, \frac{\pi}{6} \leq t \leq \pi$

Chapter 11

Applications of Integration

. . . .

This chapter presents questions related to applications of integrals: finding the area between curves, finding the volumes of a solid, and calculating the work done by a varying force. The work problems contain a variety of questions, all of which apply to a number of real-life situations and relate to questions that you may encounter in a physics class. At the end of the chapter, you answer questions related to finding the average value of a function on an interval.

The Problems You'll Work On

In this chapter, you see a variety of applications of the definite integral:

- ✔ Finding areas between curves
- ✔ Using the disk/washer method to find volumes of revolution
- ✔ Using the shell method to find volumes of revolution
- ✔ Finding volumes of solids using cross-sectional slices
- ✔ Finding the amount of work done when applying a force to an object
- ✔ Finding the average value of a continuous function on an interval

What to Watch Out For

Here are a few things to consider for the problems in this chapter:

- ✔ Make graphs for the area and volume problems to help you visualize as much as possible.
- ✔ Don't get the formulas and procedures for the disk/washer method mixed up with the shell method; it's easy to do! For example, when rotating regions about a horizontal line using disks/washers, your curve should be of the form $y = f(x)$, but if you're using shells, your curve should be of the form $x = g(y)$. When rotating a region about a vertical line and using disks/washers, your curve should be of the form $x = g(y)$, but if you're using shells, your curve should be of the form $y = f(x)$.
- ✔ Some of the volume of revolution problems can be solved using either the disk/washer method or the shell method; other problems can be solved easily only by using one method. Pay attention to which problems seem to be doable using either method and which ones do not.
- ✔ The work problems often give people a bit of a challenge, so don't worry if your first attempt isn't correct. Keep trying!

Areas between Curves

636–661 Find the area of the region bounded by the given curves. (**Tip:** It's often useful to make a rough sketch of the region.)

636. $y = x^2, y = x^4$

637. $y = x, y = \sqrt{x}$

638. $y = \cos x + 1, y = x, x = 0, x = 1$

639. $x = y^2 - y, x = 3y - y^2$

640. $x + 1 = y^2, x = \sqrt{y}, y = 0, y = 1$

641. $x = 1 + y^2, y = x - 7$

642. $x = y^2, x = 3y - 2$

643. $x = 2y^2, x + y = 1$

644. $y = 2x, y = 8 - x^2$

645. $x = 2 - y^2, x = y^2 - 2$

646. $y = 14 - x^2, y = x^2 - 4$

647. $x = y, 4x + y^2 = -3$

648. $x = 1 + \sqrt{y}, x = \dfrac{3 + y}{3}$

649. $y = x - \dfrac{\pi}{2}, y = \cos x, x = 0, x = \pi$

650. $y = x^3 - x, y = 2x$

651. $x + y = 0, x = y^2 + 4y$

652. $x = \sqrt{y + 3}, x = \dfrac{y + 3}{2}$

653. $y = \sin x,\ y = \cos x,\ x = -\frac{\pi}{4},\ x = \frac{\pi}{2}$

654. $x = y^2,\ x = \sqrt{y},\ y = 0,\ y = 2$

655. $x = y^2 - y,\ x = 4y$

656. $y = x - 1,\ y^2 = 2x + 6$

657. $y = x,\ x + 2y = 0,\ 2x + y = 3$

658. $y = \sqrt{x+4},\ y = \frac{x+4}{2}$

659. $y = |2x|,\ y = x^2 - 3$

660. $y = \cos x,\ y = \sin 2x,\ x = 0,\ x = \frac{\pi}{2}$

661. $y = 2e^{2x},\ y = 3 - 5e^x,\ x = 0$

Finding Volumes Using Disks and Washers

662–681 *Find the volume of the solid obtained by revolving the indicated region about the given line. (**Tip:** Making a rough sketch of the region that's being rotated is often useful.)*

662. The region is bounded by the curves $y = x^4$, $x = 1$, and $y = 0$ and is rotated about the x-axis.

663. The region is bounded by the curves $x = \sqrt{\sin y}$, $x = 0$, $y = 0$, and $y = \pi$ and is rotated about the y-axis.

664. The region is bounded by the curves $y = \frac{1}{x}$, $x = 3$, $x = 5$, and $y = 0$ and is rotated about the x-axis.

665. The region is bounded by the curves $y = \frac{1}{\sqrt{x}}$, $x = 1$, $x = 3$, and $y = 0$ and is rotated about the x-axis.

666. The region is bounded by the curves $y = \csc x$, $x = \frac{\pi}{4}$, $x = \frac{\pi}{2}$, and $y = 0$ and is rotated about the x-axis.

667. The region is bounded by the curves $x + 4y = 4$, $x = 0$, and $y = 0$ and is rotated about the x-axis.

668. The region is bounded by the curves $x = y^2 - y^3$ and $x = 0$ and is rotated about the y-axis.

669. The region is bounded by the curves $y = \sqrt{x-1}$, $y = 0$, and $x = 5$ and is rotated about the x-axis.

670. The region is bounded by the curves $y = 4 - \dfrac{x^2}{4}$ and $y = 2$ and is rotated about the x-axis.

671. The region is bounded by the curves $x = y^{2/3}$, $x = 0$, and $y = 8$ and is rotated about the y-axis.

672. The region is bounded by the curves $y = \sqrt{r^2 - x^2}$ and $y = 0$ and is rotated about the x-axis.

673. The region is bounded by the curves $y = \sin x$, $y = \cos x$, $x = 0$, and $x = \dfrac{\pi}{2}$ and is rotated about the x-axis.

674. The region is bounded by the curves $y = \dfrac{1}{1+x^2}$, $y = 0$, $x = 0$, and $x = 1$ and is rotated about the x-axis.

675. The region is bounded by the curves $y = 3 + 2x - x^2$ and $x + y = 3$ and is rotated about the x-axis.

676. The region is bounded by the curves $y = x^2$ and $x = y^2$ and is rotated about the y-axis.

677. The region is bounded by the curves $y = x^{2/3}$, $y = 1$, and $x = 0$ and is rotated about the line $y = 2$.

678. The region is bounded by the curves $y = x^{2/3}$, $y = 1$, and $x = 0$ and is rotated about the line $x = -1$.

679. The region is bounded by $y = \sec x$, $y = 0$, and $0 \le x \le \frac{\pi}{3}$ and is rotated about the line $y = 4$.

680. The region is bounded by the curves $x = y^2$ and $x = 4$ and is rotated about the line $x = 5$.

681. The region is bounded by the curves $y = e^{-x}$, $y = 0$, $x = 0$, and $x = 1$ and is rotated about the line $y = -1$.

Finding Volume Using Cross-Sectional Slices

682–687 Find the volume of the indicated region using the method of cross-sectional slices.

682. The base of a solid C is a circular disk that has a radius of 4 and is centered at the origin. Cross-sectional slices perpendicular to the x-axis are squares. Find the volume of the solid.

683. The base of a solid C is a circular disk that has a radius of 4 and is centered at the origin. Cross-sectional slices perpendicular to the x-axis are equilateral triangles. Find the volume of the solid.

684. The base of a solid S is an elliptical region with the boundary curve $4x^2 + 9y^2 = 36$. Cross-sectional slices perpendicular to the y-axis are squares. Find the volume of the solid.

685. The base of a solid S is triangular with vertices at $(0, 0)$, $(2, 0)$, and $(0, 4)$. Cross-sectional slices perpendicular to the y-axis are isosceles triangles with height equal to the base. Find the volume of the solid.

686. The base of a solid S is an elliptical region with the boundary curve $4x^2 + 9y^2 = 36$. Cross-sectional slices perpendicular to the x-axis are isosceles right triangles with the hypotenuse as the base. Find the volume of the solid.

687. The base of a solid S is triangular with vertices at $(0, 0)$, $(2, 0)$, and $(0, 4)$. Cross-sectional slices perpendicular to the y-axis are semicircles. Find the volume of the solid.

Finding Volumes Using Cylindrical Shells

688–711 Find the volume of the region bounded by the given functions using cylindrical shells. Give an exact answer. (**Tip:** Making a rough sketch of the region that's being rotated is often useful.)

688. The region is bounded by the curves $y = \frac{1}{x}$, $y = 0$, $x = 1$, and $x = 3$ and is rotated about the y-axis.

689. The region is bounded by the curves $y = x^2$, $y = 0$, and $x = 2$ and is rotated about the y-axis.

690. The region is bounded by the curves $x = \sqrt[3]{y}$, $x = 0$, and $y = 1$ and is rotated about the x-axis.

691. The region is bounded by the curves $y = x^2$, $y = 0$, and $x = 2$ and is rotated about the line $x = -1$.

692. The region is bounded by the curves $y = 2x$ and $y = x^2 - 4x$ and is rotated about the y-axis.

693. The region is bounded by the curves $y = x^4$, $y = 16$, and $x = 0$ and is rotated about the x-axis.

694. The region is bounded by the curves $x = 5y^2 - y^3$ and $x = 0$ and is rotated about the x-axis.

695. The region is bounded by the curves $y = x^2$ and $y = 4x - x^2$ and is rotated about the line $x = 4$.

696. The region is bounded by the curves $y = 1 + x + x^2$, $x = 0$, $x = 1$, and $y = 0$ and is rotated about the y-axis.

697. The region is bounded by the curves $y = 4x - x^2$, $x = 0$, and $y = 4$ and is rotated about the y-axis.

698. The region is bounded by the curves $y = \sqrt{9 - x}$, $x = 0$, and $y = 0$ and is rotated about the x-axis.

699. The region is bounded by the curves $y = 1 - x^2$ and $y = 0$ and is rotated about the line $x = 2$.

700. The region is bounded by the curves $y = 5 + 3x - x^2$ and $2x + y = 5$ and is rotated about the y-axis.

701. The region is bounded by the curves $x + y = 5$, $y = x$, and $y = 0$ and is rotated about the line $x = -1$.

702. The region is bounded by the curves $y = \sin(x^2)$, $x = 0$, $x = \sqrt{\pi}$, and $y = 0$ and is rotated about the y-axis.

703. The region is bounded by the curves $x = e^y$, $x = 0$, $y = 0$, and $y = 2$ and is rotated about the x-axis.

704. The region is bounded by the curves $y = e^{-x^2}$, $y = 0$, $x = 0$, and $x = 3$ and is rotated about the y-axis.

705. The region is bounded by the curves $x = y^3$ and $y = x^2$ and is rotated about the line $x = -1$.

706. The region is bounded by the curves $y = \frac{1}{x}$, $y = 0$, $x = 1$, and $x = 3$ and is rotated about the line $x = 4$.

707. The region is bounded by the curves $y = \sqrt{x}$ and $y = x^3$ and is rotated about the line $y = 1$.

708. The region is bounded by the curves $y = \sqrt{x + 2}$, $y = x$, and $y = 0$ and is rotated about the x-axis.

709. The region is bounded by the curves $y = \frac{1}{\sqrt{2\pi}} e^{-x^2/2}$, $y = 0$, $x = 0$, and $x = 1$ and is rotated about the y-axis.

710. The region is bounded by the curves $y = \sqrt{x}$, $y = \ln x$, $x = 1$, and $x = 2$ and is rotated about the y-axis.

711. The region is bounded by the curves $x = \cos y$, $y = 0$, and $y = \frac{\pi}{2}$ and is rotated about the x-axis.

Work Problems

712–735 Find the work required in each situation. Note that if the force applied is constant, work equals force times displacement (W = Fd); if the force is variable, you use the integral $W = \int_a^b f(x)dx$, where f(x) is the force on the object at x and the object moves from x = a to x = b.

712. In joules, how much work do you need to lift a 50-kilogram weight 3 meters from the floor? (*Note:* The acceleration due to gravity is 9.8 meters per second squared.)

713. In joules, how much work is done pushing a wagon a distance of 12 meters while exerting a constant force of 800 newtons in the direction of motion?

714. A heavy rope that is 30 feet long and weighs 0.75 pounds per foot hangs over the edge of a cliff. In foot-pounds, how much work is required to pull all the rope to the top of the cliff?

715. A heavy rope that is 30 feet long and weighs 0.75 pounds per foot hangs over the edge of a cliff. In foot-pounds, how much work is required to pull only half of the rope to the top of the cliff?

716. A heavy industrial cable weighing 4 pounds per foot is used to lift a 1,500-pound piece of metal up to the top of a building. In foot-pounds, how much work is required if the building is 300 feet tall?

717. A 300-pound uniform cable that's 150 feet long hangs vertically from the top of a tall building. How much work is required to lift the cable to the top of the building?

718. A container measuring 4 meters long, 2 meters wide, and 1 meter deep is full of water. In joules, how much work is required to pump the water out of the container? (*Note:* The density of water is 1,000 kilograms per cubic meter, and the acceleration due to gravity is 9.8 meters per second squared.)

719. If the work required to stretch a spring 2 feet beyond its natural length is 14 foot-pounds, how much work is required to stretch the spring 18 inches beyond its natural length? (*Note:* For a spring, force equals the spring constant k multiplied by the spring's displacement from its natural length: $F(x) = kx$.)

720. A force of 8 newtons stretches a spring 9 centimeters beyond its natural length. In joules, how much work is required to stretch the spring from 12 centimeters beyond its natural length to 22 centimeters beyond its natural length? Round the answer to the hundredths place. (*Note:* For a spring, force equals the spring constant k multiplied by the spring's displacement from its natural length: $F(x) = kx$. Also note that 1 newton-meter = 1 joule.)

721. A particle is located at a distance x meters from the origin, and a force of $2\sin\left(\frac{\pi x}{6}\right)$ newtons acts on it. In joules, how much work is done moving the particle from $x = 1$ to $x = 2$? The force is directed along the x-axis. Find an exact answer. (*Note:* 1 newton-meter = 1 joule.)

722. Five joules of work is required to stretch a spring from its natural length of 15 centimeters to a length of 25 centimeters. In joules, how much work is required to stretch the spring from a length of 30 centimeters to a length of 42 centimeters? (*Note:* For a spring, force equals the spring constant k multiplied by the spring's displacement from its natural length: $F(x) = kx$. Also note that 1 newton-meter = 1 joule.)

723. It takes a force of 15 pounds to stretch a spring 6 inches beyond its natural length. In foot-pounds, how much work is required to stretch the spring 8 inches beyond its natural length? (*Note:* For a spring, force equals the spring constant k multiplied by the spring's displacement from its natural length: $F(x) = kx$.)

724. Suppose a spring has a natural length of 10 centimeters. If a force of 30 newtons is required to stretch the spring to a length of 15 centimeters, how much work (in joules) is required to stretch the spring from 15 centimeters to 20 centimeters? (*Note:* For a spring, force equals the spring constant k multiplied by the spring's displacement from its natural length: $F(x) = kx$. Also note that 1 newton-meter = 1 joule.)

725. Suppose a 20-foot hanging chain weighs 4 pounds per foot. In foot-pounds, how much work is done in lifting the end of the chain to the top so that the chain is folded in half?

726. A 20-meter chain lying on the ground has a mass of 100 kilograms. In joules, how much work is required to raise one end of the chain to a height of 5 meters? Assume that the chain is L-shaped after being lifted with a remaining 15 meters of chain on the ground and that the chain slides without friction as its end is lifted. Also assume that the weight density of the chain is constant and is equal to $\left(\frac{100}{20} \text{ kg/m}\right)\left(9.8 \text{ m/s}^2\right) = 49 \text{ N/m}$. Round to the nearest joule. (*Note:* 1 newton-meter = 1 joule.)

727. A trough has a triangular face, and the width and height of the triangle each equal 4 meters. The trough is 10 meters long and has a 3-meter spout attached to the top of the tank. If the tank is full of water, how much work is required to empty it? Round to the nearest joule. (*Note:* The acceleration due to gravity is 9.8 meters per second squared, and the density of water is 1,000 kilograms per cubic meter.)

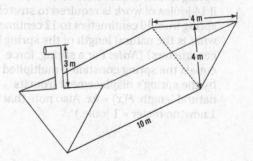

728. A cylindrical storage container has a diameter of 12 feet and a height of 8 feet. The container is filled with water to a height of 4 feet. How much work is required to pump all the water out over the side of the tank? Round to the nearest foot-pound. (*Note:* Water weighs 62.5 pounds per cubic foot.)

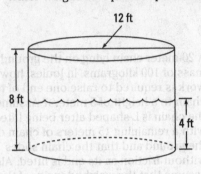

729. A thirsty farmer is using a rope of negligible weight to pull up a bucket that weighs 5 pounds from a well that is 100 feet deep. The bucket is filled with 50 pounds of water, but as the unlucky farmer pulls up the bucket at a rate of 2 feet per second, water leaks out at a constant rate and finishes draining just as the bucket reaches the top of the well. In foot-pounds, how much work has the thirsty farmer done?

730. Ten joules of work is needed to stretch a spring from 8 centimeters to 10 centimeters. If 14 joules of work is required to stretch the spring from 10 centimeters to 12 centimeters, what is the natural length of the spring in centimeters? (*Note:* For a spring, force equals the spring constant k multiplied by the spring's displacement from its natural length: $F(x) = kx$. Also note that 1 newton-meter = 1 joule.)

731. A cylindrical storage container has a diameter of 12 feet and a height of 8 feet. The container is filled with water to a distance of 4 feet from the top of the tank. Water is being pumped out, but the pump breaks after $13,500\pi$ foot-pounds of work has been completed. In feet, how far is the remaining water from the top of the tank? Round your answer to the hundredths place.

732. Twenty-five joules of work is needed to stretch a spring from 40 centimeters to 60 centimeters. If 40 joules of work is required to stretch the spring from 60 centimeters to 80 centimeters, what is the natural length of the spring in centimeters? Round the answer to two decimal places. (*Note:* For a spring, force equals the spring constant k multiplied by the spring's displacement from its natural length: $F(x) = kx$. Also note that 1 newton-meter = 1 joule.)

733. A cylindrical storage tank with a radius of 1 meter and a length of 5 meters is lying on its side and is full of water. If the top of the tank is 3 meters below ground, how much work in joules will it take to pump all the water to ground level? (*Note:* The acceleration due to gravity is 9.8 meters per second squared, and the density of water is 1,000 kilograms per cubic meter.)

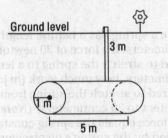

734. A tank that has the shape of a hemisphere with a radius of 4 feet is full of water. If the opening to the tank is 1 foot above the top of the tank, how much work in foot-pounds is required to empty the tank?

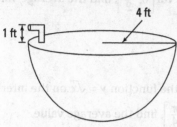

735. An open tank full of water has the shape of a right circular cone. The tank is 10 feet across the top and 6 feet high. In foot-pounds, how much work is done in emptying the tank by pumping the water over the top edge? Round to the nearest foot-pound. (**Note:** Water weighs 62.5 pounds per cubic foot.)

Average Value of a Function

736–741 Find the average value of the function on the given interval by using the formula

$$f_{avg} = \frac{1}{b-a}\int_a^b f(x)dx.$$

736. $f(x) = x^3$, $[-1, 2]$

737. $f(x) = \sin x$, $\left[0, \frac{3\pi}{2}\right]$

738. $f(x) = (\sin^3 x)(\cos x)$, $\left[0, \frac{\pi}{2}\right]$

739. $g(x) = x^2\sqrt{1+x^3}$, $[0, 2]$

740. $y = \sinh x \cosh x$, $[0, \ln 3]$

741. $f(r) = \dfrac{5}{(1+r)^2}$, $[1, 4]$

742–747 Solve the problem using the average value formula.

742. The linear density of a metal rod measuring 8 meters in length is $f(x) = \dfrac{14}{\sqrt{x+2}}$ kilograms per meter, where x is measured in meters from one end of the rod. Find the average density of the rod.

743. Find all numbers d such that the average value of $f(x) = 2 + 4x - 3x^2$ on $[0, d]$ is equal to 3.

744. Find all numbers d such that the average value of $f(x) = 3 + 6x - 9x^2$ on $[0, d]$ is equal to -33.

745. Find all values of c in the given interval such that $f_{avg} = f(c)$ for the function $f(x) = \dfrac{4(x^2 + 1)}{x^2}$ on $[1, 3]$.

746. Find all values of c in the given interval such that $f_{avg} = f(c)$ for the function $f(x) = \dfrac{1}{\sqrt{x}}$ on $[4, 9]$.

747. Find all values of c in the given interval such that $f_{avg} = f(c)$ for the function $f(x) = 5 - 3x^2$ on $[-2, 2]$.

748–749 Use the average value formula.

748. For the function $f(x) = x \sin x$ on the interval $\left[0, \dfrac{\pi}{2}\right]$, find the average value.

749. For the function $y = \sqrt{x}$ on the interval $\left[0, \dfrac{\pi}{4}\right]$, find the average value.

Chapter 12

Inverse Trigonometric Functions, Hyperbolic Functions, and L'Hôpital's Rule

· ·

This chapter looks at the very important inverse trigonometric functions and the hyperbolic functions. For these functions, you see lots of examples related to finding derivatives and integration as well. Although you don't spend much time on the hyperbolic functions in most calculus courses, the inverse trigonometric functions come up again and again; the inverse tangent function is especially important when you tackle the partial fraction problems of Chapter 14. At the end of this chapter, you experience a blast from the past: limit problems!

The Problems You'll Work On

This chapter has a variety of limit, derivative, and integration problems. Here's what you work on:

- ✔ Finding derivatives and antiderivatives using inverse trigonometric functions
- ✔ Finding derivatives and antiderivatives using hyperbolic functions
- ✔ Using L'Hôpital's rule to evaluate limits

What to Watch Out For

Here are a few things to consider for the problems in this chapter:

- ✔ The derivative questions just involve new formulas; the power, product, quotient, and chain rules still apply.
- ✔ Know the definitions of the hyperbolic functions so that if you forget any formulas, you can easily derive them. They're simply defined in terms of the exponential function, e^x.
- ✔ Although L'Hôpital's rule is great for many limit problems, make sure you have an indeterminate form before you use it, or you can get some very incorrect solutions.

Finding Derivatives Involving Inverse Trigonometric Functions

750–762 Find the derivative of the given function.

750. $y = 2\sin^{-1}(x-1)$

751. $y = 3\cos^{-1}(x^4 + x)$

752. $y = \sqrt{\tan^{-1} x}$

753. $y = \sqrt{1 - x^2}\, \sin^{-1} x$

754. $y = \tan^{-1}(\cos x)$

755. $y = e^{\sec^{-1} t}$

Note: The derivative formula for $\sec^{-1} t$ varies, depending on the definition used. For this problem, use the formula $\dfrac{d}{dt}\sec^{-1} t = \dfrac{1}{t\sqrt{t^2 - 1}}$.

756. $y = \csc^{-1} e^{2x}$

757. $y = e^{x\sec^{-1} x}$

Note: The derivative formula for $\sec^{-1} x$ varies, depending on the definition used. For this problem, use the formula $\dfrac{d}{dx}\sec^{-1} x = \dfrac{1}{x\sqrt{x^2 - 1}}$.

758. $y = 4\arccos\dfrac{x}{3}$

759. $y = x\sin^{-1} x + \sqrt{1 - x^2}$

760. $y = \cot^{-1} x + \cot^{-1}\dfrac{1}{x}$

761. $y = \tan^{-1} x + \dfrac{x}{1 + x^2}$

762. $y = \tan^{-1}\left(x - \sqrt{1 + x^2}\right)$

Finding Antiderivatives by Using Inverse Trigonometric Functions

763–774 *Find the indefinite integral or evaluate the definite integral.*

763. $\int_0^1 \frac{2}{x^2+1}\,dx$

764. $\int_{1/2}^{\sqrt{3}/2} \frac{5}{\sqrt{1-x^2}}\,dx$

765. $\int \frac{dx}{\sqrt{1-9x^2}}$

766. $\int_0^{1/2} \frac{\sin^{-1} x}{\sqrt{1-x^2}}\,dx$

767. $\int_0^{\pi/2} \frac{\cos x}{1+\sin^2 x}\,dx$

768. $\int \frac{1}{x\left(1+(\ln x)^2\right)}\,dx$

769. $\int \frac{dx}{\sqrt{x}\,(1+x)}$

770. $\int \frac{1}{x\sqrt{x^2-25}}\,dx$

771. $\int \frac{1}{(x-1)\sqrt{x^2-2x}}\,dx$

772. $\int \frac{e^{3x}}{\sqrt{1-e^{6x}}}\,dx$

773. $\int \frac{x+4}{x^2+4}\,dx$

774. $\int_2^3 \frac{2x-3}{\sqrt{4x-x^2}}\,dx$

Evaluating Hyperbolic Functions Using Their Definitions

775–779 *Use the definition of the hyperbolic functions to find the values.*

775. $\sinh 0$

776. $\cosh (\ln 2)$

777. coth (ln 6)

778. tanh 1

779. sech $\frac{1}{2}$

Finding Derivatives of Hyperbolic Functions

780–789 Find the derivative of the given function.

780. $y = \cosh^2 x$

781. $y = \sinh(x^2)$

782. $y = \frac{1}{6}\text{csch}(2x)$

783. $y = \tanh(e^x)$

784. $y = \tanh(\sinh x)$

785. $y = e^{\sinh(5x)}$

786. $y = \text{sech}^4(10x)$

787. $y = \tanh\left(\sqrt{1+t^4}\right)$

788. $y = x^3 \sinh(\ln x)$

789. $y = \ln\left(\tanh\left(\frac{x}{3}\right)\right)$

Finding Antiderivatives of Hyperbolic Functions

790–799 Find the antiderivative.

790. $\int \sinh(1-3x)\,dx$

791. $\int \cosh^2(x-3)\sinh(x-3)\,dx$

792. $\int \coth x \, dx$

793. $\int \text{sech}^2 (3x-2) \, dx$

794. $\int \dfrac{\text{sech}^2 x}{2+\tanh x} \, dx$

795. $\int x \cosh(6x) \, dx$

796. $\int x \, \text{csch}^2 \left(\dfrac{x^2}{5} \right) dx$

797. $\int \dfrac{\text{csch}\left(\dfrac{1}{x^2}\right) \coth\left(\dfrac{1}{x^2}\right)}{x^3} \, dx$

798. $\int_{\ln 2}^{\ln 3} \dfrac{\cosh x}{\cosh^2 x - 1} \, dx$

799. $\int_0^{\ln 2} \dfrac{\cosh x}{\sqrt{9-\sinh^2 x}} \, dx$

Evaluating Indeterminate Forms Using L'Hôpital's Rule

800–831 If the limit is an indeterminate form, evaluate the limit using L'Hôpital's rule. Otherwise, find the limit using any other method.

800. $\lim\limits_{x \to -1} \dfrac{x^3+1}{x+1}$

801. $\lim\limits_{x \to 1} \dfrac{x^4-1}{x^3-1}$

802. $\lim\limits_{x \to 2} \dfrac{x-2}{x^2+x-6}$

803. $\lim\limits_{x \to \left(\frac{\pi}{2}\right)^-} \dfrac{\cos x}{1-\sin x}$

804. $\lim\limits_{x \to 0} \dfrac{1-\cos x}{\tan x}$

805. $\lim\limits_{x \to \infty} \dfrac{\ln x}{x^2}$

806. $\lim\limits_{x \to 1} \dfrac{\ln x}{\cos\left(\dfrac{\pi}{2} x\right)}$

807. $\displaystyle\lim_{x\to 0}\frac{\tan^{-1}x}{x}$

808. $\displaystyle\lim_{x\to 2^+}\left(\frac{8}{x^2-4}-\frac{x}{x-2}\right)$

809. $\displaystyle\lim_{x\to 0}\frac{\sin(4x)}{2x}$

810. $\displaystyle\lim_{x\to 0^+}x\ln x$

811. $\displaystyle\lim_{x\to\infty}\frac{\ln x^4}{x^3}$

812. $\displaystyle\lim_{x\to 0}\frac{\sin^{-1}x}{x}$

813. $\displaystyle\lim_{x\to\infty}\frac{\tan^{-1}x-\dfrac{\pi}{4}}{x-1}$

814. $\displaystyle\lim_{x\to\infty}\frac{\tan^{-1}x-\dfrac{\pi}{2}}{\dfrac{1}{1+x^3}}$

815. $\displaystyle\lim_{x\to\frac{\pi}{2}^-}(\sec x-\tan x)$

816. $\displaystyle\lim_{x\to\infty}x\sin\left(\frac{1}{x}\right)$

817. $\displaystyle\lim_{x\to-\infty}x^3 e^x$

818. $\displaystyle\lim_{x\to\infty}\left(xe^{1/x}-x\right)$

819. $\displaystyle\lim_{x\to\infty}\frac{e^{3x+1}}{x^2}$

820. $\displaystyle\lim_{x\to 1}\frac{1-x+\ln x}{1+\cos\pi x}$

821. $\displaystyle\lim_{x\to 0}(1-5x)^{1/x}$

822. $\displaystyle\lim_{x\to\infty}x^{1/x^2}$

823. $\displaystyle\lim_{x\to 0^+}(\cos x)^{2/x}$

824. $\displaystyle\lim_{x\to\infty}\left(\frac{3x-1}{3x+4}\right)^{x-1}$

825. $\lim\limits_{x\to 0^+}(2x)^{x^2}$

829. $\lim\limits_{x\to 0^+}\left(e^x+x\right)^{2/x}$

826. $\lim\limits_{x\to 0^+}(\tan 3x)^x$

830. $\lim\limits_{x\to 1^+}\left(\dfrac{x^2}{x^2-1}-\dfrac{1}{\ln x}\right)$

827. $\lim\limits_{x\to 0^+}\tan x\ln x$

831. $\lim\limits_{x\to 1^+}\left(\dfrac{1}{\ln x}-\dfrac{1}{x-1}\right)$

828. $\lim\limits_{x\to\infty}\left(1+\dfrac{1}{x}\right)^x$

Chapter 13

U-Substitution and Integration by Parts

• •

In this chapter, you encounter some of the more advanced integration techniques: *u*-substitution and integration by parts. You use *u*-substitution very, very often in integration problems. For many integration problems, consider starting with a *u*-substitution if you don't immediately know the antiderivative. Another common technique is integration by parts, which comes from the product rule for derivatives. One of the difficult things about these problems is that even when you know which procedure to use, you still have some freedom in how to proceed; what to do isn't always clear, so dive in and try different things.

The Problems You'll Work On

This chapter is the start of more challenging integration problems. You work on the following skills:

- ✔ Using *u*-substitution to find definite and indefinite integrals
- ✔ Using integration by parts to find definite and indefinite integrals

What to Watch Out For

Here are a few things to keep in mind while working on the problems in this chapter:

- ✔ Even if you know you should use a substitution, there may be different substitutions to try. As a rule, start simple and make your substitution more complex if your first choice doesn't work.
- ✔ When using a *u*-substitution, don't forget to calculate *du,* the differential.
- ✔ You can algebraically manipulate both *du* and the original *u*-substitution, so play with both!
- ✔ For the integration by parts problems, if your pick of *u* and *dv* don't seem to be working, try switching them.

Using u-Substitutions

832–857 Use substitution to evaluate the integral.

832. $\int \sin(5x)\,dx$

833. $\int (x+4)^{100}\,dx$

834. $\int 3x^2 \sqrt{x^3+1}\,dx$

835. $\int \dfrac{\sec^2 \sqrt{x}}{\sqrt{x}}\,dx$

836. $\int \dfrac{6}{(4+5x)^8}\,dx$

837. $\int \sin^7 \theta \cos\theta\,d\theta$

838. $\int_{-\pi/3}^{\pi/3} \tan^3 \theta\,d\theta$

839. $\int_1^{e^2} \dfrac{(\ln x)^3}{x}\,dx$

840. $\int \dfrac{dx}{4-3x}$

841. $\int \dfrac{\tan^{-1} x}{1+x^2}\,dx$

842. $\int \tan x\,dx$

843. $\int_0^{\pi/3} e^{\cos x} \sin x\,dx$

844. $\int_0^1 \sqrt{x} \cos(1+x^{3/2})\,dx$

845. $\int \sqrt[7]{\tan x} \sec^2 x\,dx$

846. $\int \dfrac{\sin\left(\dfrac{\pi}{x}\right)}{x^2}\,dx$

847. $\int \dfrac{4+12x}{\sqrt{1+2x+3x^2}}\,dx$

848. $\int \sqrt[3]{\cot x}\,\csc^2 x\,dx$

849. $\int_0^{\sqrt{\pi}/2} x\sin\left(x^2\right)dx$

850. $\int \dfrac{3+x}{1+x^2}\,dx$

851. $\int_e^{e^2} \dfrac{dx}{x\sqrt{\ln x}}$

852. $\int \tan\theta\ln\left(\cos\theta\right)d\theta$

853. $\int x^2 e^{-x^3}dx$

854. $\int_0^3 x\sqrt{x+1}\,dx$

855. $\int_0^1 x^5\sqrt[3]{x^3+1}\,dx$

856. $\int \dfrac{x}{\sqrt[6]{x+3}}\,dx$

857. $\int x^7\sqrt{x^4+1}\,dx$

Using Integration by Parts

858–883 Use integration by parts to evaluate the integral.

858. $\int x\cos(4x)\,dx$

859. $\int xe^x dx$

860. $\int x\sinh(2x)\,dx$

861. $\int_0^3 x6^x dx$

862. $\int x \sec x \tan x \, dx$

863. $\int x \sin(5x) \, dx$

864. $\int x \csc^2 x \, dx$

865. $\int x \sin x \, dx$

866. $\int x \csc x \cot x \, dx$

867. $\int \ln(3x+1) \, dx$

868. $\int \tan^{-1} x \, dx$

869. $\int_1^4 \dfrac{\ln x}{x^2} \, dx$

870. $\int_1^9 e^{\sqrt{x}} \, dx$

871. $\int_1^9 x^{3/2} \ln x \, dx$

872. $\int \cos^{-1} x \, dx$

873. $\int \sin^{-1}(5x) \, dx$

874. $\int_0^1 \dfrac{x^2}{e^{2x}} \, dx$

875. $\int_0^1 (x^2+1) e^{-x} \, dx$

876. $\int x^3 \cos(x^2) \, dx$

877. $\int x^2 \sin(mx) \, dx$, where $m \neq 0$

878. $\int_1^3 x^5 (\ln x)^2 dx$

879. $\int e^x \cos(2x)\, dx$

880. $\int \sin x \ln(\cos x)\, dx$

881. $\int \cos \sqrt{x}\, dx$

882. $\int x^4 (\ln x)^2\, dx$

883. $\int x \tan^{-1} x\, dx$

Chapter 14

Trigonometric Integrals, Trigonometric Substitution, and Partial Fractions

• •

This chapter covers trigonometric integrals, trigonometric substitutions, and partial fractions — the remaining integration techniques you encounter in a second-semester calculus course (in addition to *u*-substitution and integration by parts; see Chapter 13). In a sense, these techniques are nothing fancy. For the trigonometric integrals, you typically use a *u*-substitution followed by a trigonometric identity, possibly throwing in a bit of algebra to solve the problem. For the trigonometric substitutions, you're often integrating a function involving a radical; by picking a clever substitution, you can often remove the radical and make the problem into a trigonometric integral and proceed from there. Last, the partial fractions technique simply decomposes a rational function into a bunch of simple fractions that are easier to integrate.

With that said, many of these problems have many steps and require you to know identities, polynomial long division, derivative formulas, and more. Many of these problems test your algebra and trigonometry skills as much as your calculus skills.

The Problems You'll Work On

This chapter finishes off the integration techniques that you see in a calculus class:

- ✔ Solving definite and indefinite integrals involving powers of trigonometric functions
- ✔ Solving definite and indefinite integrals using trigonometric substitutions
- ✔ Solving definite and indefinite integrals using partial fraction decompositions

What to Watch Out For

You can get tripped up in a lot of little places on these problems, but hopefully these tips will help:

✔ Not all of the trigonometric integrals fit into a nice mold. Try identities, *u*-substitutions, and simplifying the integral if you get stuck.

✔ You may have to use trigonometry and right triangles in the trigonometric substitution problems to recover the original variable.

✔ If you've forgotten how to do polynomial long division, you can find some examples in Chapter 1's algebra review.

✔ The trigonometric substitution problems turn into trigonometric integral problems, so make sure you can solve a variety of the latter problems!

Trigonometric Integrals

884–913 Find the antiderivative or evaluate the definite integral.

884. $\int_0^{3\pi/2} \sin^2(2\theta)\,d\theta$

885. $\int_0^{\pi/4} \cos^2\theta\,d\theta$

886. $\int \tan^2 x\,dx$

887. $\int \sec^4 t\,dt$

888. $\int \sin(3x)\sin(2x)\,dx$

889. $\int \cos(5x)\cos(2x)\,dx$

890. $\int \sin^4 x \cos x\,dx$

891. $\int \tan^2 x \sec^2 x\,dx$

892. $\int \sin(5x)\cos(4x)\,dx$

893. $\int \sec x \tan x \, dx$

901. $\int \dfrac{\cos x - \sin x}{\sin(2x)} \, dx$

894. $\int \sqrt{\csc^5 x} \, \csc x \cot x \, dx$

902. $\int \dfrac{\tan^3 x}{\cos^2 x} \, dx$

895. $\int \cos^3 x \sin^2 x \, dx$

903. $\int x \sin^2 x \, dx$

896. $\int_0^{\pi/3} \tan^3 x \sec^4 x \, dx$

904. $\int \sin^3 x \cos^3 x \, dx$

897. $\int \dfrac{\cot^3 x}{\sin^4 x} \, dx$

905. $\int \tan^3 x \sec x \, dx$

898. $\int_0^{\pi/2} \cos^3 x \, dx$

906. $\int \cot^3 x \csc x \, dx$

899. $\int \dfrac{1 - \cos x}{\sin x} \, dx$

907. $\int \cos^3 x \sqrt{\sin x} \, dx$

900. $\int (1 + \sin \theta)^2 \, d\theta$

908. $\int \cot^2 x \csc^4 x \, dx$

909. $\int \sin\theta \sin^5(\cos\theta) d\theta$

910. $\int \sin^5 x \cos^4 x \, dx$

911. $\int \sec^3 x \, dx$

912. $\int \dfrac{dx}{\sin x - 1}$

913. $\int \tan^3(4x) \sec^5(4x) dx$

Trigonometric Substitutions

914–939 Evaluate the integral using a trigonometric substitution.

914. $\int \dfrac{dx}{\sqrt{x^2 + 4}}$

915. $\int_0^{\sqrt{2}/2} \dfrac{x^2}{\sqrt{1-x^2}} dx$

916. $\int \dfrac{x}{\sqrt{x^2 - 1}} dx$

917. $\int \dfrac{1}{x^2 \sqrt{1+x^2}} dx$

918. $\int \dfrac{1}{x^2 \sqrt{x^2 - 16}} dx$

919. $\int \sqrt{9 - x^2} \, dx$

920. $\int \sqrt{1 - 25x^2} \, dx$

921. $\int \dfrac{dx}{x\sqrt{7 - x^2}}$

922. $\int \sqrt{1 - 5x^2} \, dx$

923. $\int \dfrac{dx}{x\sqrt{9 - x^2}}$

924. $\int \dfrac{\sqrt{x^2 - 4}}{x^3} dx$

925. $\int x\sqrt{1 - x^4} \, dx$

926. $\int_0^{\pi/2} \dfrac{\sin t}{\sqrt{1+\cos^2 t}}\, dt$

927. $\int \dfrac{\sqrt{x^2-1}}{x}\, dx$

928. $\int_1^2 \dfrac{1}{x^2\sqrt{x^2-1}}\, dx$

929. $\int \dfrac{x^2}{\left(4-x^2\right)^{3/2}}\, dx$

930. $\int \dfrac{1}{x^2\sqrt{4x^2-16}}\, dx$

931. $\int \dfrac{x^3}{\sqrt{x^2+16}}\, dx$

932. $\int \dfrac{x^5}{\sqrt{x^2+1}}\, dx$

933. $\int_0^3 x^3\sqrt{x^2+9}\, dx$

934. $\int x^2\sqrt{1-x^6}\, dx$

935. $\int \dfrac{dx}{\left(7-6x-x^2\right)^{3/2}}$

936. $\int \dfrac{dx}{x^2+4x+2}$

937. $\int_0^{3/2} x^3\sqrt{9-4x^2}\, dx$

938. $\int \sqrt{7+6x-x^2}\, dx$

939. $\int x^3\sqrt{16-x^2}\, dx$

Finding Partial Fraction Decompositions (without Coefficients)

940–944 Find the partial fraction decomposition without finding the coefficients.

940. $\dfrac{4x+1}{x^3\left(x+1\right)^2}$

941. $\dfrac{2x}{x^4-1}$

942. $\dfrac{5x^2 + x - 4}{(x+1)^2 (x^2 + 5)^3}$

943. $\dfrac{4x^3 + 19}{x^2 (x-1)^3 (x^2 + 17)}$

944. $\dfrac{3x^2 + 4}{(x^2 - 9)(x^4 + 2x^2 + 1)}$

Finding Partial Fraction Decompositions (Including Coefficients)

945–949 Find the partial fraction decomposition, including the coefficients.

945. $\dfrac{1}{(x+2)(x-1)}$

946. $\dfrac{x+2}{x^3 + x}$

947. $\dfrac{5x+1}{x^2 - 6x + 9}$

948. $\dfrac{x^2 + 2}{x^4 + 4x^2 + 3}$

949. $\dfrac{x^2 + 1}{(x^2 + 5)^2}$

Integrals Involving Partial Fractions

950–958 Evaluate the integral using partial fractions.

950. $\displaystyle\int \dfrac{x}{x-5}\,dx$

951. $\displaystyle\int \dfrac{x^2}{x+6}\,dx$

952. $\displaystyle\int \dfrac{x-3}{(x+4)(x-5)}\,dx$

953. $\displaystyle\int_4^5 \dfrac{1}{x^2 - 1}\,dx$

954. $\displaystyle\int \dfrac{3x+5}{x^2 + 2x + 1}\,dx$

955. $\displaystyle\int \dfrac{x^3 - 6x + 5}{x^2 - x - 6}\,dx$

956. $\displaystyle\int \dfrac{8}{x^3 + x}\,dx$

957. $\int \dfrac{6x^2+1}{x^4+6x^2+9}\,dx$

958. $\int \dfrac{1}{x^3+1}\,dx$

Rationalizing Substitutions

959–963 *Use a rationalizing substitution and partial fractions to evaluate the integral.*

959. $\int \dfrac{1}{x\sqrt{x+1}}\,dx$

960. $\int_4^9 \dfrac{\sqrt{x}}{x-1}\,dx$

961. $\int \dfrac{2}{\sqrt{x}-\sqrt[3]{x}}\,dx$

962. $\int \dfrac{1}{\sqrt[3]{x}-\sqrt[4]{x}}\,dx$

963. $\int \dfrac{e^{2x}}{e^{2x}+3e^x+2}\,dx$

Chapter 15

Improper Integrals and More Approximating Techniques

• •

The problems in this chapter involve improper integrals and two techniques to approximate definite integrals: Simpson's rule and the trapezoid rule. *Improper integrals* are definite integrals with limits thrown in, so those problems require you to make use of many different calculus techniques; they can be quite challenging. The last few problems of the chapter involve using Simpson's rule and the trapezoid rule to approximate definite integrals. When you know the formulas for these approximating techniques, the problems are more of an arithmetic chore than anything else.

The Problems You'll Work On

This chapter involves the following tasks:

✔ Solving improper integrals using definite integrals and limits

✔ Using comparison to show whether an improper integral converges or diverges

✔ Approximating definite integrals using Simpson's rule and the trapezoid rule

What to Watch Out For

Here are a few pointers to help you finish the problems in this chapter:

✔ Improper integrals involve it all: limits, l'Hôpital's rule, and any of the integration techniques.

✔ The formulas for Simpson's rule and the trapezoid rule are similar, so don't mix them up!

✔ If you're careful with the arithmetic on Simpson's rule and the trapezoid rule, you should be in good shape.

Convergent and Divergent Improper Integrals

964–987 Determine whether the integral is convergent or divergent. If the integral is convergent, give the value.

964. $\int_{1}^{\infty} \frac{1}{(x+1)^2} dx$

965. $\int_{0}^{5} \frac{1}{x\sqrt{x}} dx$

966. $\int_{0}^{2} \frac{1}{x} dx$

967. $\int_{-\infty}^{1} e^{-4x} dx$

968. $\int_{1}^{17} (x-1)^{-1/4} dx$

969. $\int_{-\infty}^{-3} \frac{1}{x+1} dx$

970. $\int_{-\infty}^{2} (x^2-5) dx$

971. $\int_{-\infty}^{\infty} (3-x^4) dx$

972. $\int_{2}^{\infty} e^{-x/3} dx$

973. $\int_{e}^{\infty} \frac{1}{x(\ln x)^2} dx$

974. $\int_{-\infty}^{\infty} \frac{x^2}{1+x^6} dx$

975. $\int_{0}^{\infty} xe^{-2x} dx$

976. $\int_{-\infty}^{\infty} x^4 e^{-x^5} dx$

977. $\int_{-\infty}^{-4} \frac{3x}{1+x^4} dx$

978. $\int_{2}^{\infty} \frac{\ln x}{x^4} dx$

979. $\int_0^\infty \frac{x}{x^2+4}\,dx$

980. $\int_1^8 \frac{1}{\sqrt[3]{x-8}}\,dx$

981. $\int_0^{\pi/2} \tan^2 x\,dx$

982. $\int_0^\infty \sin^2 x\,dx$

983. $\int_{-1}^1 \frac{e^x}{e^x-1}\,dx$

984. $\int_4^\infty \frac{1}{x^2+x-6}\,dx$

985. $\int_0^2 \frac{dx}{\sqrt{4-x^2}}$

986. $\int_{-\infty}^\infty e^{-|x|}\,dx$

987. $\int_3^5 \frac{x}{\sqrt{x-3}}\,dx$

The Comparison Test for Integrals

988–993 *Determine whether the improper integral converges or diverges using the comparison theorem for integrals.*

988. $\int_1^\infty \frac{\sin^2 x}{1+x^2}\,dx$

989. $\int_1^\infty \frac{dx}{x^4+e^{3x}}$

990. $\int_1^\infty \frac{x+2}{\sqrt{x^4-1}}\,dx$

991. $\int_1^\infty \frac{\tan^{-1}x}{x^5}\,dx$

992. $\int_2^\infty \frac{x^2}{\sqrt{x^6-1}}\,dx$

993. $\int_1^\infty \frac{5+e^{-x}}{x}\,dx$

The Trapezoid Rule

994–997 Use the trapezoid rule with the specified value of n to approximate the integral. Round to the nearest thousandth.

994. $\int_0^6 \sqrt[3]{1+x^3}\,dx$ with $n = 6$

995. $\int_1^2 \dfrac{\ln x}{1+x^2}\,dx$ with $n = 4$

996. $\int_1^3 \dfrac{\cos x}{x}\,dx$ with $n = 8$

997. $\int_0^{1/2} \sin\sqrt{x}\,dx$ with $n = 4$

Simpson's Rule

998–1,001 Use Simpson's rule with the specified value of n to approximate the integral. Round to the nearest thousandth.

998. $\int_0^6 \sqrt[3]{1+x^3}\,dx$ with $n = 6$

999. $\int_1^2 \dfrac{\ln x}{1+x^2}\,dx$ with $n = 4$

1,000. $\int_1^3 \dfrac{\cos x}{x}\,dx$ with $n = 8$

1,001. $\int_0^{1/2} \sin\sqrt{x}\,dx$ with $n = 4$

Part II
The Answers

In this part . . .

Here you get answers and explanations for all 1,001 problems. As you read the solutions, you may realize that you need a little more help. Lucky for you, the *For Dummies* series (published by Wiley) offers several excellent resources. I recommend checking out the following titles, depending on your needs:

- ✔ *Calculus For Dummies, Calculus Workbook For Dummies,* and *Calculus Essentials For Dummies,* all by Mark Ryan

- ✔ *Pre-Calculus For Dummies,* by Yang Kuang and Elleyne Kase, and *Pre-Calculus Workbook For Dummies,* by Yang Kuang and Michelle Rose Gilman

- ✔ *Trigonometry For Dummies* and *Trigonometry Workbook For Dummies,* by Mary Jane Sterling

When you're ready to step up to more advanced calculus courses, you'll find the help you need in *Calculus II For Dummies,* by Mark Zegarelli.

Visit www.dummies.com for more information.

Chapter 16

Answers and Explanations

H ere are the answer explanations for all 1,001 practice problems.

---

1. $\dfrac{5}{12}$

Get a common denominator of 12 and then perform the arithmetic in the numerator:

$$\frac{1}{2} + \frac{3}{4} - \frac{5}{6}$$

$$= \frac{1(6)}{2(6)} + \frac{3(3)}{4(3)} - \frac{5(2)}{6(2)}$$

$$= \frac{6}{12} + \frac{9}{12} - \frac{10}{12}$$

$$= \frac{6+9-10}{12}$$

$$= \frac{5}{12}$$

---

2. $\dfrac{307}{60}$

Get a common denominator of 60 and then perform the arithmetic in the numerator:

$$\frac{11}{3} + \frac{3}{5} + \frac{17}{20}$$

$$= \frac{11(20)}{3(20)} + \frac{3(12)}{5(12)} + \frac{17(3)}{20(3)}$$

$$= \frac{220}{60} + \frac{36}{60} + \frac{51}{60}$$

$$= \frac{220+36+51}{60}$$

$$= \frac{307}{60}$$

3. 11

Write each factor as a fraction and then multiply across the numerator and denominator (cancel common factors as well!):

$$\left(\frac{2}{3}\right)(21)\left(\frac{11}{14}\right)$$

$$= \left(\frac{2}{3}\right)\left(\frac{21}{1}\right)\left(\frac{11}{14}\right)$$

$$= \frac{(2)(21)(11)}{(3)(1)(14)}$$

$$= \frac{(2)(21)(11)}{(3)(1)(14)}$$

$$= \frac{11}{1} = 11$$

4. $\frac{11}{9}$

Recall that to divide by a fraction, you multiply by its reciprocal:

$\frac{a/b}{c/d} = \left(\frac{a}{b}\right)\left(\frac{d}{c}\right) = \frac{ad}{bc}$. After rewriting the fraction, cancel common factors before multiplying:

$$\frac{5/6}{15/22} = \left(\frac{5}{6}\right)\left(\frac{22}{15}\right) = \left(\frac{1}{3}\right)\left(\frac{11}{3}\right) = \frac{11}{9}$$

5. $\frac{5x + 7y + 10x^3}{x^2y}$

To add the fractions, you need a common denominator. The least common multiple of the denominators is x^2y, so multiply each fraction accordingly to get this denominator for each term. Then write the answer as a single fraction:

$$\frac{5}{yx} + \frac{7}{x^2} + \frac{10x}{y}$$

$$= \frac{5(x)}{yx(x)} + \frac{7(y)}{x^2(y)} + \frac{10x(x^2)}{y(x^2)}$$

$$= \frac{5x}{x^2y} + \frac{7y}{x^2y} + \frac{10x^3}{x^2y}$$

$$= \frac{5x + 7y + 10x^3}{x^2y}$$

Answers
1–100

6. $\dfrac{6x-4}{x^2-1}$

You need a common denominator so you can subtract the fractions. The least common multiple of the denominators is $(x-1)(x+1) = x^2-1$, so multiply each fraction accordingly to get this denominator. Then perform the arithmetic in the numerator:

$$\frac{x}{x-1} - \frac{x-4}{x+1}$$

$$= \frac{x(x+1)}{(x-1)(x+1)} - \frac{(x-4)(x-1)}{(x+1)(x-1)}$$

$$= \frac{x^2+x}{x^2-1} - \frac{x^2-5x+4}{x^2-1}$$

$$= \frac{x^2+x-\left(x^2-5x+4\right)}{x^2-1}$$

$$= \frac{x^2+x-x^2+5x-4}{x^2-1}$$

$$= \frac{6x-4}{x^2-1}$$

7. $\dfrac{(x-1)y}{x}$

Begin by factoring the expressions completely. Cancel any common factors and then simplify to get the answer:

$$\left(\frac{x^2-1}{xy^2}\right)\left(\frac{y^3}{x+1}\right)$$

$$= \left(\frac{(x+1)(x-1)}{xy^2}\right)\left(\frac{y^3}{x+1}\right)$$

$$= \frac{(x-1)}{x}\left(\frac{y}{1}\right)$$

$$= \frac{(x-1)y}{x}$$

8. $\dfrac{5x^2(x-3)}{3(x+5)y}$

Recall that to divide by a fraction, you multiply by its reciprocal:

$\dfrac{a/b}{c/d} = \left(\dfrac{a}{b}\right)\left(\dfrac{d}{c}\right) = \dfrac{ad}{bc}$. After rewriting the fraction, factor and cancel the common factors:

$$\dfrac{\dfrac{x^2-5x+6}{6xy^3}}{\dfrac{x^2+3x-10}{10x^3y^2}}$$

$$= \left(\dfrac{x^2-5x+6}{6xy^3}\right)\left(\dfrac{10x^3y^2}{x^2+3x-10}\right)$$

$$= \left(\dfrac{(x-2)(x-3)}{6xy^3}\right)\left(\dfrac{10x^3y^2}{(x+5)(x-2)}\right)$$

$$= \left(\dfrac{x-3}{3y}\right)\left(\dfrac{5x^2}{x+5}\right)$$

$$= \dfrac{5x^2(x-3)}{3(x+5)y}$$

9. $x^3y^2z^{13}$

Write each factor using positive exponents and then simplify. To start, notice that z^{-8} in the denominator becomes z^8 when it moves to the numerator:

$$\dfrac{x^3y^4z^5}{y^2z^{-8}} = \dfrac{x^3y^4z^5z^8}{y^2}$$

$$= \dfrac{x^3y^2z^{13}}{1}$$

$$= x^3y^2z^{13}$$

10. $\dfrac{(x+3)}{x^2(y+5)^3}$

To simplify, cancel the common factors:

$$\dfrac{(x+3)^2 x^4 (y+5)^{14}}{x^6(x+3)(y+5)^{17}}$$

$$= \dfrac{(x+3)}{x^2(y+5)^3}$$

11. 1

Because the entire expression is being raised to the power of zero, the answer is 1:

$$\left(\dfrac{4x^2y^{100}z^{-3}}{18x^{15}y^4z^8}\right)^0 = 1$$

There's no point in simplifying the expression inside the parentheses.

12. $z\left(x^2y^2 + z^2\right)$

Begin by factoring the numerator and then cancel the common factors:

$$\frac{x^4y^3z^2 + x^2yz^4}{x^2yz}$$

$$= \frac{x^2yz^2\left(x^2y^2 + z^2\right)}{x^2yz}$$

$$= z\left(x^2y^2 + z^2\right)$$

13. $\dfrac{x^5y^{17}}{z^8}$

For this problem, you need to know the properties of exponents. The important properties to recall here are $\left(x^m\right)^n = x^{mn}$, $(x)^0 = 1$, $x^m x^n = x^{m+n}$, $\dfrac{x^m}{x^n} = x^{m-n}$, $x^{-n} = \dfrac{1}{x^n}$, and $\dfrac{1}{x^{-n}} = x^n$.

Write all factors using positive exponents and then use the properties of exponents to simplify the expression:

$$\frac{\left(x^2y^3\right)^4\left(y^4\right)^0 z^{-2}}{\left(xz^2\right)^3 y^{-5}}$$

$$= \frac{x^8y^{12}(1)y^5}{x^3z^6z^2}$$

$$= \frac{x^5y^{17}}{z^8}$$

14. $5\sqrt{2}$

Factor the number underneath the radical and then rewrite the radical using $\sqrt{ab} = \sqrt{a}\sqrt{b}$ to simplify:

$$\sqrt{50} = \sqrt{(25)(2)}$$

$$= \sqrt{25}\sqrt{2}$$

$$= 5\sqrt{2}$$

15. $\dfrac{2\sqrt{15}}{15}$

Start by rewriting the expression so it has only one square root sign. Then reduce the fraction and use properties of radicals to simplify:

$$\frac{\sqrt{8}\sqrt{20}}{\sqrt{50}\sqrt{12}} = \sqrt{\frac{8(20)}{50(12)}}$$

$$= \sqrt{\frac{2(2)}{5(3)}}$$

$$= \frac{\sqrt{4}}{\sqrt{15}}$$

$$= \frac{2}{\sqrt{15}}$$

$$= \frac{2}{\sqrt{15}} \cdot \frac{\sqrt{15}}{\sqrt{15}}$$

$$= \frac{2\sqrt{15}}{15}$$

16. $\quad 10x^2y^4z^9\sqrt{x}$

Begin by writing the expression using a single square root sign. Then combine like bases and simplify:

$$\sqrt{20x^4y^6z^{11}}\sqrt{5xy^2z^7}$$

$$= \sqrt{\left(20x^4y^6z^{11}\right)\left(5xy^2z^7\right)}$$

$$= \sqrt{100x^5y^8z^{18}}$$

$$= \sqrt{100}\sqrt{x^5}\sqrt{y^8}\sqrt{z^{18}}$$

$$= 10x^2\sqrt{x}\,y^4z^9$$

17. $\quad x^3y^4z^5\sqrt[3]{x^2}$

Begin by writing the expression using a single cube root sign. Combine the factors with like bases and then simplify:

$$\sqrt[3]{x^4y^8z^5}\sqrt[3]{x^7y^4z^{10}}$$

$$= \sqrt[3]{x^{11}y^{12}z^{15}}$$

$$= \sqrt[3]{x^{11}}\sqrt[3]{y^{12}}\sqrt[3]{z^{15}}$$

$$= x^3\sqrt[3]{x^2}\,y^4z^5$$

18. $\quad 2xy^2\sqrt[5]{y^4}$

Simplify each root individually and then use properties of exponents to simplify further:

$$\frac{\sqrt[3]{8x^3y^6}\sqrt[5]{x^{10}y^{14}}}{\sqrt[7]{x^{14}y^{14}}}$$

$$= \frac{\left(2xy^2\right)\left(x^2y^2\sqrt[5]{y^4}\right)}{x^2y^2}$$

$$= 2xy^2\sqrt[5]{y^4}$$

19.

$$\sqrt[3]{4}\sqrt[8]{x^3}\sqrt[4]{y}\sqrt[12]{z^5}$$

Recall that $x^{m/n} = \sqrt[n]{x^m}$. The expression becomes

$$4^{1/3}\,x^{3/8}\,y^{1/4}\,z^{5/12}$$

$$= \sqrt[3]{4}\sqrt[8]{x^3}\sqrt[4]{y}\sqrt[12]{z^5}$$

20.

$$4^{1/3}\,x^{2/3}\,y^{1/3}\,z^{4/5}$$

Recall that $\sqrt[n]{x^m} = x^{m/n}$. Therefore, the expression becomes

$$\sqrt[3]{4x^2 y}\,\sqrt[5]{z^4}$$

$$= \left(4x^2 y\right)^{1/3} z^{4/5}$$

$$= 4^{1/3}\,x^{2/3}\,y^{1/3}\,z^{4/5}$$

21.

$$y = 3x + 8$$

Recall that a function is one-to-one if it satisfies the requirement that if $x_1 \neq x_2$, then $f(x_1) \neq f(x_2)$; that is, no two x coordinates have the same y value. If you consider the graph of such a function, no horizontal line would cross the graph in more than one place.

Of the given functions, only the linear function $y = 3x + 8$ passes the horizontal line test. This function is therefore one-to-one and has an inverse.

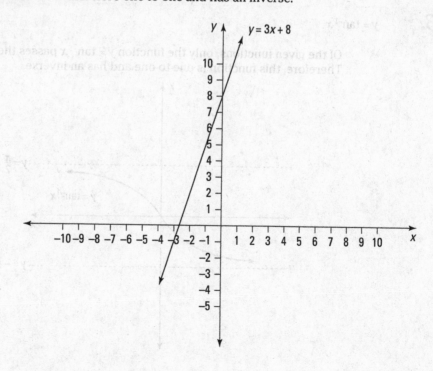

22.

$y = x^2 - 4, x \geq 0$

When determining whether a function has an inverse, consider the domain of the given function. Some functions don't pass the horizontal line test — and therefore don't have an inverse — unless the domain is restricted. The function $y = x^2 - 4$ doesn't pass the horizontal line test; however, if you restrict the domain to $x \geq 0$, then $y = x^2 - 4$ does pass the horizontal line test (and therefore has an inverse), because you're getting only the right half of the parabola.

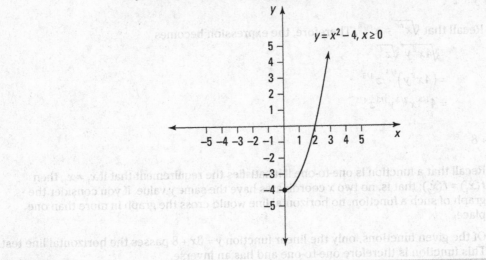

23.

$y = \tan^{-1} x$

Of the given functions, only the function $y = \tan^{-1} x$ passes the horizontal line test. Therefore, this function is one-to-one and has an inverse.

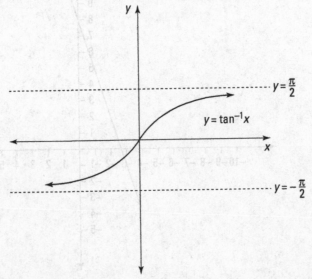

24. $f^{-1}(x) = \dfrac{4-x}{5}$

First replace $f(x)$ with y:

$$f(x) = 4 - 5x$$
$$y = 4 - 5x$$

Then replace y with x and each x with y. After making the replacements, solve for y to find the inverse:

$$x = 4 - 5y$$
$$5y = 4 - x$$
$$y = \dfrac{4-x}{5}$$

Therefore, the inverse of $f(x) = 4 - 5x$ is

$$f^{-1}(x) = \dfrac{4-x}{5}$$

25. $f^{-1}(x) = 2 + \sqrt{x+4}$

First replace $f(x)$ with y:

$$f(x) = x^2 - 4x$$
$$y = x^2 - 4x$$

Then replace y with x and each x with y. After making the replacements, solve for y to find the inverse. You need to complete the square so you can isolate y. (To complete the square, consider the quadratic expression on the right side of the equal sign; take one-half of the coefficient of the term involving y, square it, and add that value to both sides of the equation.) Then factor the right side and use square roots to solve for y:

$$x = y^2 - 4y$$
$$x + 4 = y^2 - 4y + 4$$
$$x + 4 = (y - 2)^2$$
$$\pm\sqrt{x+4} = y - 2$$
$$2 \pm \sqrt{x+4} = y$$

Notice that for the domain of the inverse to match the range of $f(x)$, you have to keep the positive root. Therefore, the inverse of $f(x) = x^2 - 4x$, where $x \geq 2$, is

$$f^{-1}(x) = 2 + \sqrt{x+4}$$

26. $f^{-1}(x) = \dfrac{8-x^2}{5}, x \geq 0$

First replace $f(x)$ with y:

$$f(x) = \sqrt{8-5x}$$
$$y = \sqrt{8-5x}$$

Then replace y with x and each x with y. After making the replacements, solve for y to find the inverse:

$$x = \sqrt{8-5y}$$
$$x^2 = 8-5y$$
$$5y = 8-x^2$$
$$y = \dfrac{8-x^2}{5}$$

Therefore, the inverse of $f(x) = \sqrt{8-5x}$ is

$$f^{-1}(x) = \dfrac{8-x^2}{5}, \quad \text{where } x \geq 0$$

The domain of the inverse is $x \geq 0$ because the range of $f(x)$ is $y \geq 0$ and the domain of $f^{-1}(x)$ is equal to the range of $f(x)$.

27. $f^{-1}(x) = \sqrt[5]{\dfrac{x-7}{3}}$

First replace $f(x)$ with y:

$$f(x) = 3x^5 + 7$$
$$y = 3x^5 + 7$$

Then replace y with x and each x with y. After making the replacements, solve for y to find the inverse:

$$x = 3y^5 + 7$$
$$x - 7 = 3y^5$$
$$\dfrac{x-7}{3} = y^5$$
$$\sqrt[5]{\dfrac{x-7}{3}} = y$$

Therefore, the inverse of $f(x) = 3x^5 + 7$ is

$$f^{-1}(x) = \sqrt[5]{\dfrac{x-7}{3}}$$

28. $f^{-1}(x) = \left(\dfrac{2-2x}{x+1}\right)^2, -1 < x \leq 1$

First replace $f(x)$ with y:

$$f(x) = \dfrac{2-\sqrt{x}}{2+\sqrt{x}}$$
$$y = \dfrac{2-\sqrt{x}}{2+\sqrt{x}}$$

Then replace y with x and each x with y. After making the replacements, solve for y to find the inverse:

$$x = \frac{2-\sqrt{y}}{2+\sqrt{y}}$$

$$x\left(2+\sqrt{y}\right) = 2-\sqrt{y}$$

$$2x + x\sqrt{y} = 2-\sqrt{y}$$

$$x\sqrt{y} + \sqrt{y} = 2-2x$$

$$\sqrt{y}(x+1) = 2-2x$$

$$\sqrt{y} = \frac{2-2x}{x+1}$$

$$y = \left(\frac{2-2x}{x+1}\right)^2$$

Note that the range of $f(x) = \frac{2-\sqrt{x}}{2+\sqrt{x}}$ is $(-1, 1]$, so the domain of the inverse function is $(-1, 1]$. Therefore, the inverse of $f(x) = \frac{2-\sqrt{x}}{2+\sqrt{x}}$ is

$$f^{-1}(x) = \left(\frac{2-2x}{x+1}\right)^2, \quad \text{where } -1 < x \le 1$$

---

29. $\qquad f^{-1}(x) = \dfrac{-4x-1}{x-2}$

First replace $f(x)$ with y:

$$f(x) = \frac{2x-1}{x+4}$$

$$y = \frac{2x-1}{x+4}$$

Then replace y with x and each x with y. After making the replacements, solve for y to find the inverse:

$$x = \frac{2y-1}{y+4}$$

$$x(y+4) = 2y-1$$

$$xy + 4x = 2y-1$$

$$xy - 2y = -4x-1$$

$$y(x-2) = -4x-1$$

$$y = \frac{-4x-1}{x-2}$$

Therefore, the inverse of $f(x) = \frac{2x-1}{x+4}$ is

$$f^{-1}(x) = \frac{-4x-1}{x-2}$$

---

30. $\qquad$ {(1, 0), (4, 3), (–6, 5)}

If the point (a, b) is on the graph of a one-to-one function, the point (b, a) is on the graph of the inverse function. Therefore, just switch the x and y coordinates to get that the set of points {(1, 0), (4, 3), (–6, 5)} belongs to the graph of $f^{-1}(x)$.

31. domain: (–1, 2); range: [–2, 4)

For a one-to-one function (which by definition has an inverse), the domain of $f(x)$ becomes the range of $f^{-1}(x)$, and the range of $f(x)$ becomes the domain of $f^{-1}(x)$. Therefore, $f^{-1}(x)$ has domain (–1, 2) and range [–2, 4).

32. $g^{-1}(x) = f^{-1}(x) - c$

Replacing x with $(x + c)$ shifts the graph c units to the left, assuming that $c > 0$. If the point (a, b) belongs to the graph of $f(x)$, then the point $(a - c, b)$ belongs to the graph of $f(x + c)$.

Now consider the inverse. The point (b, a) belongs to the graph of $f^{-1}(x)$, so the point $(b, a - c)$ belongs to the graph of $g^{-1}(x)$. Therefore, $g^{-1}(x)$ is the graph of $f^{-1}(x)$ shifted down c units so that $g^{-1}(x) = f^{-1}(x) - c$.

The same argument applies if $c < 0$.

33. $x = 2$

Put all terms involving x on one side of the equation and all constants on the other side, combining all like terms. Finally, divide by the coefficient of x to get the solution:

$$3x + 7 = 13$$
$$3x = 6$$
$$x = 2$$

34. $x = -4$

Distribute to remove the parentheses. Then put all terms involving x on one side of the equation and all constants on the other side, combining all like terms. Finally, divide by the coefficient of x to get the solution:

$$2(x + 1) = 3(x + 2)$$
$$2x + 2 = 3x + 6$$
$$-x = 4$$
$$x = -4$$

35. $x = \dfrac{5}{4}$

Distribute to remove the parentheses. Then put all terms involving x on one side of the equation and all constants on the other side, combining all like terms. Finally, divide by the coefficient of x to get the solution:

$$-4(x + 1) - 2x = 7x + 3(x - 8)$$
$$-4x - 4 - 2x = 7x + 3x - 24$$
$$-6x - 4 = 10x - 24$$
$$20 = 16x$$
$$\frac{20}{16} = x$$
$$x = \frac{5}{4}$$

36. $x = \dfrac{15}{4}$

Put all terms involving x on one side of the equation and all constants on the other side, combining all like terms. Finally, divide by the coefficient of x (that is, multiply both sides by the reciprocal of the coefficient) to get the solution:

$$\frac{5}{3}x + 5 = \frac{1}{3}x + 10$$

$$\frac{5}{3}x - \frac{1}{3}x = 10 - 5$$

$$\frac{4}{3}x = 5$$

$$x = \frac{3}{4}(5)$$

$$x = \frac{15}{4}$$

37. $x = \dfrac{10 - 3\sqrt{2}}{\sqrt{2} - \sqrt{5}}$

Distribute to remove the parentheses. Then put all terms involving x on one side of the equation and all constants on the other side, combining all like terms. Finally, divide by the coefficient of x to get the solution:

$$\sqrt{2}(x + 3) = \sqrt{5}\left(x + \sqrt{20}\right)$$

$$\sqrt{2}x + 3\sqrt{2} = \sqrt{5}x + \sqrt{100}$$

$$\sqrt{2}x - \sqrt{5}x = 10 - 3\sqrt{2}$$

$$x\left(\sqrt{2} - \sqrt{5}\right) = 10 - 3\sqrt{2}$$

$$x = \frac{10 - 3\sqrt{2}}{\sqrt{2} - \sqrt{5}}$$

38. $x = -3, 7$

This quadratic factors without too much trouble using trial and error:

$$x^2 - 4x - 21 = 0$$

$$(x - 7)(x + 3) = 0$$

Setting each factor equal to zero gives you $x - 7 = 0$ so that $x = 7$ is a solution and $x + 3 = 0$ so that $x = -3$ is a solution.

39. $x = -4 \pm \sqrt{33}$

Complete the square to solve the quadratic equation. (To complete the square, consider the quadratic expression on the left side of the equal sign; take one-half of the coefficient of the term involving x, square it, and add that value to both sides of the equation.) Then factor the left side and use square roots to solve for x:

$$x^2 + 8x - 17 = 0$$
$$x^2 + 8x = 17$$
$$x^2 + 8x + 16 = 17 + 16$$
$$(x + 4)^2 = 33$$
$$x + 4 = \pm\sqrt{33}$$
$$x = -4 \pm \sqrt{33}$$

40. $x = \dfrac{-3 \pm \sqrt{41}}{4}$

Complete the square to solve the quadratic equation. Begin by moving the constant to the right side of the equal sign and then dividing both sides of the equation by 2:

$$2x^2 + 3x - 4 = 0$$
$$2x^2 + 3x = 4$$
$$x^2 + \frac{3}{2}x = 2$$

Next, consider the quadratic expression on the left side of the equal sign; take one-half of the coefficient of the term involving x, square it, and add that value to both sides of the equation. Then factor the left side and use square roots to solve for x:

$$x^2 + \frac{3}{2}x + \frac{9}{16} = 2 + \frac{9}{16}$$
$$\left(x + \frac{3}{4}\right)^2 = \frac{41}{16}$$
$$x + \frac{3}{4} = \pm\sqrt{\frac{41}{16}}$$
$$x = \frac{-3}{4} \pm \frac{\sqrt{41}}{\sqrt{16}}$$
$$x = \frac{-3 \pm \sqrt{41}}{4}$$

41. $x = -\dfrac{4}{3}, \dfrac{1}{2}$

Factoring by trial and error gives you

$$6x^2 + 5x - 4 = 0$$
$$(2x - 1)(3x + 4) = 0$$

Setting each factor equal to zero gives you $2x - 1 = 0$ so that $x = \frac{1}{2}$ is a solution and $3x + 4 = 0$ so that $x = \frac{-4}{3}$ is a solution.

42. $x = \dfrac{-2 \pm \sqrt{10}}{3}$

The equation $3x^2 + 4x - 2 = 0$ doesn't factor nicely, so use the quadratic equation to solve for x. Here, $a = 3$, $b = 4$, and $c = -2$:

$$x = \frac{-b \pm \sqrt{b^2 - 4ac}}{2a}$$

$$= \frac{-4 \pm \sqrt{4^2 - 4(3)(-2)}}{2(3)}$$

$$= \frac{-4 \pm \sqrt{16 + 24}}{6}$$

$$= \frac{-4 \pm \sqrt{40}}{6}$$

$$= \frac{-4 \pm 2\sqrt{10}}{6}$$

$$= \frac{-2}{3} \pm \frac{\sqrt{10}}{3}$$

$$= \frac{-2 \pm \sqrt{10}}{3}$$

43. $x = \sqrt[5]{-2}, \sqrt[5]{-5}$

The equation $x^{10} + 7x^5 + 10 = 0$ isn't quadratic, but by using a substitution, you can produce a quadratic equation:

$$x^{10} + 7x^5 + 10 = 0$$

$$\left(x^5\right)^2 + 7\left(x^5\right) + 10 = 0$$

Now use the substitution $y = x^5$ to form a quadratic equation that you can easily factor:

$$y^2 + 7y + 10 = 0$$

$$(y + 2)(y + 5) = 0$$

From $y + 2 = 0$, you have the solution $y = -2$, and from $y + 5 = 0$, you have the solution $y = -5$.

Now replace y using the original substitution and solve for x to get the final answer: $x^5 = -2$ gives you $x = \sqrt[5]{-2}$, and $x^5 = -5$ gives you $x = \sqrt[5]{-5}$.

44. $x = 0, -\dfrac{5}{3}, 1$

Begin by factoring out the greatest common factor, x^2. Then factor the remaining quadratic expression:

$$3x^4 + 2x^3 - 5x^2 = 0$$

$$x^2\left(3x^2 + 2x - 5\right) = 0$$

$$x^2(3x + 5)(x - 1) = 0$$

Next, set each factor equal to zero and solve for x: $x^2 = 0$ has the solution $x = 0$, $3x + 5 = 0$ has the solution $x = -\dfrac{5}{3}$, and $x - 1 = 0$ has the solution $x = 1$.

45.

no real solutions

Factoring this polynomial by trial and error gives you the following:

$$x^8 + 12x^4 + 35 = 0$$

$$\left(x^4 + 5\right)\left(x^4 + 7\right) = 0$$

Setting the first factor equal to zero gives you $x^4 + 5 = 0$ so that $x^4 = -5$, which has no real solutions. Setting the second factor equal to zero gives you $x^4 + 7 = 0$ so that $x^4 = -7$, which also has no real solutions.

46.

$x = -1, 1$

Factor the polynomial repeatedly to get the following:

$$x^4 + 3x^2 - 4 = 0$$

$$\left(x^2 - 1\right)\left(x^2 + 4\right) = 0$$

$$(x - 1)(x + 1)\left(x^2 + 4\right) = 0$$

Now set each factor equal to zero and solve for x. Setting the first factor equal to zero gives you $x - 1 = 0$ so that $x = 1$. The second factor gives you $x + 1 = 0$ so that $x = -1$. And the last factor gives you $x^2 + 4 = 0$ so that $x^2 = -4$, which has no real solutions.

47.

$x = -3, 3$

Factor the polynomial repeatedly to get the following:

$$x^4 - 81 = 0$$

$$\left(x^2 - 9\right)\left(x^2 + 9\right) = 0$$

$$(x - 3)(x + 3)\left(x^2 + 9\right) = 0$$

Now set each factor equal to zero and solve for x. Setting the first factor equal to zero gives you $x - 3 = 0$ so that $x = 3$. Setting the second factor equal to zero gives you $x + 3 = 0$ so that $x = -3$. The last factor gives you $x^2 + 9 = 0$, or $x^2 = -9$, which has no real solutions.

48.

$x = 1, \dfrac{9}{5}$

To solve an absolute value equation of the form $|a| = b$, where $b > 0$, you must solve the two equations $a = b$ and $a = -b$. So for the equation $|5x - 7| = 2$, you have to solve $5x - 7 = 2$:

$$5x - 7 = 2$$

$$5x = 9$$

$$x = \frac{9}{5}$$

You also have to solve $5x - 7 = -2$:

$$5x - 7 = -2$$

$$5x = 5$$

$$x = 1$$

Therefore, the solutions are $x = \dfrac{9}{5}$ and $x = 1$.

49.   no real solutions

To solve an absolute value equation of the form $|a| = b$, where $b > 0$, you must solve the two equations $a = b$ and $a = -b$. However, in this case, you have $|4x - 5| + 18 = 13$, which gives you $|4x - 5| = -5$. Because the absolute value of a number can't be negative, this equation has no solutions.

50. $x = -3, 9$

To solve an absolute value equation of the form $|a| = b$, where $b > 0$, you must solve the two equations $a = b$ and $a = -b$. So for the equation $|x^2 - 6x| = 27$, you must solve $x^2 - 6x = 27$ and $x^2 - 6x = -27$.

To solve $x^2 - 6x = 27$, set the equation equal to zero and factor using trial and error:

$$x^2 - 6x - 27 = 0$$
$$(x - 9)(x + 3) = 0$$

Setting each factor equal to zero gives you $x - 9 = 0$, which has the solution $x = 9$, and $x + 3 = 0$, which has the solution $x = -3$.

Then set $x^2 - 6x = -27$ equal to zero, giving you $x^2 - 6x + 27 = 0$. This equation doesn't factor nicely, so use the quadratic equation, with $a = 1$, $b = -6$, and $c = 27$:

$$x = \frac{-b \pm \sqrt{b^2 - 4ac}}{2a}$$

$$= \frac{-(-6) \pm \sqrt{(-6)^2 - 4(1)(27)}}{2(1)}$$

$$= \frac{6 \pm \sqrt{-72}}{2}$$

The number beneath the radical is negative, so this part of the absolute value has no real solutions.

Therefore, the only real solutions to $|x^2 - 6x| = 27$ are $x = -3$ and $x = 9$.

51. $x = -3, 2$

To solve an absolute value equation of the form $|a| = |b|$, you must solve the two equations $a = b$ and $a = -b$. So for the equation $|15x - 5| = |35 - 5x|$, you have to solve $15x - 5 = 35 - 5x$ and $15x - 5 = -(35 - 5x)$.

For the first equation, you have

$$15x - 5 = 35 - 5x$$
$$20x = 40$$
$$x = 2$$

And for the second equation, you have

$$15x - 5 = -(35 - 5x)$$
$$15x - 5 = -35 + 5x$$
$$10x = -30$$
$$x = -3$$

Therefore, the solutions are $x = -3$ and $x = 2$.

52. $x = -1$

To solve a rational equation of the form $\dfrac{p(x)}{q(x)} = 0$, you need to solve the equation $p(x) = 0$. Therefore, to solve $\dfrac{x+1}{x-4} = 0$, you simply set the numerator equal to zero and solve for x:

$$x + 1 = 0$$
$$x = -1$$

53. $x = -\sqrt{2}, \sqrt{2}$

To solve the equation $\dfrac{1}{x+2} + \dfrac{1}{x} = 1$, first remove all fractions by multiplying both sides of the equation by the least common multiple of the denominators:

$$\frac{1}{x+2} + \frac{1}{x} = 1$$
$$\left[(x+2)(x)\right]\left(\frac{1}{x+2} + \frac{1}{x}\right) = \left[(x+2)(x)\right](1)$$
$$x + (x+2) = x^2 + 2x$$

In this case, multiplying leaves you with a quadratic equation to solve:

$$x + (x+2) = x^2 + 2x$$
$$2x + 2 = x^2 + 2x$$
$$2 = x^2$$
$$\pm\sqrt{2} = x$$

Check the solutions to see whether they're extraneous (incorrect) answers. In this case, you can verify that both $-\sqrt{2}$ and $\sqrt{2}$ satisfy the original equation by substituting these values into $\dfrac{1}{x+2} + \dfrac{1}{x} = 1$ and checking that you get 1 on the left side of the equation.

54. -14

To solve an equation of the form $\dfrac{a}{b} = \dfrac{c}{d}$, cross-multiply to produce the equation $ad = bc$:

$$\frac{x+5}{x+2} = \frac{x-4}{x-10}$$
$$(x+5)(x-10) = (x-4)(x+2)$$
$$x^2 - 5x - 50 = x^2 - 2x - 8$$

After cross-multiplying, you're left with a quadratic equation, which then reduces to a linear equation:

$$x^2 - 5x - 50 = x^2 - 2x - 8$$
$$-42 = 3x$$
$$-14 = x$$

You can verify that -14 is a solution of the original rational equation by substituting $x = -14$ into the equation $\dfrac{x+5}{x+2} = \dfrac{x-4}{x-10}$ and checking that you get the same value on both sides of the equation.

55.

no real solutions

Begin by factoring all denominators. Then multiply each side of the equation by the least common multiple of the denominators to remove all fractions:

$$\frac{1}{x-2} - \frac{2}{x-3} = \frac{-1}{(x-2)(x-3)}$$

$$\left[(x-2)(x-3)\right]\left(\frac{1}{x-2} - \frac{2}{x-3}\right) = \left[(x-2)(x-3)\right]\left(\frac{-1}{(x-2)(x-3)}\right)$$

$$x-3-2(x-2) = -1$$

Then simplify:

$$x-3-2(x-2) = -1$$
$$-x+1 = -1$$
$$-x = -2$$
$$x = 2$$

Because $x = 2$ makes the original equation undefined, there are no real solutions.

56.

(–4, 8)

To solve the inequality $x^2 - 4x - 32 < 0$, begin by solving the corresponding equation $x^2 - 4x - 32 = 0$. Then pick a point from each interval (determined by the solutions) to test in the original inequality.

Factoring $x^2 - 4x - 32 = 0$ gives you $(x + 4)(x - 8) = 0$. Setting the first factor equal to zero gives you $x + 4 = 0$, which has the solution $x = -4$, and setting the second factor equal to zero gives you $x - 8 = 0$, which has the solution $x = 8$.

Therefore, you need to pick a point from each of the intervals, $(-\infty, -4)$, $(-4, 8)$, and $(8, \infty)$, to test in the original inequality. Substitute each test number into the expression $x^2 - 4x - 32$ to see whether the answer is less than or greater than zero.

Using $x = -10$ to check the interval $(-\infty, -4)$ gives you

$$(-10)^2 - 4(-10) - 32 = 108$$

which is not less than zero and so doesn't satisfy the inequality.

Using $x = 0$ to check the interval $(-4, 8)$ gives you

$$0^2 - 4(0) - 32 = -32$$

which is less than zero and does satisfy the inequality.

Using $x = 10$ to check the interval $(8, \infty)$ gives you

$$(10)^2 - 4(10) - 32 = 28$$

which is not less than zero and so doesn't satisfy the inequality.

Therefore, the solution set is the interval $(-4, 8)$.

57. $(-\infty, -3] \cup \{0\} \cup [2, \infty)$

To solve the inequality $2x^4 + 2x^3 \geq 12x^2$, begin by setting one side equal to zero. Solve the corresponding equation and then pick a point from each interval (determined by the solutions) to test in the inequality.

So from $2x^4 + 2x^3 \geq 12x^2$, you have $2x^4 + 2x^3 - 12x^2 \geq 0$. Factoring the corresponding equation so you can solve for x gives you

$$2x^4 + 2x^3 - 12x^2 = 0$$

$$2x^2\left(x^2 + x - 6\right) = 0$$

$$2x^2(x+3)(x-2) = 0$$

Set each factor equal to zero and solve for x, giving you the solutions $x = 0$, $x = -3$, and $x = 2$. These values are also solutions to the inequality.

Next, pick a test point from each of the intervals, $(-\infty, -3)$, $(-3, 0)$, $(0, 2)$, and $(2, \infty)$, to see whether the answer is positive or negative. Using $x = -10$ to check the interval $(-\infty, -3)$ gives you

$$2(-10)^4 + 2(-10)^3 - 12(-10)^2$$

$$= 20,000 - 2,000 - 1,200$$

$$= 16,800$$

which is greater than zero.

Using $x = -1$ to check the interval $(-3, 0)$ gives you

$$2(-1)^4 + 2(-1)^3 - 12(-1)^2$$

$$= 2 - 2 - 12$$

$$= -12$$

which is less than zero.

Using $x = 1$ to check the interval $(0, 2)$ gives you

$$2(1)^4 + 2(1)^3 - 12(1)$$

$$= 2 + 2 - 12$$

$$= -8$$

which is less than zero.

Finally, using $x = 10$ to check the interval $(2, \infty)$ gives you

$$2(10)^4 + 2(10)^3 - 12(10)^2$$

$$= 20,000 + 2,000 - 1,200$$

$$= 20,800$$

which is greater than zero.

Therefore, the solution set is $(-\infty, -3] \cup \{0\} \cup [2, \infty)$.

Note that you could've divided the original inequality by 2 to simplify the initial inequality.

58. $(-\infty, -3) \cup (-1, 2)$

To solve the rational inequality $\dfrac{(x+1)(x-2)}{(x+3)} < 0$, first determine which values will make the numerator equal to zero and which values will make the denominator equal to zero. This helps you identify zeros (where the graph crosses the x-axis) and points where the function is not continuous.

From the numerator, you have $x + 1 = 0$, which has the solution $x = -1$. You also have $x - 2 = 0$, which has the solution $x = 2$. From the denominator, you have $x + 3 = 0$, which has the solution $x = -3$. Take a point from each interval determined by these solutions and test it in the original inequality.

You need to test a point from each of the intervals, $(-\infty, -3)$, $(-3, -1)$, $(-1, 2)$, and $(2, \infty)$.

Pick a point from each interval and substitute into the expression $\dfrac{(x+1)(x-2)}{(x+3)}$ to see whether you get a value less than or greater than zero.

Using $x = -10$ to test the interval $(-\infty, -3)$ gives you

$$\frac{(-10+1)(-10-2)}{(-10+3)}$$

$$= \frac{(-9)(-12)}{(-7)}$$

$$= -\frac{108}{7}$$

which is less than zero and so satisfies the inequality.

Using $x = -2$ to test the interval $(-3, -1)$ gives you

$$\frac{(-2+1)(-2-2)}{(-2+3)}$$

$$= \frac{(-1)(-4)}{1}$$

$$= 4$$

which is greater than zero and doesn't satisfy the inequality.

Using $x = 0$ to test the interval $(-1, 2)$ gives you

$$\frac{(0+1)(0-2)}{(0+3)}$$

$$= \frac{(1)(-2)}{3}$$

$$= -\frac{2}{3}$$

which is less than zero and so satisfies the inequality.

Using $x = 3$ to test the interval $(2, \infty)$ gives you

$$\frac{(3+1)(3-2)}{(3+3)}$$

$$= \frac{(4)(1)}{6}$$

$$= \frac{2}{3}$$

which is greater than zero and so doesn't satisfy the inequality.

Therefore, the solution set is the interval $(-\infty, -3)$ and $(-1, 2)$.

59. $\left(-1, -\frac{1}{3}\right) \cup (1, 3)$

Begin by putting the rational inequality into the form $\dfrac{p(x)}{q(x)} > 0$ by putting all terms on one side of the inequality and then getting common denominators:

$$\frac{1}{x-1} + \frac{1}{x+1} > \frac{3}{4}$$

$$\frac{1}{x-1} + \frac{1}{x+1} - \frac{3}{4} > 0$$

$$\frac{4(x+1)}{4(x-1)(x+1)} + \frac{4(x-1)}{4(x+1)(x-1)} - \frac{3(x-1)(x+1)}{4(x-1)(x+1)} > 0$$

$$\frac{4x+4+4x-4-3x^2+3}{(x-1)(x+1)} > 0$$

$$\frac{-3x^2+8x+3}{(x-1)(x+1)} > 0$$

Next, set the numerator equal to zero and solve for x. Setting the numerator equal to zero gives you a quadratic equation that you can factor by trial and error:

$$-3x^2 + 8x + 3 = 0$$

$$3x^2 - 8x - 3 = 0$$

$$(3x+1)(x-3) = 0$$

The solutions are $x = -\dfrac{1}{3}$ and $x = 3$.

Also set each factor from the denominator equal to zero and solve for x. This gives you both $(x-1) = 0$, which has the solution $x = 1$, and $(x+1) = 0$, which has the solution $x = -1$.

Next, take a point from each of the intervals, $(-\infty, -1)$, $\left(-1, -\frac{1}{3}\right)$, $\left(-\frac{1}{3}, 1\right)$, $(1, 3)$, and $(3, \infty)$, and test it in the expression $\dfrac{-3x^2+8x+3}{(x-1)(x+1)}$ to see whether the answer is positive or negative; or equivalently, you can use the expression $\dfrac{-(3x+1)(x-3)}{(x-1)(x+1)}$.

Using $x = -10$ to test the interval $(-\infty, 1)$ gives you

$$\frac{-(3(-10)+1)(-10-3)}{(-10-1)(-10+1)}$$

$$= \frac{-(-29)(-13)}{(-11)(-9)}$$

$$= \frac{-377}{99}$$

which is less than zero and so doesn't satisfy the inequality.

Using $x = -\dfrac{1}{2}$ to test the interval $\left(-1, -\frac{1}{3}\right)$ gives you

$$\frac{-\left(3\left(-\frac{1}{2}\right)+1\right)\left(-\frac{1}{2}-3\right)}{\left(-\frac{1}{2}-1\right)\left(-\frac{1}{2}+1\right)}$$

$$=\frac{-\left(-\frac{1}{2}\right)\left(-\frac{7}{2}\right)}{\left(-\frac{3}{2}\right)\left(\frac{1}{2}\right)}$$

$$=\frac{7}{3}$$

which is greater than zero and so satisfies the inequality.

Using $x = 0$ to test the interval $\left(-\frac{1}{3}, 1\right)$ gives you

$$\frac{-(3(0)+1)(0-3)}{(0-1)(0+1)}$$

$$=\frac{-1(-3)}{(-1)(1)}$$

$$=-3$$

which is less than zero and so doesn't satisfy the inequality.

Using $x = 2$ to test the interval $(1, 3)$ gives you

$$\frac{-(3(2)+1)(2-3)}{(2-1)(2+1)}$$

$$=\frac{-(7)(-1)}{(1)(3)}$$

$$=\frac{7}{3}$$

which is greater than zero and so satisfies the inequality.

And using $x = 10$ to test the interval $(3, \infty)$ gives you

$$\frac{-(3(10)+1)(10-3)}{(10-1)(10+1)}$$

$$=\frac{-(31)(7)}{(9)(11)}$$

$$=\frac{-217}{99}$$

which is less than zero and so doesn't satisfy the inequality.

Therefore, the solution set is the intervals $\left(-1, -\frac{1}{3}\right)$ and $(1, 3)$.

60. $\left(-\frac{3}{2}, \frac{5}{2}\right)$

To solve an absolute inequality of the form $|a| < b$, where $b > 0$, solve the compound inequality $-b < a < b$. So for the inequality $|2x - 1| < 4$, you have

$$-4 < 2x - 1 < 4$$
$$-3 < 2x < 5$$
$$-\frac{3}{2} < x < \frac{5}{2}$$

The solution set is $\left(-\frac{3}{2}, \frac{5}{2}\right)$.

61. $(-\infty, 1) \cup \left(\frac{9}{5}, \infty\right)$

To solve an absolute value inequality of the form $|a| > b$, where $b > 0$, you have to solve the corresponding inequalities $a > b$ and $a < -b$. Therefore, for the inequality $|5x - 7| > 2$, you have to solve $5x - 7 > 2$ and $5x - 7 < -2$. For the first inequality, you have

$$5x - 7 > 2$$
$$5x > 9$$
$$x > \frac{9}{5}$$

And for the second inequality, you have

$$5x - 7 < -2$$
$$5x < 5$$
$$x < 1$$

Therefore, the solution set is the intervals $(-\infty, 1)$ and $\left(\frac{9}{5}, \infty\right)$.

62. $\left[-\frac{4}{3}, 2\right]$

To solve an absolute inequality of the form $|a| \le b$, where $b > 0$, solve the compound inequality $-b \le a \le b$. For the inequality $|-3x + 1| \le 5$, you have

$$-5 \le -3x + 1 \le 5$$
$$-6 \le -3x \le 4$$
$$2 \ge x \ge -\frac{4}{3}$$

Therefore, the solution is the interval $\left[-\frac{4}{3}, 2\right]$.

63. $\frac{7}{4}$

The slope of a line that goes through the points (x_1, y_1) and (x_2, y_2) is given by $m = \frac{y_2 - y_1}{x_2 - x_1}$. Therefore, the slope of the line that passes through $(1, 2)$ and $(5, 9)$ is

$$m = \frac{9 - 2}{5 - 1} = \frac{7}{4}$$

64. $y = 4x + 5$

The graph of $y = mx + b$ is a line with slope of m and a y-intercept at $(0, b)$. Therefore, using $m = 4$ and $b = 5$ gives you the equation $y = 4x + 5$.

65. $y = \frac{5}{6}x + \frac{14}{3}$

The equation of a line that goes through the point (x_1, y_1) and has a slope of m is given by the point-slope formula: $y - y_1 = m(x - x_1)$.

The slope of the line that goes through $(-2, 3)$ and $(4, 8)$ is

$$m = \frac{y_2 - y_1}{x_2 - x_1} = \frac{8 - 3}{4 - (-2)} = \frac{5}{6}$$

Now use the slope and the point $(4, 8)$ in the point-slope formula:

$$y - y_1 = m(x - x_1)$$
$$y - 8 = \frac{5}{6}(x - 4)$$
$$y - 8 = \frac{5}{6}x - \frac{20}{6}$$
$$y = \frac{5}{6}x - \frac{10}{3} + \frac{8}{1}$$
$$y = \frac{5}{6}x - \frac{10}{3} + \frac{24}{3}$$
$$y = \frac{5}{6}x + \frac{14}{3}$$

66. $y = \frac{3}{4}x + \frac{17}{4}$

The equation of a line that goes through the point (x_1, y_1) and has a slope of m is given by the point-slope formula: $y - y_1 = m(x - x_1)$.

The two lines are parallel, so the slopes are the same. Therefore, the slope of the line you're trying to find is $m = \frac{3}{4}$.

Using the slope and the point $(1, 5)$ in the point-slope formula gives you

$$y - y_1 = m(x - x_1)$$
$$y - 5 = \frac{3}{4}(x - 1)$$
$$y - 5 = \frac{3}{4}x - \frac{3}{4}$$
$$y = \frac{3}{4}x - \frac{3}{4} + \frac{5}{1}$$
$$y = \frac{3}{4}x - \frac{3}{4} + \frac{20}{4}$$
$$y = \frac{3}{4}x + \frac{17}{4}$$

67.

$$y = \frac{3}{2}x - \frac{17}{2}$$

The equation of a line that goes through the point (x_1, y_1) and has a slope of m is given by the point-slope formula: $y - y_1 = m(x - x_1)$.

The slope of the line that goes through the points $(-6, 2)$ and $(3, -4)$ is

$$m = \frac{y_2 - y_1}{x_2 - x_1} = \frac{-4 - 2}{3 - (-6)} = \frac{-6}{9} = -\frac{2}{3}$$

The slopes of perpendicular lines are opposite reciprocals of each other, so the slope of the line you want to find is $m = \frac{3}{2}$ (you flip the fraction and change the sign).

Using the slope and the point $(3, -4)$ in the point-slope formula gives you

$$y - y_1 = m(x - x_1)$$

$$y - (-4) = \frac{3}{2}(x - 3)$$

$$y + 4 = \frac{3}{2}(x - 3)$$

$$y = \frac{3}{2}x - \frac{9}{2} - \frac{4}{1}$$

$$y = \frac{3}{2}x - \frac{9}{2} - \frac{8}{2}$$

$$y = \frac{3}{2}x - \frac{17}{2}$$

Here's what the perpendicular lines look like:

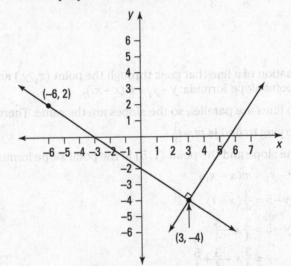

68.

$$y = 2\sqrt{x - 3} + 4$$

Stretching the graph of $y = \sqrt{x}$ vertically by a factor of 2 produces the equation $y = 2\sqrt{x}$. To shift the graph of $y = 2\sqrt{x}$ to the right 3 units, replace x with $(x - 3)$ to get $y = 2\sqrt{x - 3}$. Last, to move the graph of $y = 2\sqrt{x - 3}$ up 4 units, add 4 to the right side to get $y = 2\sqrt{x - 3} + 4$.

69.

$$y = \frac{3}{2}(x+2)^2 - 4$$

The vertex form of a parabola is given by $y = a(x - h)^2 + k$, with the vertex at the point (h, k). Using the vertex $(-2, -4)$ gives you the following equation:

$$y = a\left(x - (-2)\right)^2 - 4$$
$$= a(x+2)^2 - 4$$

Now use the point $(0, 2)$ to solve for a:

$$2 = a(0+2)^2 - 4$$
$$2 = 4a - 4$$
$$6 = 4a$$
$$\frac{3}{2} = a$$

Therefore, the equation of the parabola is

$$y = \frac{3}{2}(x+2)^2 - 4$$

70.

$$y = -(x+1)^2 + 6$$

The vertex form of a parabola is given by $y = a(x - h)^2 + k$, with the vertex at the point (h, k). Using the vertex $(-1, 6)$, you have the following equation:

$$y = a\left(x - (-1)\right)^2 + 6$$
$$= a(x+1)^2 + 6$$

Now use the point $(1, 2)$ to solve for a:

$$2 = a(1+1)^2 + 6$$
$$2 = 4a + 6$$
$$-4 = 4a$$
$$-1 = a$$

Therefore, the vertex form of the parabola is

$$y = -(x+1)^2 + 6$$

Here's the graph of $y = -(x + 1)^2 + 6$:

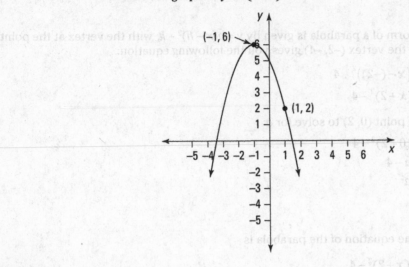

71.

$y = 3(x - 5)^2 + 2$

To translate the parabola 6 units to the right, you replace x with the quantity $(x - 6)$. And to translate the parabola 2 units down, you subtract 2 from the original expression for y:

$$y = 3\big((x-6)+1\big)^2 + 4 - 2$$

$$= 3(x - 5)^2 + 2$$

Another option is simply to count and determine that the new vertex is at (5, 2); then use the vertex form $y = a(x - h)^2 + k$, where the vertex is at the point (h, k).

72.

$y = e^{-2x} - 5$

To compress the graph of $y = e^x$ horizontally by a factor of 2, replace x with $2x$ to get $y = e^{2x}$. To reflect the graph of $y = e^{2x}$ across the y-axis, replace x with $-x$ to get $y = e^{2(-x)}$, or $y = e^{-2x}$. Last, to shift the graph of $y = e^{-2x}$ down 5 units, subtract 5 from the right side of the equation to get $y = e^{-2x} - 5$.

Here's the graph of $y = e^{-2x} - 5$:

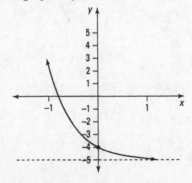

73. $\quad y = -\left|\dfrac{1}{5}x\right| + 3$

To stretch the graph of $y = |x|$ horizontally by a factor of 5, replace x with $\dfrac{1}{5}x$ to get $y = \left|\dfrac{1}{5}x\right|$. Next, to reflect the graph of $y = \left|\dfrac{1}{5}x\right|$ across the x-axis, multiply the right side of the equation $y = \left|\dfrac{1}{5}x\right|$ by -1 to get $y = -\left|\dfrac{1}{5}x\right|$. Last, to shift the graph of $y = -\left|\dfrac{1}{5}x\right|$ up 3 units, add 3 to the right side of the equation to get

$$y = -\left|\dfrac{1}{5}x\right| + 3$$

74. $\quad f(x) = -\dfrac{3}{8}x^3 - \dfrac{15}{8}x^2 - \dfrac{3}{4}x + 3$

The polynomial has x intercepts at $x = -4$, $x = -2$, and $x = 1$, so you know that the polynomial has the factors $(x + 4)$, $(x + 2)$, and $(x - 1)$. Therefore, you can write

$$f(x) = a(x + 4)(x + 2)(x - 1)$$

Use the point $(0, 3)$ to solve for a:

$$3 = a(0 + 4)(0 + 2)(0 - 1)$$
$$3 = -8a$$
$$-\dfrac{3}{8} = a$$

Now you can enter the value of a in the equation of the polynomial and simplify:

$$f(x) = -\dfrac{3}{8}(x + 4)(x + 2)(x - 1)$$
$$= -\dfrac{3}{8}\left(x^2 + 6x + 8\right)(x - 1)$$
$$= -\dfrac{3}{8}\left(x^3 - x^2 + 6x^2 - 6x + 8x - 8\right)$$
$$= -\dfrac{3}{8}\left(x^3 + 5x^2 + 2x - 8\right)$$
$$= -\dfrac{3}{8}x^3 - \dfrac{15}{8}x^2 - \dfrac{3}{4}x + 3$$

Here's the graph of $f(x) = -\dfrac{3}{8}x^3 - \dfrac{15}{8}x^2 - \dfrac{3}{4}x + 3$:

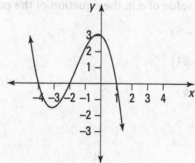

75.

$$f(x) = -\frac{1}{2}x^4 + \frac{7}{2}x^3 - \frac{13}{2}x^2 - \frac{3}{2}x + 9$$

The polynomial is fourth-degree and has the x-intercepts at $x = -1$, 2, and 3, where 3 is a repeated root. Therefore, the polynomial has the factors $(x + 1)$, $(x - 2)$, and $(x - 3)^2$, so you can write

$$f(x) = a(x+1)(x-2)(x-3)^2$$

Use the point (1, 4) to solve for a:

$$4 = a(1+1)(1-2)(1-3)^2$$
$$4 = a(2)(-1)(4)$$
$$-\frac{1}{2} = a$$

Now you can enter the value of a in the equation of the polynomial and simplify:

$$f(x) = -\frac{1}{2}(x+1)(x-2)(x-3)^2$$
$$= -\frac{1}{2}(x^2 - x - 2)(x^2 - 6x + 9)$$
$$= -\frac{1}{2}(x^4 - 6x^3 + 9x^2 - x^3 + 6x^2 - 9x - 2x^2 + 12x - 18)$$
$$= -\frac{1}{2}(x^4 - 7x^3 + 13x^2 + 3x - 18)$$
$$= -\frac{1}{2}x^4 + \frac{7}{2}x^3 - \frac{13}{2}x^2 - \frac{3}{2}x + 9$$

76.

$$y = -\frac{1}{3}x^2 + \frac{2}{3}x + 8$$

The parabola has x-intercepts at $x = -4$ and $x = 6$, so you know that $(x + 4)$ and $(x - 6)$ are factors of the parabola. Therefore, you can write

$$y = a(x+4)(x-6)$$

Use the point (0, 8) to solve for a:

$$8 = a(0+4)(0-6)$$
$$8 = -24a$$
$$-\frac{1}{3} = a$$

Now you can enter the value of a in the equation of the parabola and simplify:

$$y = -\frac{1}{3}(x+4)(x-6)$$
$$= -\frac{1}{3}(x^2 - 2x - 24)$$
$$= -\frac{1}{3}x^2 + \frac{2}{3}x + 8$$

77. $f(x) = \frac{3}{2}x^2 + 15x + 24$

The parabola has x-intercepts at $x = -8$ and $x = -2$, so you know that $(x + 8)$ and $(x + 2)$ are factors of the parabola. Therefore, you can write

$$f(x) = a(x + 8)(x + 2)$$

Use the point $(-4, -12)$ to solve for a:

$$-12 = a(-4 + 8)(-4 + 2)$$
$$-12 = a(4)(-2)$$
$$-12 = -8a$$
$$\frac{3}{2} = a$$

Now you can enter the value of a in the equation of the parabola and simplify:

$$f(x) = \frac{3}{2}(x + 8)(x + 2)$$
$$= \frac{3}{2}\left(x^2 + 10x + 16\right)$$
$$= \frac{3}{2}x^2 + 15x + 24$$

78.   domain: $[-2, \infty)$; range: $[1, \infty)$

The function is continuous on its domain.

The lowest x coordinate occurs at the point $(-2, 3)$, and then the graph extends indefinitely to the right; therefore, the domain is $[-2, \infty)$. The lowest y coordinate occurs at the point $(-1, 1)$, and then the graph extends indefinitely upward; therefore, the range is $[1, \infty)$.

Note that the brackets in the interval notation indicate that the value -2 is included in the domain and that 1 is included in the range.

79.   domain: $[-5, 3]$; range: $[-2, 2]$

Notice that the function is continuous on its domain.

The smallest x coordinate occurs at the point $(-5, 0)$, and the largest x coordinate occurs at the point $(3, 0)$; therefore, the domain is $[-5, 3]$. The smallest y coordinate occurs at the point $(1, -2)$, and the largest y coordinate occurs at the point $(-3, 2)$; therefore, the range is $[-2, 2]$.

Note that the brackets in the interval notation indicate that the values -5 and 3 are included in the domain and that the values -2 and 2 are included in the range.

80.
domain: $(-\infty, -1) \cup [2, \infty)$; range: $(-3, -1) \cup [2, 4)$

Notice that the function is continuous on its domain.

For the portion of the graph that's to the left of the y-axis, there's no largest x value due to the hole at $(-1, -1)$; instead, the x values get arbitrarily close to –1. The graph then extends indefinitely to the left so that the domain for the left side of the graph is $(-\infty, -1)$. There's also no largest y value due to the hole at $(-1, -1)$; the y values get arbitrarily close to –1. The graph then decreases as it approaches the horizontal asymptote at $y = -3$, so the range for this portion of the graph is $(-3, -1)$.

For the portion of the graph that's to the right of the y-axis, the smallest x value is at the point $(2, 2)$. The graph then extends indefinitely to the right, so the domain for the right side of the graph is $[2, \infty)$. The smallest y value also occurs at the point $(2, 2)$, and then the graph increases as it approaches the horizontal asymptote at $y = 4$; therefore, the range of this portion of the graph is $[2, 4)$.

Putting the two parts together, the domain of the function is $(-\infty, -1) \cup [2, \infty)$, and the range is $(-3, -1) \cup [2, 4)$.

81.
$\lim\limits_{x \to -\infty} \left(3x^6 - 40x^5 + 33\right) = \infty$, $\lim\limits_{x \to \infty} \left(3x^6 - 40x^5 + 33\right) = \infty$

To determine the end behavior of a polynomial, you just have to determine the end behavior of the highest-powered term.

As x approaches $-\infty$, you have

$$\lim\limits_{x \to -\infty} \left(3x^6 - 40x^5 + 33\right) = \lim\limits_{x \to -\infty} \left(3x^6\right)$$
$$= \infty$$

Note that even though the limit is approaching negative infinity, the limit is still positive due to the even exponent.

As x approaches ∞, you have

$$\lim\limits_{x \to \infty} \left(3x^6 - 40x^5 + 33\right) = \lim\limits_{x \to \infty} \left(3x^6\right)$$
$$= \infty$$

82.
$\lim\limits_{x \to -\infty} \left(-7x^9 + 33x^8 - 51x^7 + 19x^4 - 1\right) = \infty$, $\lim\limits_{x \to \infty} \left(-7x^9 + 33x^8 - 51x^7 + 19x^4 - 1\right) = -\infty$

To determine the end behavior of a polynomial, determine the end behavior of the highest-powered term.

As x approaches $-\infty$, you have

$$\lim\limits_{x \to -\infty} \left(-7x^9 + 33x^8 - 51x^7 + 19x^4 - 1\right) = \lim\limits_{x \to -\infty} \left(-7x^9\right)$$
$$= \infty$$

Note that the limit is positive because a negative number raised to an odd power is negative, but after multiplying by –7, the answer becomes positive.

As x approaches ∞, you have

$$\lim_{x \to \infty}\left(-7x^9 + 33x^8 - 51x^7 + 19x^4 - 1\right) = \lim_{x \to \infty}\left(-7x^9\right)$$
$$= -\infty$$

Likewise, the limit here is negative because a positive number raised to an odd power (or any power for that matter!) is positive, but after multiplying by –7, the answer becomes negative.

83. $3x + 12$

Remove the parentheses and combine like terms:

$$(5x + 6) + (-2x + 6)$$
$$= 5x + 6 - 2x + 6$$
$$= 3x + 12$$

84. $3x - 2$

Remove the parentheses and combine like terms:

$$\left(2x^2 - x + 7\right) + \left(-2x^2 + 4x - 9\right)$$
$$= 2x^2 - x + 7 - 2x^2 + 4x - 9$$
$$= 3x - 2$$

85. $x^3 - x^2 + 2x + 14$

Remove the parentheses and combine like terms:

$$\left(x^3 - 5x^2 + 6\right) + \left(4x^2 + 2x + 8\right)$$
$$= x^3 - 5x^2 + 6 + 4x^2 + 2x + 8$$
$$= x^3 - x^2 + 2x + 14$$

86. $-2x^4 + 3x + 8$

Remove the parentheses and combine like terms:

$$\left(3x + x^4 + 2\right) + \left(-3x^4 + 6\right)$$
$$= 3x + x^4 + 2 - 3x^4 + 6$$
$$= -2x^4 + 3x + 8$$

87. $x^4 + 5x^3 - 3x^2$

Remove the parentheses and combine like terms:

$$\left(x^4 - 6x^2 + 3\right) + \left(5x^3 + 3x^2 - 3\right)$$
$$= x^4 - 6x^2 + 3 + 5x^3 + 3x^2 - 3$$
$$= x^4 + 5x^3 - 3x^2$$

88. $3x - 7$

Begin by removing the parentheses, being careful to distribute the –1 to the second term. Then collect like terms to simplify:

$$(5x - 3) - (2x + 4)$$
$$= 5x - 3 - 2x - 4$$
$$= 3x - 7$$

89. $6x^2 - 5x + 5$

Begin by removing the parentheses, being careful to distribute the –1 to the second term. Then collect like terms to simplify:

$$\left(x^2 - 3x + 1\right) - \left(-5x^2 + 2x - 4\right)$$
$$= x^2 - 3x + 1 + 5x^2 - 2x + 4$$
$$= 6x^2 - 5x + 5$$

90. $4x^3 + 5x^2 - 8x + 14$

Begin by removing the parentheses, being careful to distribute the –1 to the second term. Then collect like terms to simplify:

$$\left(8x^3 + 5x^2 - 3x + 2\right) - \left(4x^3 + 5x - 12\right)$$
$$= 8x^3 + 5x^2 - 3x + 2 - 4x^3 - 5x + 12$$
$$= 4x^3 + 5x^2 - 8x + 14$$

91. $2x^2 + 3x + 1$

Begin by removing the parentheses, being careful to distribute the –1 to the second and third terms. Then collect like terms to simplify:

$$(x + 3) - \left(x^2 + 3x - 4\right) - \left(-3x^2 - 5x + 6\right)$$
$$= x + 3 - x^2 - 3x + 4 + 3x^2 + 5x - 6$$
$$= 2x^2 + 3x + 1$$

92. $10x^4 - 7x^3 - 9x^2 - 8x + 10$

Begin by removing the parentheses, being careful to distribute the –1 to the second term. Then collect like terms to simplify:

$$\left(10x^4 - 6x^3 + x^2 + 6\right) - \left(x^3 + 10x^2 + 8x - 4\right)$$
$$= 10x^4 - 6x^3 + x^2 + 6 - x^3 - 10x^2 - 8x + 4$$
$$= 10x^4 - 7x^3 - 9x^2 - 8x + 10$$

93. $5x^3 - 15x^2$

Distribute to get

$$5x^2(x-3)$$
$$= 5x^3 - 15x^2$$

94. $3x^2 + 7x - 20$

Distribute each term in the first factor to each term in the second factor. Then collect like terms:

$$(x+4)(3x-5)$$
$$= 3x^2 - 5x + 12x - 20$$
$$= 3x^2 + 7x - 20$$

95. $x^2y - xy^2 + 6xy$

Multiply each term in the first factor by xy:

$$(x-y+6)(xy)$$
$$= x^2y - xy^2 + 6xy$$

96. $2x^3 - 3x^2 + 9x - 4$

Distribute each term in the first factor to each term in the second factor and then collect like terms:

$$(2x-1)(x^2 - x + 4)$$
$$= 2x^3 - 2x^2 + 8x - x^2 + x - 4$$
$$= 2x^3 - 3x^2 + 9x - 4$$

97. $-x^6 - 3x^5 - 3x^4 - 9x^3 - 2x^2 - 6x$

Begin by distributing $-x$ to each term in the second factor:

$$-x(x^4 + 3x^2 + 2)(x+3)$$
$$= (-x^5 - 3x^3 - 2x)(x+3)$$

Next, multiply each term in the first factor by each term in the second factor:

$$(-x^5 - 3x^3 - 2x)(x+3)$$
$$= -x^6 - 3x^5 - 3x^4 - 9x^3 - 2x^2 - 6x$$

98. $x + 6 + \dfrac{18}{x-2}$

Using polynomial long division gives you

$$
\begin{array}{r}
x+6 \\
(x-2)\overline{\smash{\big)}\,x^2 + 4x + 6} \\
-\underline{(x^2 - 2x)} \\
6x + 6 \\
-\underline{(6x - 12)} \\
18 \text{ (Remainder)}
\end{array}
$$

Remember to put the remainder over the divisor when writing the answer:

$$\frac{x^2 + 4x + 6}{x - 2} = x + 6 + \frac{18}{x - 2}$$

99. $2x - 11 + \dfrac{52}{x+4}$

Using polynomial long division gives you

$$
\begin{array}{r}
2x - 11 \\
(x+4)\overline{\smash{\big)}\,2x^2 - 3x + 8} \\
-\underline{(2x^2 + 8x)} \\
-11x + 8 \\
-\underline{(-11x - 44)} \\
52 \text{ (Remainder)}
\end{array}
$$

When writing the answer, put the remainder over the divisor:

$$\frac{2x^2 - 3x + 8}{x + 4} = 2x - 11 + \frac{52}{x + 4}$$

100. $x^2 + 3x + 7 + \dfrac{27}{x-3}$

First add a placeholder for the missing x^2 term in the numerator. Rewriting $x^3 - 2x + 6$ as $x^3 + 0x^2 - 2x + 6$ will make all the like terms line up when you do the long division, making the subtraction a bit easier to follow.

Then use polynomial long division:

$$
\begin{array}{r}
x^2 + 3x + 7 \\
(x-3)\overline{\smash{\big)}\,x^3 + 0x^2 - 2x + 6} \\
-\underline{(x^3 - 3x^2)} \\
3x^2 - 2x \\
-\underline{(3x^2 - 9x)} \\
7x + 6 \\
-\underline{(7x - 21)} \\
27 \text{ (Remainder)}
\end{array}
$$

When you write the answer, put the remainder over the divisor:

$$\frac{x^3 - 2x + 6}{x - 3} = x^2 + 3x + 7 + \frac{27}{x - 3}$$

101.

$$3x^3 + 4x^2 - 15x - 21 + \frac{75x + 106}{x^2 + 5}$$

Begin by filling in all the missing terms so that everything will line up when you perform the long division. Here, add $0x^3$ and $0x$ as placeholders in the numerator and put $0x$ in the denominator:

$$\frac{3x^5 + 4x^4 - x^2 + 1}{x^2 + 5} = \frac{3x^5 + 4x^4 + 0x^3 - x^2 + 0x + 1}{x^2 + 0x + 5}$$

Then use polynomial long division:

$$
\begin{array}{r}
3x^3 + 4x^2 - 15x - 21 \\
(x^2 + 0x + 5)\overline{)\,3x^5 + 4x^4 + 0x^3 - x^2 + 0x + 1} \\
-(3x^5 + 0x^4 + 15x^3) \\
\hline
4x^4 - 15x^3 - x^2 \\
-(4x^4 + 0x^3 + 20x^2) \\
\hline
-15x^3 - 21x^2 + 0x \\
-(-15x^3 + 0x^2 - 75x) \\
\hline
-21x^2 + 75x + 1 \\
-(-21x^2 + 0x - 105) \\
\hline
75x + 106 \text{ (Remainder)}
\end{array}
$$

Put the remainder over the divisor when writing your answer:

$$\frac{3x^5 + 4x^4 - x^2 + 1}{x^2 + 5} = 3x^3 + 4x^2 - 15x - 21 + \frac{75x + 106}{x^2 + 5}$$

102.

$$3x^3 - 8x^2 + 15x - 41 + \frac{114x^2 - 60x + 166}{x^3 + 2x^2 + 4}$$

Begin by filling in all the missing terms so that everything will line up when you perform the long division:

$$\frac{3x^6 - 2x^5 - x^4 + x^3 + 2}{x^3 + 2x^2 + 4} = \frac{3x^6 - 2x^5 - x^4 + x^3 + 0x^2 + 0x + 2}{x^3 + 2x^2 + 0x + 4}$$

Using polynomial long division gives you

$$
\begin{array}{r}
3x^3 - 8x^2 + 15x - 41 \\
(x^3 + 2x^2 + 0x + 4)\overline{)\,3x^4 - 2x^5 - x^4 + x^3 + 0x^2 + 0x + 2} \\
-(3x^4 + 6x^5 + 0x^4 + 12x^3) \\
\hline
-8x^5 - 1x^4 - 11x^3 + 0x^2 \\
-(-8x^5 - 16x^4 + 0x^3 - 32x^2) \\
\hline
15x^4 - 11x^3 + 32x^2 + 0x \\
-(15x^4 + 30x^3 + 0x^2 + 60x) \\
\hline
-41x^3 + 32x^2 - 60x + 2 \\
-(-41x^3 + 82x^2 + 0x - 164) \\
\hline
114x^2 - 60x + 166 \text{ (Remainder)}
\end{array}
$$

Then write your answer, putting the remainder over the divisor:

$$\frac{3x^6 - 2x^5 - x^4 + x^3 + 2}{x^3 + 2x^2 + 4} = 3x^3 - 8x^2 + 15x - 41 + \frac{114x^2 - 60x + 166}{x^3 + 2x^2 + 4}$$

103. $\sin\theta = \frac{7\sqrt{65}}{65}$; $\cos\theta = \frac{4\sqrt{65}}{65}$; $\tan\theta = \frac{7}{4}$

When considering the sides of the right triangle, the values of the trigonometric functions are given by $\sin\theta = \frac{\text{opposite}}{\text{hypotenuse}}$, $\cos\theta = \frac{\text{adjacent}}{\text{hypotenuse}}$, and $\tan\theta = \frac{\text{opposite}}{\text{adjacent}}$; therefore, $\sin\theta = \frac{7}{\sqrt{65}} = \frac{7\sqrt{65}}{65}$, $\cos\theta = \frac{4}{\sqrt{65}} = \frac{4\sqrt{65}}{65}$, and $\tan\theta = \frac{7}{4}$.

104. $\sin\theta = \frac{4\sqrt{17}}{17}$; $\cos\theta = \frac{\sqrt{17}}{17}$; $\tan\theta = 4$

When considering the sides of the right triangle, the values of the trigonometric functions are given by $\sin\theta = \frac{\text{opposite}}{\text{hypotenuse}}$, $\cos\theta = \frac{\text{adjacent}}{\text{hypotenuse}}$, and $\tan\theta = \frac{\text{opposite}}{\text{adjacent}}$; therefore, $\sin\theta = \frac{8}{2\sqrt{17}} = \frac{4\sqrt{17}}{17}$, $\cos\theta = \frac{2}{2\sqrt{17}} = \frac{\sqrt{17}}{17}$, and $\tan\theta = \frac{8}{2} = 4$.

105. $\frac{-2\sqrt{10}}{3}$

You know the value of $\sin\theta$, so if you can find the value of $\cos\theta$, you can evaluate $\cot\theta$ using $\cot\theta = \frac{\cos\theta}{\sin\theta}$. To find $\cos\theta$, use the identity $\sin^2\theta + \cos^2\theta = 1$:

$$\left(\frac{3}{7}\right)^2 + \cos^2\theta = 1$$

$$\cos^2\theta = 1 - \frac{9}{49}$$

$$\cos^2\theta = \frac{40}{49}$$

Next, take the square root of both sides, keeping the negative solution for cosine because $\frac{\pi}{2} < \theta < \pi$:

$$\cos^2\theta = \frac{40}{49}$$

$$\cos\theta = -\sqrt{\frac{40}{49}}$$

$$\cos\theta = -\frac{2\sqrt{10}}{7}$$

Therefore, using $\cot\theta = \frac{\cos\theta}{\sin\theta}$, you have

$$\cot\theta = \frac{-\dfrac{2\sqrt{10}}{7}}{\dfrac{3}{7}} = \frac{-2\sqrt{10}}{3}$$

106. $\dfrac{-4\sqrt{7}}{7}$

To find the value of $\csc\theta$, you can find the value of $\sin\theta$ and then use $\csc\theta = \dfrac{1}{\sin\theta}$. Use the identity $\sin^2\theta + \cos^2\theta = 1$ to solve for $\sin\theta$:

$$\sin^2\theta + \left(\dfrac{3}{4}\right)^2 = 1$$

$$\sin^2\theta = 1 - \dfrac{9}{16}$$

$$\sin^2\theta = \dfrac{7}{16}$$

Next, take the square root of both sides, keeping the negative solution for sine because $\dfrac{3\pi}{2} < \theta < 2\pi$:

$$\sin^2\theta = \dfrac{7}{16}$$

$$\sin\theta = -\dfrac{\sqrt{7}}{4}$$

Therefore, using $\csc\theta = \dfrac{1}{\sin\theta}$, you have

$$\csc\theta = \dfrac{1}{-\dfrac{\sqrt{7}}{4}} = \dfrac{-4}{\sqrt{7}} = \dfrac{-4\sqrt{7}}{7}$$

107. $-\dfrac{80}{89}$

To find the value of $\sin(2\theta)$, you can use the identity $\sin(2\theta) = 2\sin\theta\cos\theta$. Notice that because $\sin\theta > 0$ and $\cos\theta < 0$, angle θ must be in the second quadrant. Using $\tan\theta = -\dfrac{8}{5}$, you can make a right triangle and find the missing side using the Pythagorean theorem. When making the triangle, you can neglect the negative sign:

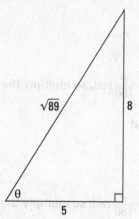

$$h = \sqrt{8^2 + 5^2} = \sqrt{89}$$

Because θ is in the second quadrant, you have $\sin\theta = \dfrac{8}{\sqrt{89}}$ and $\cos\theta = -\dfrac{5}{\sqrt{89}}$. Enter these values in the identity $\sin(2\theta) = 2\sin\theta\cos\theta$ and solve:

$$\sin(2\theta) = 2\sin\theta\cos\theta$$
$$= 2\left(\frac{8}{\sqrt{89}}\right)\left(-\frac{5}{\sqrt{89}}\right)$$
$$= -\frac{80}{89}$$

108. $\dfrac{77}{85}$

To find the value of $\cos(2\theta)$, you can use the identity $\cos(2\theta) = 1 - 2\sin^2\theta$. Using $\cot\theta = -\dfrac{9}{2}$, you can make a right triangle and find the missing side using the Pythagorean theorem. When making the triangle, you can neglect the negative sign:

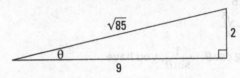

$$h = \sqrt{9^2 + 2^2} = \sqrt{85}$$

Now use the identity $\cos(2\theta) = 1 - 2\sin^2\theta$. The sine is negative, so you have $\sin\theta = -\dfrac{2}{\sqrt{85}}$:

$$\cos(2\theta) = 1 - 2\sin^2\theta$$
$$= 1 - 2\left(-\frac{2}{\sqrt{85}}\right)^2$$
$$= 1 - \frac{8}{85}$$
$$= \frac{77}{85}$$

109. $\dfrac{3\pi}{4}$ rad

Because $180° = \pi$ rad, you have $1° = \dfrac{\pi}{180}$ rad, so multiply the number of degrees by this value:

$$135° = 135\left(\frac{\pi}{180}\right) \text{ rad} = \frac{3\pi}{4} \text{ rad}$$

110. $-\dfrac{14\pi}{9}$ rad

Because $180° = \pi$ rad, you have $1° = \dfrac{\pi}{180}$ rad, so multiply the number of degrees by this value:

$$-280° = -280\left(\frac{\pi}{180}\right) \text{ rad} = -\frac{14\pi}{9} \text{ rad}$$

111. $\frac{\pi}{5}$ rad

Because $180° = \pi$ rad, you have $1° = \frac{\pi}{180}$ rad, so multiply the number of degrees by this value:

$$36° = 36\left(\frac{\pi}{180}\right) \text{ rad} = \frac{\pi}{5} \text{ rad}$$

112. $\frac{-7\pi}{4}$ rad

Because $180° = \pi$ rad, you have $1° = \frac{\pi}{180}$ rad, so multiply the number of degrees by this value:

$$-315° = -315\left(\frac{\pi}{180}\right) \text{ rad} = -\frac{7\pi}{4} \text{ rad}$$

113. $210°$

Because π rad $= 180°$, you have 1 rad $= \left(\frac{180}{\pi}\right)°$, so multiply the number of radians by this value:

$$\frac{7\pi}{6} \text{ rad} = \frac{7\pi}{6}\left(\frac{180}{\pi}\right)° = 210°$$

114. $165°$

Because π rad $= 180°$, you have 1 rad $= \left(\frac{180}{\pi}\right)°$, so multiply the number of radians by this value:

$$\frac{11\pi}{12} \text{ rad} = \frac{11\pi}{12}\left(\frac{180}{\pi}\right)° = 165°$$

115. $-108°$

Because π rad $= 180°$, you have 1 rad $= \left(\frac{180}{\pi}\right)°$, so multiply the number of radians by this value:

$$-\frac{3\pi}{5} \text{ rad} = -\frac{3\pi}{5}\left(\frac{180}{\pi}\right)° = -108°$$

116. $-630°$

Because π rad $= 180°$, you have 1 rad $= \left(\frac{180}{\pi}\right)°$, so multiply the number of radians by this value:

$$-\frac{7\pi}{2} \text{ rad} = -\frac{7\pi}{2}\left(\frac{180}{\pi}\right)° = -630°$$

117. $\dfrac{3\pi}{4}$

Because θ is in Quadrant II and the angle is measured counterclockwise, you have $\dfrac{\pi}{2} \le \theta \le \pi$. Therefore, $\dfrac{3\pi}{4}$ is the most appropriate choice.

118. $-\dfrac{2\pi}{3}$

Because θ is in Quadrant III and the angle is measured clockwise, you have $-\pi \le \theta \le -\dfrac{\pi}{2}$. Therefore, $-\dfrac{2\pi}{3}$ is the most appropriate choice.

119. $\dfrac{11\pi}{6}$

Because θ is in Quadrant IV and the angle is measured counterclockwise, you have $\dfrac{3\pi}{2} \le \theta \le 2\pi$. Therefore, $\dfrac{11\pi}{6}$ is the most appropriate choice.

120. $\sin\theta = \dfrac{\sqrt{2}}{2}$; $\cos\theta = \dfrac{\sqrt{2}}{2}$; $\tan\theta = 1$

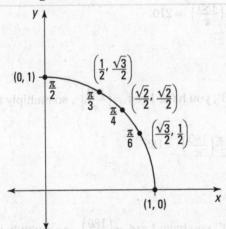

The given angle measure is $\theta = \dfrac{\pi}{4}$. Using the first quadrant of the unit circle, you have the following:

$$\sin\theta = \frac{\sqrt{2}}{2}$$

$$\cos\theta = \frac{\sqrt{2}}{2}$$

$$\tan\theta = \frac{\sin\theta}{\cos\theta} = \frac{\sqrt{2}/2}{\sqrt{2}/2} = 1$$

121. $\sin\theta = \frac{1}{2}$; $\cos\theta = -\frac{\sqrt{3}}{2}$; $\tan\theta = -\frac{\sqrt{3}}{3}$

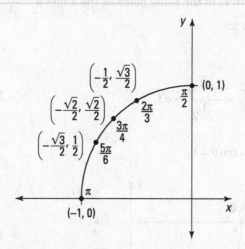

The given angle measure is $\theta = \frac{5\pi}{6}$. Using the second quadrant of the unit circle, you have the following:

$\sin\theta = \frac{1}{2}$

$\cos\theta = -\frac{\sqrt{3}}{2}$

$\tan\theta = \frac{\sin\theta}{\cos\theta} = \frac{1/2}{-\sqrt{3}/2} = -\frac{1}{\sqrt{3}} = -\frac{\sqrt{3}}{3}$

122. $\sin\theta = -\frac{\sqrt{3}}{2}$; $\cos\theta = -\frac{1}{2}$; $\tan\theta = \sqrt{3}$

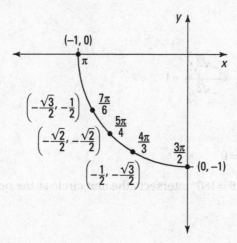

Use the third quadrant of the unit circle. Note that because the angle $\theta = \dfrac{-2\pi}{3}$ touches the unit circle in the same place as the angle $\theta = \dfrac{4\pi}{3}$, you have the following:

$$\sin\theta = -\frac{\sqrt{3}}{2}$$

$$\cos\theta = -\frac{1}{2}$$

$$\tan\theta = \frac{\sin\theta}{\cos\theta} = \frac{-\sqrt{3}/2}{-1/2} = \sqrt{3}$$

123. $\sin\theta = -\dfrac{\sqrt{2}}{2}$; $\cos\theta = -\dfrac{\sqrt{2}}{2}$; $\tan\theta = 1$

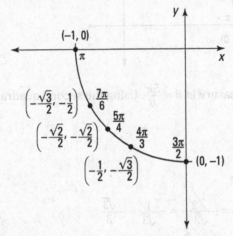

Use the third quadrant of the unit circle. The angle $-135° = \dfrac{-3\pi}{4}$ rad touches the unit circle in the same place as the angle $\dfrac{5\pi}{4}$, so you have the following:

$$\sin\theta = -\frac{\sqrt{2}}{2}$$

$$\cos\theta = -\frac{\sqrt{2}}{2}$$

$$\tan\theta = \frac{\sin\theta}{\cos\theta} = \frac{-\sqrt{2}/2}{-\sqrt{2}/2} = 1$$

124. $\sin\theta = 0$; $\cos\theta = -1$; $\tan\theta = 0$

Because the angle $\theta = 180°$ intersects the unit circle at the point $(-1, 0)$, you have the following:

$$\sin\theta = 0$$

$$\cos\theta = -1$$

$$\tan\theta = \frac{\sin\theta}{\cos\theta} = \frac{0}{-1} = 0$$

125. $\cos\theta$

Simply rewrite cotangent and simplify:

$$\sin\theta\cot\theta$$

$$=\sin\theta\left(\frac{\cos\theta}{\sin\theta}\right)$$

$$=\cos\theta$$

126. $\sin x \tan x$

Begin by writing $\sec x$ as $\dfrac{1}{\cos x}$. Then get common denominators and simplify:

$$\sec x - \cos x$$

$$=\frac{1}{\cos x}-\frac{\cos x}{1}$$

$$=\frac{1-\cos^2 x}{\cos x}$$

$$=\frac{\sin^2 x}{\cos x}$$

$$=\sin x\left(\frac{\sin x}{\cos x}\right)$$

$$=\sin x \tan x$$

127. $1 + \sin 2x$

Expand the given expression and use the identity $\sin^2 x + \cos^2 x = 1$ along with $2\sin x\cos x = \sin(2x)$:

$$\left(\sin x + \cos x\right)^2$$

$$=\left(\sin x + \cos x\right)\left(\sin x + \cos x\right)$$

$$=\sin^2 x + 2\cos x\sin x + \cos^2 x$$

$$=1 + 2\cos x\sin x$$

$$=1 + \sin 2x$$

128. $\sin x$

Use the identity $\sin(A - B) = \sin A\cos B - \cos A\sin B$ on the expression $\sin(\pi - x)$:

$$\sin(\pi - x)$$

$$=\sin\pi\cos x - \cos\pi\sin x$$

$$=0\cos x - (-1)\sin x$$

$$=\sin x$$

129. $\cos x$

Use the identity $\cos A \cos B + \sin A \sin B = \cos(A - B)$ to get

$$\sin x \sin 2x + \cos x \cos 2x$$
$$= \cos 2x \cos x + \sin 2x \sin x$$
$$= \cos(2x - x)$$
$$= \cos x$$

Note that you can also use the following:

$$\sin x \sin 2x + \cos x \cos 2x$$
$$= \cos x \cos 2x + \sin x \sin 2x$$
$$= \cos(x - 2x)$$
$$= \cos(-x)$$
$$= \cos x$$

130. $2\csc^2\theta$

First get common denominators:

$$\frac{1}{1-\cos\theta} + \frac{1}{1+\cos\theta}$$
$$= \frac{(1+\cos\theta)}{(1-\cos\theta)(1+\cos\theta)} + \frac{(1-\cos\theta)}{(1-\cos\theta)(1+\cos\theta)}$$
$$= \frac{2}{1-\cos^2\theta}$$

Then use the identity $\sin^2 x = 1 - \cos^2 x$ in the denominator:

$$\frac{2}{1-\cos^2\theta}$$
$$= \frac{2}{\sin^2\theta}$$
$$= 2\csc^2\theta$$

131. $\csc x + \cot x$

Begin by multiplying the numerator and denominator by the conjugate of the denominator, $1 + \cos x$. Then start to simplify:

$$\frac{\sin x}{1-\cos x}$$
$$= \frac{\sin x}{(1-\cos x)} \frac{(1+\cos x)}{(1+\cos x)}$$
$$= \frac{\sin x(1+\cos x)}{1-\cos^2 x}$$

Answers
101–200

Continue, using the identity $1 - \cos^2 x = \sin^2 x$ to simplify the denominator:

$$= \frac{\sin x(1 + \cos x)}{\sin^2 x}$$

$$= \frac{1 + \cos x}{\sin x}$$

$$= \frac{1}{\sin x} + \frac{\cos x}{\sin x}$$

$$= \csc x + \cot x$$

132. $4\cos^3 \theta - 3\cos\theta$

Use the identity $\cos(A + B) = \cos A \cos B - \sin A \sin B$:

$$\cos(3\theta)$$
$$= \cos(\theta + 2\theta)$$
$$= \cos\theta \cos 2\theta - \sin\theta \sin 2\theta$$

Then use the identities $\cos(2\theta) = 2\cos^2\theta - 1$ and $\sin(2\theta) = 2\sin\theta\cos\theta$ and simplify:

$$\cos\theta \cos 2\theta - \sin\theta \sin 2\theta$$

$$= \cos\theta\left(2\cos^2\theta - 1\right) - \sin\theta(2\sin\theta\cos\theta)$$

$$= 2\cos^3\theta - \cos\theta - 2\sin^2\theta\cos\theta$$

$$= 2\cos^3\theta - \cos\theta - 2\left(1 - \cos^2\theta\right)\cos\theta$$

$$= 2\cos^3\theta - \cos\theta - 2\cos\theta + 2\cos^3\theta$$

$$= 4\cos^3\theta - 3\cos\theta$$

133. $\dfrac{\pi}{6}, \dfrac{5\pi}{6}$

Solve for $\sin x$:

$$2\sin x - 1 = 0$$

$$\sin x = \frac{1}{2}$$

The solutions are $x = \dfrac{\pi}{6}$ and $\dfrac{5\pi}{6}$ in the interval $[0, 2\pi]$.

134. $0, \pi, 2\pi$

Make one side of the equation zero, use the identity $\tan x = \dfrac{\sin x}{\cos x}$, and then factor:

$$\sin x = \tan x$$

$$\sin x - \tan x = 0$$

$$\sin x - \frac{\sin x}{\cos x} = 0$$

$$\sin x\left(1 - \frac{1}{\cos x}\right) = 0$$

Setting the first factor equal to zero gives you $\sin x = 0$, which has the solutions $x = 0$, π, and 2π. Setting the second factor equal to zero gives you $1 - \dfrac{1}{\cos x} = 0$ so that $\cos x = 1$, which has the solutions $x = 0$ and 2π.

135. $\frac{\pi}{3}, \pi, \frac{5\pi}{3}$

Factor the equation:

$$2\cos^2 x + \cos x - 1 = 0$$

$$(2\cos x - 1)(\cos x + 1) = 0$$

Setting the first factor equal to zero gives you $2\cos x - 1 = 0$ so that $\cos x = \frac{1}{2}$, which has the solutions $x = \frac{\pi}{3}$ and $\frac{5\pi}{3}$. Setting the second factor equal to zero gives you $\cos x + 1 = 0$ so that $\cos x = -1$, which has the solution $x = \pi$.

136. $\frac{\pi}{4}, \frac{3\pi}{4}, \frac{5\pi}{4}, \frac{7\pi}{4}$

To solve the equation $|\tan x| = 1$, you must find the solutions to two equations: $\tan x = 1$ and $\tan x = -1$. For the equation $\tan x = 1$, you have the solutions $x = \frac{\pi}{4}$ and $\frac{5\pi}{4}$. For the equation $\tan x = -1$, you have the solutions $x = \frac{3\pi}{4}$ and $\frac{7\pi}{4}$.

137. $\frac{7\pi}{6}, \frac{11\pi}{6}$

Factor the equation $2\sin^2 x - 5\sin x - 3 = 0$ to get

$$(2\sin x + 1)(\sin x - 3) = 0$$

Setting the first factor equal to zero gives you $2\sin x + 1 = 0$ so that $\sin x = -\frac{1}{2}$, which has the solutions $x = \frac{7\pi}{6}$ and $\frac{11\pi}{6}$. Setting the second factor equal to zero gives you $\sin x - 3 = 0$ so that $\sin x = 3$. Because 3 is outside the range of the sine function, this equation from the second factor has no solution.

138. $\frac{\pi}{2}, \frac{3\pi}{2}$

Begin by making one side of the equation equal to zero. Then use the identity $\cot x = \frac{\cos x}{\sin x}$ and factor:

$$\cos x = \cot x$$

$$\cos x - \cot x = 0$$

$$\cos x - \frac{\cos x}{\sin x} = 0$$

$$\cos x\left(1 - \frac{1}{\sin x}\right) = 0$$

Setting the first factor equal to zero gives you $\cos x = 0$, which has the solutions $x = \frac{\pi}{2}$ and $\frac{3\pi}{2}$. Setting the second factor equal to zero gives you $1 - \frac{1}{\sin x} = 0$ so that $\sin x = 1$, which has the solution $x = \frac{\pi}{2}$.

139. $\dfrac{\pi}{12}, \dfrac{5\pi}{12}, \dfrac{13\pi}{12}, \dfrac{17\pi}{12}$

Begin by making the substitution $2x = y$. Because $0 \le x \le 2\pi$, it follows that $0 \le 2x \le 4\pi$ and that $0 \le y \le 4\pi$.

Solving the equation $\sin(y) = \dfrac{1}{2}$ gives you the solutions $y = \dfrac{\pi}{6}, \dfrac{5\pi}{6}, \dfrac{13\pi}{6}$, and $\dfrac{17\pi}{6}$ in the interval $[0, 4\pi]$.

Next, take each solution, set it equal to $2x$, and solve for x: $2x = \dfrac{\pi}{6}$ so that $x = \dfrac{\pi}{12}$; $2x = \dfrac{5\pi}{6}$ so that $x = \dfrac{5\pi}{12}$; $2x = \dfrac{13\pi}{6}$ so that $x = \dfrac{13\pi}{12}$; and $2x = \dfrac{17\pi}{6}$ so that $x = \dfrac{17\pi}{12}$.

140. $\dfrac{\pi}{6}, \dfrac{\pi}{2}, \dfrac{5\pi}{6}, \dfrac{3\pi}{2}$

Begin by making one side of the equation zero. Then use the identity $\sin 2x = 2 \sin x \cos x$ and factor:

$$\sin 2x = \cos x$$
$$\sin 2x - \cos x = 0$$
$$2 \sin x \cos x - \cos x = 0$$
$$\cos x (2 \sin x - 1) = 0$$

Setting the first factor equal to zero gives you $\cos x = 0$, which has the solutions $x = \dfrac{\pi}{2}$ and $\dfrac{3\pi}{2}$. Setting the second factor equal to zero gives you $2 \sin x - 1 = 0$ so that $\sin x = \dfrac{1}{2}$, which has the solutions $x = \dfrac{\pi}{6}$ and $\dfrac{5\pi}{6}$.

141. $\dfrac{\pi}{2}, \dfrac{3\pi}{2}$

Begin by using the identity $\sin 2x = 2 \sin x \cos x$ and then factor:

$$2 \cos x + \sin 2x = 0$$
$$2 \cos x + 2 \sin x \cos x = 0$$
$$2 \cos x (1 + \sin x) = 0$$

Set each factor equal to zero and solve for x. The equation $\cos x = 0$ has the solutions $x = \dfrac{\pi}{2}$ and $\dfrac{3\pi}{2}$, and the equation $1 + \sin x = 0$ gives you $\sin x = -1$, which has the solution $x = \dfrac{3\pi}{2}$.

142. $\dfrac{2\pi}{3}, \pi, \dfrac{4\pi}{3}$

Use the identity $\cos 2x = 2 \cos^2 x - 1$, make one side of the equation zero, and factor:
$$2 + \cos 2x = -3 \cos x$$
$$2 + 2 \cos^2 x - 1 = -3 \cos x$$
$$2 \cos^2 x + 3 \cos x + 1 = 0$$
$$(2 \cos x + 1)(\cos x + 1) = 0$$

Set each factor equal to zero and solve for x. Setting the first factor equal to zero gives you $2\cos x + 1 = 0$ so that $\cos x = -\frac{1}{2}$, which has the solutions $x = \frac{2\pi}{3}$ and $\frac{4\pi}{3}$. Setting the second factor equal to zero gives you $\cos x + 1 = 0$ so that $\cos x = -1$, which has the solution $x = \pi$.

143. $\dfrac{\pi}{4}, \dfrac{7\pi}{12}, \dfrac{11\pi}{12}, \dfrac{5\pi}{4}, \dfrac{19\pi}{12}, \dfrac{23\pi}{12}$

Begin by making the substitution $y = 3x$. Because $0 \le x \le 2\pi$, you have $0 \le 3x \le 6\pi$ so that $0 \le y \le 6\pi$. Finding all solutions of the equation $\tan y = -1$ in the interval $[0, 6\pi]$ gives you $y = \dfrac{3\pi}{4}, \dfrac{7\pi}{4}, \dfrac{11\pi}{4}, \dfrac{15\pi}{4}, \dfrac{19\pi}{4}$, and $\dfrac{23\pi}{4}$.

Next, take each solution, set it equal to $3x$, and solve for x: $3x = \dfrac{3\pi}{4}$ so that $x = \dfrac{\pi}{4}$; $3x = \dfrac{7\pi}{4}$ so that $x = \dfrac{7\pi}{12}$; $3x = \dfrac{11\pi}{4}$ so that $x = \dfrac{11\pi}{12}$; $3x = \dfrac{15\pi}{4}$ so that $x = \dfrac{5\pi}{4}$; $3x = \dfrac{19\pi}{4}$ so that $x = \dfrac{19\pi}{12}$; and $3x = \dfrac{23\pi}{4}$ so that $x = \dfrac{23\pi}{12}$.

144. $\dfrac{\pi}{4}, \dfrac{3\pi}{4}, \dfrac{5\pi}{4}, \dfrac{7\pi}{4}$

Begin by making one side of the equation equal to zero and then factor:

$$\cos(2x) = \cot(2x)$$
$$\cos 2x - \cot 2x = 0$$
$$\cos 2x - \frac{\cos 2x}{\sin 2x} = 0$$
$$\cos 2x \left(1 - \frac{1}{\sin 2x}\right) = 0$$

Next, make the substitution $y = 2x$:

$$\cos y \left(1 - \frac{1}{\sin y}\right) = 0$$

Because $0 \le x \le 2\pi$, it follows that $0 \le 2x \le 4\pi$ so that $0 \le y \le 4\pi$.

Setting the first factor equal to zero gives you $\cos y = 0$, which has the solutions $y = \dfrac{\pi}{2}$, $\dfrac{3\pi}{2}, \dfrac{5\pi}{2}$, and $\dfrac{7\pi}{2}$. Take each solution, set it equal to $2x$, and solve for x: $2x = \dfrac{\pi}{2}$ so that $x = \dfrac{\pi}{4}$; $2x = \dfrac{3\pi}{2}$ so that $x = \dfrac{3\pi}{4}$; $2x = \dfrac{5\pi}{2}$ so that $x = \dfrac{5\pi}{4}$; and $2x = \dfrac{7\pi}{2}$ so that $x = \dfrac{7\pi}{4}$.

Proceed in a similar manner for the second factor: $1 - \dfrac{1}{\sin y} = 0$ so that $\sin y = 1$, which has the solutions $y = \dfrac{\pi}{2}$ and $\dfrac{5\pi}{2}$. Now take each solution, set it equal to $2x$, and solve for x: $2x = \dfrac{\pi}{2}$ so that $x = \dfrac{\pi}{4}$, and $2x = \dfrac{5\pi}{2}$ so that $x = \dfrac{5\pi}{4}$.

145. amplitude: $\frac{1}{2}$; period: 2π; phase shift: $-\frac{\pi}{2}$; midline $y = 0$

For the function $f(x) = A\sin\left(B\left(x - \frac{C}{B}\right)\right) + D$, the amplitude is $|A|$, the

period is $\left|\frac{2\pi}{B}\right|$, the phase shift is $\frac{C}{B}$, and the midline is $y = D$. Writing

$f(x) = \frac{1}{2}\sin\left(x + \frac{\pi}{2}\right) = \frac{1}{2}\sin\left(1\left(x - \left(-\frac{\pi}{2}\right)\right)\right) + 0$ gives you the amplitude as $\left|\frac{1}{2}\right| = \frac{1}{2}$,

the period as $\left|\frac{2\pi}{1}\right| = 2\pi$, the phase shift as $-\frac{\pi}{2}$, and the midline as $y = 0$.

146. amplitude: $\frac{1}{4}$; period: 2; phase shift: $\frac{4}{\pi}$; midline $y = 0$

For the function $f(x) = A\cos\left(B\left(x - \frac{C}{B}\right)\right) + D$, the amplitude is $|A|$, the period is

$\left|\frac{2\pi}{B}\right|$, the phase shift is $\frac{C}{B}$, and the midline is $y = D$. Writing

$f(x) = -\frac{1}{4}\cos(\pi x - 4) = -\frac{1}{4}\cos\left(\pi\left(x - \frac{4}{\pi}\right)\right) + 0$ gives you the amplitude as

$\left|-\frac{1}{4}\right| = \frac{1}{4}$, the period as $\left|\frac{2\pi}{\pi}\right| = 2$, the phase shift as $\frac{4}{\pi}$, and the midline as $y = 0$.

147. amplitude: 3; period: 2; phase shift: $\frac{6}{\pi}$; midline $y = 2$

For the function $f(x) = A\cos\left(B\left(x - \frac{C}{B}\right)\right) + D$, the amplitude is $|A|$, the period

is $\left|\frac{2\pi}{B}\right|$, the phase shift is $\frac{C}{B}$, and the midline is $y = D$. Writing

$f(x) = 2 - 3\cos(\pi x - 6) = -3\cos\left(\pi\left(x - \frac{6}{\pi}\right)\right) + 2$ gives you the amplitude as $|-3| = 3$,

the period as $\left|\frac{2\pi}{\pi}\right| = 2$, the phase shift as $\frac{6}{\pi}$, and the midline as $y = 2$.

148. amplitude: 1; period: 4π; phase shift: $-\pi$; midline $y = \frac{1}{2}$

For the function $f(x) = A\sin\left(B\left(x - \frac{C}{B}\right)\right) + D$, the amplitude is $|A|$, the

period is $\left|\frac{2\pi}{B}\right|$, the phase shift is $\frac{C}{B}$, and the midline is $y = D$. Writing

$f(x) = \frac{1}{2} - \sin\left(\frac{1}{2}x + \frac{\pi}{2}\right) = -1\sin\left(\frac{1}{2}(x - (-\pi))\right) + \frac{1}{2}$ gives you the amplitude as

$|-1| = 1$, the period as $\left|\frac{2\pi}{1/2}\right| = 4\pi$, the phase shift as $-\pi$, and the midline as $y = \frac{1}{2}$.

149. $f(x) = 2\sin(2x)$

For the function $f(x) = A\sin\left(B\left(x - \frac{C}{B}\right)\right) + D$, the amplitude is $|A|$, the period is $\left|\frac{2\pi}{B}\right|$, the phase shift is $\frac{C}{B}$, and the midline is $y = D$. The function has a period of π, so one possible value of B is $B = 2$. If you use a sine function to describe the graph, there's no phase shift, so you can use $C = 0$. The line $y = 0$ is the midline of the function so that $D = 0$. Last, the amplitude is 2, so you can use $A = 2$ because the function increases for values of x that are slightly larger than zero. Therefore, the function $f(x) = 2\sin(2x)$ describes the given graph. Note that you can also use other functions to describe this graph.

150. $f(x) = 2\cos(2x)$

For the function $f(x) = A\cos\left(B\left(x - \frac{C}{B}\right)\right) + D$, the amplitude is $|A|$, the period is $\left|\frac{2\pi}{B}\right|$, the phase shift is $\frac{C}{B}$, and the midline is $y = D$. The function has a period of π, so one possible value of B is $B = 2$. If you use a cosine function to describe the graph, there's no phase shift; therefore, you can use $C = 0$. The line $y = 0$ is the midline of the function, so $D = 0$. Last, the amplitude is 2; you can use $A = 2$ because the function decreases for values of x that are slightly larger than zero. Therefore, the function $f(x) = 2\cos(2x)$ describes the given graph. Note that you can also use other functions to describe this graph.

151. $f(x) = -2\cos\left(\frac{1}{2}x\right)$

For the function $f(x) = A\cos\left(B\left(x - \frac{C}{B}\right)\right) + D$, the amplitude is $|A|$, the period is $\left|\frac{2\pi}{B}\right|$, the phase shift is $\frac{C}{B}$, and the midline is $y = D$. The function has a period of 4π, so one possible value of B is $B = \frac{1}{2}$. If you use a cosine function to describe the graph, there's no phase shift, so you can use $C = 0$. The line $y = 0$ is the midline of the function, so $D = 0$. Last, the amplitude is 2, so you can use $A = -2$ because the function increases for values of x that are slightly larger than zero. Therefore, the function $f(x) = -2\cos\left(\frac{1}{2}x\right)$ describes the given graph. Note that you can also use other functions to describe this graph.

152. $f(x) = 2\cos(2x) + 1$

For the function $f(x) = A\cos\left(B\left(x - \frac{C}{B}\right)\right) + D$, the amplitude is $|A|$, the period is $\left|\frac{2\pi}{B}\right|$, the phase shift is $\frac{C}{B}$, and the midline is $y = D$. The function has a period of π, so one possible value of B is $B = 2$. If you use a cosine function to describe the graph, there's no phase shift, so you can use $C = 0$. The line $y = 1$ is the midline of the function, so $D = 1$. Last, the amplitude is 2, so you can use $A = 2$ because the function decreases for values of x that are slightly larger than zero. Therefore, the function $f(x) = 2\cos(2x) + 1$ describes the given graph. Note that you can also use other functions to describe this graph.

153. $f(x) = -2\cos\left(\frac{1}{2}x - \frac{\pi}{4}\right)$

For the function $f(x) = A\cos\left(B\left(x - \frac{C}{B}\right)\right) + D$, the amplitude is $|A|$, the period is $\left|\frac{2\pi}{B}\right|$, the phase shift is $\frac{C}{B}$, and the midline is $y = D$. The function has a period of 4π, so one possible value of B is $B = \frac{1}{2}$. Think of the graph as a cosine graph that's flipped about the x-axis and shifted to the right; that means there's a phase shift of $\frac{\pi}{2}$ to the right, so $\frac{C}{B} = \frac{\pi}{2}$. The line $y = 0$ is the midline of the function, so $D = 0$. Last, the amplitude is 2, so you can use $A = -2$ because the function increases for values of x that are slightly larger than $\frac{\pi}{2}$. Therefore, the following function describes the graph:

$$f(x) = -2\cos\left(\frac{1}{2}\left(x - \frac{\pi}{2}\right)\right)$$

$$= -2\cos\left(\frac{1}{2}x - \frac{\pi}{4}\right)$$

Note that you can also use other functions to describe this graph.

154. $f(x) = -2\cos\left(\frac{1}{2}x - \frac{\pi}{4}\right) - 1$

For the function $f(x) = A\cos\left(B\left(x - \frac{C}{B}\right)\right) + D$, the amplitude is $|A|$, the period is $\left|\frac{2\pi}{B}\right|$, the phase shift is $\frac{C}{B}$, and the midline is $y = D$. The function has a period of 4π, so one possible value of B is $B = \frac{1}{2}$. Think of the graph as a cosine graph that's flipped about the x-axis and shifted to the right and down; there's a phase shift of $\frac{\pi}{2}$ to the right so that $\frac{C}{B} = \frac{\pi}{2}$. The line $y = -1$ is the midline of the function, so $D = -1$. Last, the amplitude is 2, so you can use $A = -2$ because the function increases for values of x that are slightly larger than $\frac{\pi}{2}$. Therefore, the following function describes the given graph.

$$f(x) = -2\cos\left(\frac{1}{2}\left(x - \frac{\pi}{2}\right)\right) - 1$$

$$= -2\cos\left(\frac{1}{2}x - \frac{\pi}{4}\right) - 1$$

Note that you can also use other functions to describe this graph.

155. $\frac{\pi}{3}$

To evaluate $\sin^{-1}\left(\frac{\sqrt{3}}{2}\right) = x$, find the solution of $\frac{\sqrt{3}}{2} = \sin x$, where $-\frac{\pi}{2} \le x \le \frac{\pi}{2}$. Because $\sin\frac{\pi}{3} = \frac{\sqrt{3}}{2}$, you have $\sin^{-1}\left(\frac{\sqrt{3}}{2}\right) = \frac{\pi}{3}$.

156. $-\dfrac{\pi}{4}$

To evaluate $\arctan(-1) = x$, find the solution of $-1 = \tan x$, where $-\dfrac{\pi}{2} \le x \le \dfrac{\pi}{2}$. Because $\tan\left(-\dfrac{\pi}{4}\right) = -1$, you have $\arctan(-1) = -\dfrac{\pi}{4}$.

157. $\dfrac{\sqrt{3}}{2}$

To evaluate $\cos\left(\sin^{-1}\left(\dfrac{1}{2}\right)\right)$, first find the value of $\sin^{-1}\left(\dfrac{1}{2}\right)$. To evaluate $\sin^{-1}\left(\dfrac{1}{2}\right) = x$, where $-\dfrac{\pi}{2} \le x \le \dfrac{\pi}{2}$, find the solution of $\dfrac{1}{2} = \sin x$. Because $\sin\left(\dfrac{\pi}{6}\right) = \dfrac{1}{2}$, you have $\sin^{-1}\left(\dfrac{1}{2}\right) = \dfrac{\pi}{6}$. Therefore, $\cos\left(\sin^{-1}\left(\dfrac{1}{2}\right)\right) = \cos\left(\dfrac{\pi}{6}\right) = \dfrac{\sqrt{3}}{2}$.

158. $\dfrac{\sqrt{3}}{3}$

To evaluate $\tan\left(\cos^{-1}\left(\dfrac{\sqrt{3}}{2}\right)\right)$, first find the value of $\cos^{-1}\left(\dfrac{\sqrt{3}}{2}\right)$. To evaluate $\cos^{-1}\left(\dfrac{\sqrt{3}}{2}\right) = x$, where $0 \le x \le \pi$, find the solution of $\dfrac{\sqrt{3}}{2} = \cos x$. Because $\cos\left(\dfrac{\pi}{6}\right) = \dfrac{\sqrt{3}}{2}$, you have $\cos^{-1}\left(\dfrac{\sqrt{3}}{2}\right) = \dfrac{\pi}{6}$. Therefore,

$$\tan\left(\cos^{-1}\left(\dfrac{\sqrt{3}}{2}\right)\right) = \tan\left(\dfrac{\pi}{6}\right) = \dfrac{\sqrt{3}}{3}.$$

159. $\dfrac{5}{3}$

The value of $\arccos\dfrac{4}{5}$ probably isn't something you've memorized, so to evaluate $\csc\left(\arccos\dfrac{4}{5}\right)$, you can create a right triangle.

Let $\arccos\dfrac{4}{5} = \theta$ so that $\dfrac{4}{5} = \cos\theta$. Using $\dfrac{4}{5} = \cos\theta$, create the right triangle; then use the Pythagorean theorem to find the missing side:

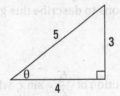

By the substitution, you have $\csc\left(\arccos\dfrac{4}{5}\right) = \csc(\theta)$, and from the right triangle, you have $\csc(\theta) = \dfrac{5}{3}$.

160. $\dfrac{\sqrt{2}}{2}$

To evaluate $\sin\left(\tan^{-1}(2)+\tan^{-1}(3)\right)$, first create two right triangles using the substitutions $\tan^{-1}(2)=\alpha$ and $\tan^{-1}(3)=\beta$.

To make the first right triangle, use $\tan^{-1}(2)=\alpha$. You know that $2=\tan\alpha$, and you can find the missing side of the first right triangle using the Pythagorean theorem:

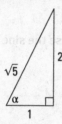

As for the second right triangle, because $\tan^{-1}(3)=\beta$, you know that $3=\tan\beta$. Again, you can use the Pythagorean theorem to find the missing side of the right triangle:

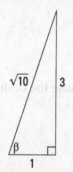

The substitutions give you $\sin\left(\tan^{-1}(2)+\tan^{-1}(3)\right)=\sin(\alpha+\beta)$. Using a trigonometric identity, you know that $\sin(\alpha+\beta)=\sin\alpha\cos\beta+\cos\alpha\sin\beta$. From the right triangles, you can read off each of the values to get the following:

$$
\begin{aligned}
\sin\alpha\cos\beta+\cos\alpha\sin\beta &= \left(\frac{2}{\sqrt{5}}\right)\left(\frac{1}{\sqrt{10}}\right)+\left(\frac{1}{\sqrt{5}}\right)\left(\frac{3}{\sqrt{10}}\right) \\
&= \frac{2}{\sqrt{50}}+\frac{3}{\sqrt{50}} \\
&= \frac{5}{\sqrt{50}} \\
&= \frac{5}{5\sqrt{2}} \\
&= \frac{1}{\sqrt{2}} \\
&= \frac{\sqrt{2}}{2}
\end{aligned}
$$

161. $x = 0.412, \pi - 0.412$

To solve $\sin x = 0.4$ for x over the interval $[0, 2\pi]$, begin by taking the inverse sine of both sides:

$$\sin x = 0.4$$
$$x = \sin^{-1}(0.4)$$
$$\approx 0.412$$

The other solution belongs to Quadrant II because the sine function also has positive values there:

$$x = \pi - 0.412$$

162. $x = 2.465, 2\pi - 2.456$

To solve $\cos x = -0.78$ for x over the interval $[0, 2\pi]$, begin by taking the inverse cosine of both sides:

$$\cos x = -0.78$$
$$x = \cos^{-1}(-0.78)$$
$$\approx 2.465$$

The other solution belongs to Quadrant III because the cosine function also has negative values there:

$$2\pi - 2.465$$

163. $x = 0.322, \frac{\pi}{2} - 0.322, \pi + 0.322, \frac{3\pi}{2} - 0.322$

To solve $5 \sin(2x) + 1 = 4$ for x over the interval $[0, 2\pi]$, begin by isolating the term involving sine:

$$5\sin(2x) + 1 = 4$$
$$5\sin(2x) = 3$$
$$\sin(2x) = \frac{3}{5}$$

You can also use the substitution $y = 2x$ to help simplify. Because $0 \le x \le 2\pi$, it follows that $0 \le 2x \le 4\pi$ so that $0 \le y \le 4\pi$. Use the substitution and take the inverse sine of both sides:

$$\sin(y) = \frac{3}{5}$$
$$y = \sin^{-1}\left(\frac{3}{5}\right)$$
$$\approx 0.644$$

It follows that in the interval $[0, 4\pi]$, $2\pi + 0.644$ is also a solution because when you add 2π, the resulting angle lies at the same place on the unit circle.

Likewise, there's a solution to the equation $\sin(y) = \frac{3}{5}$ in the second quadrant because the function also has positive values there, namely $\pi - 0.644$; adding 2π gives you the other solution, $3\pi - 0.644$. Therefore, $y = 0.644$, $\pi - 0.644$, $2\pi + 0.644$, and $3\pi - 0.644$ are all solutions.

Last, substitute $2x$ into the equations and divide to get the solutions for x:

$$2x = 0.644$$
$$x = 0.322$$

$$2x = \pi - 0.644$$
$$x = \frac{\pi}{2} - 0.322$$

$$2x = 2\pi + 0.644$$
$$x = \pi + 0.322$$

$$2x = 3\pi - 0.644$$
$$x = \frac{3\pi}{2} - 0.322$$

The solutions are $x = 0.322$, $\frac{\pi}{2} - 0.322$, $\pi + 0.322$, and $\frac{3\pi}{2} - 0.322$.

164. $x = 0.321$, $\frac{2\pi}{3} - 0.321$, $\frac{2\pi}{3} + 0.321$, $\frac{4\pi}{3} - 0.321$, $\frac{4\pi}{3} + 0.321$, $2\pi - 0.321$

To solve $7\cos(3x) - 1 = 3$ for x over the interval $[0, 2\pi]$, first isolate the term involving cosine:

$$7\cos(3x) - 1 = 3$$
$$\cos(3x) = \frac{4}{7}$$

You can also use the substitution $y = 3x$ to simplify the equation. Because $0 \le x \le 2\pi$, it follows that $0 \le 3x \le 6\pi$ so that $0 \le y \le 6\pi$. Use the substitution and take the inverse cosine of both sides:

$$\cos y = \frac{4}{7}$$
$$y = \cos^{-1}\left(\frac{4}{7}\right)$$
$$\approx 0.963$$

It follows that in the interval $[0, 6\pi]$, $y = 2\pi + 0.963$ and $y = 4\pi + 0.963$ are also solutions because adding multiples of 2π makes the resulting angles fall at the same places on the unit circle.

Likewise, there's a solution to the equation $\cos y = \frac{4}{7}$ in the fourth quadrant because cosine also has positive values there, namely $y = 2\pi - 0.963$. Because $y = 2\pi - 0.963$ is a solution, it follows that $y = 4\pi - 0.963$ and $y = 6\pi - 0.963$ are also solutions.

Therefore, you have $y = 0.963$, $2\pi - 0.963$, $2\pi + 0.963$, $4\pi - 0.963$, $4\pi + 0.963$, and $6\pi - 0.963$ as solutions to $\cos y = \frac{4}{7}$. Last, substitute $3x$ into the equations and divide to solve for x:

$$3x = 0.963$$

$$x = 0.321$$

$$3x = 2\pi - 0.963$$

$$x = \frac{2\pi}{3} - 0.321$$

$$3x = 2\pi + 0.963$$

$$x = \frac{2\pi}{3} + 0.321$$

$$3x = 4\pi - 0.963$$

$$x = \frac{4\pi}{3} - 0.321$$

$$3x = 4\pi + 0.963$$

$$x = \frac{4\pi}{3} + 0.321$$

$$3x = 6\pi - 0.963$$

$$x = 2\pi - 0.321$$

Therefore, the solutions are $x = 0.321$, $\frac{2\pi}{3} - 0.321$, $\frac{2\pi}{3} + 0.321$, $\frac{4\pi}{3} - 0.321$, $\frac{4\pi}{3} + 0.321$, and $2\pi - 0.321$.

165. $x = \pi + 0.887, 2\pi - 0.887$

To solve $2\sin^2 x + 8\sin x + 5 = 0$ for x over the interval $[0, 2\pi]$, first use the quadratic formula:

$$\sin x = \frac{-8 \pm \sqrt{8^2 - 4(2)(5)}}{2(2)}$$

$$= \frac{-8 \pm \sqrt{24}}{2(2)}$$

$$= -\frac{8}{4} \pm \frac{\sqrt{24}}{4}$$

Simplifying gives you $-2 + \frac{\sqrt{24}}{4} \approx -0.775$ and $-2 - \frac{\sqrt{24}}{4} \approx -3.225$.

You now need to find solutions to $\sin x = -0.775$ and $\sin x = -3.225$. Notice that $\sin x = -3.225$ has no solutions because -3.225 is outside the range of the sine function.

To solve $\sin x = -0.775$, take the inverse sine of both sides:

$$\sin x = -0.775$$

$$x = \sin^{-1}(-0.775)$$

$$\approx -0.887$$

Note that this solution isn't in the desired interval, $[0, 2\pi]$. The solutions in the given interval belong to Quadrants III and IV, respectively, because in those quadrants, the sine function has negative values; those solutions are $x = \pi + 0.887$ and $x = 2\pi - 0.887$.

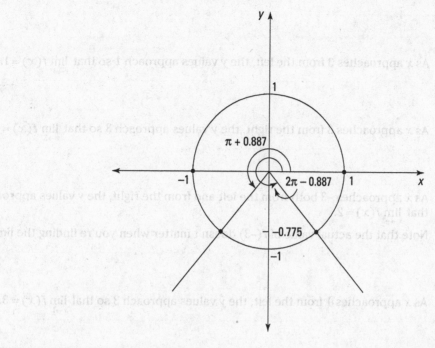

166. $x = \pi - 0.322, 2\pi - 0.322, \dfrac{3\pi}{4}, \dfrac{7\pi}{4}$

To solve $3\sec^2 x + 4\tan x = 2$ for x over the interval $[0, 2\pi]$, begin by using the identity $\sec^2 x = 1 + \tan^2 x$ and make one side of the equation equal to zero:

$$3\left(1 + \tan^2 x\right) + 4\tan x = 2$$
$$3 + 3\tan^2 x + 4\tan x = 2$$
$$3\tan^2 x + 4\tan x + 1 = 0$$

Next, use the quadratic formula to find $\tan x$:

$$\tan x = \frac{-4 \pm \sqrt{4^2 - 4(3)(1)}}{2(3)}$$
$$= \frac{-4 \pm \sqrt{4}}{2(3)}$$
$$= \frac{-4 \pm 2}{2(3)}$$

Therefore, $\tan x = -1$ and $-\dfrac{1}{3}$.

The solutions to the equation $\tan x = -1$ are $x = \dfrac{3\pi}{4}$ and $x = \dfrac{7\pi}{4}$. To solve $\tan x = -\dfrac{1}{3}$, take the inverse tangent of both sides:

$$\tan x = -\frac{1}{3}$$
$$x = \tan^{-1}\left(-\frac{1}{3}\right)$$
$$\approx -0.322$$

Note that this solution isn't in the given interval. The solutions that are in the given interval and belong to Quadrants II and IV (where the tangent function is negative) are $x = \pi - 0.322$ and $x = 2\pi - 0.322$.

167. 1

As x approaches 3 from the left, the y values approach 1 so that $\lim\limits_{x \to 3^-} f(x) = 1$.

168. 3

As x approaches 3 from the right, the y values approach 3 so that $\lim\limits_{x \to 3^+} f(x) = 3$.

169. 2

As x approaches −3 both from the left and from the right, the y values approach 2 so that $\lim\limits_{x \to -3} f(x) = 2$.

Note that the actual value of $f(-3)$ doesn't matter when you're finding the limit.

170. 3

As x approaches 1 from the left, the y values approach 3 so that $\lim\limits_{x \to 1^-} f(x) = 3$.

171. 3

As x approaches 1 from the right, the y values approach 3 so that $\lim\limits_{x \to 1^+} f(x) = 3$.

172. 5

As x approaches −2 both from the left and from the right, the y values approach 5 so that $\lim\limits_{x \to -2} f(x) = 5$.

173. 4

Note that substituting the limiting value, 3, into the function $\dfrac{x^2 - 2x - 3}{x - 3}$ gives you the indeterminate form $\dfrac{0}{0}$.

To find the limit, first factor the numerator and simplify:

$$\lim_{x \to 3} \frac{x^2 - 2x - 3}{x - 3}$$

$$= \lim_{x \to 3} \frac{(x - 3)(x + 1)}{(x - 3)}$$

$$= \lim_{x \to 3} (x + 1)$$

Then substitute 3 for x:

$$\lim_{x \to 3} (x + 1) = 3 + 1 = 4$$

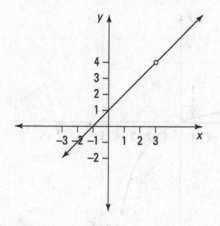

174. $-\dfrac{7}{4}$

Note that substituting the limiting value, 2, into the function $\dfrac{x^2+3x-10}{x^2-8x+12}$ gives you the indeterminate form $\dfrac{0}{0}$.

To find the limit, first factor the numerator and denominator and simplify:

$$\lim_{x\to 2}\frac{x^2+3x-10}{x^2-8x+12}$$

$$=\lim_{x\to 2}\frac{(x-2)(x+5)}{(x-2)(x-6)}$$

$$=\lim_{x\to 2}\frac{(x+5)}{(x-6)}$$

Then substitute 2 for x:

$$\lim_{x\to 2}\frac{(x+5)}{(x-6)}=\frac{2+5}{2-6}=-\frac{7}{4}$$

175. $\dfrac{1}{2}$

Note that substituting the limiting value, –5, into the function $\dfrac{x^2+5x}{x^2-25}$ gives you the indeterminate form $\dfrac{0}{0}$.

To find the limit, first factor the numerator and denominator and simplify:

$$\lim_{x\to -5}\frac{x^2+5x}{x^2-25}$$

$$=\lim_{x\to -5}\frac{x(x+5)}{(x-5)(x+5)}$$

$$=\lim_{x\to -5}\frac{x}{x-5}$$

Then substitute –5 for x:

$$\lim_{x\to -5}\frac{x}{x-5}=\frac{-5}{-5-5}=\frac{1}{2}$$

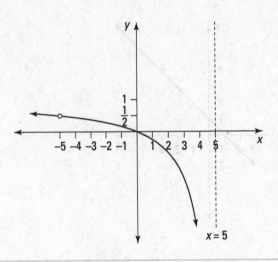

$x = 5$

176.   4

Note that substituting the limiting value, 4, into the function $\dfrac{4-x}{2-\sqrt{x}}$ gives you the indeterminate form $\dfrac{0}{0}$.

To find the limit, first factor the numerator and simplify:

$$\lim_{x \to 4} \frac{4-x}{2-\sqrt{x}}$$

$$= \lim_{x \to 4} \frac{\left(2-\sqrt{x}\right)\left(2+\sqrt{x}\right)}{\left(2-\sqrt{x}\right)}$$

$$= \lim_{x \to 4} \left(2+\sqrt{x}\right)$$

Then substitute 4 for x:

$$\lim_{x \to 4}\left(2+\sqrt{x}\right) = 2+\sqrt{4} = 4$$

177.   1

Note that substituting in the limiting value, 0, gives you an indeterminate form. For example, as x approaches 0 from the right, you have the indeterminate form $\infty - \infty$, and as x approaches 0 from the left, you have the indeterminate form $-\infty + \infty$.

To find the limit, first get common denominators and simplify:

$$\lim_{x \to 0}\left(\frac{1}{x} - \frac{1}{x^2+x}\right)$$

$$= \lim_{x \to 0}\left(\frac{1}{x} - \frac{1}{x(x+1)}\right)$$

$$= \lim_{x \to 0}\left(\frac{1(x+1)}{x(x+1)} - \frac{1}{x(x+1)}\right)$$

$$= \lim_{x \to 0}\frac{x}{x(x+1)}$$

$$= \lim_{x \to 0}\frac{1}{x+1}$$

Then substitute 0 for x:

$$\lim_{x \to 0} \frac{1}{x+1} = \frac{1}{0+1} = 1$$

178. 0

The given function is continuous everywhere, so you can simply substitute in the limiting value:

$$\lim_{x \to 4} |x-4| = |4-4| = 0$$

179. 3

Note that substituting the limiting value, -1, into the function $\frac{x^3+1}{x+1}$ gives you the indeterminate form $\frac{0}{0}$.

To find the limit, first factor the numerator and simplify:

$$\lim_{x \to -1} \frac{x^3+1}{x+1}$$

$$= \lim_{x \to -1} \frac{(x+1)(x^2-x+1)}{(x+1)}$$

$$= \lim_{x \to -1} (x^2-x+1)$$

Then substitute -1 for x:

$$\lim_{x \to -1} (x^2-x+1) = (-1)^2 - (-1) + 1 = 3$$

180. $\frac{1}{4}$

Note that substituting the limiting value, 0, into the function $\frac{\sqrt{4+h}-2}{h}$ gives you the indeterminate form $\frac{0}{0}$.

To find the limit, first multiply the numerator and denominator by the conjugate of the numerator and then simplify:

$$\lim_{h \to 0} \frac{\sqrt{4+h}-2}{h}$$

$$= \lim_{h \to 0} \frac{\left(\sqrt{4+h}-2\right)}{h} \frac{\left(\sqrt{4+h}+2\right)}{\left(\sqrt{4+h}+2\right)}$$

$$= \lim_{h \to 0} \frac{4+h-4}{h\left(\sqrt{4+h}+2\right)}$$

$$= \lim_{h \to 0} \frac{h}{h\left(\sqrt{4+h}+2\right)}$$

$$= \lim_{h \to 0} \frac{1}{\sqrt{4+h}+2}$$

Then substitute 0 for h:

$$\lim_{h \to 0} \frac{1}{\sqrt{4+h}+2} = \frac{1}{\sqrt{4+0}+2} = \frac{1}{4}$$

181. 108

Note that substituting the limiting value, 3, into the function $\dfrac{x^4-81}{x-3}$ gives you the indeterminate form $\dfrac{0}{0}$.

To find the limit, first factor the numerator and simplify:

$$\lim_{x\to3}\frac{x^4-81}{x-3}$$

$$=\lim_{x\to3}\frac{\left(x^2-9\right)\left(x^2+9\right)}{x-3}$$

$$=\lim_{x\to3}\frac{(x-3)(x+3)\left(x^2+9\right)}{x-3}$$

$$=\lim_{x\to3}(x+3)\left(x^2+9\right)$$

Then substitute 3 for x:

$$\lim_{x\to3}(x+3)\left(x^2+9\right)$$

$$=(3+3)\left(3^2+9\right)$$

$$=(6)(18)$$

$$=108$$

182. ∞

Note that substituting in the limiting value gives you an indeterminate form.

Because x is approaching 0 from the right, you have $x > 0$ so that $|x| = x$. Therefore, the limit becomes

$$\lim_{x\to0^+}\left(\frac{2}{x^2}+\frac{2}{|x|}\right)=\lim_{x\to0^+}\left(\frac{2}{x^2}+\frac{2}{x}\right)$$

$$=\lim_{x\to0^+}\left(\frac{2+2x}{x^2}\right)$$

As $x \to 0^+$, you have $(2 + 2x) \to 2$ and $x^2 \to 0^+$ so that $\left(\dfrac{2+2x}{x^2}\right) \to \dfrac{2}{0^+} \to \infty$. The limit is positive infinity because dividing 2 by a very small positive number close to zero gives you a very large positive number.

183. ∞

Note that substituting in the limiting value gives you the indeterminate form $\infty - \infty$.

Because x is approaching 0 from the left, you have $x < 0$ so that $|x| = -x$. Therefore, the limit becomes

$$\lim_{x \to 0^-} \left(\frac{2}{x^2} - \frac{2}{|x|} \right) = \lim_{x \to 0^-} \left(\frac{2}{x^2} - \frac{2}{-x} \right)$$

$$= \lim_{x \to 0^-} \left(\frac{2}{x^2} + \frac{2}{x} \right)$$

$$= \lim_{x \to 0^-} \left(\frac{2 + 2x}{x^2} \right)$$

Now consider the numerator and denominator as $x \to 0^-$. As $x \to 0^-$, you have

$(2 + 2x) \to 2$ and $x^2 \to 0^+$ so that $\left(\dfrac{2+2x}{x^2} \right) \to \dfrac{2}{0^+} \to \infty$. The limit is positive infinity

because dividing 2 by a very small positive number close to zero gives you a very large positive number.

---

184.   limit does not exist

Note that substituting in the limiting value gives you the indeterminate form $\frac{0}{0}$.

Examine both the left-hand limit and right-hand limit to determine whether the limits are equal. To find the left-hand limit, consider values that are slightly smaller than $\frac{4}{3}$ and substitute into the limit. Notice that to simplify the absolute value in the denominator of the fraction, you replace the absolute values bars with parentheses and add a negative sign, because substituting in a value less than $\frac{4}{3}$ will make the number in the parentheses negative; the extra negative sign will make the value positive again.

$$\lim_{x \to \frac{4}{3}^-} \frac{3x^2 - 4x}{|3x - 4|} = \lim_{x \to \frac{4}{3}^-} \frac{x(3x - 4)}{-(3x - 4)}$$

$$= \lim_{x \to \frac{4}{3}^-} \frac{x}{-1}$$

$$= \frac{\frac{4}{3}}{-1}$$

$$= -\frac{4}{3}$$

You deal with the right-hand limit similarly. Here, when removing the absolute value bars, you simply replace them with parentheses; you don't need the negative sign because the value in the parentheses is positive when you're substituting in a value larger than $\frac{4}{3}$.

$$\lim_{x \to \frac{4}{3}^+} \frac{3x^2 - 4x}{|3x - 4|} = \lim_{x \to \frac{4}{3}^+} \frac{x(3x - 4)}{(3x - 4)}$$

$$= \lim_{x \to \frac{4}{3}^+} \frac{x}{1}$$

$$= \frac{4}{3}$$

Because the right-hand limit doesn't equal the left-hand limit, the limit doesn't exist.

185. $\dfrac{5}{7}$

Note that substituting the limiting value, 5, into the function $\dfrac{x^2-25}{3x^2-16x+5}$ gives you the indeterminate form $\dfrac{0}{0}$.

To find the limit, first factor the numerator and denominator and simplify:

$$\lim_{x \to 5} \frac{x^2-25}{3x^2-16x+5}$$

$$=\lim_{x \to 5} \frac{(x-5)(x+5)}{(x-5)(3x-1)}$$

$$=\lim_{x \to 5} \frac{x+5}{3x-1}$$

Then substitute 5 for x:

$$\lim_{x \to 5} \frac{x+5}{3x-1} = \frac{5+5}{3(5)-1} = \frac{5}{7}$$

186. $-\dfrac{1}{6\sqrt{6}}$

Note that substituting in the limiting value gives you the indeterminate form $\dfrac{0}{0}$.

Begin by multiplying the numerator and denominator of the fraction by the conjugate of the numerator:

$$\lim_{x \to 3} \frac{\sqrt{x+3}-\sqrt{2x}}{x^2-3x}$$

$$=\lim_{x \to 3} \frac{\left(\sqrt{x+3}-\sqrt{2x}\right)}{x(x-3)} \frac{\left(\sqrt{x+3}+\sqrt{2x}\right)}{\left(\sqrt{x+3}+\sqrt{2x}\right)}$$

$$=\lim_{x \to 3} \frac{x+3-2x}{x(x-3)\left(\sqrt{x+3}+\sqrt{2x}\right)}$$

$$=\lim_{x \to 3} \frac{3-x}{x(x-3)\left(\sqrt{x+3}+\sqrt{2x}\right)}$$

Next, factor –1 from the numerator and continue simplifying:

$$\lim_{x \to 3} \frac{3-x}{x(x-3)\left(\sqrt{x+3}+\sqrt{2x}\right)}$$

$$=\lim_{x \to 3} \frac{-1(x-3)}{x(x-3)\left(\sqrt{x+3}+\sqrt{2x}\right)}$$

$$=\lim_{x \to 3} \frac{-1}{x\left(\sqrt{x+3}+\sqrt{2x}\right)}$$

To find the limit, substitute 3 for x:

$$\lim_{x \to 3} \frac{-1}{x\left(\sqrt{x+3}+\sqrt{2x}\right)}$$

$$= \frac{-1}{3\left(\sqrt{3+3}+\sqrt{2(3)}\right)}$$

$$= \frac{-1}{3\left(2\sqrt{6}\right)}$$

$$= \frac{-1}{6\sqrt{6}}$$

187. 12

Note that substituting the limiting value, 0, into the function $\dfrac{(2+h)^3-8}{h}$ gives you the indeterminate form $\dfrac{0}{0}$.

Expand the numerator and simplify:

$$\lim_{h \to 0} \frac{(2+h)^3-8}{h}$$

$$= \lim_{h \to 0} \frac{h^3+6h^2+12h+8-8}{h}$$

$$= \lim_{h \to 0} \frac{h^3+6h^2+12h}{h}$$

$$= \lim_{h \to 0} \frac{h\left(h^2+6h+12\right)}{h}$$

$$= \lim_{h \to 0} \left(h^2+6h+12\right)$$

To find the limit, substitute 0 for h:

$$\lim_{h \to 0} \left(h^2+6h+12\right) = 0^2+6(0)+12 = 12$$

188. 1

Note that substituting in the limiting value, 4, gives you the indeterminate form $\dfrac{0}{0}$.

Because x is approaching 4 from the right, you have $x > 4$ so that $|x-4| = (x-4)$. Therefore, the limit becomes

$$\lim_{x \to 4^+} \frac{x-4}{|x-4|} = \lim_{x \to 4^+} \frac{x-4}{x-4}$$

$$= \lim_{x \to 4^+} 1$$

$$= 1$$

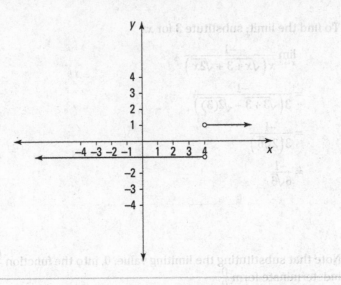

189. −1

Note that substituting in the limiting value gives you the indeterminate form $\frac{0}{0}$.

Because x is approaching 5 from the left, you have $x < 5$ so that $|x-5| = -(x-5)$. Therefore, the limit becomes

$$\lim_{x\to 5^-}\frac{x-5}{|x-5|} = \lim_{x\to 5^-}\frac{x-5}{-(x-5)}$$

$$= \lim_{x\to 5^-}(-1)$$

$$= -1$$

190. $-\frac{1}{25}$

Note that substituting in the limiting value, −5, into the function $\dfrac{\frac{1}{5}+\frac{1}{x}}{x+5}$ gives you the indeterminate form $\frac{0}{0}$.

Begin by writing the two fractions in the numerator as a single fraction by getting common denominators. Then simplify:

$$\lim_{x\to -5}\frac{\frac{1}{5}+\frac{1}{x}}{x+5}$$

$$= \lim_{x\to -5}\frac{\frac{1(x)}{5(x)}+\frac{1(5)}{x(5)}}{x+5}$$

$$= \lim_{x\to -5}\frac{\frac{x+5}{5x}}{\frac{x+5}{1}}$$

$$= \lim_{x\to -5}\left(\frac{x+5}{5x}\right)\left(\frac{1}{x+5}\right)$$

$$= \lim_{x\to -5}\frac{1}{5x}$$

To find the limit, substitute –5 for x:

$$\lim_{x \to -5} \frac{1}{5x} = \frac{1}{5(-5)} = -\frac{1}{25}$$

191. $-\frac{1}{2}$

Note that substituting in the limiting value gives you an indeterminate form. For example, as x approaches 0 from the right, you have the indeterminate form $\infty - \infty$, and as x approaches 0 from the left, you have the indeterminate form $-\infty + \infty$.

Begin by getting common denominators:

$$\lim_{x \to 0} \left(\frac{1}{x\sqrt{1+x}} - \frac{1}{x} \right)$$

$$= \lim_{x \to 0} \left(\frac{1}{x\sqrt{1+x}} - \frac{\sqrt{1+x}}{x\sqrt{1+x}} \right)$$

$$= \lim_{x \to 0} \frac{\left(1 - \sqrt{1+x} \right)}{x\sqrt{1+x}}$$

Next, multiply the numerator and denominator of the fraction by the conjugate of the numerator and simplify:

$$\lim_{x \to 0} \frac{\left(1 - \sqrt{1+x} \right)}{x\sqrt{1+x}}$$

$$= \lim_{x \to 0} \frac{\left(1 - \sqrt{1+x} \right)}{\left(x\sqrt{1+x} \right)} \frac{\left(1 + \sqrt{1+x} \right)}{\left(1 + \sqrt{1+x} \right)}$$

$$= \lim_{x \to 0} \frac{1 - (1+x)}{x\sqrt{1+x}\left(1 + \sqrt{1+x} \right)}$$

$$= \lim_{x \to 0} \frac{-x}{x\sqrt{1+x}\left(1 + \sqrt{1+x} \right)}$$

$$= \lim_{x \to 0} \frac{-1}{\sqrt{1+x}\left(1 + \sqrt{1+x} \right)}$$

To find the limit, substitute 0 for x:

$$\lim_{x \to 0} \frac{-1}{\sqrt{1+x}\left(1 + \sqrt{1+x} \right)}$$

$$= \frac{-1}{\sqrt{1+0}\left(1 + \sqrt{1+0} \right)}$$

$$= \frac{-1}{1(2)}$$

$$= -\frac{1}{2}$$

192. $-\dfrac{1}{16}$

Note that substituting in the limiting value, 0, gives you the indeterminate form $\dfrac{0}{0}$.

Begin by rewriting the two terms in the numerator using positive exponents. After that, get common denominators in the numerator and simplify:

$$\lim_{h\to0}\frac{(4+h)^{-1}-4^{-1}}{h}$$

$$=\lim_{h\to0}\frac{\dfrac{1}{4+h}-\dfrac{1}{4}}{h}$$

$$=\lim_{h\to0}\frac{\dfrac{4}{4(4+h)}-\dfrac{1(4+h)}{4(4+h)}}{h}$$

$$=\lim_{h\to0}\frac{\dfrac{4-(4+h)}{4(4+h)}}{\dfrac{h}{1}}$$

$$=\lim_{h\to0}\left(\frac{-h}{4(4+h)}\frac{1}{h}\right)$$

$$=\lim_{h\to0}\frac{-1}{4(4+h)}$$

To find the limit, substitute 0 for h:

$$\lim_{h\to0}\frac{-1}{4(4+h)}=\frac{-1}{4(4+0)}=-\frac{1}{16}$$

193.   5

Recall that the Squeeze Theorem states: If $f(x)\le g(x)\le h(x)$ when x is near a (except possibly at a) and if $\lim\limits_{x\to a}f(x)=\lim\limits_{x\to a}h(x)=L$, then $\lim\limits_{x\to a}g(x)=L$.

Note that $\lim\limits_{x\to2^+}5=5$ and that $\lim\limits_{x\to2^+}(x^2+3x-5)=2^2+3(2)-5=5$. Therefore, by the Squeeze Theorem, $\lim\limits_{x\to2^+}f(x)=5$.

194.   4

Note that $\lim\limits_{x\to0}(x^2+4)=0^2+4=4$ and that $\lim\limits_{x\to0}(4+\sin x)=4+\sin0=4$. Therefore, by the squeeze theorem, $\lim\limits_{x\to0}f(x)=4$.

195.   2

Note that $\lim\limits_{x\to1}2x=2(1)=2$ and that $\lim\limits_{x\to1}(x^3+1)=1^3+1=2$. Therefore, by the squeeze theorem, $\lim\limits_{x\to1}f(x)=2$.

196.

0

Notice that for all values of x except for $x = 0$, you have $-1 \le \cos\left(\dfrac{2}{x^2}\right) \le 1$ due to the range of the cosine function. So for all values of x except for $x = 0$, you have

$$-1\left(x^4\right) \le \left(x^4\right)\cos\left(\dfrac{2}{x^2}\right) \le 1\left(x^4\right)$$

Because x is approaching 0 (but isn't equal to 0), you can apply the squeeze theorem:

$$\lim_{x \to 0}\left(-x^4\right) \le \lim_{x \to 0} x^4 \cos\left(\dfrac{2}{x^2}\right) \le \lim_{x \to 0}\left(x^4\right)$$

$$0 \le \lim_{x \to 0} x^4 \cos\left(\dfrac{2}{x^2}\right) \le 0$$

Therefore, you can conclude that $\lim\limits_{x \to 0} x^4 \cos\left(\dfrac{2}{x^2}\right) = 0$.

197.

0

Notice that for all values of $x > 0$, you have $-1 \le \sin\left(\dfrac{2}{\sqrt{x}}\right) \le 1$ due to the range of the sine function. So for all values of $x > 0$, you have

$$-1\left(x^2\right) \le x^2 \sin\left(\dfrac{2}{\sqrt{x}}\right) \le 1\left(x^2\right)$$

Because x is approaching 0 (but isn't equal to 0), you can apply the squeeze theorem:

$$\lim_{x \to 0^+}\left(-x^2\right) \le \lim_{x \to 0^+} x^2 \sin\left(\dfrac{2}{\sqrt{x}}\right) \le \lim_{x \to 0^+}\left(x^2\right)$$

$$0 \le \lim_{x \to 0^+} x^2 \sin\left(\dfrac{2}{\sqrt{x}}\right) \le 0$$

Therefore, you can conclude that $\lim\limits_{x \to 0^+} x^2 \sin\left(\dfrac{2}{\sqrt{x}}\right) = 0$.

198.

0

Notice that for all values of x except for $x = 0$, you have $-1 \le \sin^2\left(\dfrac{\pi}{x}\right) \le 1$ due to the range of the sine function. So for all values of x except for $x = 0$, you have the following (after multiplying by -1):

$$-1 \le -\sin^2\left(\dfrac{\pi}{x}\right) \le 1$$

You need to make the center expression match the given one, $\sqrt[3]{x}\left(3 - \sin^2\left(\dfrac{\pi}{x}\right)\right)$, so do a little algebra. Adding 3 gives you

$$-1 + 3 \le 3 - \sin^2\left(\dfrac{\pi}{x}\right) \le 1 + 3$$

$$2 \le 3 - \sin^2\left(\dfrac{\pi}{x}\right) \le 4$$

Now multiply by $\sqrt[3]{x}$:

$$2\sqrt[3]{x} \le \sqrt[3]{x}\left(3 - \sin^2\left(\frac{\pi}{x}\right)\right) \le 4\sqrt[3]{x}$$

Note that $\sqrt[3]{x} > 0$ for values of x greater than 0, so you don't have to flip the inequalities.

Because the limit is approaching 0 from the right (but isn't equal to 0), you can apply the squeeze theorem to get

$$\lim_{x \to 0^+} 2\sqrt[3]{x} \le \lim_{x \to 0^+} \sqrt[3]{x}\left(3 - \sin^2\left(\frac{\pi}{x}\right)\right) \le \lim_{x \to 0^+} 4\sqrt[3]{x}$$

$$0 \le \lim_{x \to 0^+} \sqrt[3]{x}\left(3 - \sin^2\left(\frac{\pi}{x}\right)\right) \le 0$$

Therefore, you can conclude that $\lim\limits_{x \to 0^+} \sqrt[3]{x}\left(3 - \sin^2\left(\frac{\pi}{x}\right)\right) = 0$.

199.   5

To use $\lim\limits_{x \to 0} \dfrac{\sin x}{x} = 1$, you need the denominator of the function to match the argument of the sine. Begin by multiplying the numerator and denominator by 5. Then simplify:

$$\lim_{x \to 0} \frac{\sin(5x)}{x} = \lim_{x \to 0} \frac{5\sin(5x)}{5x}$$

$$= 5\lim_{x \to 0} \frac{\sin(5x)}{5x}$$

$$= 5(1)$$

$$= 5$$

200.   0

Factor the 2 from the numerator and then multiply the numerator and denominator by $\dfrac{1}{x}$. Then simplify:

$$\lim_{x \to 0} \frac{2\cos x - 2}{\sin x} = \lim_{x \to 0} \frac{2(\cos x - 1)}{\sin x}$$

$$= \lim_{x \to 0} \frac{2\dfrac{\cos x - 1}{x}}{\dfrac{\sin x}{x}}$$

$$= \frac{2(0)}{1}$$

$$= 0$$

201. $-\sqrt{2}$

Begin by using a trigonometric identity in the numerator and then factor:

$$\lim_{x \to \frac{\pi}{4}} \frac{\cos 2x}{\sin x - \cos x}$$

$$= \lim_{x \to \frac{\pi}{4}} \frac{\cos^2 x - \sin^2 x}{\sin x - \cos x}$$

$$= \lim_{x \to \frac{\pi}{4}} \frac{(\cos x - \sin x)(\cos x + \sin x)}{\sin x - \cos x}$$

Next, factor out a –1 from the numerator and simplify:

$$\lim_{x \to \frac{\pi}{4}} \frac{(\cos x - \sin x)(\cos x + \sin x)}{\sin x - \cos x}$$

$$= \lim_{x \to \frac{\pi}{4}} \frac{\left[(-1)(-\cos x + \sin x) \right](\cos x + \sin x)}{\sin x - \cos x}$$

$$= \lim_{x \to \frac{\pi}{4}} \frac{(-1)(\sin x - \cos x)(\cos x + \sin x)}{\sin x - \cos x}$$

$$= \lim_{x \to \frac{\pi}{4}} (-1)(\cos x + \sin x)$$

$$= (-1)\left(\cos \frac{\pi}{4} + \sin \frac{\pi}{4} \right)$$

$$= (-1)\left(\frac{\sqrt{2}}{2} + \frac{\sqrt{2}}{2} \right)$$

$$= -\sqrt{2}$$

202. $\dfrac{5}{9}$

You want to rewrite the expression so you can use $\lim_{x \to 0} \dfrac{\sin x}{x} = 1$. Begin by multiplying

the numerator and denominator by $\dfrac{1}{x}$ and rewrite the expression as the product of two fractions:

$$\lim_{x \to 0} \frac{\sin(5x)}{\sin(9x)}$$

$$= \lim_{x \to 0} \frac{\dfrac{\sin(5x)}{x}}{\dfrac{\sin(9x)}{x}}$$

$$= \lim_{x \to 0} \left(\frac{\sin(5x)}{x} \cdot \frac{x}{\sin(9x)} \right)$$

Now you want a 5 in the denominator of the first fraction and a 9 in the numerator of the second fraction. Multiplying by $\frac{5}{5}$ and also by $\frac{9}{9}$ gives you

$$= \lim_{x \to 0}\left(\left(\frac{\sin(5x)}{5x} \cdot \frac{9x}{\sin(9x)}\right)\left(\frac{5}{9}\right)\right)$$

$$= (1)(1)\left(\frac{5}{9}\right)$$

$$= \frac{5}{9}$$

203. $\frac{7}{3}$

Begin by rewriting $\tan(7x)$ to get

$$\lim_{x \to 0} \frac{\tan(7x)}{\sin(3x)}$$

$$= \lim_{x \to 0} \frac{\frac{\sin(7x)}{\cos(7x)}}{\sin(3x)}$$

$$= \lim_{x \to 0}\left(\frac{\sin(7x)}{\cos(7x)} \cdot \frac{1}{\sin(3x)}\right)$$

$$= \lim_{x \to 0}\left(\frac{\sin(7x)}{1} \cdot \frac{1}{\cos(7x)} \cdot \frac{1}{\sin(3x)}\right)$$

Now you want $7x$ in the denominator of the first fraction and $3x$ in the numerator of the third fraction. Therefore, multiply by $\frac{x}{x}$, $\frac{3}{3}$, and $\frac{7}{7}$ to get

$$\lim_{x \to 0}\left(\frac{\sin(7x)}{7x} \cdot \frac{1}{\cos(7x)} \cdot \frac{3x}{\sin(3x)} \cdot \frac{7}{3}\right)$$

$$= (1)\left(\frac{1}{\cos 0}\right)(1)\left(\frac{7}{3}\right)$$

$$= \frac{7}{3}$$

204. 8

Begin by breaking up the fraction as

$$\lim_{x \to 0} \frac{\sin^3(2x)}{x^3}$$

$$= \lim_{x \to 0}\left(\frac{\sin(2x)}{x} \cdot \frac{\sin(2x)}{x} \cdot \frac{\sin(2x)}{x}\right)$$

Next, you want each fraction to have a denominator of $2x$, so multiply by $\frac{2}{2}$, $\frac{2}{2}$, and $\frac{2}{2}$ — or equivalently, by $\frac{8}{8}$ — to get

$$\lim_{x \to 0}\left(\frac{\sin(2x)}{2x} \cdot \frac{\sin(2x)}{2x} \cdot \frac{\sin(2x)}{2x} \cdot \frac{8}{1}\right)$$

$$= (1)(1)(1)(8)$$

$$= 8$$

205. $\frac{1}{5}$

Begin by factoring the denominator and rewriting the limit:

$$\lim_{x\to2}\frac{\sin(x-2)}{x^2+x-6}$$

$$=\lim_{x\to2}\frac{\sin(x-2)}{(x-2)(x+3)}$$

$$=\lim_{x\to2}\frac{\sin(x-2)}{(x-2)}\cdot\lim_{x\to2}\frac{1}{x+3}$$

Notice that you can rewrite the first limit using the substitution $\theta=x-2$ so that as $x\to2$, you have $\theta\to0$; this step isn't necessary, but it clarifies how to use $\lim_{x\to0}\frac{\sin x}{x}=1$ in this problem.

Replacing $(x-2)$ with θ and replacing $x\to2$ with $\theta\to0$ in the limit $\lim_{x\to2}\frac{\sin(x-2)}{(x-2)}$ gives you the following:

$$\lim_{x\to2}\frac{\sin(x-2)}{(x-2)}\cdot\lim_{x\to2}\frac{1}{x+3}$$

$$=\lim_{\theta\to0}\frac{\sin(\theta)}{(\theta)}\cdot\lim_{x\to2}\frac{1}{x+3}$$

$$=(1)\left(\frac{1}{2+3}\right)$$

$$=\frac{1}{5}$$

206. $\frac{1}{2}$

Begin by rewriting $\tan x$ and then multiply the numerator and denominator by $\frac{1}{x}$. Then simplify:

$$\lim_{x\to0}\frac{\sin x}{x+\tan x}$$

$$=\lim_{x\to0}\frac{\sin x}{x+\frac{\sin x}{\cos x}}$$

$$=\lim_{x\to0}\frac{\frac{1}{x}(\sin x)}{\frac{1}{x}\left(x+\frac{\sin x}{\cos x}\right)}$$

$$=\lim_{x\to0}\frac{\frac{\sin x}{x}}{1+\frac{\sin x}{x\cos x}}$$

$$=\lim_{x\to0}\frac{\frac{\sin x}{x}}{1+\frac{\sin x}{x}\left(\frac{1}{\cos x}\right)}$$

Then substitute 0 into the equation and solve:

$$\lim_{x \to 0} \frac{\dfrac{\sin x}{x}}{1 + \dfrac{\sin x}{x}\left(\dfrac{1}{\cos x}\right)}$$

$$= \frac{1}{1 + 1\left(\dfrac{1}{\cos 0}\right)}$$

$$= \frac{1}{1 + 1(1)}$$

$$= \frac{1}{2}$$

207. ∞

As x approaches 3 from the left, the y values approach ∞ so that $\lim\limits_{x \to 3^-} f(x) = \infty$.

208. −∞

As x approaches 3 from the right, the y values approach −∞ so that $\lim\limits_{x \to 3^+} f(x) = -\infty$.

209. ∞

As x approaches 5 from the left, the y values approach ∞ so that $\lim\limits_{x \to 5^-} f(x) = \infty$.

210. −∞

As x approaches 5 from the right, the y values approach −∞ so that $\lim\limits_{x \to 5^+} f(x) = -\infty$.

211. limit does not exist

As x approaches 5 from the left, the y values approach ∞. However, as x approaches 5 from the right, the y values approach −∞. Because the left-hand limit doesn't equal the right-hand limit, the limit doesn't exist.

212. ∞

As x approaches 1 from the right, you have $(x - 1) \to 0^+$ so that

$$\frac{3}{x - 1} \to \frac{3}{0^+} \to \infty$$

Note that the limit is positive infinity because dividing 3 by a small positive number close to zero gives you a large positive number.

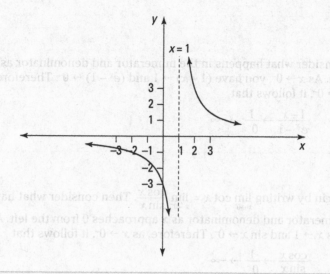

213. $-\infty$

As x approaches 1 from the left, you have $(x - 1) \to 0^-$ so that

$$\frac{3}{x-1} \to \frac{3}{0^-} \to -\infty$$

Note that the limit is negative infinity because dividing 3 by a small negative number close to zero gives you a negative number whose absolute value is large.

214. $-\infty$

Begin by writing the limit as $\lim\limits_{x\to\frac{\pi}{2}^+}(\tan x) = \lim\limits_{x\to\frac{\pi}{2}^+}\frac{\sin x}{\cos x}$. Then consider what happens in the numerator and the denominator as x approaches $\frac{\pi}{2}$ from the right.

As $x \to \frac{\pi}{2}^+$, you have $\sin x \to 1$ and $\cos x \to 0^-$. Therefore, as $x \to \frac{\pi}{2}^+$, it follows that

$$\frac{\sin x}{\cos x} \to \frac{1}{0^-} \to -\infty$$

215. ∞

Consider what happens in the numerator and denominator as x approaches π from the left. As $x \to \pi^-$, you have $x^2 \to \pi^2$ and $\sin x \to 0^+$. Therefore, as $x \to \pi^-$, it follows that

$$\frac{x^2}{\sin x} \to \frac{\pi^2}{0^+} \to \infty$$

216. $-\infty$

Consider what happens in the numerator and denominator as x approaches 5 from the left. As $x \to 5^-$, you have $(x + 3) \to 8$ and $(x - 5) \to 0^-$. Therefore, as $x \to 5^-$, it follows that

$$\frac{x+3}{x-5} \to \frac{8}{0^-} \to -\infty$$

217. $-\infty$

Consider what happens in the numerator and denominator as x approaches 0 from the left. As $x \to 0^-$, you have $(1-x) \to 1$ and $(e^x - 1) \to 0^-$. Therefore, as $x \to 0^-$, it follows that

$$\frac{1-x}{e^x - 1} \to \frac{1}{0^-} \to -\infty$$

218. $-\infty$

Begin by writing $\displaystyle\lim_{x \to 0^-} \cot x = \lim_{x \to 0^-} \frac{\cos x}{\sin x}$. Then consider what happens in the numerator and denominator as x approaches 0 from the left. As $x \to 0^-$, you have $\cos x \to 1$ and $\sin x \to 0^-$. Therefore, as $x \to 0^-$, it follows that

$$\frac{\cos x}{\sin x} \to \frac{1}{0^-} \to -\infty$$

219. ∞

Consider what happens in the numerator and denominator as x approaches 2. As $x \to 2$, you have $4e^x \to 4e^2$ and $|2-x| \to 0^+$. Therefore, as $x \to 2$, it follows that

$$\frac{4e^x}{|2-x|} \to \frac{4e^2}{0^+} \to \infty$$

220. $-\infty$

Consider what happens to each factor in the numerator and denominator as x approaches $\frac{1}{2}$ from the right. As $x \to \frac{1}{2}^+$, you have $(x^2 + 1) \to \frac{5}{4}$ and $\cos(\pi x) \to \cos\left(\frac{\pi}{2}^+\right) \to 0^-$. Therefore, as $x \to \frac{1}{2}^+$, it follows that

$$\frac{x^2 + 1}{x\cos(\pi x)} \to \frac{\frac{5}{4}}{\frac{1}{2}(0^-)} \to \frac{\frac{5}{4}}{0^-} \to -\infty$$

221. $-\infty$

Consider what happens in the numerator and denominator as x approaches 5. As $x \to 5$, you have $\sin x \to \sin 5$, and because $\pi < 5 < 2\pi$, it follows that $\sin 5 < 0$. And as $x \to 5$, you also have $(5-x)^4 \to 0^+$. Therefore, as $x \to 5$, it follows that

$$\frac{\sin x}{(5-x)^4} \to \frac{\sin 5}{0^+} \to -\infty$$

Note that the limit is negative infinity because $\sin(5)$ is negative, and dividing $\sin(5)$ by a small positive number close to zero gives you a negative number whose absolute value is large.

222. $-\infty$

Consider what happens to each factor in the numerator and denominator as x approaches 0. You need to examine both the left-hand limit and the right-hand limit.

For the left-hand limit, as $x \to 0^-$, you have $(x + 5) \to 5$, $x^4 \to 0^+$, and $(x - 6) \to -6$. Therefore, as $x \to 0^-$, it follows that

$$\frac{x+5}{x^4(x-6)} \to \frac{5}{0^+(-6)} \to \frac{5}{0^-} \to -\infty$$

For the right-hand limit, you have $(x + 5) \to 5$, $x^4 \to 0^+$, and $(x - 6) \to -6$. Therefore, as $x \to 0^+$, it follows that

$$\frac{x+5}{x^4(x-6)} \to \frac{5}{0^+(-6)} \to \frac{5}{0^-} \to -\infty$$

Because the left-hand limit is equal to the right-hand limit, $\lim\limits_{x \to 0} \dfrac{x+5}{x^4(x-6)} = -\infty$.

223. limit does not exist

Begin by examining the left-hand limit and the right-hand limit to determine whether they're equal.

To find the limits, consider what happens in the numerator and denominator as x approaches 1. Using the left-hand limit, as $x \to 1^-$, you have $(3x) \to 3$ and $(e^x - e) \to 0^-$. Therefore, as $x \to 1^-$, it follows that

$$\frac{3x}{e^x-e} \to \frac{3}{0^-} \to -\infty$$

Using the right-hand limit, as $x \to 1^+$, you have $(3x) \to 3$ and $(e^x - e) \to 0^+$. So as $x \to 1^+$, it follows that

$$\frac{3x}{e^x-e} \to \frac{3}{0^+} \to \infty$$

Because the left-hand limit doesn't equal the right-hand limit, the limit doesn't exist.

224. $-\infty$

Consider what happens to each factor in the numerator and denominator as x approaches 0 from the right. You have $(x - 1) \to -1$, $x^2 \to 0^+$, and $(x + 2) \to 2$. Therefore, as $x \to 0^+$, you get the following:

$$\frac{x-1}{x^2(x+2)} \to \frac{-1}{0^+(2)} \to \frac{-1}{0^+} \to -\infty$$

225. $-\infty$

Consider what happens in the numerator and denominator as x approaches e from the left. As $x \to e^-$, you have $x^3 \to e^3$ and $(\ln x - 1) \to (\ln e^- - 1) \to 0^-$. Therefore, as $x \to e^-$, it follows that

$$\frac{x^3}{\ln x-1} \to \frac{e^3}{0^-} \to -\infty$$

226. limit does not exist

Begin by examining both the left-hand limit and the right-hand limit to determine whether they're equal.

To find the limits, consider what happens in the numerator and denominator as x approaches e^2. For the left-hand limit, as $x \to e^{2-}$, you have $-x \to -e^2$ and $(\ln x - 2) \to 0^-$. Therefore, as $x \to e^{2-}$, it follows that

$$\frac{-x}{\ln x - 2} \to \frac{-e^2}{0^-} \to \infty$$

Note that the limit is positive infinity because dividing $-e^2$ by a small negative number close to zero gives you a large positive number.

For the right-hand limit, as $x \to e^{2+}$, you have $-x \to -e^2$ and $(\ln x - 2) \to 0^+$. Therefore, as $x \to e^{2+}$, it follows that

$$\frac{-x}{\ln x - 2} \to \frac{-e^2}{0^+} \to -\infty$$

Because the left-hand limit doesn't equal the right-hand limit, the limit doesn't exist.

227. limit does not exist

Begin by examining the left-hand limit and the right-hand limit to determine whether they're equal.

To find the limits, consider what happens in the numerator and denominator as x approaches 2. For the left-hand limit, as $x \to 2^-$, you have $(x + 2) \to 4$ and $(x^2 - 4) \to 0^-$. Therefore, as $x \to 2^-$, it follows that

$$\frac{x+2}{x^2 - 4} \to \frac{4}{0^-} \to -\infty$$

For the right-hand limit, as $x \to 2^+$, you have $(x + 2) \to 4$ and $(x^2 - 4) \to 0^+$. Therefore, as $x \to 2^+$, it follows that

$$\frac{x+2}{x^2 - 4} \to \frac{4}{0^+} \to \infty$$

Because the left-hand limit doesn't equal the right-hand limit, the limit doesn't exist.

228. limit does not exist

Begin by examining the left-hand limit and the right-hand limit to determine whether they're equal.

To find the limits, consider what happens in the numerator and denominator as x approaches 25. For the left-hand limit, as $x \to 25^-$, you have $\left(5 + \sqrt{x}\right) \to 5 + \sqrt{25} = 10$ and $(x - 25) \to 0^-$. Therefore, as $x \to 25^-$, it follows that

$$\frac{5 + \sqrt{x}}{x - 25} \to \frac{10}{0^-} \to -\infty$$

For the right-hand limit, as $x \to 25^+$, you have $\left(5 + \sqrt{x}\right) \to 5 + \sqrt{25} = 10$ and $(x - 25) \to 0^+$.

Therefore, as $x \to 25^+$, you have

$$\frac{5+\sqrt{x}}{x-25} \to \frac{10}{0^+} \to \infty$$

Because the left-hand limit doesn't equal the right-hand limit, the limit doesn't exist.

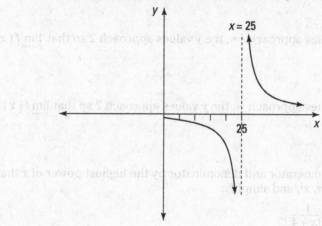

229. $-\infty$

Consider what happens to each factor in the numerator and denominator as x approaches 0. As $x \to 0$, you have $(x^2 + 4) \to 4$, $x^2 \to 0^+$ and $(x - 1) \to -1$. Therefore, as $x \to 0$, it follows that

$$\frac{x^2 + 4}{x^2(x-1)} \to \frac{4}{0^+(-1)} \to \frac{4}{0^-} \to -\infty$$

230. $-\infty$

Consider what happens to each factor in the numerator and denominator as x approaches 0 from the left. You have $(x - 1) \to -1$, $x^2 \to 0^+$, and $(x + 2) \to 2$. Therefore, as $x \to 0^-$, you get the following:

$$\frac{x-1}{x^2(x+2)} \to \frac{-1}{0^+(2)} \to \frac{-1}{0^+} \to -\infty$$

231. ∞

Consider what happens in the numerator and denominator as x approaches 3. As $x \to 3$, you have $(3 + x) \to 6$ and $|3 - x| \to 0^+$. Therefore, as $x \to 3$, it follows that

$$\frac{3+x}{|3-x|} \to \frac{6}{0^+} \to \infty$$

232. $-\dfrac{\pi}{2}$

As the x values approach $-\infty$, the y values approach $-\frac{\pi}{2}$ so that $\lim\limits_{x \to -\infty} f(x) = -\frac{\pi}{2}$.

233. $\dfrac{\pi}{2}$

As the x values approach ∞, the y values approach $\dfrac{\pi}{2}$ so that $\lim\limits_{x\to\infty} f(x) = \dfrac{\pi}{2}$.

234. 2

As the x values approach $-\infty$, the y values approach 2 so that $\lim\limits_{x\to-\infty} f(x) = 2$.

235. 2

As the x values approach ∞, the y values approach 2 so that $\lim\limits_{x\to\infty} f(x) = 2$.

236. 0

Divide the numerator and denominator by the highest power of x that appears in the denominator, x^1, and simplify:

$$\lim_{x\to\infty} \frac{1}{3x+4}$$

$$= \lim_{x\to\infty} \frac{\dfrac{1}{x}}{\dfrac{3x}{x}+\dfrac{4}{x}}$$

$$= \lim_{x\to\infty} \frac{\dfrac{1}{x}}{3+\dfrac{4}{x}}$$

Then apply the limit:

$$= \frac{0}{3+0} = 0$$

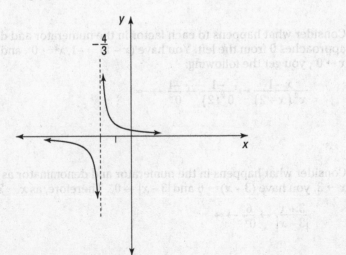

237. $-\infty$

Begin by multiplying to get

$$\lim_{x \to -\infty} \left[x^2 (x+1)(3-x) \right]$$

$$= \lim_{x \to -\infty} \left[-x^4 + 2x^3 + 3x^2 \right]$$

For a polynomial function, the end behavior is determined by the term containing the largest power of x, so you have

$$\lim_{x \to -\infty} \left[-x^4 + 2x^3 + 3x^2 \right]$$

$$= \lim_{x \to -\infty} \left(-x^4 \right)$$

$$= -\infty$$

238. 3

Divide the numerator and denominator by the highest power of x that appears in the denominator, x^1, and simplify:

$$\lim_{x \to \infty} \frac{3x+4}{x-7}$$

$$= \lim_{x \to \infty} \frac{\dfrac{3x}{x} + \dfrac{4}{x}}{\dfrac{x}{x} - \dfrac{7}{x}}$$

$$= \lim_{x \to \infty} \frac{3 + \dfrac{4}{x}}{1 - \dfrac{7}{x}}$$

Then apply the limit:

$$= \frac{3+0}{1-0} = 3$$

239. limit does not exist

Because $\cos x$ doesn't approach a specific value as x approaches ∞, the limit doesn't exist.

240. 0

Begin by expanding the denominator:

$$\lim_{x \to \infty} \frac{5x^4 + 5}{\left(x^2 - 1 \right)\left(2x^3 + 3 \right)}$$

$$= \lim_{x \to \infty} \frac{5x^4 + 5}{2x^5 - 2x^3 + 3x^2 - 3}$$

Next, divide the numerator and denominator by the highest power of x that appears in the denominator, x^5, and simplify:

$$\lim_{x \to \infty} \frac{5x^4 + 5}{2x^5 - 2x^3 + 3x^2 - 3}$$

$$= \lim_{x \to \infty} \frac{\dfrac{5x^4}{x^5} + \dfrac{5}{x^5}}{\dfrac{2x^5}{x^5} - \dfrac{2x^3}{x^5} + \dfrac{3x^2}{x^5} - \dfrac{3}{x^5}}$$

$$= \lim_{x \to \infty} \frac{\dfrac{5}{x} + \dfrac{5}{x^5}}{2 - \dfrac{2}{x^2} + \dfrac{3}{x^3} - \dfrac{3}{x^5}}$$

Then apply the limit:

$$= \frac{0 + 0}{2 - 0 + 0 - 0} = 0$$

241. $\quad -1$

Begin by multiplying the numerator and denominator by $\dfrac{1}{x}$ so that you can simplify the expression underneath the square root:

$$\lim_{x \to -\infty} \frac{x}{\sqrt{x^2 + 1}}$$

$$= \lim_{x \to -\infty} \frac{\dfrac{1}{x}(x)}{\dfrac{1}{x}\sqrt{x^2 + 1}}$$

Because x is approaching $-\infty$, you know that $x < 0$. So as you take the limit, you need to use the substitution $\dfrac{1}{x} = -\sqrt{\dfrac{1}{x^2}}$ in the denominator. Therefore, you have

$$\lim_{x \to -\infty} \frac{\dfrac{1}{x}(x)}{\dfrac{1}{x}\sqrt{x^2 + 1}}$$

$$= \lim_{x \to -\infty} \frac{1}{-\sqrt{\dfrac{1}{x^2}}\sqrt{x^2 + 1}}$$

$$= \lim_{x \to -\infty} \frac{1}{-\sqrt{\dfrac{x^2}{x^2} + \dfrac{1}{x^2}}}$$

$$= \lim_{x \to -\infty} \frac{1}{-\sqrt{1 + \dfrac{1}{x^2}}}$$

Now apply the limit:

$$= \frac{1}{-\sqrt{1+0}} = -1$$

242. ∞

Divide the numerator and denominator by the highest power of x that appears in the denominator, x^2, and simplify:

$$\lim_{x \to -\infty} \frac{8x^4 + 3x - 5}{x^2 + 1}$$

$$= \lim_{x \to -\infty} \frac{\dfrac{8x^4}{x^2} + \dfrac{3x}{x^2} - \dfrac{5}{x^2}}{\dfrac{x^2}{x^2} + \dfrac{1}{x^2}}$$

$$= \lim_{x \to -\infty} \frac{8x^2 + \dfrac{3}{x} - \dfrac{5}{x^2}}{1 + \dfrac{1}{x^2}}$$

Then apply the limit. Consider what happens in the numerator and the denominator as $x \to -\infty$. In the numerator, $\left(8x^2 + \dfrac{3}{x} - \dfrac{5}{x^2}\right) \to \infty$, and in the denominator, $\left(1 + \dfrac{1}{x^2}\right) \to 1$. Therefore, you have

$$\lim_{x \to -\infty} \frac{8x^2 + \dfrac{3}{x} - \dfrac{5}{x^2}}{1 + \dfrac{1}{x^2}} = \infty$$

243. −3

To simplify the expression underneath the radical, begin by multiplying the numerator and denominator by $\dfrac{1}{x^5}$:

$$\lim_{x \to -\infty} \frac{\sqrt{9x^{10} - x}}{x^5 + 1}$$

$$= \lim_{x \to -\infty} \frac{\dfrac{1}{x^5}\sqrt{9x^{10} - x}}{\dfrac{1}{x^5}\left(x^5 + 1\right)}$$

Because x is approaching $-\infty$, you know that $x < 0$. So as you take the limit, you need to use the substitution $\dfrac{1}{x^5} = -\sqrt{\dfrac{1}{x^{10}}}$ in the numerator:

$$\lim_{x \to -\infty} \frac{\dfrac{1}{x^5}\sqrt{9x^{10}-x}}{\dfrac{1}{x^5}\left(x^5+1\right)}$$

$$= \lim_{x \to -\infty} \frac{-\sqrt{\dfrac{1}{x^{10}}}\sqrt{9x^{10}-x}}{\dfrac{x^5}{x^5}+\dfrac{1}{x^5}}$$

$$= \lim_{x \to -\infty} \frac{-\sqrt{\dfrac{9x^{10}}{x^{10}}-\dfrac{x}{x^{10}}}}{1+\dfrac{1}{x^5}}$$

$$= \lim_{x \to -\infty} \frac{-\sqrt{9-\dfrac{1}{x^9}}}{1+\dfrac{1}{x^5}}$$

Now apply the limit:

$$= \frac{-\sqrt{9-0}}{1+0} = -3$$

244. 3

To simplify the expression underneath the radical, begin by multiplying the numerator and denominator by $\dfrac{1}{x^5}$:

$$\lim_{x \to \infty} \frac{\sqrt{9x^{10}-x}}{x^5+1}$$

$$= \lim_{x \to \infty} \frac{\dfrac{1}{x^5}\sqrt{9x^{10}-x}}{\dfrac{1}{x^5}\left(x^5+1\right)}$$

$$= \lim_{x \to \infty} \frac{\sqrt{\dfrac{1}{x^{10}}}\sqrt{9x^{10}-x}}{\dfrac{x^5}{x^5}+\dfrac{1}{x^5}}$$

$$= \lim_{x \to \infty} \frac{\sqrt{\dfrac{9x^{10}}{x^{10}}-\dfrac{x}{x^{10}}}}{1+\dfrac{1}{x^5}}$$

$$= \lim_{x \to \infty} \frac{\sqrt{9-\dfrac{1}{x^9}}}{1+\dfrac{1}{x^5}}$$

Now apply the limit:

$$= \frac{\sqrt{9-0}}{1+0} = 3$$

245. $-\infty$

Notice that if you consider the limit immediately, you have the indeterminate form $\infty - \infty$.

Begin by factoring to get

$$\lim_{x \to \infty} \left(\sqrt{x} - x \right)$$
$$= \lim_{x \to \infty} x^{1/2} \left(1 - x^{1/2} \right)$$
$$= \left(\lim_{x \to \infty} x^{1/2} \right) \left(\lim_{x \to \infty} \left(1 - x^{1/2} \right) \right)$$
$$= \infty \left(-\infty \right)$$
$$= -\infty$$

246. $\dfrac{3}{2}$

Notice that if you consider the limit immediately, you have the indeterminate form $\infty - \infty$.

Create a fraction and multiply the numerator and denominator by the conjugate of the expression $\left(\sqrt{x^4 + 3x^2} - x^2 \right)$. The conjugate is $\left(\sqrt{x^4 + 3x^2} + x^2 \right)$, so you have the following:

$$\lim_{x \to \infty} \left(\sqrt{x^4 + 3x^2} - x^2 \right)$$
$$= \lim_{x \to \infty} \frac{\left(\sqrt{x^4 + 3x^2} - x^2 \right)}{1} \left(\frac{\sqrt{x^4 + 3x^2} + x^2}{\sqrt{x^4 + 3x^2} + x^2} \right)$$
$$= \lim_{x \to \infty} \frac{3x^2}{\sqrt{x^4 + 3x^2} + x^2}$$

Next, multiply the numerator and denominator by $\dfrac{1}{x^2}$ so you can simplify the expression underneath the radical:

$$\lim_{x \to \infty} \frac{\dfrac{1}{x^2} \left(3x^2 \right)}{\dfrac{1}{x^2} \left(\sqrt{x^4 + 3x^2} + x^2 \right)}$$
$$= \lim_{x \to \infty} \frac{3}{\left(\dfrac{1}{x^2} \sqrt{x^4 + 3x^2} + \dfrac{x^2}{x^2} \right)}$$
$$= \lim_{x \to \infty} \frac{3}{\left(\sqrt{\dfrac{1}{x^4} \left(x^4 + 3x^2 \right)} + \dfrac{x^2}{x^2} \right)}$$
$$= \lim_{x \to \infty} \frac{3}{\left(\sqrt{\dfrac{x^4}{x^4} + \dfrac{3x^2}{x^4}} + 1 \right)}$$
$$= \lim_{x \to \infty} \frac{3}{\left(\sqrt{1 + \dfrac{3}{x^2}} + 1 \right)}$$

Now apply the limit:

$$= \frac{3}{\left(\sqrt{1 + 0} + 1 \right)} = \frac{3}{2}$$

247. $-\dfrac{5}{2}$

Notice that if you consider the limit immediately, you have the indeterminate form $-\infty + \infty$.

Create a fraction and multiply the numerator and denominator by the conjugate of the expression $\left(x + \sqrt{x^2 + 5x}\right)$. The conjugate is $\left(x - \sqrt{x^2 + 5x}\right)$, so you have

$$\lim_{x \to -\infty} \left(x + \sqrt{x^2 + 5x}\right)$$

$$= \lim_{x \to -\infty} \frac{\left(x + \sqrt{x^2 + 5x}\right)}{1} \left(\frac{x - \sqrt{x^2 + 5x}}{x - \sqrt{x^2 + 5x}}\right)$$

$$= \lim_{x \to -\infty} \frac{-5x}{x - \sqrt{x^2 + 5x}}$$

Next, multiply the numerator and denominator by $\dfrac{1}{x}$ so that you can simplify the expression underneath the square root:

$$\lim_{x \to -\infty} \frac{-5x}{x - \sqrt{x^2 + 5x}}$$

$$= \lim_{x \to -\infty} \frac{\frac{1}{x}(-5x)}{\frac{1}{x}\left(x - \sqrt{x^2 + 5x}\right)}$$

$$= \lim_{x \to -\infty} \frac{-5}{\frac{1}{x}(x) - \frac{1}{x}\sqrt{x^2 + 5x}}$$

$$= \lim_{x \to -\infty} \frac{-5}{1 - \frac{1}{x}\sqrt{x^2 + 5x}}$$

Because x is approaching $-\infty$, you know that $x < 0$. So as you take the limit, you need to use the substitution $\dfrac{1}{x} = -\sqrt{\dfrac{1}{x^2}}$ in the denominator. Therefore, you have

$$\lim_{x \to -\infty} \frac{-5}{1 - \frac{1}{x}\sqrt{x^2 + 5x}}$$

$$= \lim_{x \to -\infty} \frac{-5}{1 - \left(-\frac{1}{\sqrt{x^2}}\sqrt{x^2 + 5x}\right)}$$

$$= \lim_{x \to -\infty} \frac{-5}{1 + \sqrt{\frac{x^2}{x^2} + \frac{5x}{x^2}}}$$

$$= \lim_{x \to -\infty} \frac{-5}{1 + \sqrt{1 + \frac{5}{x}}}$$

Now apply the limit:

$$= \frac{-5}{1 + \sqrt{1 + 0}} = -\frac{5}{2}$$

248. $y = \dfrac{3}{5}$

To find any horizontal asymptotes of the function $y = \dfrac{1+3x^4}{x+5x^4}$, you need to consider the limit of the function as $x \to \infty$ and as $x \to -\infty$. For the limit as $x \to \infty$, begin by multiplying the numerator and denominator by $\dfrac{1}{x^4}$:

$$\lim_{x \to \infty} \frac{1+3x^4}{x+5x^4} = \lim_{x \to \infty} \frac{\dfrac{1}{x^4} + \dfrac{3x^4}{x^4}}{\dfrac{x}{x^4} + \dfrac{5x^4}{x^4}}$$

$$= \lim_{x \to \infty} \frac{\dfrac{1}{x^4} + 3}{\dfrac{1}{x^3} + 5}$$

$$= \frac{0+3}{0+5}$$

$$= \frac{3}{5}$$

To find the limit as $x \to -\infty$, you can proceed in the same way in order to simplify:

$$\lim_{x \to -\infty} \frac{1+3x^4}{x+5x^4} = \lim_{x \to -\infty} \frac{\dfrac{1}{x^4} + \dfrac{3x^4}{x^4}}{\dfrac{x}{x^4} + \dfrac{5x^4}{x^4}}$$

$$= \lim_{x \to -\infty} \frac{\dfrac{1}{x^4} + 3}{\dfrac{1}{x^3} + 5}$$

$$= \frac{0+3}{0+5}$$

$$= \frac{3}{5}$$

Therefore, the only horizontal asymptote is $y = \dfrac{3}{5}$.

249. $y = -1$

To find any horizontal asymptotes of the function $y = \dfrac{5-x^2}{5+x^2}$, you need to consider the limit of the function as $x \to \infty$ and as $x \to -\infty$. For the limit as $x \to \infty$, begin by multiplying the numerator and denominator by $\dfrac{1}{x^2}$:

$$\lim_{x \to \infty} \frac{5-x^2}{5+x^2} = \lim_{x \to \infty} \frac{\dfrac{5}{x^2} - \dfrac{x^2}{x^2}}{\dfrac{5}{x^2} + \dfrac{x^2}{x^2}}$$

$$= \lim_{x \to \infty} \frac{\dfrac{5}{x^2} - 1}{\dfrac{5}{x^2} + 1}$$

$$= \frac{0-1}{0+1}$$

$$= -1$$

To find the limit as $x \to -\infty$, proceed in the same way:

$$\lim_{x \to -\infty} \frac{5-x^2}{5+x^2} = \lim_{x \to -\infty} \frac{\dfrac{5}{x^2} - \dfrac{x^2}{x^2}}{\dfrac{5}{x^2} + \dfrac{x^2}{x^2}}$$

$$= \lim_{x \to -\infty} \frac{\dfrac{5}{x^2} - 1}{\dfrac{5}{x^2} + 1}$$

$$= \frac{0-1}{0+1}$$

$$= -1$$

Therefore, the only horizontal asymptote is $y = -1$.

250. $y = \dfrac{1}{3}$

In order to find any horizontal asymptotes of the function $y = \dfrac{\sqrt{x^4 + x}}{3x^2}$, you need to consider the limit of the function as $x \to \infty$ and as $x \to -\infty$. For the limit as $x \to \infty$, begin by multiplying the numerator and denominator by $\dfrac{1}{x^2}$:

$$\lim_{x \to \infty} \frac{\sqrt{x^4 + x}}{3x^2} = \lim_{x \to \infty} \frac{\dfrac{1}{x^2}\sqrt{x^4 + x}}{\dfrac{1}{x^2}\left(3x^2\right)}$$

$$= \lim_{x \to \infty} \frac{\sqrt{\dfrac{1}{x^4}}\sqrt{x^4 + x}}{3}$$

$$= \lim_{x \to \infty} \frac{\sqrt{\dfrac{x^4}{x^4} + \dfrac{x}{x^4}}}{3}$$

$$= \lim_{x \to \infty} \frac{\sqrt{1 + \dfrac{1}{x^3}}}{3}$$

$$= \frac{\sqrt{1+0}}{3}$$

$$= \frac{1}{3}$$

In order to find the limit as $x \to -\infty$, proceed in the same way, noting that as $x \to -\infty$, you still use $\dfrac{1}{x^2} = \sqrt{\dfrac{1}{x^4}}$:

$$\lim_{x \to -\infty} \frac{\sqrt{x^4 + x}}{3x^2} = \lim_{x \to -\infty} \frac{\frac{1}{x^2}\sqrt{x^4 + x}}{\frac{1}{x^2}\left(3x^2\right)}$$

$$= \lim_{x \to -\infty} \frac{\sqrt{\frac{1}{x^4}}\sqrt{x^4 + x}}{3}$$

$$= \lim_{x \to -\infty} \frac{\sqrt{\frac{x^4}{x^4} + \frac{x}{x^4}}}{3}$$

$$= \lim_{x \to -\infty} \frac{\sqrt{1 + \frac{1}{x^3}}}{3}$$

$$= \frac{\sqrt{1 + 0}}{3}$$

$$= \frac{1}{3}$$

Therefore, the only horizontal asymptote is $y = \frac{1}{3}$.

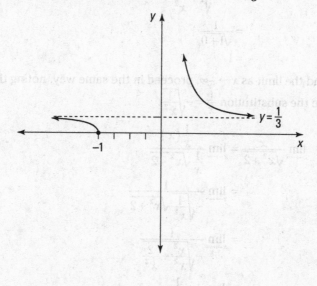

251. $y = 1$ and $y = -1$

To find any horizontal asymptotes of the function $y = \dfrac{x}{\sqrt{x^2 + 2}}$, you need to consider the limit of the function as $x \to \infty$ and as $x \to -\infty$. For the limit as $x \to \infty$, begin by multiplying the numerator and denominator by $\dfrac{1}{x}$:

$$\lim_{x \to \infty} \frac{x}{\sqrt{x^2 + 2}} = \lim_{x \to \infty} \frac{\frac{1}{x}(x)}{\frac{1}{x}\sqrt{x^2 + 2}}$$

$$= \lim_{x \to \infty} \frac{1}{\sqrt{\frac{1}{x^2}}\sqrt{x^2 + 2}}$$

$$= \lim_{x \to \infty} \frac{1}{\sqrt{\frac{x^2}{x^2} + \frac{2}{x^2}}}$$

$$= \lim_{x \to \infty} \frac{1}{\sqrt{1 + \frac{2}{x^2}}}$$

$$= \frac{1}{\sqrt{1 + 0}}$$

$$= 1$$

To find the limit as $x \to -\infty$, proceed in the same way, noting that as $x \to -\infty$, you need to use the substitution $\dfrac{1}{x} = -\sqrt{\dfrac{1}{x^2}}$:

$$\lim_{x \to -\infty} \frac{x}{\sqrt{x^2 + 2}} = \lim_{x \to -\infty} \frac{\frac{1}{x}(x)}{\frac{1}{x}\sqrt{x^2 + 2}}$$

$$= \lim_{x \to -\infty} \frac{1}{-\sqrt{\frac{1}{x^2}}\sqrt{x^2 + 2}}$$

$$= \lim_{x \to -\infty} \frac{1}{-\sqrt{\frac{x^2}{x^2} + \frac{2}{x^2}}}$$

$$= \lim_{x \to -\infty} \frac{1}{-\sqrt{1 + \frac{2}{x^2}}}$$

$$= \frac{1}{-\sqrt{1 + 0}}$$

$$= -1$$

Therefore, the function has the horizontal asymptotes $y = 1$ and $y = -1$.

252.   removable discontinuity at $x = -3$, jump discontinuity at $x = 3$

The limit exists at $x = -3$ but isn't equal to $f(-3)$, which corresponds to a removable discontinuity.

At $x = 3$, the left-hand limit doesn't equal the right-hand limit (both limits exist as finite values); this corresponds to a jump discontinuity.

253. removable discontinuity at $x = 1$, infinite discontinuity at $x = 5$

The limit exists at $x = 1$, but $f(1)$ is undefined, which corresponds to a removable discontinuity.

At $x = 5$, the left-hand limit is ∞ and the right-hand limit is $-\infty$, so an infinite discontinuity exists at $x = 5$.

254. jump discontinuity at $x = -2$, jump discontinuity at $x = 3$

At $x = -2$, the left-hand limit doesn't equal the right-hand limit (both limits exist as finite values), which corresponds to a jump discontinuity.

At $x = 3$, the left-hand limit again doesn't equal the right-hand limit (both limits exist as finite values), so this also corresponds to a jump discontinuity.

255. removable discontinuity at $x = -1$, jump discontinuity at $x = 4$, infinite discontinuity at $x = 6$

At $x = -1$, the left-hand limit equals the right-hand limit, but the limit doesn't equal $f(-1)$, which is undefined. Therefore, a removable discontinuity is at $x = -1$.

At $x = 4$, the left-hand limit doesn't equal the right-hand limit (both limits exist as finite values), so a jump discontinuity is at $x = 4$.

At $x = 6$, both the left and right-hand limits equal ∞, so an infinite discontinuity is at $x = 6$.

256. not continuous, infinite discontinuity

A function $f(x)$ is continuous at $x = a$ if it satisfies the equation $\lim\limits_{x \to a} f(x) = f(a)$.
The left-hand limit at a is given by $\lim\limits_{x \to 2^-} \dfrac{1}{x-2}$. As $x \to 2^-$, you have $(x-2) \to 0^-$ so that $\dfrac{1}{x-2} \to \dfrac{1}{0^-} \to -\infty$. Because the discontinuity is infinite, you don't need to examine the right-hand limit; you can conclude that the function is not continuous.

257. continuous, $f(a) = 2$

A function $f(x)$ is continuous at $x = a$ if it satisfies the equation $\lim\limits_{x \to a} f(x) = f(a)$.
The left-hand limit at a is $\lim\limits_{x \to 1^-} \left(1 + x^2\right) = 1 + 1^2 = 2$, and the right-hand limit at a is $\lim\limits_{x \to 1^+} \left(4\sqrt{x} - 2\right) = 4\sqrt{1} - 2 = 2$. The left-hand and right-hand limits match, so the limit at a exists and is equal to 2.

The value at a is $f(a) = f(1) = 4\sqrt{1} - 2 = 2$. Because the function satisfies the definition of continuity, you can conclude that that function is continuous at $a = 1$.

258. continuous, $f(a) = 5$

A function $f(x)$ is continuous at $x = a$ if it satisfies the equation $\lim\limits_{x \to a} f(x) = f(a)$.
The left-hand limit at a is

$$\lim_{x \to 3^-} \frac{x^2 - x - 6}{x - 3} = \lim_{x \to 3^-} \frac{(x-3)(x+2)}{x-3}$$

$$= \lim_{x \to 3^-} (x + 2)$$

$$= 3 + 2$$

$$= 5$$

And the right-hand limit at a is

$$\lim_{x \to 3^+} \frac{x^2 - x - 6}{x - 3} = \lim_{x \to 3^+} \frac{(x-3)(x+2)}{x-3}$$

$$= \lim_{x \to 3^+} (x + 2)$$

$$= 3 + 2$$

$$= 5$$

Note that in this case, you could have simply evaluated the limit as x approaches 3 instead of examining the left-hand limit and right-hand limit separately.

The left-hand and right-hand limits match, so the limit exists and is equal to 5. Because $f(a) = f(3) = 5$, the definition of continuity is satisfied and the function is continuous.

---

259.   continuous, $f(a) = \frac{1}{8}$

A function $f(x)$ is continuous at $x = a$ if it satisfies the equation $\lim_{x \to a} f(x) = f(a)$. The left-hand limit at a is

$$\lim_{x \to 16^-} \frac{4 - \sqrt{x}}{16 - x} = \lim_{x \to 16^-} \frac{4 - \sqrt{x}}{\left(4 - \sqrt{x}\right)\left(4 + \sqrt{x}\right)}$$

$$= \lim_{x \to 16^-} \frac{1}{\left(4 + \sqrt{x}\right)}$$

$$= \frac{1}{4 + \sqrt{16}}$$

$$= \frac{1}{8}$$

The right-hand limit at a is

$$\lim_{x \to 16^+} \frac{4 - \sqrt{x}}{16 - x} = \lim_{x \to 16^+} \frac{4 - \sqrt{x}}{\left(4 - \sqrt{x}\right)\left(4 + \sqrt{x}\right)}$$

$$= \lim_{x \to 16^+} \frac{1}{\left(4 + \sqrt{x}\right)}$$

$$= \frac{1}{4 + \sqrt{16}}$$

$$= \frac{1}{8}$$

The left-hand and right-hand limits match, so the limit exists.

Note that in this case, you could have simply evaluated the limit as x approaches 16 instead of examining the left-hand limit and right-hand limit separately.

The value at a is given by $f(a) = f(16) = \frac{1}{8}$, which matches the limit. Because the definition of continuity is satisfied, you can conclude that that function is continuous at $a = 16$.

---

260.   not continuous, jump discontinuity

A function $f(x)$ is continuous at $x = a$ if it satisfies the equation $\lim\limits_{x \to a} f(x) = f(a)$. The left-hand limit at a is

$$\lim_{x \to -6^-} \frac{x+6}{|x+6|} = \lim_{x \to -6^-} \frac{x+6}{-(x+6)}$$

$$= \lim_{x \to -6^-} (-1)$$

$$= -1$$

The right-hand limit at a is

$$\lim_{x \to -6^+} \frac{x+6}{|x+6|} = \lim_{x \to -6^+} \frac{x+6}{(x+6)}$$

$$= \lim_{x \to -6^+} (1)$$

$$= 1$$

Because the left- and right-hand limits exist but aren't equal to each other, there's a jump discontinuity at $a = -6$.

---

261.   not continuous, removable discontinuity

A function $f(x)$ is continuous at $x = a$ if it satisfies the equation $\lim\limits_{x \to a} f(x) = f(a)$. The left-hand limit at a is

$$\lim_{x \to -1^-} \frac{x^3+1}{x+1} = \lim_{x \to -1^-} \frac{(x+1)(x^2-x+1)}{(x+1)}$$

$$= \lim_{x \to -1^-} (x^2 - x + 1)$$

$$= (-1)^2 - (-1) + 1$$

$$= 3$$

The right-hand limit at a is

$$\lim_{x \to -1^+} \frac{x^3+1}{x+1} = \lim_{x \to -1^+} \frac{(x+1)(x^2-x+1)}{(x+1)}$$

$$= \lim_{x \to -1^+} (x^2 - x + 1)$$

$$= (-1)^2 - (-1) + 1$$

$$= 3$$

The left-hand and right-hand limits match, so the limit exists and is equal to 3.

Note that in this case, you could have simply evaluated the limit as x approaches -1 instead of examining the left-hand limit and right-hand limit separately.

However, because $f(a) = f(-1) = 2$, the function is not continuous; it has a removable discontinuity.

262. continuous at $a = 0$ and at $a = \pi$

A function $f(x)$ is continuous at $x = a$ if it satisfies the equation $\lim_{x \to a} f(x) = f(a)$.

To determine whether the function is continuous at $a = 0$, see whether it satisfies the equation $\lim_{x \to 0} f(x) = f(0)$. The left-hand limit at $a = 0$ is $\lim_{x \to 0^-} 2 + x^2 = 2 + (0)^2 = 2$, and the right-hand limit at $a = 0$ is $\lim_{x \to 0^+} 2\cos x = 2\cos(0) = 2$. Because $f(0) = 2\cos(0) = 2$, the function is continuous at $a = 0$.

Likewise, decide whether the function satisfies the equation $\lim_{x \to \pi} f(x) = f(\pi)$.

The left-hand limit at $a = \pi$ is $\lim_{x \to \pi^-} 2\cos x = 2\cos(\pi) = 2(-1) = -2$, and the right-hand limit at $a = \pi$ is $\lim_{x \to \pi^+} (\sin x - 2) = \sin\pi - 2 = 0 - 2 = -2$. Because $f(\pi) = 2\cos x = 2(-1) = -2$, the function is also continuous at $a = \pi$.

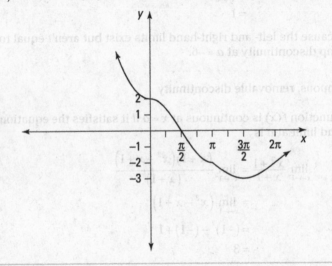

263. jump discontinuity at $a = 1$ and at $a = 3$

A function $f(x)$ is continuous at $x = a$ if it satisfies the equation $\lim_{x \to a} f(x) = f(a)$.

To determine whether the function is continuous at $a = 1$, see whether it satisfies the equation $\lim_{x \to 1} f(x) = f(1)$. The left-hand limit at $a = 1$ is $\lim_{x \to 1^-} (x + 2) = 1 + 2 = 3$, and the right-hand limit at $a = 1$ is $\lim_{x \to 1^+} 2x^2 = 2(1)^2 = 2$. The limits differ, so there's a jump discontinuity.

Likewise, decide whether the function satisfies the equation $\lim_{x \to 3} f(x) = f(3)$.

The left-hand limit at $a = 3$ is $\lim_{x \to 3^-} 2x^2 = 2(3)^2 = 18$, and the right-hand limit at $a = 3$ is $\lim_{x \to 3^+} x^3 = (3)^3 = 27$, so there's another jump discontinuity.

264.

continuous at $a = 2$, jump discontinuity at $a = 3$

A function $f(x)$ is continuous at $x = a$ if it satisfies the equation $\lim_{x \to a} f(x) = f(a)$.

To determine whether the function is continuous at $a = 2$, see whether it satisfies the equation $\lim_{x \to 2} f(x) = f(2)$. The left-hand limit at $a = 2$ is $\lim_{x \to 2^-} \sqrt{2 - x} = \sqrt{2 - 2} = 0$, and the right-hand limit at $a = 2$ is $\lim_{x \to 2^+} (x^2 - 4) = (2)^2 - 4 = 0$. Because $f(2) = \sqrt{2 - 2} = 0$, the function is continuous at $a = 2$.

Likewise, decide whether the function satisfies the equation $\lim_{x \to 3} f(x) = f(3)$. The left-hand limit at $a = 3$ is $\lim_{x \to 3^-} (x^2 - 4) = (3)^2 - 4 = 5$, and the right-hand limit at $a = 3$ is $\lim_{x \to 3^+} \left(\frac{1}{x + 5} \right) = \frac{1}{3 + 5} = \frac{1}{8}$. The limits don't match, so there's a jump discontinuity at $a = 3$.

265.

infinite discontinuity at $a = 0$, continuous at $a = 4$

A function $f(x)$ is continuous at $x = a$ if it satisfies the equation $\lim_{x \to a} f(x) = f(a)$.

To determine whether the function is continuous at $a = 0$, see whether it satisfies the equation $\lim_{x \to 0} f(x) = f(0)$. The left-hand limit at $a = 0$ is $\lim_{x \to 0^-} \cos x = \cos 0 = 1$, and the right-hand limit at $a = 0$ is $\lim_{x \to 0^+} \frac{1}{x} \to \frac{1}{0^+} \to \infty$, so there's an infinite discontinuity at $a = 0$.

Likewise, decide whether the function satisfies the equation $\lim_{x \to 4} f(4) = f(4)$. The left-hand limit at $a = 4$ is $\lim_{x \to 4^-} \frac{1}{x} = \frac{1}{4}$, and the right-hand limit at $a = 4$ is

$\lim_{x \to 4^+} \frac{2}{x + 4} = \frac{2}{4 + 4} = \frac{2}{8} = \frac{1}{4}$. Because $f(4) = \frac{1}{4}$, the function is continuous at $a = 4$.

266.

$c = -\frac{3}{2}$

To determine the value of c, you must satisfy the definition of a continuous function: $\lim_{x \to 2} f(x) = f(2)$.

The left-hand limit at $x = 2$ is given by $\lim_{x \to 2^-} (cx - 2) = c(2) - 2 = 2c - 2$, and the right-hand limit is given by $\lim_{x \to 2^+} (cx^2 + 1) = c(2)^2 + 1 = 4c + 1$. Also note that $f(2) = 2c - 2$.

The left-hand limit must equal the right-hand limit, so set them equal to each other and solve for c:

$$2c - 2 = 4c + 1$$
$$-3 = 2c$$
$$-\frac{3}{2} = c$$

Therefore, $c = -\frac{3}{2}$ is the solution.

267.

$c = 2$

To determine the value of c, you must satisfy the definition of a continuous function: $\lim_{x \to 4} f(x) = f(4)$.

The left-hand limit at $x = 4$ is given by $\lim\limits_{x \to 4^-}\left(x^2 + c^2\right) = \left(4\right)^2 + c^2 = 16 + c^2$, and the

right-hand limit is given by $\lim\limits_{x \to 4^+}\left(cx + 12\right) = 4c + 12$. Also note that $f\left(4\right) = 16 + c^2$.

The left-hand limit must equal the right-hand limit, so set them equal to each other:

$$16 + c^2 = 4c + 12$$

$$c^2 - 4c + 4 = 0$$

$$\left(c - 2\right)^2 = 0$$

Therefore, $c - 2 = 0$, which gives you the solution $c = 2$.

---

268. [1, 2]

Recall the intermediate value theorem: Suppose that f is continuous on the closed interval $[a, b]$, and let N be any number between $f(a)$ and $f(b)$, where $f(a) \neq f(b)$. Then a number c exists in (a, b) such that $f(c) = N$.

Notice that $f(x) = x^2 - \dfrac{3}{2}$ is a polynomial that's continuous everywhere, so the intermediate value theorem applies. Checking the endpoints of the interval $[1, 2]$ gives you

$$f(1) = 1^2 - \frac{3}{2} = -\frac{1}{2}$$

$$f(2) = 2^2 - \frac{3}{2} = 4 - \frac{3}{2} = \frac{5}{2}$$

Because the function changes signs on this interval, there's at least one root in the interval by the intermediate value theorem.

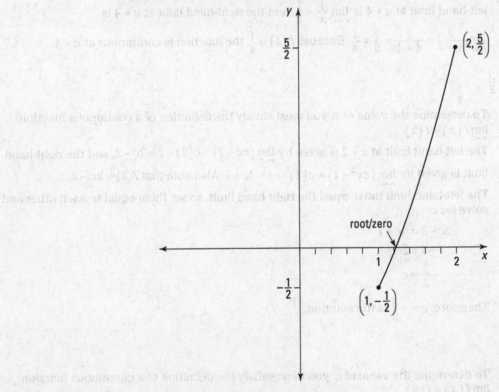

269. [16, 25]

Notice that $f(x) = 3\sqrt{x} - x + 5$ is a continuous function for $x \geq 0$, so the intermediate value theorem applies. Checking the endpoints of the interval [16, 25] gives you

$$f(16) = 3\sqrt{16} - 16 + 5 = 1$$
$$f(25) = 3\sqrt{25} - 25 + 5 = -5$$

Because the function changes signs on this interval, there's at least one root in the interval by the intermediate value theorem.

270. [2, 3]

The function $f(x) = 2(3^x) + x^2 - 4$ is continuous everywhere, so the intermediate value theorem applies. Checking the endpoints of the interval [2, 3] gives you

$$f(2) = 2\left(3^2\right) + 2^2 - 4 = 18$$
$$f(3) = 2\left(3^3\right) + 3^2 - 4 = 59$$

The number 32 is between 18 and 59, so by the intermediate value theorem, there exists at least one point c in the interval [2, 3] such that $f(c) = 32$.

271. [3, 4]

The function $f(x) = 4|2x - 3| + 5$ is continuous everywhere, so the intermediate value theorem applies. Checking the endpoints of the interval [3, 4] gives you

$$f(3) = 4|2(3) - 3| + 5 = 17$$
$$f(4) = 4|2(4) - 3| + 5 = 25$$

The number 22 is between 17 and 25, so by the intermediate value theorem, there exists at least one point c in the interval [3, 4] such that $f(c) = 22$.

272. $x = 1$

The graph has a point of discontinuity at $x = 1$, so the graph isn't differentiable there. However, the rest of the graph is smooth and continuous, so the derivative exists for all other points.

273. differentiable everywhere

Because the graph is continuous and smooth everywhere, the function is differentiable everywhere.

274. $x = -2, x = 0, x = 2$

The graph of the function has sharp corners at $x = -2$, $x = 0$, and $x = 2$, so the function isn't differentiable there.

You can also note that for each of the points $x = -2$, $x = 0$, and $x = 2$, the slopes of the tangent lines jump from 1 to –1 or from –1 to 1. This jump in slopes is another way to recognize values of x where the function is not differentiable.

275. $x = \dfrac{\pi}{2} + \pi n$, where n is an integer

Because $y = \tan x$ has points of discontinuity at $x = \dfrac{\pi}{2} + \pi n$, where n is an integer, the function isn't differentiable at those points.

276. $x = 0$

The tangent line would be vertical at $x = 0$, so the function isn't differentiable there.

277. 2

Use the derivative definition with $f(x) = 2x - 1$ and $f(x + h) = 2(x + h) - 1 = 2x + 2h - 1$:

$$\lim_{h \to 0} \frac{2x + 2h - 1 - (2x - 1)}{h}$$
$$= \lim_{h \to 0} \frac{2x + 2h - 1 - 2x + 1}{h}$$
$$= \lim_{h \to 0} \frac{2h}{h}$$
$$= \lim_{h \to 0} 2$$
$$= 2$$

278. $2x$

Use the derivative definition with $f(x) = x^2$ and $f(x + h) = (x + h)^2 = (x + h)(x + h) = x^2 + 2xh + h^2$:

$$\lim_{h \to 0} \frac{x^2 + 2xh + h^2 - x^2}{h}$$

$$= \lim_{h \to 0} \frac{2xh + h^2}{h}$$

$$= \lim_{h \to 0} \frac{h(2x + h)}{h}$$

$$= \lim_{h \to 0} (2x + h)$$

$$= 2x + 0$$

$$= 2x$$

279. 1

Use the derivative definition with $f(x) = 2 + x$ and $f(x + h) = 2 + (x + h)$:

$$\lim_{h \to 0} \frac{2 + x + h - (2 + x)}{h}$$

$$= \lim_{h \to 0} \frac{h}{h} = 1$$

280. $-\dfrac{1}{x^2}$

Use the derivative definition with $f(x) = \dfrac{1}{x}$ and with $f(x + h) = \dfrac{1}{x + h}$ to get the following:

$$\lim_{h \to 0} \frac{\frac{1}{x + h} - \frac{1}{x}}{h} = \lim_{h \to 0} \frac{\frac{x - (x + h)}{x(x + h)}}{\frac{h}{1}}$$

$$= \lim_{h \to 0} \frac{-h}{hx(x + h)}$$

$$= \lim_{h \to 0} \frac{-1}{x(x + h)}$$

$$= \frac{-1}{x(x + 0)}$$

$$= -\frac{1}{x^2}$$

281. $3x^2 - 2x$

Using the derivative definition with $f(x) = x^3 - x^2$ and with

$$f(x+h) = (x+h)^3 - (x+h)^2$$
$$= x^3 + 3x^2h + 3xh^2 + h^3 - (x^2 + 2xh + h^2)$$

gives you the following:

$$\lim_{h \to 0} \frac{x^3 + 3x^2h + 3xh^2 + h^3 - x^2 - 2xh - h^2 - (x^3 - x^2)}{h}$$

$$= \lim_{h \to 0} \frac{3x^2h + 3xh^2 + h^3 - 2xh - h^2}{h}$$

$$= \lim_{h \to 0} \frac{h(3x^2 + 3xh + h^2 - 2x - h)}{h}$$

$$= \lim_{h \to 0} (3x^2 + 3xh + h^2 - 2x - h)$$

$$= 3x^2 + 0 + 0 - 2x - 0$$

$$= 3x^2 - 2x$$

282. $6x + 4$

Using the derivative definition with $f(x) = 3x^2 + 4x$ and with

$$f(x+h) = 3(x+h)^2 + 4(x+h)$$
$$= 3(x^2 + 2xh + h^2) + 4x + 4h$$
$$= 3x^2 + 6xh + 3h^2 + 4x + 4h$$

gives you the following:

$$\lim_{h \to 0} \frac{3x^2 + 6xh + 3h^2 + 4x + 4h - (3x^2 + 4x)}{h}$$

$$= \lim_{h \to 0} \frac{6xh + 3h^2 + 4h}{h}$$

$$= \lim_{h \to 0} \frac{h(6x + 3h + 4)}{h}$$

$$= \lim_{h \to 0} (6x + 3h + 4)$$

$$= 6x + 3(0) + 4$$

$$= 6x + 4$$

283. $\dfrac{1}{2\sqrt{x}}$

Using the derivative definition with $f(x) = \sqrt{x}$ and with $f(x+h) = \sqrt{x+h}$ gives you the following:

$$\lim_{h \to 0} \frac{\sqrt{x+h} - \sqrt{x}}{h}$$

$$= \lim_{h \to 0} \frac{\left(\sqrt{x+h} - \sqrt{x}\right)\left(\sqrt{x+h} + \sqrt{x}\right)}{h\left(\sqrt{x+h} + \sqrt{x}\right)}$$

$$= \lim_{h \to 0} \frac{x+h-x}{h\left(\sqrt{x+h} + \sqrt{x}\right)}$$

$$= \lim_{h \to 0} \frac{h}{h\left(\sqrt{x+h} + \sqrt{x}\right)}$$

$$= \lim_{h \to 0} \frac{1}{\sqrt{x+h} + \sqrt{x}}$$

$$= \frac{1}{\sqrt{x+0} + \sqrt{x}}$$

$$= \frac{1}{2\sqrt{x}}$$

284. $\dfrac{-5}{2\sqrt{2-5x}}$

Using the derivative definition with $f(x) = \sqrt{2-5x}$ and with

$f(x+h) = \sqrt{2 - 5(x+h)} = \sqrt{2 - 5x - 5h}$ gives you the following:

$$\lim_{h \to 0} \frac{\sqrt{2-5x-5h} - \sqrt{2-5x}}{h}$$

$$= \lim_{h \to 0} \frac{\sqrt{2-5x-5h} - \sqrt{2-5x}}{h} \left(\frac{\sqrt{2-5x-5h} + \sqrt{2-5x}}{\sqrt{2-5x-5h} + \sqrt{2-5x}} \right)$$

$$= \lim_{h \to 0} \frac{(2-5x-5h) - (2-5x)}{h\left(\sqrt{2-5x-5h} + \sqrt{2-5x}\right)}$$

$$= \lim_{h \to 0} \frac{-5h}{h\left(\sqrt{2-5x-5h} + \sqrt{2-5x}\right)}$$

$$= \lim_{h \to 0} \frac{-5}{\left(\sqrt{2-5x-5h} + \sqrt{2-5x}\right)}$$

$$= \frac{-5}{\left(\sqrt{2-5x-0} + \sqrt{2-5x}\right)}$$

$$= \frac{-5}{2\sqrt{2-5x}}$$

285. $\dfrac{-1}{2(x-1)^{3/2}}$

Use the derivative definition with $f(x) = \dfrac{1}{\sqrt{x-1}}$ and with $f(x+h) = \dfrac{1}{\sqrt{x+h-1}}$:

$$\lim_{h \to 0} \frac{\dfrac{1}{\sqrt{x+h-1}} - \dfrac{1}{\sqrt{x-1}}}{h}$$

$$= \lim_{h \to 0} \frac{\dfrac{\sqrt{x-1} - \sqrt{x+h-1}}{\sqrt{x-1}\sqrt{x+h-1}}}{\dfrac{h}{1}}$$

$$= \lim_{h \to 0} \left(\frac{\sqrt{x-1} - \sqrt{x+h-1}}{h\sqrt{x-1}\sqrt{x+h-1}} \right) \left(\frac{\sqrt{x-1} + \sqrt{x+h-1}}{\sqrt{x-1} + \sqrt{x+h-1}} \right)$$

$$= \lim_{h \to 0} \frac{(x-1) - (x+h-1)}{\left(h\sqrt{x-1}\sqrt{x+h-1} \right)\left(\sqrt{x-1} + \sqrt{x+h-1} \right)}$$

$$= \lim_{h \to 0} \frac{-h}{\left(h\sqrt{x-1}\sqrt{x+h-1} \right)\left(\sqrt{x-1} + \sqrt{x+h-1} \right)}$$

$$= \lim_{h \to 0} \frac{-1}{\left(\sqrt{x-1}\sqrt{x+h-1} \right)\left(\sqrt{x-1} + \sqrt{x+h-1} \right)}$$

$$= \frac{-1}{\left(\sqrt{x-1}\sqrt{x+0-1} \right)\left(\sqrt{x-1} + \sqrt{x+0-1} \right)}$$

$$= \frac{-1}{(x-1)(2\sqrt{x-1})}$$

$$= \frac{-1}{2(x-1)^{3/2}}$$

286. $3x^2 + 3$

Use the derivative definition with $f(x) = x^3 + 3x$ and with $f(x+h) = (x+h)^3 + 3(x+h) = x^3 + 3x^2h + 3xh^2 + h^3 + 3x + 3h$:

$$\lim_{h \to 0} \frac{x^3 + 3x^2h + 3xh^2 + h^3 + 3x + 3h - \left(x^3 + 3x \right)}{h}$$

$$= \lim_{h \to 0} \frac{3x^2h + 3xh^2 + h^3 + 3h}{h}$$

$$= \lim_{h \to 0} \frac{h\left(3x^2 + 3xh + h^2 + 3 \right)}{h}$$

$$= \lim_{h \to 0} \left(3x^2 + 3xh + h^2 + 3 \right)$$

$$= 3x^2 + 0 + 0 + 3$$

$$= 3x^2 + 3$$

287. $\dfrac{-2x}{\left(x^2+5\right)^2}$

Using the derivative definition with $f(x)=\dfrac{1}{x^2+5}$ and with

$f(x+h)=\dfrac{1}{(x+h)^2+5}=\dfrac{1}{x^2+2xh+h^2+5}$ gives you the following:

$$\lim_{h\to0}\frac{\dfrac{1}{x^2+2xh+h^2+5}-\dfrac{1}{x^2+5}}{\dfrac{h}{1}}$$

$$=\lim_{h\to0}\frac{\dfrac{\left(x^2+5\right)-\left(x^2+2xh+h^2+5\right)}{\left(x^2+2xh+h^2+5\right)\left(x^2+5\right)}}{\dfrac{h}{1}}$$

$$=\lim_{h\to0}\frac{-2xh-h^2}{h\left(x^2+2xh+h^2+5\right)\left(x^2+5\right)}$$

$$=\lim_{h\to0}\frac{h(-2x-h)}{h\left(x^2+2xh+h^2+5\right)\left(x^2+5\right)}$$

$$=\lim_{h\to0}\frac{-2x-h}{\left(x^2+2xh+h^2+5\right)\left(x^2+5\right)}$$

$$=\frac{-2x}{\left(x^2+0+0+5\right)\left(x^2+5\right)}$$

$$=\frac{-2x}{\left(x^2+5\right)^2}$$

288. $\dfrac{7}{\left(x+4\right)^2}$

Using the derivative definition with $f(x)=\dfrac{2x+1}{x+4}$ and with $f(x+h)=\dfrac{2(x+h)+1}{x+h+4}$ gives you the following:

$$\lim_{h\to0}\frac{\dfrac{2x+2h+1}{x+h+4}-\dfrac{2x+1}{x+4}}{\dfrac{h}{1}}$$

$$=\lim_{h\to0}\frac{\dfrac{(2x+2h+1)(x+4)-(2x+1)(x+h+4)}{(x+4)(x+h+4)}}{\dfrac{h}{1}}$$

$$=\lim_{h\to0}\frac{7h}{h(x+4)(x+h+4)}$$

$$= \lim_{h \to 0} \frac{7}{(x+4)(x+h+4)}$$

$$= \frac{7}{(x+4)(x+0+4)}$$

$$= \frac{7}{(x+4)^2}$$

289. $\dfrac{x^2+6x-1}{(x+3)^2}$

Using the derivative definition with $f(x) = \dfrac{x^2+1}{x+3}$ and with

$f(x+h) = \dfrac{(x+h)^2+1}{x+h+3} = \dfrac{x^2+2xh+h^2+1}{x+h+3}$ gives you the following:

$$\lim_{h \to 0} \frac{\dfrac{x^2+2xh+h^2+1}{x+h+3} - \dfrac{x^2+1}{x+3}}{\dfrac{h}{1}}$$

$$= \lim_{h \to 0} \frac{\dfrac{(x^2+2xh+h^2+1)(x+3)-(x^2+1)(x+h+3)}{(x+h+3)(x+3)}}{\dfrac{h}{1}}$$

$$= \lim_{h \to 0} \frac{hx^2+6xh+h^2x+3h^2-h}{h(x+h+3)(x+3)}$$

$$= \lim_{h \to 0} \frac{h(x^2+6x+hx+3h-1)}{h(x+h+3)(x+3)}$$

$$= \frac{x^2+6x+0+0-1}{(x+0+3)(x+3)}$$

$$= \frac{x^2+6x-1}{(x+3)^2}$$

290. $\dfrac{2}{3(2x+1)^{2/3}}$

Using the derivative definition with $f(x) = \sqrt[3]{2x+1}$ and with

$f(x+h) = \sqrt[3]{2(x+h)+1} = \sqrt[3]{2x+2h+1}$ gives you

$$\lim_{h \to 0} \frac{\sqrt[3]{2x+2h+1} - \sqrt[3]{2x+1}}{h} = \lim_{h \to 0} \frac{(2x+2h+1)^{1/3} - (2x+1)^{1/3}}{h}$$

To rationalize the numerator, consider the formula $a^3 - b^3 = (a-b)(a^2 + ab + b^2)$. If you let $a = (2x + 2h + 1)^{1/3}$ and $b = (2x + 1)^{1/3}$, you have $(a - b)$ in the numerator; that means you can rationalize the numerator by multiplying by $(a^2 + ab + b^2)$:

$$\lim_{h \to 0} \frac{(2x+2h+1)^{1/3} - (2x+1)^{1/3}}{h}$$

$$= \lim_{h \to 0} \frac{\left[(2x+2h+1)^{1/3} - (2x+1)^{1/3}\right]}{h} \frac{\left[(2x+2h+1)^{2/3} + (2x+2h+1)^{1/3}(2x+1)^{1/3} + (2x+1)^{2/3}\right]}{\left[(2x+2h+1)^{2/3} + (2x+2h+1)^{1/3}(2x+1)^{1/3} + (2x+1)^{2/3}\right]}$$

$$= \lim_{h \to 0} \frac{(2x+2h+1) - (2x+1)}{h\left[(2x+2h+1)^{2/3} + (2x+2h+1)^{1/3}(2x+1)^{1/3} + (2x+1)^{2/3}\right]}$$

$$= \lim_{h \to 0} \frac{2h}{h\left[(2x+2h+1)^{2/3} + (2x+2h+1)^{1/3}(2x+1)^{1/3} + (2x+1)^{2/3}\right]}$$

$$= \lim_{h \to 0} \frac{2}{\left[(2x+2h+1)^{2/3} + (2x+2h+1)^{1/3}(2x+1)^{1/3} + (2x+1)^{2/3}\right]}$$

$$= \frac{2}{\left[(2x+2(0)+1)^{2/3} + (2x+2(0)+1)^{1/3}(2x+1)^{1/3} + (2x+1)^{2/3}\right]}$$

$$= \frac{2}{(2x+1)^{2/3} + (2x+1)^{2/3} + (2x+1)^{2/3}}$$

$$= \frac{2}{3(2x+1)^{2/3}}$$

291. 0

The tangent line at $x = 3$ is horizontal, so the slope is zero. Therefore, $f'(3) = 0$.

292. –1

The slope of the tangent line at $x = -1$ is equal to -1, so $f'(-1) = -1$.

293. 1

The slope of the tangent line at $x = -3$ is equal to 1, so $f'(-3) = 1$.

294. 3

The slope of the tangent line at any point on the graph of $y = 3x + 4$ is equal to 3, so $f'(-22\pi^3) = 3$.

295. $f'(1) < f'(-2) < f'(-3)$

The tangent line at $x = -3$ has a positive slope, the slope of the tangent line at $x = -2$ is equal to zero, and the slope of the tangent line at $x = 1$ is negative. Therefore, $f'(1) < f'(-2) < f'(-3)$.

296. $f'(1) < f'(2) < 0.1 < f'(5)$

The slope of the tangent line at $x = 1$ is negative, the slope of the tangent line at $x = 2$ is equal to zero, and the slope of the tangent line at $x = 5$ is clearly larger than 0.1. Therefore, $f'(1) < f'(2) < 0.1 < f'(5)$.

297. 5

Use basic derivative rules to get $f'(x) = 5$.

298. $2x + 3$

Apply the power rule to each term, recalling that the derivative of a constant is zero: $f'(x) = 2x + 3$.

299. $4x + 7$

Begin by multiplying the two factors together:

$$f(x) = (x+4)(2x-1)$$
$$= 2x^2 + 7x - 4$$

Then apply the power rule to get the derivative:

$$f'(x) = 2(2x) + 7$$
$$= 4x + 7$$

300. 0

Because π^3 is constant, $f'(x) = 0$.

301. $\dfrac{\sqrt{5}}{2\sqrt{x}}$

Split up the radical and rewrite the power on the variable using exponential notation:

$$f(x) = \sqrt{5x} = \sqrt{5}\sqrt{x} = \sqrt{5}x^{1/2}$$

Then apply the power rule to find the derivative:

$$f'(x) = \sqrt{5}\left(\frac{1}{2}x^{-1/2}\right) = \frac{\sqrt{5}}{2x^{1/2}} = \frac{\sqrt{5}}{2\sqrt{x}}$$

302. $\dfrac{-3}{x^2} - \dfrac{2}{x^3} + \dfrac{12}{x^4}$

Begin by breaking up the fraction:

$$f(x) = \frac{3x^2 + x - 4}{x^3}$$
$$= \frac{3x^2}{x^3} + \frac{x}{x^3} - \frac{4}{x^3}$$
$$= 3x^{-1} + x^{-2} - 4x^{-3}$$

Then apply the power rule to find the derivative:

$$f'(x) = -3x^{-2} - 2x^{-3} + 12x^{-4}$$

$$= \frac{-3}{x^2} - \frac{2}{x^3} + \frac{12}{x^4}$$

303. $\frac{7}{2}x^{5/2} + \frac{1}{2\sqrt{x}}$

Begin by multiplying the factors together:

$$f(x) = \sqrt{x}\left(x^3 + 1\right)$$

$$= x^{1/2}\left(x^3 + 1\right)$$

$$= x^{7/2} + x^{1/2}$$

Then apply the power rule to find the derivative:

$$f'(x) = \frac{7}{2}x^{5/2} + \frac{1}{2}x^{-1/2}$$

$$= \frac{7}{2}x^{5/2} + \frac{1}{2\sqrt{x}}$$

304. $\frac{-4\sqrt{3}}{x^5} + 2.2x^{0.1}$

Begin by rewriting the function using a negative exponent:

$$f(x) = \frac{\sqrt{3}}{x^4} + 2x^{1.1} = \sqrt{3}x^{-4} + 2x^{1.1}$$

Then apply the power rule to each term to get the derivative:

$$f'(x) = \sqrt{3}\left(-4x^{-5}\right) + 2\left(1.1x^{0.1}\right)$$

$$= \frac{-4\sqrt{3}}{x^5} + 2.2x^{0.1}$$

305. $\frac{-12}{7x^{10/7}} + 8$

Apply the power rule to the first two terms of $f(x) = 4x^{-3/7} + 8x + \sqrt{5}$, recalling that the derivative of a constant is zero:

$$f'(x) = 4\left(-\frac{3}{7}x^{-10/7}\right) + 8$$

$$= \frac{-12}{7x^{10/7}} + 8$$

306. $-\frac{1}{x^2} - \frac{2}{x^3} - 2$

Multiply the factors and rewrite the function using exponential notation:

$$f(x) = \left(\frac{1}{x^3} - \frac{2}{x}\right)\left(x^2 + x\right)$$

$$= \frac{1}{x} + \frac{1}{x^2} - 2x - 2$$

$$= x^{-1} + x^{-2} - 2x - 2$$

Next, apply the power rule to find the derivative:

$$f'(x) = -1x^{-2} - 2x^{-3} - 2$$
$$= -\frac{1}{x^2} - \frac{2}{x^3} - 2$$

307. $-\dfrac{5}{x^6} + \dfrac{2}{x^3} - 20$

Begin by multiplying the factors together:

$$f(x) = \left(x^{-3} + 4\right)\left(x^{-2} - 5x\right)$$
$$= x^{-5} - 5x^{-2} + 4x^{-2} - 20x$$
$$= x^{-5} - x^{-2} - 20x$$

Then find the derivative using the power rule:

$$f'(x) = -5x^{-6} + 2x^{-3} - 20$$
$$= -\frac{5}{x^6} + \frac{2}{x^3} - 20$$

308. $16x^3 - 2x + 8$

Apply the power rule to each term, recalling that the derivative of a constant is zero:
$f'(x) = 16x^3 - 2x + 8$.

309. $\dfrac{1}{2x^{1/2}} + \dfrac{1}{4x^{5/4}}$

Rewrite the function using exponential notation:

$$f(x) = \sqrt{x} - \frac{1}{\sqrt[4]{x}}$$
$$= x^{1/2} - \frac{1}{x^{1/4}}$$
$$= x^{1/2} - x^{-1/4}$$

Then apply the power rule to each term to find the derivative:

$$f'(x) = \frac{1}{2}x^{-1/2} - \left(-\frac{1}{4}x^{-5/4}\right)$$
$$= \frac{1}{2x^{1/2}} + \frac{1}{4x^{5/4}}$$

310. $-\dfrac{1}{3}, 1$

Begin by finding the derivative of the function $f(x) = x^3 - x^2 - x + 1$:

$$f'(x) = 3x^2 - 2x - 1$$

A horizontal tangent line has a slope of zero, so set the derivative equal to zero, factor, and solve for x:

$$3x^2 - 2x - 1 = 0$$
$$(3x + 1)(x - 1) = 0$$

Setting each factor equal to zero gives you $3x + 1 = 0$, which has the solution $x = -\frac{1}{3}$, and gives you $x - 1 = 0$, which has the solution $x = 1$.

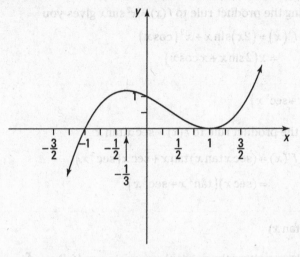

311.

$\pm\dfrac{1}{3\sqrt{2}}$

Begin by finding the derivative of the function:

$$f'(x) = 18x^2 + 5$$

Next, set the derivative equal to 6 and solve for x:

$$18x^2 + 5 = 6$$
$$x^2 = \frac{1}{18}$$
$$x = \pm\sqrt{\frac{1}{18}}$$
$$x = \pm\frac{1}{3\sqrt{2}}$$

312.

$16x^7 + 5x^4 - 8x^3 - 1$

Recall that the product rule states

$$\frac{d}{dx}\left[f(x)g(x)\right] = f'(x)g(x) + f(x)g'(x)$$

You can multiply out the expression $f(x) = (2x^3 + 1)(x^5 - 1)$ first and then avoid using the product rule, but here's how to find the derivative using the product rule:

$$f'(x) = \left(6x^2\right)\left(x^5 - x\right) + \left(2x^3 + 1\right)\left(5x^4 - 1\right)$$
$$= 6x^7 - 6x^3 + 10x^7 - 2x^3 + 5x^4 - 1$$
$$= 16x^7 + 5x^4 - 8x^3 - 1$$

313.

$x(2\sin x + x\cos x)$

Applying the product rule to $f(x) = x^2 \sin x$ gives you

$$f'(x) = (2x)\sin x + x^2 (\cos x)$$
$$= x(2\sin x + x\cos x)$$

314.

$(\sec x)(\tan^2 x + \sec^2 x)$

Apply the product rule to $f(x) = \sec x \tan x$:

$$f'(x) = (\sec x \tan x)\tan x + \sec x(\sec^2 x)$$
$$= (\sec x)(\tan^2 x + \sec^2 x)$$

315.

$(\sec x)(1 + x\tan x)$

Begin by rewriting the original expression as $f(x) = \dfrac{x}{\cos x} = x\sec x$ and then apply the product rule:

$$f'(x) = (1)\sec x + x\sec x \tan x$$
$$= (\sec x)(1 + x\tan x)$$

316.

$4(\csc x)(1 - x\cot x)$

Apply the product rule to $f(x) = 4x\csc x$ to get

$$f'(x) = 4(\csc x) + 4x(-\csc x \cot x)$$
$$= 4(\csc x)(1 - x\cot x)$$

317.

12

According to the product rule,

$$(fg)'(x) = f'(x)g(x) + f(x)g'(x)$$

To find $(fg)'(4)$, enter the numbers and solve:

$$(fg)'(4) = f'(4)g(4) + f(4)g'(4)$$
$$= (2)(-6) + (3)(8)$$
$$= 12$$

318.

$-\dfrac{3g(x)}{x^4} + \dfrac{g'(x)}{x^3}$

Recall that the product rule states

$$\frac{d}{dx}[f(x)g(x)] = f'(x)g(x) + f(x)g'(x)$$

You can apply the quotient rule directly, or you can rewrite the original function as $f(x) = \dfrac{g(x)}{x^3} = x^{-3}[g(x)]$ and then apply the product rule to get the following:

$$f'(x) = (-3x^{-4})[g(x)] + x^{-3}[g'(x)]$$

$$= -\frac{3g(x)}{x^4} + \frac{g'(x)}{x^3}$$

319.

$x \sec x \tan x + 2 \sec^3 x$

Apply the product rule to $f(x) = \sec x(x + \tan x)$ as follows:

$$f'(x) = \sec x \tan x(x + \tan x) + \sec x(1 + \sec^2 x)$$

$$= x \sec x \tan x + \sec x \tan^2 x + \sec x + \sec^3 x$$

$$= x \sec x \tan x + (\tan^2 x + 1)\sec x + \sec^3 x$$

$$= x \sec x \tan x + \sec^3 x + \sec^3 x$$

$$= x \sec x \tan x + 2 \sec^3 x$$

320.

$(2x+1)\csc x - (x^2 + x)(\csc x \cot x)$

Apply the product rule to $f(x) = (x^2 + x)\csc x$ to get

$$f'(x) = (2x+1)\csc x + (x^2 + x)(-\csc x \cot x)$$

$$= (2x+1)\csc x - (x^2 + x)(\csc x \cot x)$$

321.

$4x^2 (\sec x)(3 + x \tan x)$

Applying the product rule to $f(x) = 4x^3 \sec x$ gives you

$$f'(x) = (12x^2)\sec x + 4x^3(\sec x \tan x)$$

$$= 4x^2(\sec x)(3 + x \tan x)$$

322.

$-\dfrac{\cot x}{2x^{3/2}} - \dfrac{\csc^2 x}{x^{1/2}}$

You can apply the quotient rule directly, or you can rewrite the original function as $f(x) = \dfrac{\cot x}{x^{1/2}} = x^{-1/2} \cot x$ and then apply the product rule as follows:

$$f'(x) = -\frac{1}{2}x^{-3/2}\cot x + x^{-1/2}(-\csc^2 x)$$

$$= -\frac{\cot x}{2x^{3/2}} - \frac{\csc^2 x}{x^{1/2}}$$

323. $x\left[2g(x)+xg'(x)\right]$

Using the product rule on $f(x) = x^2g(x)$ and then factoring gives you the derivative as follows:

$$f'(x) = 2xg(x) + x^2\left[g'(x)\right]$$
$$= x\left[2g(x) + xg'(x)\right]$$

324. $g(x)+xg'(x)-\dfrac{1}{x^2}$

Begin by breaking up the fraction and simplifying:

$$f(x) = \frac{1+x^2g(x)}{x}$$
$$= \frac{1}{x} + \frac{x^2g(x)}{x}$$
$$= x^{-1} + xg(x)$$

Next, apply the power rule to the first term and the product rule to the second term:

$$f'(x) = -1x^{-2} + \left[1g(x) + xg'(x)\right]$$
$$= g(x) + xg'(x) - \frac{1}{x^2}$$

325. -46

The product rule tells you that

$$(fg)'(x) = f'(x)g(x) + f(x)g'(x)$$

To find $(fg)'(3)$, enter the numbers and solve:

$$(fg)'(3) = f'(3)g(3) + f(3)g'(3)$$
$$= (4)(-8) + (-2)(7)$$
$$= -46$$

326. $2x\cos x\sin x - x^2\sin^2 x + x^2\cos^2 x$

Recall that the product rule states

$$\frac{d}{dx}\left[f(x)g(x)\right] = f'(x)g(x) + f(x)g'(x)$$

You can group the factors however you want and then apply the product rule within the product rule. If you group together the trigonometric functions and apply the product rule, you have $f(x) = x^2(\cos x\sin x)$ so that

$$f'(x) = (2x)(\cos x\sin x) + x^2\left((-\sin x)(\sin x) + (\cos x)(\cos x)\right)$$
$$= 2x\cos x\sin x - x^2\sin^2 x + x^2\cos^2 x$$

327. $-\dfrac{1}{x^{3/2}}+\dfrac{g(x)}{2x^{1/2}}+x^{1/2}g'(x)$

Begin by simplifying the given expression:

$$f(x)=\dfrac{2+xg(x)}{\sqrt{x}}$$

$$=2x^{-1/2}+x^{1/2}g(x)$$

Then apply the product rule to the second term and the power rule to the first term:

$$f'(x)=2\left(-\dfrac{1}{2}x^{-3/2}\right)+\dfrac{1}{2}x^{-1/2}g(x)+x^{1/2}g'(x)$$

$$=-\dfrac{1}{x^{3/2}}+\dfrac{g(x)}{2x^{1/2}}+x^{1/2}g'(x)$$

328. $\left(\dfrac{-2}{x^{3}}+\dfrac{2}{x^{2}}\right)\tan x+\left(\dfrac{1}{x^{2}}-\dfrac{2}{x}\right)\sec^{2}x$

Rewrite the original expression:

$$f(x)=\left(\dfrac{1}{x^{2}}-\dfrac{2}{x}\right)(\tan x)=\left(x^{-2}-2x^{-1}\right)(\tan x)$$

Then apply the product rule to get the derivative:

$$f'(x)=\left(-2x^{-3}+2x^{-2}\right)\tan x+\left(x^{-2}-2x^{-1}\right)\left(\sec^{2}x\right)$$

$$=\left(\dfrac{-2}{x^{3}}+\dfrac{2}{x^{2}}\right)\tan x+\left(\dfrac{1}{x^{2}}-\dfrac{2}{x}\right)\sec^{2}x$$

329. $-\dfrac{1}{x^{2}}+\dfrac{4\cos x}{3x^{5/3}}+\dfrac{2\sin x}{x^{2/3}}$

First simplify the given function:

$$f(x)=\dfrac{x-2x\sqrt[3]{x}\cos x}{x^{2}}$$

$$=\dfrac{x}{x^{2}}-\dfrac{2x\sqrt[3]{x}\cos x}{x^{2}}$$

$$=x^{-1}-2x^{-2/3}\cos x$$

Then apply the product rule to find the derivative:

$$f'(x)=-x^{-2}-\left(-\dfrac{4}{3}x^{-5/3}\cos x+2x^{-2/3}(-\sin x)\right)$$

$$=-\dfrac{1}{x^{2}}+\dfrac{4\cos x}{3x^{5/3}}+\dfrac{2\sin x}{x^{2/3}}$$

330.

$$\frac{-20g(x)}{x^6}+\frac{4g'(x)}{x^5}$$

You can use the quotient rule directly, or you can rewrite the original expression as $f(x)=\frac{4g(x)}{x^5}=4x^{-5}g(x)$ and then apply the product rule as follows:

$$f'(x)=-20x^{-6}g(x)+4x^{-5}g'(x)$$

$$=\frac{-20g(x)}{x^6}+\frac{4g'(x)}{x^5}$$

331.

$$(\sin x)g'(x)h(x)+(\sin x)g(x)h'(x)+(\cos x)g(x)h(x)$$

To find the derivative of $f(x)=[g(x)h(x)]\sin x$, use the product rule within the product rule:

$$f'(x)=[g'(x)h(x)+g(x)h'(x)]\sin x+[g(x)h(x)]\cos x$$

$$=(\sin x)g'(x)h(x)+(\sin x)g(x)h'(x)+(\cos x)g(x)h(x)$$

332.

$$\frac{5}{(3x+4)^2}$$

Recall that the quotient rule states

$$\frac{d}{dx}\left(\frac{f(x)}{g(x)}\right)=\frac{g(x)f'(x)-f(x)g'(x)}{[g(x)]^2}$$

Apply the quotient rule to $f(x)=\frac{2x+1}{3x+4}$:

$$f'(x)=\frac{(3x+4)(2)-(2x+1)(3)}{(3x+4)^2}$$

$$=\frac{5}{(3x+4)^2}$$

333.

$$\frac{10-2x^2}{\left(5+x^2\right)^2}$$

Apply the quotient rule to $f(x)=\frac{2x}{5+x^2}$ to get the derivative:

$$f'(x)=\frac{\left(5+x^2\right)(2)-(2x)(2x)}{\left(5+x^2\right)^2}$$

$$=\frac{10-2x^2}{\left(5+x^2\right)^2}$$

334.

$$\frac{\cos x-\sin x+1}{(\cos x+1)^2}$$

Apply the quotient rule to $f(x)=\frac{\sin x-1}{\cos x+1}$:

$$f'(x) = \frac{(\cos x + 1)(\cos x) - (\sin x - 1)(-\sin x)}{(\cos x + 1)^2}$$

$$= \frac{\cos^2 x + \cos x + \sin^2 x - \sin x}{(\cos x + 1)^2}$$

$$= \frac{\cos x - \sin x + 1}{(\cos x + 1)^2}$$

335. $\dfrac{-4x^5 - 2x^3 + 1}{\left(x^5 + x^3 + 1\right)^2}$

Apply the quotient rule to $f(x) = \dfrac{x}{x^5 + x^3 + 1}$ to get the following:

$$f'(x) = \frac{\left(x^5 + x^3 + 1\right)(1) - x\left(5x^4 + 3x^2\right)}{\left(x^5 + x^3 + 1\right)^2}$$

$$= \frac{-4x^5 - 2x^3 + 1}{\left(x^5 + x^3 + 1\right)^2}$$

336. $-\dfrac{19}{16}$

The quotient rule gives you

$$\left(\frac{f}{g}\right)'(x) = \frac{g(x)f'(x) - f(x)g'(x)}{[g(x)]^2}$$

To find $\left(\dfrac{f}{g}\right)'(4)$, enter the numbers and solve:

$$\left(\frac{f}{g}\right)'(4) = \frac{g(4)f'(4) - f(4)g'(4)}{[g(4)]^2}$$

$$= \frac{(8)(-7) - (5)(4)}{8^2}$$

$$= -\frac{19}{16}$$

337. $\dfrac{-1}{(3x+5)^2}$

Recall that the quotient rule states

$$\frac{d}{dx}\left(\frac{f(x)}{g(x)}\right) = \frac{g(x)f'(x) - f(x)g'(x)}{[g(x)]^2}$$

Apply the quotient rule to $f(x) = \dfrac{x+2}{3x+5}$:

$$f'(x) = \frac{(3x+5)(1) - (x+2)(3)}{(3x+5)^2}$$

$$= \frac{-1}{(3x+5)^2}$$

338. $\dfrac{-2x^2 + 2x}{\left(3x^2 - 2x + 1\right)^2}$

Apply the quotient rule to $f(x) = \dfrac{x^2}{3x^2 - 2x + 1}$ to find the derivative:

$$f'(x) = \frac{\left(3x^2 - 2x + 1\right)2x - x^2(6x - 2)}{\left(3x^2 - 2x + 1\right)^2}$$

$$= \frac{-2x^2 + 2x}{\left(3x^2 - 2x + 1\right)^2}$$

339. $\dfrac{x\cos x - 3\sin x}{x^4}$

Apply the quotient rule to $f(x) = \dfrac{\sin x}{x^3}$ to get the derivative:

$$f'(x) = \frac{x^3 \cos x - \sin x\left(3x^2\right)}{\left(x^3\right)^2}$$

$$= \frac{x^2(x\cos x - 3\sin x)}{x^6}$$

$$= \frac{x\cos x - 3\sin x}{x^4}$$

340. $\dfrac{-3x^2 - 2x - 7}{\left(x^2 - 2x - 3\right)^2}$

First, simplify the numerator and denominator:

$$f(x) = \frac{(x-1)(x+2)}{(x-3)(x+1)} = \frac{x^2 + x - 2}{x^2 - 2x - 3}$$

Then apply the quotient rule to get

$$f'(x) = \frac{\left(x^2 - 2x - 3\right)(2x + 1) - \left(x^2 + x - 2\right)(2x - 2)}{\left(x^2 - 2x - 3\right)^2}$$

$$= \frac{-3x^2 - 2x - 7}{\left(x^2 - 2x - 3\right)^2}$$

341. $\dfrac{\sec x \tan x}{\left(1 + \sec x\right)^2}$

Apply the quotient rule to $f(x) = \dfrac{\sec x}{1 + \sec x}$ to get the following:

$$f'(x) = \frac{(1 + \sec x)(\sec x \tan x) - \sec x(\sec x \tan x)}{(1 + \sec x)^2}$$

$$= \frac{\sec x \tan x}{(1 + \sec x)^2}$$

342.

$$\frac{8x\sin x+8x\cos x-4x^2\cos x+4x^2\sin x}{\left(\sin x+\cos x\right)^2}$$

Apply the quotient rule to $f(x)=\dfrac{4x^2}{\sin x+\cos x}$:

$$f'(x)=\frac{(\sin x+\cos x)(8x)-\left(4x^2\right)(\cos x-\sin x)}{(\sin x+\cos x)^2}$$

$$=\frac{8x\sin x+8x\cos x-4x^2\cos x+4x^2\sin x}{(\sin x+\cos x)^2}$$

343.

$$-\frac{38}{49}$$

The quotient rule says that

$$\left(\frac{f}{g}\right)'(x)=\frac{g(x)f'(x)-f(x)g'(x)}{\left[g(x)\right]^2}$$

To find $\left(\dfrac{f}{g}\right)'(5)$, enter the numbers and solve:

$$\left(\frac{f}{g}\right)'(5)=\frac{g(5)f'(5)-f(5)g'(5)}{\left[g(5)\right]^2}$$

$$=\frac{(-7)(2)-(-4)(-6)}{(-7)^2}$$

$$=-\frac{38}{49}$$

344.

$$\frac{4x+3x^{3/2}}{2\left(1+x^{1/2}\right)^2}$$

Recall that the quotient rule states

$$\frac{d}{dx}\left(\frac{f(x)}{g(x)}\right)=\frac{g(x)f'(x)-f(x)g'(x)}{\left[g(x)\right]^2}$$

Apply the quotient rule to $f(x)=\dfrac{x^2}{1+\sqrt{x}}$ to get

$$f'(x)=\frac{\left(1+x^{1/2}\right)(2x)-x^2\left(\frac{1}{2}x^{-1/2}\right)}{\left(1+x^{1/2}\right)^2}$$

$$=\frac{2x+2x^{3/2}-\frac{1}{2}x^{3/2}}{\left(1+x^{1/2}\right)^2}$$

$$=\frac{2x+\frac{3}{2}x^{3/2}}{\left(1+x^{1/2}\right)^2}$$

$$=\frac{4x+3x^{3/2}}{2\left(1+x^{1/2}\right)^2}$$

Answers
301–400

345. $\dfrac{1}{\left(\cos x+\sin x\right)^{2}}$

Apply the quotient rule to $f(x)=\dfrac{\sin x}{\cos x+\sin x}$ to get the following:

$$f'(x)=\frac{(\cos x+\sin x)(\cos x)-\sin x(-\sin x+\cos x)}{(\cos x+\sin x)^{2}}$$

$$=\frac{\cos^{2}x+\cos x\sin x+\sin^{2}x-\cos x\sin x}{(\cos x+\sin x)^{2}}$$

$$=\frac{1}{(\cos x+\sin x)^{2}}$$

346. $\dfrac{6x+5x^{4/3}}{3\left(1+x^{1/3}\right)^{2}}$

Apply the quotient rule to $f(x)=\dfrac{x^{2}}{1+\sqrt[3]{x}}$:

$$f'(x)=\frac{\left(1+x^{1/3}\right)(2x)-x^{2}\left(\dfrac{1}{3}x^{-2/3}\right)}{\left(1+x^{1/3}\right)^{2}}$$

$$=\frac{2x+\dfrac{5}{3}x^{4/3}}{\left(1+x^{1/3}\right)^{2}}$$

$$=\frac{6x+5x^{4/3}}{3\left(1+x^{1/3}\right)^{2}}$$

347. $\dfrac{-1}{2\sqrt{x}\left(\sqrt{x}+1\right)^{2}}$

Apply the quotient rule to $f(x)=\dfrac{\sqrt{x}+2}{\sqrt{x}+1}$ to get

$$f'(x)=\frac{\left(\sqrt{x}+1\right)\left(\dfrac{1}{2}x^{-1/2}\right)-\left(\sqrt{x}+2\right)\left(\dfrac{1}{2}x^{-1/2}\right)}{\left(\sqrt{x}+1\right)^{2}}$$

$$=\frac{-\dfrac{1}{2}x^{-1/2}}{\left(\sqrt{x}+1\right)^{2}}$$

$$=\frac{-1}{2\sqrt{x}\left(\sqrt{x}+1\right)^{2}}$$

348.

$$\frac{(\sec x)(1+\sec x+\tan x)}{(\sec x+1)^2}$$

Apply the quotient rule to $f(x) = \dfrac{\tan x - 1}{\sec x + 1}$ as follows:

$$f'(x) = \frac{(\sec x+1)\sec^2 x - (\tan x - 1)(\sec x \tan x)}{(\sec x + 1)^2}$$

$$= \frac{\sec^3 x + \sec^2 x - \tan^2 x \sec x + \sec x \tan x}{(\sec x + 1)^2}$$

$$= \frac{(\sec x)(\sec^2 x + \sec x - \tan^2 x + \tan x)}{(\sec x + 1)^2}$$

$$= \frac{(\sec x)(1 + \sec x + \tan x)}{(\sec x + 1)^2}$$

Note that the identity $\sec^2 x - \tan^2 x = 1$ was used to simplify the final expression.

349.

$$\frac{8x}{\left(x^2 + 4\right)^2}$$

First multiply the numerator and denominator by x:

$$f(x) = \frac{x}{x + \dfrac{4}{x}}$$

$$= \frac{x^2}{x^2 + 4}$$

Then use the quotient rule to find the derivative:

$$f'(x) = \frac{\left(x^2 + 4\right)(2x) - x^2(2x)}{\left(x^2 + 4\right)^2}$$

$$= \frac{8x}{\left(x^2 + 4\right)^2}$$

350.

$$\frac{xg'(x) - 3g(x)}{x^4}$$

Applying the quotient rule to $f(x) = \dfrac{g(x)}{x^3}$, followed by factoring and simplifying, gives you the following:

$$f'(x) = \frac{x^3 g'(x) - g(x)\left[3x^2\right]}{\left(x^3\right)^2}$$

$$= \frac{x^2\left[xg'(x) - 3g(x)\right]}{x^6}$$

$$= \frac{xg'(x) - 3g(x)}{x^4}$$

351. $-\dfrac{g(x)\sin x + g'(x)\cos x}{\left[g(x)\right]^2}$

Applying the quotient rule to $f(x) = \dfrac{\cos x}{g(x)}$ gives you

$$f'(x) = \frac{g(x)[-\sin x]-(\cos x)g'(x)}{\left[g(x)\right]^2}$$

$$= -\frac{g(x)\sin x + g'(x)\cos x}{\left[g(x)\right]^2}$$

352. $100(2x+3)\left(x^2+3x\right)^{99}$

Recall that the chain rule states

$$\frac{d}{dx}f(g(x)) = f'(g(x))g'(x)$$

Applying the chain rule to $f(x) = \left(x^2+3x\right)^{100}$ gives you

$$f'(x) = 100\left(x^2+3x\right)^{99}(2x+3)$$

$$= 100(2x+3)\left(x^2+3x\right)^{99}$$

353. $4\cos(4x)$

Apply the chain rule to $f(x) = \sin(4x)$:

$$f'(x) = \left[\cos(4x)\right](4)$$

$$= 4\cos(4x)$$

354. $\dfrac{\sec x \tan x}{3(1+\sec x)^{2/3}}$

Rewrite the function using exponential notation:

$$f(x) = \sqrt[3]{1+\sec x} = (1+\sec x)^{1/3}$$

Then apply the chain rule:

$$f'(x) = \frac{1}{3}(1+\sec x)^{-2/3}(\sec x \tan x)$$

$$= \frac{\sec x \tan x}{3(1+\sec x)^{2/3}}$$

355. $\dfrac{-5(2x-1)}{\left(x^2-x\right)^6}$

Rewrite the function as

$$f(x) = \frac{1}{\left(x^2-x\right)^5} = \left(x^2-x\right)^{-5}$$

Then apply the chain rule to get the derivative:

$$f'(x) = -5\left(x^2 - x\right)^{-6}(2x-1)$$

$$= \frac{-5(2x-1)}{\left(x^2 - x\right)^6}$$

---

356. $\dfrac{2\csc\left(\dfrac{1}{x^2}\right)\cot\left(\dfrac{1}{x^2}\right)}{x^3}$

First rewrite the function:

$$f(x) = \csc\left(\frac{1}{x^2}\right) = \csc\left(x^{-2}\right)$$

Then apply the chain rule:

$$f'(x) = \left(-\csc\left(x^{-2}\right)\cot\left(x^{-2}\right)\right)\left(-2x^{-3}\right)$$

$$= \frac{2\csc\left(\dfrac{1}{x^2}\right)\cot\left(\dfrac{1}{x^2}\right)}{x^3}$$

---

357. $5\left(1 - 2\cos x \sin x\right)\left(x + \cos^2 x\right)^4$

Apply the chain rule to $f(x) = \left(x + \cos^2 x\right)^5$ to get

$$f'(x) = 5\left(x + \cos^2 x\right)^4\left[1 + 2\cos x(-\sin x)\right]$$

$$= 5(1 - 2\cos x \sin x)\left(x + \cos^2 x\right)^4$$

---

358. $\dfrac{\pi}{\left(\cos(\pi x) + \sin(\pi x)\right)^2}$

Recall that the chain rule states

$$\frac{d}{dx}f(g(x)) = f'(g(x))g'(x)$$

and that the quotient rule states

$$\frac{d}{dx}\left(\frac{f(x)}{g(x)}\right) = \frac{g(x)f'(x) - f(x)g'(x)}{[g(x)]^2}$$

To find the derivative of $f(x) = \dfrac{\sin(\pi x)}{\cos(\pi x) + \sin(\pi x)}$, apply the quotient rule along with the chain rule:

$$f'(x) = \frac{(\cos(\pi x) + \sin(\pi x))(\cos(\pi x))\pi - \sin(\pi x)((-\sin \pi x)\pi + (\cos(\pi x))\pi)}{\left(\cos(\pi x) + \sin(\pi x)\right)^2}$$

$$= \frac{\pi\left(\cos^2(\pi x) + \sin^2(\pi x)\right)}{\left(\cos(\pi x) + \sin(\pi x)\right)^2}$$

$$= \frac{\pi}{\left(\cos(\pi x) + \sin(\pi x)\right)^2}$$

359. $-(\sin x + x\cos x)\sin(x\sin x)$

Recall that the chain rule states

$$\frac{d}{dx}f(g(x)) = f'(g(x))g'(x)$$

and that the product rule states

$$\frac{d}{dx}[f(x)g(x)] = f'(x)g(x) + f(x)g'(x)$$

To find the derivative of $f(x) = \cos(x\sin x)$, apply the chain rule and the product rule:

$$f'(x) = [-\sin(x\sin x)][1\sin x + x\cos x]$$
$$= -1(\sin x + x\cos x)\sin(x\sin x)$$

360. $\sin\sqrt[3]{x} + \frac{1}{3}\sqrt[3]{x}\cos\sqrt[3]{x}$

Rewrite the function using exponential notation:

$$f(x) = x\sin\sqrt[3]{x}$$
$$= x\sin\left(x^{1/3}\right)$$

Now apply the product rule, being careful to use the chain rule when taking the derivative of the second factor:

$$f'(x) = 1\sin\left(x^{1/3}\right) + x\left(\cos\left(x^{1/3}\right)\right)\left(\frac{1}{3}x^{-2/3}\right)$$
$$= \sin\left(x^{1/3}\right) + \left(\frac{1}{3}x^{1/3}\right)\left(\cos\left(x^{1/3}\right)\right)$$
$$= \sin\sqrt[3]{x} + \frac{1}{3}\sqrt[3]{x}\cos\sqrt[3]{x}$$

361. $\dfrac{3-x}{(3-2x)^{3/2}}$

Recall that the chain rule states

$$\frac{d}{dx}f(g(x)) = f'(g(x))g'(x)$$

and that the quotient rule states

$$\frac{d}{dx}\left(\frac{f(x)}{g(x)}\right) = \frac{g(x)f'(x) - f(x)g'(x)}{[g(x)]^2}$$

Rewrite the function using exponential notation:

$$f(x) = \frac{x}{\sqrt{3-2x}}$$
$$= \frac{x}{(3-2x)^{1/2}}$$

Next, apply the quotient rule, making sure to use the chain rule when taking the derivative of the denominator:

$$f'(x) = \frac{(3-2x)^{1/2}(1) - x\left(\frac{1}{2}(3-2x)^{-1/2}(-2)\right)}{\left((3-2x)^{1/2}\right)^2}$$

$$= \frac{(3-2x)^{-1/2}\left((3-2x)+x\right)}{(3-2x)}$$

$$= \frac{3-x}{(3-2x)^{3/2}}$$

362. $4\sec^2 x \tan x$

First rewrite the function:

$$f(x) = \sec^2 x + \tan^2 x = (\sec x)^2 + (\tan x)^2$$

Applying the chain rule gives you the derivative:

$$f'(x) = 2(\sec x)\sec x \tan x + 2(\tan x)\sec^2 x$$
$$= 4\sec^2 x \tan x$$

363. $\dfrac{1}{\left(1+x^2\right)^{3/2}}$

Recall that the chain rule states

$$\frac{d}{dx}f(g(x)) = f'(g(x))g'(x)$$

and that the quotient rule states

$$\frac{d}{dx}\left(\frac{f(x)}{g(x)}\right) = \frac{g(x)f'(x) - f(x)g'(x)}{[g(x)]^2}$$

Rewrite the function using exponential notation:

$$f(x) = \frac{x}{\sqrt{1+x^2}}$$

$$= \frac{x}{\left(1+x^2\right)^{1/2}}$$

Apply the quotient rule and the chain rule to get the derivative:

$$f'(x) = \frac{\left(1+x^2\right)^{1/2}(1) - x\left(\frac{1}{2}\left(1+x^2\right)^{-1/2}(2x)\right)}{\left(\left(1+x^2\right)^{1/2}\right)^2}$$

$$= \frac{\left(1+x^2\right)^{-1/2}\left(\left(1+x^2\right) - x^2\right)}{\left(1+x^2\right)}$$

$$= \frac{1}{\left(1+x^2\right)^{3/2}}$$

364. $\dfrac{x\left(11x^3+3x+40\right)}{4\left(x^3+5\right)^{3/4}}$

Recall that the chain rule states

$$\frac{d}{dx}f\big(g(x)\big)=f'\big(g(x)\big)g'(x)$$

and that the product rule states

$$\frac{d}{dx}\big[f(x)g(x)\big]=f'(x)g(x)+f(x)g'(x)$$

Rewrite the function using exponential notation:

$$f(x)=\left(x^2+1\right)\sqrt[4]{x^3+5}$$

$$=\left(x^2+1\right)\left(x^3+5\right)^{1/4}$$

Then apply the product rule and also use the chain rule on the second factor to get the derivative:

$$f'(x)=(2x)\left(x^3+5\right)^{1/4}+\left(x^2+1\right)\left(\tfrac{1}{4}\left(x^3+5\right)^{-3/4}\left(3x^2\right)\right)$$

$$=\left(x^3+5\right)^{-3/4}\left(2x\left(x^3+5\right)+\tfrac{3}{4}x^2\left(x^2+1\right)\right)$$

$$=\frac{\tfrac{11}{4}x^4+\tfrac{3}{4}x^2+10x}{\left(x^3+5\right)^{3/4}}$$

$$=\frac{x\left(11x^3+3x+40\right)}{4\left(x^3+5\right)^{3/4}}$$

365. $6x^3\left(x^4-1\right)^2\left(x^5+1\right)^5\left(7x^5-5x+2\right)$

To find the derivative of $f(x)=\left(x^4-1\right)^3\left(x^5+1\right)^6$, use the product rule along with the chain rule and then factor:

$$f'(x)=\left(3\left(x^4-1\right)^2\left(4x^3\right)\right)\left(x^5+1\right)^6+\left(x^4-1\right)^3\left(6\left(x^5+1\right)^5\left(5x^4\right)\right)$$

$$=6x^3\left(x^4-1\right)^2\left(x^5+1\right)^5\left(2\left(x^5+1\right)+5x\left(x^4-1\right)\right)$$

$$=6x^3\left(x^4-1\right)^2\left(x^5+1\right)^5\left(7x^5-5x+2\right)$$

366.
$$\dfrac{(x-1)^2\left(-7x^2+8x+5\right)}{\left(x^2+x\right)^6}$$

Recall that the chain rule states

$$\dfrac{d}{dx}f\left(g(x)\right)=f'\left(g(x)\right)g'(x)$$

and that the quotient rule states

$$\dfrac{d}{dx}\left(\dfrac{f(x)}{g(x)}\right)=\dfrac{g(x)f'(x)-f(x)g'(x)}{\left[g(x)\right]^2}$$

Applying the quotient rule and the chain rule to $f(x)=\dfrac{(x-1)^3}{\left(x^2+x\right)^5}$ gives you the following derivative:

$$f'(x)=\dfrac{\left(x^2+x\right)^5\left(3(x-1)^2\right)-(x-1)^3\left(5\left(x^2+x\right)^4(2x+1)\right)}{\left(\left(x^2+x\right)^5\right)^2}$$

$$=\dfrac{(x-1)^2\left(x^2+x\right)^4\left(3\left(x^2+x\right)-5(x-1)(2x+1)\right)}{\left(x^2+x\right)^{10}}$$

$$=\dfrac{(x-1)^2\left(-7x^2+8x+5\right)}{\left(x^2+x\right)^6}$$

367.
$$\dfrac{-1}{(x+1)^{1/2}(x-1)^{3/2}}$$

First rewrite the function using exponential notation:

$$f(x)=\sqrt{\dfrac{x+1}{x-1}}=\left(\dfrac{x+1}{x-1}\right)^{1/2}$$

Then apply both the chain rule and the quotient rule:

$$f'(x)=\dfrac{1}{2}\left(\dfrac{x+1}{x-1}\right)^{-1/2}\left[\dfrac{(x-1)(1)-(x+1)(1)}{(x-1)^2}\right]$$

$$=\dfrac{1}{2}\dfrac{(x+1)^{-1/2}}{(x-1)^{-1/2}}\left[\dfrac{-2}{(x-1)^2}\right]$$

$$=\dfrac{(x-1)^{1/2}}{(x+1)^{1/2}}\left(\dfrac{-1}{(x-1)^2}\right)$$

$$=\dfrac{-1}{(x+1)^{1/2}(x-1)^{3/2}}$$

368. $2x\Big(\cos\big(x^2\big)\Big)\Big(\cos\big(\sin\big(x^2\big)\big)\Big)\Big(\cos\big(\sin\big(\sin\big(x^2\big)\big)\big)\Big)$

To find the derivative of $f(x) = \sin\Big(\sin\big(\sin\big(x^2\big)\big)\Big)$, apply the chain rule repeatedly:

$$f'(x) = \Big(\cos\big(\sin\big(\sin\big(x^2\big)\big)\big)\Big)\Big(\cos\big(\sin\big(x^2\big)\big)\Big)\Big(\cos\big(x^2\big)\Big)(2x)$$

$$= 2x\Big(\cos\big(x^2\big)\Big)\Big(\cos\big(\sin\big(x^2\big)\big)\Big)\Big(\cos\big(\sin\big(\sin\big(x^2\big)\big)\big)\Big)$$

369. $\dfrac{2x^{1/2}+1}{6x^{1/2}\big(x+x^{1/2}\big)^{2/3}}$

Rewrite the function using exponential notation:

$$f(x) = \sqrt[3]{x+\sqrt{x}}$$

$$= \Big(x+(x)^{1/2}\Big)^{1/3}$$

Then apply the chain rule repeatedly to get the derivative:

$$f'(x) = \frac{1}{3}\big(x+x^{1/2}\big)^{-2/3}\Big(1+\frac{1}{2}x^{-1/2}\Big)$$

$$= \frac{1+\frac{1}{2}x^{-1/2}}{3\big(x+x^{1/2}\big)^{2/3}}$$

$$= \frac{2x^{1/2}+1}{6x^{1/2}\big(x+x^{1/2}\big)^{2/3}}$$

370. $(1+5x)^3\big(2+x-x^2\big)^6\big(-90x^2+41x+47\big)$

Recall that the chain rule states

$$\frac{d}{dx}f(g(x)) = f'(g(x))g'(x)$$

and that the product rule states

$$\frac{d}{dx}[f(x)g(x)] = f'(x)g(x)+f(x)g'(x)$$

Applying the product rule and chain rule to $f(x) = (1+5x)^4(2+x-x^2)^7$ and then factoring (factoring can be the tricky part!) gives you the following:

$$f'(x) = \Big(4(1+5x)^3(5)\Big)\big(2+x-x^2\big)^7 + (1+5x)^4\Big(7\big(2+x-x^2\big)^6(1-2x)\Big)$$

$$= (1+5x)^3\big(2+x-x^2\big)^6\Big(20\big(2+x-x^2\big)+7(1+5x)(1-2x)\Big)$$

$$= (1+5x)^3\big(2+x-x^2\big)^6\big(-90x^2+41x+47\big)$$

371. $x = 0, \pi, 2\pi$

Recall that the chain rule states

$$\frac{d}{dx} f(g(x)) = f'(g(x)) g'(x)$$

Begin by finding the derivative of the function $f(x) = 2\cos x + \sin^2 x$:

$$f'(x) = 2(-\sin x) + 2\sin x \cos x$$
$$= 2\sin x(\cos x - 1)$$

Then set the derivative equal to zero to find the x values where the function has a horizontal tangent line:

$$2\sin x(\cos x - 1) = 0$$

Next, set each factor equal to zero and solve for x: $\sin x = 0$ has the solutions $x = 0, \pi, 2\pi$, and $\cos x - 1 = 0$, or $\cos x = 1$, has the solutions $x = 0, 2\pi$. The slope of the tangent line is zero at each of these x values, so the solutions are $x = 0, \pi, 2\pi$.

372. $\dfrac{6[H(x)]^2}{x}$

To find the derivative of $F(x) = [H(x)]^3$, use the chain rule:

$$F'(x) = 3[H(x)]^2 H'(x)$$
$$= 3[H(x)]^2 \frac{2}{x}$$
$$= \frac{6[H(x)]^2}{x}$$

373. 28

Because the chain rule gives you $F'(x) = [f'(g(x))] g'(x)$, it follows that

$$F'(2) = [f'(g(2))] g'(2)$$
$$= [f'(-2)](4)$$
$$= (7)(4)$$
$$= 28$$

374. −40

Because the chain rule gives you $F'(x) = [f'(f(x))] f'(x)$, it follows that

$$F'(2) = [f'(f(2))] f'(2)$$
$$= [f'(-2)](-5)$$
$$= (8)(-5)$$
$$= -40$$

375. $\dfrac{6}{x}$

Recall that the chain rule states

$$\frac{d}{dx}f(g(x)) = f'(g(x))g'(x)$$

Using the chain rule on $f(x) = H(x^3)$ gives you the following:

$$f'(x) = \left[H'(x^3)\right](3x^2)$$

Because $H'(x) = \dfrac{2}{x}$, you know that that $H'(x^3) = \dfrac{2}{x^3}$, so the derivative becomes

$$f'(x) = \left[H'(x^3)\right](3x^2)$$

$$= \frac{2}{x^3}(3x^2)$$

$$= \frac{6}{x}$$

376. 80

Because the chain rule gives you $F'(x) = \left[f'(g(x))\right]g'(x)$, it follows that

$$F'(4) = \left[f'(g(4))\right]g'(4)$$

$$= \left[f'(6)\right](8)$$

$$= (10)(8)$$

$$= 80$$

377. $\dfrac{2}{x}$

Use properties of logarithms to rewrite the function:

$$f(x) = \ln x^2 = 2\ln x$$

Then take the derivative:

$$f'(x) = 2\frac{1}{x} = \frac{2}{x}$$

378. $\dfrac{4(\ln x)^3}{x}$

To find the derivative of $f(x) = (\ln x)^4$, apply the chain rule:

$$f'(x) = 4(\ln x)^3\left(\frac{1}{x}\right)$$

$$= \frac{4(\ln x)^3}{x}$$

379.
$$\frac{x}{x^2+5}$$

Begin by using properties of logarithms to rewrite the function:

$$f(x) = \ln\sqrt{x^2+5}$$
$$= \ln\left(x^2+5\right)^{1/2}$$
$$= \frac{1}{2}\ln\left(x^2+5\right)$$

Then apply the chain rule to find the derivative:

$$f'(x) = \frac{1}{2}\left(\frac{1}{x^2+5}\right)(2x)$$
$$= \frac{x}{x^2+5}$$

380.
$$\frac{1}{(\ln 10)\sqrt{6+x^2}}$$

Begin by writing the function using exponential notation:

$$f(x) = \log_{10}\left(x+\sqrt{6+x^2}\right)$$
$$= \log_{10}\left(x+\left(6+x^2\right)^{1/2}\right)$$

Then use the chain rule to find the derivative:

$$f'(x) = \frac{1}{\ln 10}\left(\frac{1}{x+\sqrt{6+x^2}}\right)\left(1+\frac{1}{2}\left(6+x^2\right)^{-1/2}(2x)\right)$$
$$= \frac{1}{\ln 10}\left(\frac{1}{x+\sqrt{6+x^2}}\right)\left(1+\frac{x}{\sqrt{6+x^2}}\right)$$
$$= \frac{1}{\ln 10}\left(\frac{1}{x+\sqrt{6+x^2}}\right)\left(\frac{\sqrt{6+x^2}}{\sqrt{6+x^2}}+\frac{x}{\sqrt{6+x^2}}\right)$$
$$= \frac{1}{(\ln 10)\sqrt{6+x^2}}$$

381.
$$\frac{2\cos x}{(\ln 6)(-1+2\sin x)} - \frac{3\cos x}{(\ln 6)(4+3\sin x)}$$

To make the derivative easier to find, use properties of logarithms to break up the function:

$$f(x) = \log_6\left|\frac{-1+2\sin x}{4+3\sin x}\right|$$
$$= \log_6\left|-1+2\sin x\right| - \log_6\left|4+3\sin x\right|$$

Now apply the chain rule to each term to get the derivative:

$$f'(x) = \frac{1}{(\ln 6)(-1+2\sin x)}(2\cos x) - \frac{1}{(\ln 6)(4+3\sin x)}(3\cos x)$$

$$= \frac{2\cos x}{(\ln 6)(-1+2\sin x)} - \frac{3\cos x}{(\ln 6)(4+3\sin x)}$$

382. $\dfrac{1}{(\ln 7)x\ln x}$

Begin by rewriting the function using properties of logarithms:

$$f(x) = \log_7\left(\log_8 x^5\right)$$

$$= \log_7\left(5\log_8 x\right)$$

$$= \log_7 5 + \log_7\left(\log_8 x\right)$$

The derivative becomes

$$f'(x) = 0 + \frac{1}{\ln 7(\log_8 x)}\left(\frac{1}{(\ln 8)x}\right)$$

$$= \frac{1}{(\ln 7)(\ln 8)(x)(\log_8 x)}$$

Note that $\log_7 5$ is a constant, so its derivative is equal to zero.

You can further simplify by using the change of base formula to write $\log_8 x = \dfrac{\ln x}{\ln 8}$:

$$f'(x) = \frac{1}{(\ln 7)(\ln 8)(x)(\log_8 x)}$$

$$= \frac{1}{(\ln 7)(\ln 8)(x)\left(\dfrac{\ln x}{\ln 8}\right)}$$

$$= \frac{1}{(\ln 7)x\ln x}$$

383. $\sec x$

Applying the chain rule to $f(x) = \ln|\sec x + \tan x|$ gives you

$$f'(x) = \frac{1}{\sec x + \tan x}\left(\sec x\tan x + \sec^2 x\right)$$

$$= \frac{(\sec x)(\tan x + \sec x)}{\sec x + \tan x}$$

$$= \sec x$$

384. $\dfrac{\sqrt{1+x^2}}{x^2}$

To find the derivative of $f(x) = -\dfrac{\sqrt{x^2+1}}{x} + \ln\left(x + \sqrt{x^2+1}\right)$, apply the quotient rule and chain rule to the first term and apply the chain rule to the second term:

$$f'(x) = \frac{x\left(-\frac{1}{2}\left(x^2+1\right)^{-1/2}(2x) + \left(x^2+1\right)^{1/2}(1)\right)}{x^2} + \frac{1}{x+\left(x^2+1\right)^{1/2}}\left(1 + \frac{1}{2}\left(x^2+1\right)^{-1/2}(2x)\right)$$

$$= \frac{\left(x^2+1\right)^{-1/2}\left[-x^2 + \left(x^2+1\right)\right]}{x^2} + \frac{1}{x+\left(x^2+1\right)^{1/2}}\left(1 + \frac{x}{\left(x^2+1\right)^{1/2}}\right)$$

$$= \frac{1}{x^2\left(x^2+1\right)^{1/2}} + \left(\frac{1}{x+\left(x^2+1\right)^{1/2}}\right)\left(\frac{\left(x^2+1\right)^{1/2}+x}{\left(x^2+1\right)^{1/2}}\right)$$

$$= \frac{1}{x^2\left(x^2+1\right)^{1/2}} + \frac{1}{\left(x^2+1\right)^{1/2}}$$

$$= \frac{\left(1+x^2\right)}{x^2\left(x^2+1\right)^{1/2}}$$

$$= \frac{\sqrt{1+x^2}}{x^2}$$

385. $\dfrac{2(x+1) - 6x\ln x}{(\ln 5)x(x+1)^4}$

Begin by using properties of logarithms to rewrite the function:

$$f(x) = \frac{\log_5 x^2}{(x+1)^3}$$

$$= \frac{2\log_5 x}{(x+1)^3}$$

Then apply the quotient rule and the chain rule:

$$f'(x) = \frac{(x+1)^3(2)\dfrac{1}{(\ln 5)x} - \left(2\log_5 x\right)\left(3(x+1)^2\right)}{(x+1)^6}$$

$$= \frac{(x+1)^2\left[\dfrac{2}{x\ln 5}(x+1) - 6\log_5 x\right]}{(x+1)^6}$$

$$= \frac{\dfrac{2(x+1)}{x\ln 5} - 6\log_5 x}{(x+1)^4}$$

$$= \frac{2(x+1) - (6\ln 5)x\left(\log_5 x\right)}{(\ln 5)x(x+1)^4}$$

You can now further simplify by using the change of base formula to write $\log_5 x = \dfrac{\ln x}{\ln 5}$ so that you have

$$\frac{2(x+1)-(6\ln 5)x\left(\log_5 x\right)}{(\ln 5)x(x+1)^4}$$

$$=\frac{2(x+1)-(6\ln 5)x\left(\dfrac{\ln x}{\ln 5}\right)}{(\ln 5)x(x+1)^4}$$

$$=\frac{2(x+1)-6x\ln x}{(\ln 5)x(x+1)^4}$$

386. $x^{\tan x}\left[\left(\sec^2 x\right)\ln x+\left(\tan x\right)\left(\dfrac{1}{x}\right)\right]$

Begin by rewriting the function and taking the natural logarithm of each side:

$$y = x^{\tan x}$$

$$\ln(y) = \ln\left(x^{\tan x}\right)$$

$$\ln(y) = (\tan x)\ln x$$

Take the derivative of each side with respect to x:

$$\frac{1}{y}\frac{dy}{dx} = \left(\sec^2 x\right)\ln x+\left(\tan x\right)\left(\frac{1}{x}\right)$$

Then multiply both sides by y:

$$\frac{dy}{dx} = y\left[\left(\sec^2 x\right)\ln x+\left(\tan x\right)\left(\frac{1}{x}\right)\right]$$

Finally, replacing y with $x^{\tan x}$ gives you the answer:

$$\frac{dy}{dx} = x^{\tan x}\left[\left(\sec^2 x\right)\ln x+\left(\tan x\right)\left(\frac{1}{x}\right)\right]$$

387. $(\ln x)^{\cos x}\left[(-\sin x)\left[\ln(\ln x)\right]+\dfrac{\cos x}{x\ln x}\right]$

Begin by rewriting the function and taking the natural logarithm of each side:

$$y = (\ln x)^{\cos x}$$

$$\ln(y) = \ln\left[(\ln x)^{\cos x}\right]$$

$$\ln y = (\cos x)\left[\ln(\ln x)\right]$$

Then take the derivative of each side with respect to x:

$$\frac{1}{y}\frac{dy}{dx} = (-\sin x)\left[\ln(\ln x)\right]+(\cos x)\left(\frac{1}{\ln x}\left(\frac{1}{x}\right)\right)$$

Multiplying both sides by y produces

$$\frac{dy}{dx} = y\left[(-\sin x)\left[\ln(\ln x)\right]+(\cos x)\left(\frac{1}{\ln x}\left(\frac{1}{x}\right)\right)\right]$$

And replacing y with $(\ln x)^{\cos x}$ gives you the answer:

$$\frac{dy}{dx} = (\ln x)^{\cos x}\left[(-\sin x)[\ln(\ln x)] + \frac{\cos x}{x\ln x}\right]$$

388. $\dfrac{-8x}{\sqrt{x^2-4}\left(x^2+4\right)^{3/2}}$

Begin by rewriting the function and taking the natural logarithm of each side:

$$y = \left(\frac{x^2-4}{x^2+4}\right)^{1/2}$$

$$\ln(y) = \ln\left[\left(\frac{x^2-4}{x^2+4}\right)^{1/2}\right]$$

Using properties of logarithms to expand gives you

$$\ln\left[\left(\frac{x^2-4}{x^2+4}\right)^{1/2}\right] = \frac{1}{2}\left[\ln\left(x^2-4\right) - \ln\left(x^2+4\right)\right]$$

Next, take the derivative of each side with respect to x:

$$\frac{1}{y}\frac{dy}{dx} = \frac{1}{2}\left(\frac{1}{x^2-4}(2x) - \frac{1}{x^2+4}(2x)\right)$$

Multiplying both sides by y produces the following:

$$\frac{dy}{dx} = y\left[\frac{x}{x^2-4} - \frac{x}{x^2+4}\right]$$

Replacing y with $\left(\dfrac{x^2-4}{x^2+4}\right)^{1/2}$ and simplifying gives you the solution:

$$\frac{dy}{dx} = \sqrt{\frac{x^2-4}{x^2+4}}\left[\frac{x}{x^2-4} - \frac{x}{x^2+4}\right]$$

$$= \frac{\sqrt{x^2-4}}{\sqrt{x^2+4}}\left[\frac{x^3+4x}{\left(x^2-4\right)\left(x^2+4\right)} - \frac{x^3-4x}{\left(x^2-4\right)\left(x^2+4\right)}\right]$$

$$= \frac{\sqrt{x^2-4}}{\sqrt{x^2+4}}\left[\frac{-8x}{\left(x^2-4\right)\left(x^2+4\right)}\right]$$

$$= \frac{-8x}{\sqrt{x^2-4}\left(x^2+4\right)^{3/2}}$$

Note that you can find the derivative without logarithmic differentiation, but using it makes the math easier.

389. $\dfrac{(x-1)\sin x}{\sqrt{x+2}\,(x+4)^{3/2}}\left[\dfrac{1}{x-1}+\cot x-\dfrac{1}{2(x+2)}-\dfrac{3}{2(x+4)}\right]$

Begin by rewriting the function and taking the natural logarithm of each side:

$$y=\dfrac{(x-1)\sin x}{\sqrt{x+2}\,(x+4)^{3/2}}$$

$$\ln(y)=\ln\left[\dfrac{(x-1)\sin x}{\sqrt{x+2}(x+4)^{3/2}}\right]$$

Use properties of logarithms to expand:

$$\ln y=\ln(x-1)+\ln(\sin x)-\dfrac{1}{2}\ln(x+2)-\dfrac{3}{2}\ln(x+4)$$

Then take the derivative of each side with respect to x:

$$\dfrac{1}{y}\dfrac{dy}{dx}=\dfrac{1}{x-1}+\dfrac{1}{\sin x}(\cos x)-\dfrac{1}{2(x+2)}-\dfrac{3}{2(x+4)}$$

Multiplying both sides by y produces

$$\dfrac{dy}{dx}=y\left(\dfrac{1}{x-1}+\dfrac{1}{\sin x}(\cos x)-\dfrac{1}{2(x+2)}-\dfrac{3}{2(x+4)}\right)$$

Replacing with y with $\dfrac{(x-1)\sin x}{\sqrt{x+2}(x+4)^{3/2}}$ gives you the answer:

$$\dfrac{dy}{dx}=\dfrac{(x-1)\sin x}{\sqrt{x+2}(x+4)^{3/2}}\left[\dfrac{1}{x-1}+\cot x-\dfrac{1}{2(x+2)}-\dfrac{3}{2(x+4)}\right]$$

Note that you can find the derivative without logarithmic differentiation, but using this technique makes the calculations easier.

390. $5e^{5x}$

Apply the chain rule to find the derivative of $f(x)=e^{5x}$:

$$f'(x)=\left(e^{5x}\right)(5)$$
$$=5e^{5x}$$

391. $e^{\sin x+x^4}\left(\cos x+4x^3\right)$

Applying the chain rule to $f(x)=e^{\sin x+x^4}$ gives you

$$f'(x)=e^{\sin x+x^4}\left(\cos x+4x^3\right)$$

392. $2^x\left[3x^2+(\ln 2)\left(x^3+1\right)\right]$

Applying the product rule to $f(x)=\left(x^3+1\right)2^x$ gives you the derivative as follows:

$$f'(x)=\left(3x^2\right)2^x+\left(x^3+1\right)\left(2^x\ln 2\right)$$
$$=2^x\left[3x^2+(\ln 2)\left(x^3+1\right)\right]$$

393.

$2x$

Use properties of logarithms to simplify the function:

$$f(x) = \log_5 5^{x^2+3} = x^2 + 3$$

The derivative is simply equal to

$$f'(x) = 2x$$

394.

$e^{x^3}\left[3x^2(\sin x + \cos x) + \cos x - \sin x\right]$

To find the derivative of $f(x) = e^{x^3}(\sin x + \cos x)$, apply the product rule while also applying the chain rule to the first factor:

$$f'(x) = \left[e^{x^3}(3x^2)\right](\sin x + \cos x) + e^{x^3}(\cos x - \sin x)$$

$$= e^{x^3}\left[3x^2(\sin x + \cos x) + \cos x - \sin x\right]$$

395.

$5^{\sqrt{x}}\left[\dfrac{(\ln 5)\sin x}{2\sqrt{x}} + \cos x\right]$

Put the radical in exponential form:

$$f(x) = 5^{\sqrt{x}}\sin x$$

$$= 5^{x^{1/2}}\sin x$$

Then apply the product rule as well as the chain rule to the first factor to get the derivative:

$$f'(x) = \left[(\ln 5)5^{x^{1/2}}\frac{1}{2}x^{-1/2}\right]\sin x + 5^{x^{1/2}}\cos x$$

$$= 5^{\sqrt{x}}\left[\frac{(\ln 5)\sin x}{2\sqrt{x}} + \cos x\right]$$

396.

$3(\ln 4)\left(4^{-x} + 4^x\right)^2\left(4^x - 4^{-x}\right)$

Applying the chain rule to $f(x) = \left(4^{-x} + 4^x\right)^3$ gives you the derivative as follows:

$$f'(x) = 3\left(4^{-x} + 4^x\right)^2\left(4^{-x}(\ln 4)(-1) + 4^x(\ln 4)\right)$$

$$= 3(\ln 4)\left(4^{-x} + 4^x\right)^2\left(4^x - 4^{-x}\right)$$

397.

$\dfrac{e^x}{1+e^x} + \dfrac{e^x}{1-e^x}$

Use properties of logarithms to break up the function:

$$f(x) = \ln\left(\frac{1+e^x}{1-e^x}\right)$$

$$= \ln\left(1+e^x\right) - \ln\left(1-e^x\right)$$

Then apply the chain rule to each term to get the derivative:

$$f'(x) = \frac{1}{1+e^x}(e^x) - \frac{1}{1-e^x}(-e^x)$$

$$= \frac{e^x}{1+e^x} + \frac{e^x}{1-e^x}$$

398. $\dfrac{(x^2+1)\big((\ln 6)6^x+1\big)-2x\big(6^x+x\big)}{\big(x^2+1\big)^2}$

Apply the quotient rule to find the derivative of $f(x) = \dfrac{6^x+x}{x^2+1}$:

$$f'(x) = \frac{(x^2+1)\big(6^x(\ln 6)+1\big)-\big(6^x+x\big)(2x)}{\big(x^2+1\big)^2}$$

$$= \frac{(x^2+1)\big((\ln 6)6^x+1\big)-2x\big(6^x+x\big)}{\big(x^2+1\big)^2}$$

399. $\dfrac{-4\big(e^x-e^{-x}\big)}{\big(e^x+e^{-x}\big)^2}$

First rewrite the function:

$$f(x) = \frac{4}{e^x+e^{-x}} = 4\big(e^x+e^{-x}\big)^{-1}$$

Applying the chain rule gives you the derivative as follows:

$$f'(x) = -4\big(e^x+e^{-x}\big)^{-2}\big(e^x+e^{-x}(-1)\big)$$

$$= \frac{-4\big(e^x-e^{-x}\big)}{\big(e^x+e^{-x}\big)^2}$$

400. $\dfrac{\big(8^{x^2+1}\big)\big[(2\ln 8)x\cos x+\sin x\big]}{\big(\cos x\big)^2}$

To find the derivative of $f(x) = \dfrac{8^{x^2+1}}{\cos x}$, apply the quotient rule while applying the chain rule to the numerator:

$$f'(x) = \frac{\cos x\big(8^{x^2+1}(\ln 8)(2x)\big)-8^{x^2+1}(-\sin x)}{\big(\cos x\big)^2}$$

$$= \frac{\big(8^{x^2+1}\big)\big[(2\ln 8)x\cos x+\sin x\big]}{\big(\cos x\big)^2}$$

401. $6^{-5x}\left[1-(5\ln 6)x\right]$

To find the derivative of $f(x)=x(6^{-5x})$, apply the product rule while applying the chain rule to the second factor:

$$f'(x)=(1)\left(6^{-5x}\right)+x\left(6^{-5x}(\ln 6)(-5)\right)$$
$$=6^{-5x}\left[1-(5\ln 6)x\right]$$

402. $y = \pi x + 3$

You can begin by finding the y value, because it isn't given:

$$f(0)=3\cos(0)+\pi(0)=3$$

Next, find the derivative of the function:

$$f'(x)=-3\sin x+\pi$$

Substitute in the given x value to find the slope of the tangent line:

$$f'(0)=-3\sin(0)+\pi=\pi$$

Now use the point-slope formula for a line to get the tangent line at $x = 0$:

$$y-3=\pi(x-0)$$
$$y=\pi x+3$$

403. $y = x + 1$

Begin by finding the derivative of the function $f(x) = x^2 - x + 2$:

$$f'(x)=2x-1$$

Substitute in the given x value to find the slope of the tangent line:

$$f'=2(1)-1=1$$

Now use the point-slope formula for a line to get the tangent line at (1, 2):

$$y-2=1(x-1)$$
$$y=x+1$$

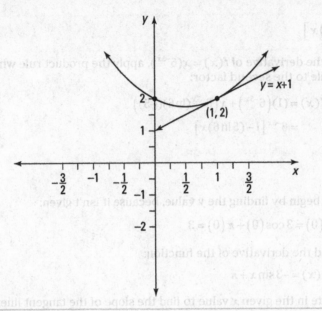

404. $y = \dfrac{7e^4}{4}x - 3e^4$

You can begin by finding the y value, because it isn't given:

$$f(2) = \frac{e^{2^2}}{2} = \frac{e^4}{2}$$

Next, find the derivative of the function:

$$f'(x) = \frac{x\left(e^{x^2}(2x)\right) - e^{x^2}(1)}{x^2}$$

Substitute in the given x value to find the slope of the tangent line:

$$f'(2) = \frac{(2)\left(e^{(2)^2}(2(2))\right) - e^{(2)^2}(1)}{2^2} = \frac{7e^4}{4}$$

Now use the point-slope formula for a line to get the tangent line at $x = 2$:

$$y - \frac{e^4}{2} = \frac{7e^4}{4}(x - 2)$$

$$y = \frac{7e^4}{4}x - 3e^4$$

405. $y = -\dfrac{1}{19}x + \dfrac{535}{19}$

The normal line is perpendicular to the tangent line. Begin by finding the derivative of the function $f(x) = 3x^2 + x - 2$:

$$f'(x) = 6x + 1$$

Then substitute in the given x value to find the slope of the tangent line:

$$f'(3) = 6(3) + 1 = 19$$

To find the slope of the normal line, take the opposite reciprocal of the slope of the tangent line to get $-\frac{1}{19}$.

Now use the point-slope formula for a line to get the normal line at (3, 28):

$$y - 28 = -\frac{1}{19}(x - 3)$$

$$y = -\frac{1}{19}x + \frac{535}{19}$$

406. $\quad y = -x + \frac{\pi + 2}{4}$

The normal line is perpendicular to the tangent line. You can begin by finding the y value at $x = \frac{\pi}{4}$, because it isn't given:

$$f\left(\frac{\pi}{4}\right) = \left(\sin\frac{\pi}{4}\right)^2 = \left(\frac{\sqrt{2}}{2}\right)^2 = \frac{1}{2}$$

Next, find the derivative of the function:

$$f'(x) = 2\sin x \cos x$$

Then substitute in the given x value to find the slope of the tangent line:

$$f'\left(\frac{\pi}{4}\right) = 2\sin\frac{\pi}{4}\cos\frac{\pi}{4} = 2\left(\frac{\sqrt{2}}{2}\right)\left(\frac{\sqrt{2}}{2}\right) = 1$$

To find the slope of the normal line, take the opposite reciprocal of the slope of the tangent line to get –1.

Now use the point-slope formula for a line to get the normal line at $x = \frac{\pi}{4}$:

$$y - \frac{1}{2} = -1\left(x - \frac{\pi}{4}\right)$$

$$y = -x + \frac{\pi + 2}{4}$$

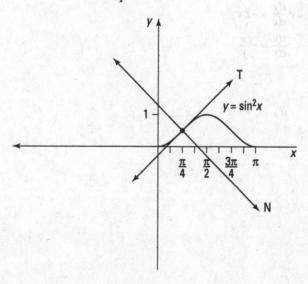

407. $y = -\dfrac{e^2}{4}x + \dfrac{e^4 + 40}{4}$

The normal line is perpendicular to the tangent line. You can begin by finding the y value at $x = e^2$, because it isn't given:

$$f\left(e^2\right) = 4\ln\left(e^2\right) + 2 = 4(2) + 2 = 10$$

Next, find the derivative of the function:

$$f'(x) = 4\left(\frac{1}{x}\right) = \frac{4}{x}$$

Then substitute in the given x value to find the slope of the tangent line:

$$f'\left(e^2\right) = \frac{4}{e^2}$$

To find the slope of the normal line, take the opposite reciprocal of the slope of the tangent line to get $-\dfrac{e^2}{4}$.

Now use the point-slope formula for a line to get the normal line at $x = e^2$:

$$y - 10 = -\frac{e^2}{4}\left(x - e^2\right)$$

$$y = -\frac{e^2}{4}x + \frac{e^4 + 40}{4}$$

408. $-\dfrac{x}{y}$

Taking the derivative of both sides of $x^2 + y^2 = 9$ and solving for $\dfrac{dy}{dx}$ gives you the following:

$$2x + 2y\frac{dy}{dx} = 0$$

$$2y\frac{dy}{dx} = -2x$$

$$\frac{dy}{dx} = -\frac{x}{y}$$

409.

$$\frac{2xy - 2xy^3}{5y^4 + 3x^2y^2 - x^2}$$

Take the derivative of both sides of $y^5 + x^2y^3 = 2 + x^2y$ and solve for $\frac{dy}{dx}$:

$$5y^4 \frac{dy}{dx} + \left(2xy^3 + x^2\left(3y^2 \frac{dy}{dx}\right)\right) = 2xy + x^2 \frac{dy}{dx}$$

$$5y^4 \frac{dy}{dx} + 3x^2y^2 \frac{dy}{dx} - x^2 \frac{dy}{dx} = 2xy - 2xy^3$$

$$\frac{dy}{dx}\left(5y^4 + 3x^2y^2 - x^2\right) = 2xy - 2xy^3$$

$$\frac{dy}{dx} = \frac{2xy - 2xy^3}{5y^4 + 3x^2y^2 - x^2}$$

410.

$$\frac{-3x^2y^3 - \cos y}{3x^3y^2 - x\sin y}$$

Take the derivative of both sides of $x^3y^3 + x\cos(y) = 7$ and solve for $\frac{dy}{dx}$:

$$3x^2y^3 + x^3\left(3y^2 \frac{dy}{dx}\right) + 1\cos(y) + x\left(-\sin y \frac{dy}{dx}\right) = 0$$

$$3x^3y^2 \frac{dy}{dx} - x\sin y \frac{dy}{dx} = -3x^2y^3 = -\cos y$$

$$\frac{dy}{dx}\left(3x^3y^2 - x\sin y\right) = -3x^2y^3 - \cos y$$

$$\frac{dy}{dx} = \frac{-3x^2y^3 - \cos y}{3x^3y^2 - x\sin y}$$

411.

$$\frac{-1}{1 + 4y\sqrt{x+y}\,\sin\left(y^2\right)}$$

Take the derivative of both sides of $\sqrt{x+y} = \cos\left(y^2\right)$ and solve for $\frac{dy}{dx}$:

$$\frac{1}{2}(x+y)^{-1/2}\left(1 + \frac{dy}{dx}\right) = \left(-\sin\left(y^2\right)\right)\left(2y \frac{dy}{dx}\right)$$

$$\frac{1}{2\sqrt{x+y}} + \frac{1}{2\sqrt{x+y}} \frac{dy}{dx} = \left(-2y\sin\left(y^2\right)\right)\frac{dy}{dx}$$

$$\frac{1}{2\sqrt{x+y}} \frac{dy}{dx} + 2y\sin\left(y^2\right)\frac{dy}{dx} = \frac{-1}{2\sqrt{x+y}}$$

$$\frac{dy}{dx}\left[\frac{1}{2\sqrt{x+y}} + 2y\sin\left(y^2\right)\right] = \frac{-1}{2\sqrt{x+y}}$$

$$\frac{dy}{dx} = \frac{\dfrac{-1}{2\sqrt{x+y}}}{\dfrac{1}{2\sqrt{x+y}} + 2y\sin\left(y^2\right)}$$

$$\frac{dy}{dx} = \frac{-1}{1 + 4y\sqrt{x+y}\,\sin\left(y^2\right)}$$

412.
$$\frac{y\csc^2\left(\dfrac{y}{x}\right)-2x^3}{x^2+x\csc^2\left(\dfrac{y}{x}\right)}$$

Take the derivative of both sides of $\cot\left(\dfrac{y}{x}\right)=x^2+y$ and solve for $\dfrac{dy}{dx}$:

$$-\csc^2\left(\frac{y}{x}\right)\left[\frac{x\dfrac{dy}{dx}-y(1)}{x^2}\right]=2x+(1)\frac{dy}{dx}$$

$$-\csc^2\left(\frac{y}{x}\right)\left[\frac{1}{x}\frac{dy}{dx}-\frac{y}{x^2}\right]=2x+\frac{dy}{dx}$$

$$-\frac{1}{x}\csc^2\left(\frac{y}{x}\right)\frac{dy}{dx}+\frac{y}{x^2}\csc^2\left(\frac{y}{x}\right)=2x+\frac{dy}{dx}$$

$$\frac{y}{x^2}\csc^2\left(\frac{y}{x}\right)-2x=\frac{dy}{dx}+\frac{1}{x}\csc^2\left(\frac{y}{x}\right)\frac{dy}{dx}$$

$$\frac{y}{x^2}\csc^2\left(\frac{y}{x}\right)-2x=\frac{dy}{dx}\left(1+\frac{1}{x}\csc^2\left(\frac{y}{x}\right)\right)$$

$$\frac{\dfrac{y}{x^2}\csc^2\left(\dfrac{y}{x}\right)-2x}{1+\dfrac{1}{x}\csc^2\left(\dfrac{y}{x}\right)}=\frac{dy}{dx}$$

$$\frac{x^2}{x^2}\left(\frac{\dfrac{y}{x^2}\csc^2\left(\dfrac{y}{x}\right)-2x}{1+\dfrac{1}{x}\csc^2\left(\dfrac{y}{x}\right)}\right)=\frac{dy}{dx}$$

$$\frac{y\csc^2\left(\dfrac{y}{x}\right)-2x^3}{x^2+x\csc^2\left(\dfrac{y}{x}\right)}=\frac{dy}{dx}$$

413.
$$\frac{-2xy-\left(1+x^2\right)^2 y\sec(xy)\tan(xy)}{\left(1+x^2\right)\left[\left(1+x^2\right)x\sec(xy)\tan(xy)-1\right]}$$

Take the derivative of both sides of $\sec(xy)=\dfrac{y}{1+x^2}$ and solve for $\dfrac{dy}{dx}$:

$$\sec(xy)\tan(xy)\left[(1)y+x\frac{dy}{dx}\right]=\frac{\left(1+x^2\right)\dfrac{dy}{dx}-y(2x)}{\left(1+x^2\right)^2}$$

$$y\sec(xy)\tan(xy)+x\sec(xy)\tan(xy)\frac{dy}{dx}=\frac{1}{1+x^2}\frac{dy}{dx}-\frac{2xy}{\left(1+x^2\right)^2}$$

$$\frac{dy}{dx}\left[x\sec(xy)\tan(xy)-\frac{1}{1+x^2}\right]=-\frac{2xy}{\left(1+x^2\right)^2}-y\sec(xy)\tan(xy)$$

$$\frac{dy}{dx}=\frac{-\dfrac{2xy}{\left(1+x^2\right)^2}-y\sec(xy)\tan(xy)}{x\sec(xy)\tan(xy)-\dfrac{1}{1+x^2}}$$

$$\frac{dy}{dx}=\frac{-2xy-\left(1+x^2\right)^2 y\sec(xy)\tan(xy)}{\left(1+x^2\right)^2 x\sec(xy)\tan(xy)-\left(1+x^2\right)}$$

$$\frac{dy}{dx}=\frac{-2xy-\left(1+x^2\right)^2 y\sec(xy)\tan(xy)}{\left(1+x^2\right)\left[\left(1+x^2\right)x\sec(xy)\tan(xy)-1\right]}$$

414.
$$\frac{-64}{y^3}$$

Begin by finding the first derivative of $8x^2+y^2=8$:

$$16x+2y\frac{dy}{dx}=0$$

$$2y\frac{dy}{dx}=-16x$$

$$\frac{dy}{dx}=\frac{-8x}{y}$$

Next, find the second derivative:

$$\frac{d^2y}{dx^2}=\frac{y(-8)-(-8x)\dfrac{dy}{dx}}{y^2}$$

Substituting in the value of $\frac{dy}{dx} = \frac{-8x}{y}$ gives you

$$\frac{d^2y}{dx^2} = \frac{y(-8)-(-8x)\frac{dy}{dx}}{y^2}$$

$$= \frac{-8y + 8x\left(\frac{-8x}{y}\right)}{y^2}$$

$$= \frac{-8y^2 - 64x^2}{y^3}$$

$$= \frac{-8\left(8x^2 + y^2\right)}{y^3}$$

And now using $8x^2 + y^2 = 8$ gives you the answer:

$$\frac{d^2y}{dx^2} = \frac{-8(8)}{y^3} = \frac{-64}{y^3}$$

415. $\dfrac{-4x^3}{y^9}$

Begin by finding the first derivative of $x^5 + y^5 = 1$:

$$5x^4 + 5y^4\frac{dy}{dx} = 0$$

$$\frac{dy}{dx} = -\frac{x^4}{y^4}$$

Next, find the second derivative:

$$\frac{d^2y}{dx^2} = \frac{y^4\left(-4x^3\right)-\left(-x^4\right)\left(4y^3\right)\frac{dy}{dx}}{\left(y^4\right)^2}$$

Substituting in the value of $\frac{dy}{dx} = -\frac{x^4}{y^4}$ gives you

$$\frac{d^2y}{dx^2} = \frac{-4x^3y^4 + 4x^4y^3\left(-\frac{x^4}{y^4}\right)}{y^8}$$

$$= \frac{-4x^3y^4 - \frac{4x^8}{y}}{y^8}$$

$$= \frac{-4x^3y^5 - 4x^8}{y^9}$$

$$= \frac{-4x^3\left(y^5 + x^5\right)}{y^9}$$

And now using $x^5 + y^5 = 1$ gives you the answer:

$$\frac{d^2y}{dx^2} = \frac{-4x^3(1)}{y^9} = \frac{-4x^3}{y^9}$$

416. $\dfrac{-10x}{y^5}$

Begin by finding the first derivative of $x^3 + y^3 = 5$:

$$3x^2 + 3y^2 \frac{dy}{dx} = 0$$

$$\frac{dy}{dx} = \frac{-x^2}{y^2}$$

Next, find the second derivative:

$$\frac{d^2y}{dx^2} = \frac{y^2(-2x) - (-x^2)\left(2y\dfrac{dy}{dx}\right)}{\left(y^2\right)^2}$$

Substituting in the value of $\dfrac{dy}{dx} = \dfrac{-x^2}{y^2}$ gives you

$$\frac{d^2y}{dx^2} = \frac{y^2(-2x) - (-x^2)\left(2y\dfrac{dy}{dx}\right)}{\left(y^2\right)^2}$$

$$= \frac{-2xy^2 + 2x^2y\left(\dfrac{-x^2}{y^2}\right)}{y^4}$$

$$= \frac{-2xy^2 - \dfrac{2x^4}{y}}{y^4}$$

$$= \frac{-2xy^3 - 2x^4}{y^5}$$

$$= \frac{-2x\left(y^3 + x^3\right)}{y^5}$$

And now using $x^3 + y^3 = 5$ gives you the answer:

$$\frac{d^2y}{dx^2} = \frac{-2x(5)}{y^5} = \frac{-10x}{y^5}$$

417. $\dfrac{1}{2x^{3/2}}$

Begin by finding the first derivative of $\sqrt{x} + \sqrt{y} = 1$:

$$\frac{1}{2}x^{-1/2} + \frac{1}{2}y^{-1/2}\frac{dy}{dx} = 0$$

$$\frac{1}{2}y^{-1/2}\frac{dy}{dx} = -\frac{1}{2}x^{-1/2}$$

$$\frac{dy}{dx} = -\frac{y^{1/2}}{x^{1/2}}$$

Next, find the second derivative:

$$\frac{d^2y}{dx^2} = \frac{x^{1/2}\left(-\frac{1}{2}y^{-1/2}\frac{dy}{dx}\right) - \left(-y^{1/2}\right)\left(\frac{1}{2}x^{-1/2}\right)}{\left(x^{1/2}\right)^2}$$

Substituting in the value of $\frac{dy}{dx} = \frac{-y^{1/2}}{x^{1/2}}$ gives you

$$\frac{d^2y}{dx^2} = \frac{-\frac{1}{2}x^{1/2}y^{-1/2}\left(\frac{-y^{1/2}}{x^{1/2}}\right) + \frac{y^{1/2}}{2x^{1/2}}}{x}$$

$$= \frac{\frac{1}{2} + \frac{y^{1/2}}{2x^{1/2}}}{x}$$

$$= \frac{x^{1/2} + y^{1/2}}{2x^{3/2}}$$

And now using $\sqrt{x} + \sqrt{y} = 1$ gives you the answer:

$$\frac{d^2y}{dx^2} = \frac{1}{2x^{3/2}}$$

418. $y = -x + 2$

You know a point on the tangent line, so you just need to find the slope. Begin by finding the derivative of $x^2 + xy + y^2 = 3$:

$$2x + \left(1y + x\frac{dy}{dx}\right) + 2y\frac{dy}{dx} = 0$$

$$2x + y + x\frac{dy}{dx} + 2y\frac{dy}{dx} = 0$$

Next, enter the values $x = 1$ and $y = 1$ and solve for the slope $\frac{dy}{dx}$:

$$2(1) + 1 + (1)\frac{dy}{dx} + 2(1)\frac{dy}{dx} = 0$$

$$3 + 3\frac{dy}{dx} = 0$$

$$\frac{dy}{dx} = -1$$

The tangent line has a slope of -1 and passes through $(1, 1)$, so its equation is

$$y - 1 = -1(x - 1)$$

$$y = -x + 2$$

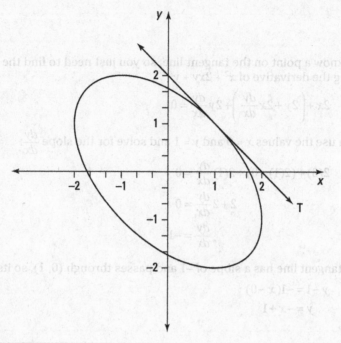

419. $y = -\dfrac{2}{11}x + \dfrac{15}{11}$

You know a point on the tangent line, so you just need to find the slope. Begin by finding the derivative of $3(x^2 + y^2)^2 = 25(x^2 - y^2)$:

$$3\left[2\left(x^2 + y^2\right)\left(2x + 2y\dfrac{dy}{dx}\right)\right] = 25\left(2x - 2y\dfrac{dy}{dx}\right)$$

Then enter the values $x = 2$ and $y = 1$ and solve for the slope $\dfrac{dy}{dx}$:

$$6\left(2^2 + 1^2\right)\left(2(2) + 2(1)\dfrac{dy}{dx}\right) = 25\left(2(2) - 2(1)\dfrac{dy}{dx}\right)$$

$$6(5)\left(4 + 2\dfrac{dy}{dx}\right) = 25\left(4 - 2\dfrac{dy}{dx}\right)$$

$$110\dfrac{dy}{dx} = -20$$

$$\dfrac{dy}{dx} = -\dfrac{2}{11}$$

The tangent line has a slope of $-\dfrac{2}{11}$ and passes through $(2, 1)$, so its equation is

$$y - 1 = -\dfrac{2}{11}(x - 2)$$

$$y = -\dfrac{2}{11}x + \dfrac{15}{11}$$

420.

$y = -x + 1$

You know a point on the tangent line, so you just need to find the slope. Begin by taking the derivative of $x^2 + 2xy + y^2$:

$$2x + \left(2y + 2x\frac{dy}{dx}\right) + 2y\frac{dy}{dx} = 0$$

Then use the values $x = 0$ and $y = 1$ and solve for the slope $\frac{dy}{dx}$:

$$2(0) + (2(1) + 0) + 2(1)\frac{dy}{dx} = 0$$

$$2 + 2\frac{dy}{dx} = 0$$

$$\frac{dy}{dx} = -1$$

The tangent line has a slope of –1 and passes through (0, 1), so its equation is

$$y - 1 = -1(x - 0)$$

$$y = -x + 1$$

421.

$y = \left(2 - \frac{\pi}{2}\right)x + (\pi - 2)$

You know a point on the tangent line, so you just need to find the slope. Begin by taking the derivative of $\cos(xy) + x^2 = \sin y$:

$$-\sin(xy)\left[1y + x\frac{dy}{dx}\right] + 2x = (\cos y)\frac{dy}{dx}$$

Then use the values $x = 1$ and $y = \frac{\pi}{2}$ and solve for the slope $\frac{dy}{dx}$:

$$-\sin\left(\frac{\pi}{2}\right)\left[\frac{\pi}{2} + 1\frac{dy}{dx}\right] + 2 = 0$$

$$\frac{dy}{dx} = 2 - \frac{\pi}{2}$$

The tangent line has a slope of $2 - \frac{\pi}{2}$ and passes through $\left(1, \frac{\pi}{2}\right)$, so its equation is

$$y - \frac{\pi}{2} = \left(2 - \frac{\pi}{2}\right)(x - 1)$$

$$y = \left(2 - \frac{\pi}{2}\right)x + (\pi - 2)$$

422. $y = 1$

You know a point on the tangent line, so you just need to find the slope. Begin by simplifying the left side of the equation.

$$y^2(y^2 - 1) = x^2 \tan y$$
$$y^4 - y^2 = x^2 \tan y$$

Next, find the derivative:

$$4y^3 \frac{dy}{dx} - 2y \frac{dy}{dx} = 2x(\tan y) + x^2 \left(\sec^2 y \frac{dy}{dx} \right)$$

Use the values $x = 0$ and $y = 1$ and solve for the slope $\frac{dy}{dx}$:

$$4(1)^3 \frac{dy}{dx} - 2(1) \frac{dy}{dx} = 0$$
$$2 \frac{dy}{dx} = 0$$
$$\frac{dy}{dx} = 0$$

The tangent line has a slope of 0 and passes through (0, 1), so its equation is

$$y - 1 = 0 (x - 0)$$
$$y = 1$$

423. $\frac{1}{3}$

Use the formula for the differential, $dy = f'(x)\, dx$:

$$y = x^2 - 4x$$
$$dy = (2x - 4)\, dx$$

So for the given values, $x = 3$ and $dx = \frac{1}{6}$, find dy:

$$dy = (2(3) - 4)\frac{1}{6}$$
$$= (2)\frac{1}{6} = \frac{1}{3}$$

424. $\frac{1}{20}$

Use the formula for the differential, $dy = f'(x)\, dx$:

$$y = \frac{1}{x^2 + 1}$$
$$dy = (-1(x^2 + 1)^{-2}(2x))\, dx$$
$$dy = \frac{-2x}{(x^2 + 1)^2}\, dx$$

So for the given values, $x = 1$ and $dx = -0.1$, find dy:

$$dy = \frac{-2(1)}{(1^2 + 1)^2}(-0.1)$$
$$= \frac{-1}{2}\left(\frac{-1}{10}\right)$$
$$= \frac{1}{20}$$

425. $\frac{-\sqrt{3}}{100}$

Use the formula for the differential, $dy = f'(x)\, dx$:

$$y = \cos^2 x$$
$$dy = 2(\cos x)(-\sin x)\, dx$$
$$dy = -2\cos x \sin x\, dx$$

So for the given values, $x = \frac{\pi}{3}$ and $dx = 0.02$, find dy:

$$dy = \left(-2\cos\frac{\pi}{3}\sin\frac{\pi}{3}\right)(0.02)$$
$$= \left(-2\left(\frac{1}{2}\right)\left(\frac{\sqrt{3}}{2}\right)\right)\left(\frac{2}{100}\right)$$
$$= \frac{-\sqrt{3}}{100}$$

426. $L(x) = 6x - 3$

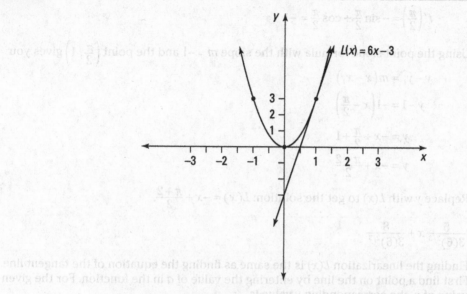

Finding the linearization $L(x)$ is the same as finding the equation of the tangent line. First find a point on the line by entering the value of a in the function. For the given value of a, the corresponding y value is

$$f(1) = 3(1)^2 = 3$$

Next, find the derivative of the function to get the slope of the line:

$$f'(x) = 6x$$
$$f'(1) = 6(1) = 6$$

Using the point-slope formula with the slope $m = 6$ and the point $(1, 3)$ gives you

$$y - y_1 = m(x - x_1)$$
$$y - 3 = 6(x - 1)$$
$$y = 6x - 6 + 3$$
$$y = 6x - 3$$

Replace y with $L(x)$ to get the solution: $L(x) = 6x - 3$.

427. $L(x) = -x + \dfrac{\pi + 2}{2}$

Finding the linearization $L(x)$ is the same as finding the equation of the tangent line. First find a point on the line by entering the value of a in the function. For the given value of a, the corresponding y value is

$$f\left(\frac{\pi}{2}\right) = \cos\frac{\pi}{2} + \sin\frac{\pi}{2} = 1$$

Next, find the derivative of the function to get the slope of the line:

$$f'(x) = -\sin x + \cos x$$

$$f'\left(\frac{\pi}{2}\right) = -\sin\frac{\pi}{2} + \cos\frac{\pi}{2} = -1$$

Using the point-slope formula with the slope $m = -1$ and the point $\left(\frac{\pi}{2}, 1\right)$ gives you

$$y - y_1 = m(x - x_1)$$

$$y - 1 = -1\left(x - \frac{\pi}{2}\right)$$

$$y = -x + \frac{\pi}{2} + 1$$

$$y = -x + \frac{\pi + 2}{2}$$

Replace y with $L(x)$ to get the solution: $L(x) = -x + \frac{\pi + 2}{2}$.

428. $L(x) = \dfrac{5}{3(6)^{2/3}}x + \dfrac{8}{3(6)^{2/3}}$

Finding the linearization $L(x)$ is the same as finding the equation of the tangent line. First find a point on the line by entering the value of a in the function. For the given value of a, the corresponding y value is

$$f(2) = \sqrt[3]{2^2 + 2} = \sqrt[3]{6} = 6^{1/3}$$

Next, find the derivative of the function to get the slope of the line:

$$f'(x) = \frac{1}{3}\left(x^2 + x\right)^{-2/3}(2x + 1) = \frac{2x + 1}{3\left(x^2 + x\right)^{2/3}}$$

$$f'(2) = \frac{2(2) + 1}{3\left(2^2 + 2\right)^{2/3}} = \frac{5}{3(6)^{2/3}}$$

Using the point-slope formula with the slope $m = \dfrac{5}{3(6)^{2/3}}$ and the point $(2, 6^{1/3})$ gives you

$$y - y_1 = m(x - x_1)$$

$$y - 6^{1/3} = \frac{5}{3(6)^{2/3}}(x - 2)$$

$$y = \frac{5}{3(6)^{2/3}}x - \frac{10}{3(6)^{2/3}} + 6^{1/3}$$

$$y = \frac{5}{3(6)^{2/3}}x - \frac{10}{3(6)^{2/3}} + \frac{18}{3(6)^{2/3}}$$

$$y = \frac{5}{3(6)^{2/3}}x + \frac{8}{3(6)^{2/3}}$$

Replace y with $L(x)$ to get the solution: $L(x) = \dfrac{5}{3(6)^{2/3}}x + \dfrac{8}{3(6)^{2/3}}$.

429.   3.987

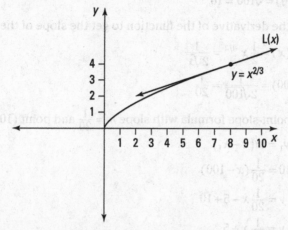

To estimate the value, you can find a linearization by using $f(x) = x^{2/3}$ and $a = 8$ and then substitute 7.96 into the linearization.

Find a point on the line by entering the value of a in the function. For the given value of a, the corresponding y value is

$$f(8) = 8^{2/3} = \left(8^{1/3}\right)^2 = 4$$

Next, find the derivative of the function to get the slope of the line:

$$f'(x) = \frac{2}{3}x^{-1/3} = \frac{2}{3\sqrt[3]{x}}$$

$$f'(8) = \frac{2}{3\sqrt[3]{8}} = \frac{2}{3(2)} = \frac{1}{3}$$

Using the point-slope formula with the slope $m = \frac{1}{3}$ and the point $(8, 4)$ gives you

$$y - y_1 = m(x - x_1)$$

$$y - 4 = \frac{1}{3}(x - 8)$$

$$y = \frac{1}{3}x - \frac{8}{3} + 4$$

$$y = \frac{1}{3}x + \frac{4}{3}$$

Replacing y with $L(x)$ gives you the linearization $L(x) = \frac{1}{3}x + \frac{4}{3}$.

Finally, substitute in the value $x = 7.96$ to find the estimate:

$$L(7.96) = \frac{1}{3}(7.96) + \frac{4}{3} \approx 3.987$$

430.   10.1

To estimate the value, you can find a linearization by using $f(x) = \sqrt{x} = x^{1/2}$ and $a = 100$ and then substitute 102 into the linearization.

Find a point on the line by entering the value of a in the function. For the given value of a, the corresponding y value is

$$f(100) = \sqrt{100} = 10$$

Next, find the derivative of the function to get the slope of the line:

$$f'(x) = \frac{1}{2}x^{-1/2} = \frac{1}{2\sqrt{x}}$$

$$f'(100) = \frac{1}{2\sqrt{100}} = \frac{1}{20}$$

Using the point-slope formula with slope $m = \frac{1}{20}$ and point $(100, 10)$ gives you

$$y - y_1 = m(x - x_1)$$

$$y - 10 = \frac{1}{20}(x - 100)$$

$$y = \frac{1}{20}x - 5 + 10$$

$$y = \frac{1}{20}x + 5$$

Replacing y with $L(x)$ gives you the linearization $L(x) = \frac{1}{20}x + 5$.

Finally, substitute in the value $x = 102$ to find the estimate:

$$L(102) = \frac{1}{20}(102) + 5$$

$$= 10.1$$

431.   1.035

To estimate the value of $\tan 46°$, you can find a linearization using $f(x) = \tan x$ and $a = 45° = \frac{\pi}{4}$ rad and then substitute the value $46° = \frac{46\pi}{180}$ rad into the linearization.

Find a point on the line by entering the value of a in the function. For the given value of a, the corresponding y value is

$$f\left(\frac{\pi}{4}\right) = \tan\left(\frac{\pi}{4}\right) = 1$$

Next, find the derivative of the function to get the slope of the line:

$$f'(x) = \sec^2 x$$

$$f'\left(\frac{\pi}{4}\right) = \left(\sec\frac{\pi}{4}\right)^2 = \left(\sqrt{2}\right)^2 = 2$$

Using the point-slope formula with slope $m = 2$ and point $\left(\frac{\pi}{4}, 1\right)$ gives you

$$y - y_1 = m(x - x_1)$$

$$y - 1 = 2\left(x - \frac{\pi}{4}\right)$$

$$y = 2\left(x - \frac{\pi}{4}\right) + 1$$

Replacing y with $L(x)$ gives you the linearization $L(x) = 1 + 2\left(x - \frac{\pi}{4}\right)$.

Finally, substitute in the value of $x = \frac{46\pi}{180}$ radians (that is, $46°$) to find the estimate of $\tan 46°$:

$$L\left(\frac{46\pi}{180}\right) = 1 + 2\left(\frac{46\pi}{180} - \frac{\pi}{4}\right)$$

$$= 1 + 2\left(\frac{46\pi}{180} - \frac{45\pi}{180}\right)$$

$$= 1 + 2\left(\frac{\pi}{180}\right)$$

$$= 1 + \frac{\pi}{90} \approx 1.035$$

432. $\quad 4\pi r^2 \dfrac{dr}{dt}$

The volume of a sphere is $V = \frac{4}{3}\pi r^3$. To find the rate of expansion, take the derivative of each side with respect to time:

$$V = \frac{4}{3}\pi r^3$$

$$\frac{dV}{dt} = \frac{4}{3}\pi\left(3r^2\frac{dr}{dt}\right)$$

$$= 4\pi r^2\frac{dr}{dt}$$

433. $\quad 8\pi$ m²/s

The area of a circle is $A = \pi r^2$. To find the rate of increase, take the derivative of each side with respect to time:

$$A = \pi r^2$$

$$\frac{dA}{dt} = \pi\left(2r\frac{dr}{dt}\right)$$

$$= 2\pi r\frac{dr}{dt}$$

The circle increases at a rate of $\frac{dr}{dt} = 1$ meter per second, so when radius $r = 4$ meters, the area increases at the following rate:

$$\frac{dA}{dt} = 2\pi(4)(1) = 8\pi \text{ m}^2/\text{s}$$

434. $\quad 508$

Take the derivative of both sides with respect to t:

$$y = x^4 + 3x^2 + x$$

$$\frac{dy}{dt} = 4x^3\frac{dx}{dt} + 6x\frac{dx}{dt} + 1\frac{dx}{dt}$$

Then substitute in the given values, $x = 3$ and $\frac{dx}{dt} = 4$:

$$\frac{dy}{dt} = 4(3)^3(4) + 6(3)(4) + 1(4) = 508$$

435. $\frac{4}{9}\sqrt[3]{15}$

Take the derivative of both sides with respect to t:

$$z^3 = x^2 - y^2$$

$$3z^2\left(\frac{dz}{dt}\right) = 2x\left(\frac{dx}{dt}\right) - 2y\left(\frac{dy}{dt}\right)$$

To find the value of z, enter $x = 4$ and $y = 1$ in the original equation:

$$z^3 = 4^2 - 1^2$$

$$z = \sqrt[3]{15}$$

$$z = 15^{1/3}$$

Then substitute in the given values along with the value of z and solve for $\frac{dz}{dt}$:

$$3z^2\left(\frac{dz}{dt}\right) = 2x\frac{dx}{dt} - 2y\left(\frac{dy}{dt}\right)$$

$$3\left((15)^{1/3}\right)^2\left(\frac{dz}{dt}\right) = 2(4)(3) - 2(1)(2)$$

$$\frac{dz}{dt} = \frac{4}{9}\sqrt[3]{15}$$

436. $2.49\ \text{m}^2/\text{s}$

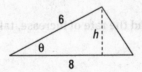

To find the area of the triangle, you need the base and the height. If you let the length of the base equal 8 meters, then one side of the triangle equals 6 meters. Therefore, if the height equals h, then $\frac{h}{6} = \sin\theta$ so that $h = 6\sin\theta$.

Because the area of a triangle is $A = \frac{1}{2}bh$, you can write the area as

$$A = \frac{1}{2}(8)6\sin\theta$$

$$A = 24\sin\theta$$

This equation now involves θ, so you can take the derivative of both sides with respect to t to get the rate of increase:

$$\frac{dA}{dt} = (24\cos\theta)\left(\frac{d\theta}{dt}\right)$$

Substitute in the given values, $\theta = \frac{\pi}{6}$ and $\frac{d\theta}{dt} = 0.12$ radians/second, and simplify:

$$\frac{dA}{dt} = \left(24\cos\frac{\pi}{6}\right)(0.12)$$

$$= (24)\left(\frac{\sqrt{3}}{2}\right)(0.12)$$

$$\approx 2.49 \text{ m}^2/\text{s}$$

When the angle between the sides is $\frac{\pi}{6}$, the area is increasing at a rate of about 2.49 square meters per second.

---

437. $\frac{3}{4}$ rad/s

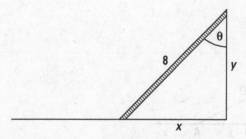

The problem tells you how quickly the bottom of the ladder slides away from the wall $\left(\frac{dx}{dt}\right)$, so first write an expression for x, the distance from the wall. Assuming that the ground and wall meet at a right angle, you can write

$$\sin\theta = \frac{x}{8}$$

$$8\sin\theta = x$$

Taking the derivative with respect to time, t, gives you the rate at which the angle is changing:

$$8\cos\theta\frac{d\theta}{dt} = \frac{dx}{dt}$$

$$\frac{d\theta}{dt} = \frac{\frac{dx}{dt}}{8\cos\theta}$$

Substitute in the given information, where $\theta = \frac{\pi}{3}$ and $\frac{dx}{dt} = 3$ feet per second:

$$\frac{d\theta}{dt} = \frac{3}{8\cos\left(\frac{\pi}{3}\right)} = \frac{3}{4} \text{ rad/s}$$

When the angle is $\frac{\pi}{3}$, the angle is increasing at a rate of $\frac{3}{4}$ radians per second.

---

438. 72 cm²/min

The area of a triangle is $A = \frac{1}{2}bh$, and the problem tells you how quickly the base and height are changing. To find the rate of the change in area, take the derivative of both sides of the equation with respect to time, t:

$$A = \frac{1}{2}bh$$

$$\frac{dA}{dt} = \frac{1}{2}\left[\left(1\frac{db}{dt}\right)h + b\left(1\frac{dh}{dt}\right)\right]$$

Note that you have to use the product rule to find the derivative of the right side of the equation.

Substitute in the given information, $\frac{db}{dt} = 2$ centimeters per minute, $h = 32$ centimeters, $b = 20$ centimeters, and $\frac{dh}{dt} = 4$ centimeters per minute:

$$\frac{dA}{dt} = \frac{1}{2}[(2)(32) + 20(4)]$$

$$= 72 \text{ cm}^2/\text{min}$$

439. 13.56 km/h

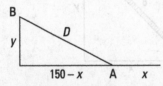

Let x be the distance sailed by Ship A and let y be the distance sailed by Ship B. Using the Pythagorean theorem, the distance between the ships is

$$D^2 = (150 - x)^2 + y^2$$

From the given information, you have $\frac{dx}{dt} = 20$ kilometers per hour and $\frac{dy}{dt} = 35$ kilometers per hour. Taking the derivative of both sides of the equation with respect to time gives you

$$2D\frac{dD}{dt} = 2(150 - x)\left(-1\frac{dx}{dt}\right) + 2y\frac{dy}{dt}$$

Notice that after 3 hours have elapsed, $x = 60$ and $y = 105$. Therefore, using the Pythagorean theorem, you can deduce that $D = \sqrt{19,125} \approx 138.29$ kilometers. Substitute in all these values and solve for $\frac{dD}{dt}$:

$$2(138.29)\frac{dD}{dt} = 2(150 - 60)(-20) + 2(105)(35)$$

$$\frac{dD}{dt} \approx 13.56 \text{ km/h}$$

At 3 p.m., the ships are moving apart at a rate of about 13.56 kilometers per hour.

440. 4.83 cm/s

From the Pythagorean theorem, the distance from the origin is $D^2 = x^2 + y^2$. Using the particle's path, $y = \sqrt[3]{x} + 1$, you get

$$D^2 = x^2 + \left(x^{1/3} + 1\right)^2$$

$$D^2 = x^2 + x^{2/3} + 2x^{1/3} + 1$$

Taking the derivative of both sides with respect to time gives you

$$2D\frac{dD}{dt} = 2x\frac{dx}{dt} + \frac{2}{3}x^{-1/3}\frac{dx}{dt} + \frac{2}{3}x^{-2/3}\frac{dx}{dt}$$

From the given values $x = 8$ centimeters and $y = 3$ centimeters, you find that the distance from the origin is $D = \sqrt{73}$ centimeters. Use these values along with $\frac{dx}{dt} = 5$ centimeters per second and solve for $\frac{dD}{dt}$:

$$2\sqrt{73}\frac{dD}{dt} = 2(8)(5) + \frac{2}{3}(8)^{-1/3}(5) + \frac{2}{3}(8)^{-2/3}(5)$$

$$\frac{dD}{dt} \approx 4.83 \text{ cm/s}$$

---

441.   2.95 mi/h

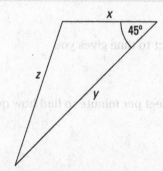

Let x be the distance that the person who is walking west has traveled, let y be the distance that the person traveling southwest has traveled, and let z be the distance between them.

Use the law of cosines to relate x, y, and z:

$$z^2 = x^2 + y^2 - 2xy\cos(45°)$$
$$z^2 = x^2 + y^2 - \sqrt{2}xy$$

Take the derivative of both sides of the equation with respect to time:

$$2z\frac{dz}{dt} = 2x\frac{dx}{dt} + 2y\frac{dy}{dt} - \sqrt{2}\left(\frac{dx}{dt}(y) + x\frac{dy}{dt}\right)$$

After 40 minutes, or $\frac{2}{3}$ hour, $x = \frac{4}{3}$ and $y = \frac{8}{3}$. Using the initial equation, you find that $z \approx 1.96$ miles at this time. Substitute these values into the derivative and solve for $\frac{dz}{dt}$:

$$2(1.96)\frac{dz}{dt} = 2\left(\frac{4}{3}\right)(2) + 2\left(\frac{8}{3}\right)(4) - \sqrt{2}\left((2)\left(\frac{8}{3}\right) + \frac{4}{3}(4)\right)$$

$$\frac{dz}{dt} \approx 2.95 \text{ mi/h}$$

After 40 minutes, the distance between these people is increasing at a rate of about 2.95 miles per hour.

442. $\frac{4}{25}$ ft/min

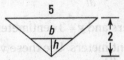

The water in the trough has a volume of $V = \frac{1}{2}b(h)(20)$. As the water level rises, the base and height of the triangle shapes increase. By similar triangles, you have $\frac{5}{2} = \frac{b}{h}$ so that $b = \frac{5}{2}h$. Therefore, the volume is

$$V = \frac{1}{2}b(h)(20)$$

$$V = 10\left(\frac{5}{2}h\right)h$$

$$V = 25h^2$$

Taking the derivative with respect to time gives you

$$\frac{dV}{dt} = 50h\frac{dh}{dt}$$

Use $h = 1$ foot and $\frac{dV}{dt} = 8$ cubic feet per minute to find how quickly the water level is rising:

$$8 = 50(1)\frac{dh}{dt}$$

$$\frac{dh}{dt} = \frac{8}{50}$$

$$= \frac{4}{25} \text{ ft/min}$$

When the water is 1 foot deep, the water level is rising at a rate of $\frac{4}{25}$ feet per minute.

443. 698.86 km/h

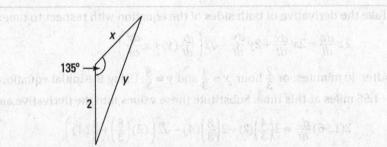

Let x be the distance that the jet travels, and let y be the distance between the jet and the radar station.

Use the law of cosines to relate x and y:

$$y^2 = x^2 + 2^2 - 2(2)x\cos(135°)$$

$$y^2 = x^2 + 4 - 4x\left(-\frac{\sqrt{2}}{2}\right)$$

$$y^2 = x^2 + 2\sqrt{2}x + 4$$

Taking the derivative of both sides of the equation with respect to time gives you

$$2y\frac{dy}{dt} = 2x\frac{dx}{dt} + 2\sqrt{2}\frac{dx}{dt}$$

$$\frac{dy}{dt} = \frac{2x + 2\sqrt{2}}{2y}\frac{dx}{dt}$$

$$\frac{dy}{dt} = \frac{x + \sqrt{2}}{y}\frac{dx}{dt}$$

After 2 minutes, the jet has traveled $x = 2\left(\frac{700}{60}\right) = \frac{70}{3}$ kilometers, so the distance from the radar station is

$$y = \sqrt{\left(\frac{70}{3}\right)^2 + 2\sqrt{2}\left(\frac{70}{3}\right) + 4}$$

Enter the value of y in the $\frac{dy}{dt}$ equation and solve:

$$\frac{dy}{dt} = \frac{\frac{70}{3} + \sqrt{2}}{\sqrt{\left(\frac{70}{3}\right)^2 + 2\sqrt{2}\left(\frac{70}{3}\right) + 4}}(700) \approx 698.86 \text{ km/h}$$

After 2 minutes, the jet is moving away from the radar station at a rate of about 698.86 kilometers per hour.

444. $\frac{348\pi}{5}$ km/min

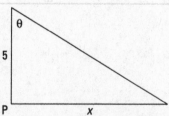

Let x be the distance from the Point P to the spot on the shore where the light is shining. From the diagram, you have $\frac{x}{5} = \tan\theta$, or $x = 5\tan\theta$. Taking the derivative of both sides of the equation with respect to time gives you

$$\frac{dx}{dt} = 5\sec^2\theta\left(\frac{d\theta}{dt}\right)$$

When $x = 2$, $\tan\theta = \frac{2}{5}$. Using a trigonometric identity, you have

$$\sec^2\theta = 1 + \left(\frac{2}{5}\right)^2 = \frac{29}{25}$$

From the given information, you know that $\frac{d\theta}{dt} = 6$ revolutions per minute. Because 2π radians are in one revolution, you can convert as follows: $\frac{d\theta}{dt} = 6(2\pi) = 12\pi$ radians per minute:

$$\frac{dx}{dt} = 5\left(\frac{29}{25}\right)(12\pi)$$

$$= \frac{348\pi}{5}$$

$$\approx 218.65 \text{ km/min}$$

445. $\dfrac{5}{36\pi}$ ft/min

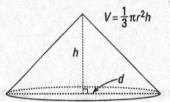

$$V = \frac{1}{3}\pi r^2 h$$

The volume of a cone is $V = \frac{1}{3}\pi r^2 h$. The problem tells you that $\dfrac{dV}{dt} = 20$ cubic feet per minute and that $d = 2h$. Because the diameter is twice the radius, you have $2r = d = 2h$ so that $r = h$; therefore, the volume becomes

$$V = \frac{1}{3}\pi h^2 h$$

$$V = \frac{1}{3}\pi h^3$$

Take the derivative with respect to time:

$$\frac{dV}{dt} = \left(\frac{1}{3}\pi\right)\left(3h^2\right)\frac{dh}{dt}$$

Substitute in the given information and solve for $\dfrac{dh}{dt}$:

$$20 = \left(\frac{1}{3}\pi\right)\left(3(12)^2\right)\frac{dh}{dt}$$

$$\frac{dh}{dt} = \frac{20}{\left(\frac{1}{3}\pi\right)\left(3(12)^2\right)}$$

$$= \frac{5}{36\pi}$$

When the pile is 12 feet high, the height of the pile is increasing at a rate of $\dfrac{5}{36\pi}$ (about 0.04) feet per minute.

446. no absolute maximum; absolute minimum: $y = -1$; no local maxima; local minimum: $(1, -1)$

There's no absolute maximum because the graph doesn't attain a largest y value. There are also no local maxima. The absolute minimum is $y = -1$. The point $(1, -1)$ is a local minimum.

447. absolute maximum: $y = 4$; absolute minimum: $y = 0$; local maximum: $(5, 4)$; local minima: $(1, 0), (7, 2)$

The absolute maximum value is 4, which the graph attains at the point $(5, 4)$. The absolute minimum is 0, which is attained at the point $(1, 0)$. The point $(5, 4)$ also corresponds to a local maximum, and the local minima occur at $(1, 0)$ and $(7, 2)$.

448. absolute maximum: $y = 3$; no absolute minimum; local maxima: (3, 3), (5, 3); local minimum: (4, 1)

The absolute maximum value is 3, which the graph attains at the points (3, 3) and (5, 3). The function approaches the x-axis but doesn't cross or touch it, so there's no absolute minimum. The points (3, 3) and (5, 3) also correspond to local maxima, and the local minimum occurs at (4, 1).

449. no maxima or minima

The graph has no absolute maximum or minimum because the graph approaches ∞ on the left and $-\infty$ on the right. Likewise, there are no local maxima or minima because (2, 1) and (5, 3) are points of discontinuity.

450. I and II

Because Point A satisfies the definition of both a local maximum and a local minimum, it's both. The graph decreases to negative infinity on the left side, so Point A is not an absolute minimum.

451. absolute maximum: 5; absolute minimum: –7

Begin by finding the derivative of the function; then find any critical numbers on the given interval by determining where the derivative equals zero or is undefined. Note that by finding critical numbers, you're finding potential turning points or cusp points of the graph.

The derivative of $f(x) = 3x^2 - 12x + 5$ is

$$f'(x) = 6x - 12$$
$$= 6(x - 2)$$

Setting the derivative equal to zero and solving gives you the only critical number, $x = 2$. Next, substitute the endpoints of the interval and the critical number into the original function and pick the largest and smallest values:

$$f(0) = 3(0)^2 - 12(0) + 5 = 5$$
$$f(2) = 3(2)^2 - 12(2) + 5 = -7$$
$$f(3) = 3(3)^2 - 12(3) + 5 = -4$$

Therefore, the absolute maximum is 5, and the absolute minimum is –7.

452. absolute maximum: 67; absolute minimum: 3

Begin by finding the derivative of the function; then find any critical numbers on the given interval. The derivative of $f(x) = x^4 - 2x^2 + 4$ is

$$f'(x) = 4x^3 - 4x$$
$$= 4x(x^2 - 1)$$

Setting the derivative equal to zero and solving gives you the critical numbers $x = 0$ and $x = \pm 1$, all of which fall within the given interval. Next, substitute the endpoints of the interval along with the critical numbers into the original function and pick the largest and smallest values:

$$f(-2) = (-2)^4 - 2(-2)^2 + 4 = 12$$
$$f(-1) = (-1)^4 - 2(-1)^2 + 4 = 3$$
$$f(0) = 0^4 - 2(0) + 4 = 4$$
$$f(1) = 1^4 - 2(1)^2 + 4 = 3$$
$$f(3) = 3^4 - 2(3)^2 + 4 = 67$$

Therefore, the absolute maximum is 67, and the absolute minimum is 3.

453. absolute maximum: $\frac{1}{2}$; absolute minimum: 0

Begin by finding the derivative of the function; then find any critical numbers on the given interval. The derivative of $f(x) = \dfrac{x}{x^2 + 1}$ is

$$f'(x) = \frac{(x^2 + 1)(1) - x(2x)}{(x^2 + 1)^2} = \frac{1 - x^2}{(x^2 + 1)^2}$$

Next, find the critical numbers by setting the numerator equal to zero and solving for x (note that the denominator will never be zero):

$$1 - x^2 = 0$$
$$1 = x^2$$
$$\pm 1 = x$$

Only $x = 1$ is in the given interval, so don't use $x = -1$. Next, substitute the endpoints of the interval along with the critical number into the original function and pick the largest and smallest values:

$$f(0) = 0$$
$$f(1) = \frac{1}{2}$$
$$f(3) = \frac{3}{10}$$

Therefore, the absolute maximum is $\frac{1}{2}$, and the absolute minimum is 0.

454. absolute maximum: 2; absolute minimum: $-\sqrt{3}$

Begin by finding the derivative of the function; then find any critical numbers on the given interval. The derivative of $f(t) = t\sqrt{4 - t^2} = t\left(4 - t^2\right)^{1/2}$ is

$$f'(t) = 1\left(4 - t^2\right)^{1/2} + \frac{1}{2}t\left(4 - t^2\right)^{-1/2}(-2t)$$

$$= \left(4 - t^2\right)^{1/2} - \frac{t^2}{\left(4 - t^2\right)^{1/2}}$$

$$= \frac{4 - 2t^2}{\left(4 - t^2\right)^{1/2}}$$

Next, find any critical numbers of the function by setting the numerator and denominator of the derivative equal to zero and solving for t.

From the numerator of the derivative, you have $4 - 2t^2 = 0$ so that $4 = 2t^2$, or $2 = t^2$, which has the solutions $t = \pm\sqrt{2}$. From the denominator, you have $4 - t^2 = 0$, or $4 = t^2$, which has the solutions $t = \pm 2$.

Substitute the endpoints along with the critical points that fall within the interval into the original function and pick the largest and smallest values:

$$f(-1) = -1\sqrt{4-1} = -\sqrt{3}$$

$$f\left(\sqrt{2}\right) = \sqrt{2}\sqrt{4 - \sqrt{2}^2} = \sqrt{2}\sqrt{2} = 2$$

$$f(2) = 2\sqrt{4-4} = 0$$

Therefore, the absolute maximum is 2, and the absolute minimum is $-\sqrt{3}$.

455.   absolute maximum: $\pi + 2$; absolute minimum: $-\dfrac{\pi}{6} - \sqrt{3}$

Begin by finding the derivative of the function; then find any critical numbers on the given interval. The derivative of $f(x) = x - 2\cos x$ is

$$f'(x) = 1 + 2\sin x$$

Setting the derivative equal to zero in order to find the critical numbers gives you $1 + 2$ $\sin x = 0$, or $\sin x = -\dfrac{1}{2}$, which has the solutions $x = \dfrac{-5\pi}{6}$ and $x = \dfrac{-\pi}{6}$ in the given interval of $[-\pi, \pi]$.

Substitute these critical numbers along with the endpoints of the interval into the original function and pick the largest and smallest values:

$$f(-\pi) = -\pi - 2\cos(-\pi) = -\pi + 2$$

$$f\left(\frac{-5\pi}{6}\right) = \frac{-5\pi}{6} - 2\cos\left(\frac{-5\pi}{6}\right) = \frac{-5\pi}{6} + \sqrt{3}$$

$$f\left(\frac{-\pi}{6}\right) = -\frac{\pi}{6} - 2\cos\left(\frac{-\pi}{6}\right) = \frac{-\pi}{6} - \sqrt{3}$$

$$f(\pi) = \pi - 2\cos\pi = \pi + 2$$

Therefore, the absolute maximum is $\pi + 2$, and the absolute minimum is $-\dfrac{\pi}{6} - \sqrt{3}$.

456.   increasing on $(-\infty, -2)$ and $(2, \infty)$; decreasing on $(-2, 2)$

Begin by finding the derivative of the function; then find any critical numbers on the given interval by determining where the derivative equals zero or is undefined. Note that by finding critical numbers, you're finding potential turning points or cusp points of the graph.

The derivative of the function $f(x) = 2x^3 - 24x + 1$ is

$$f'(x) = 6x^2 - 24$$

$$= 6\left(x^2 - 4\right)$$

Setting the derivative equal to zero in order to find the critical points gives you $x^2 - 4 = 0$, or $x^2 = 4$, so that $x = \pm 2$.

To determine where the function is increasing or decreasing, substitute a test point inside each interval into the derivative to see whether the derivative is positive or negative.

To test $(-\infty, -2)$, you can use $x = -3$. In that case, $f'(-3) = 6(-3)^2 - 24 = 30$. The derivative is positive, so the function is increasing on $(-\infty, -2)$. Proceed in a similar manner for the intervals $(-2, 2)$ and $(2, \infty)$. When $x = 0$, you have $f'(0) = 6(0)^2 - 24 = -24$, so the function is decreasing on $(-2, 2)$. And if $x = 3$, then $f'(3) = 6(3)^2 - 24 = 30$, so the function is increasing on $(2, \infty)$.

457. increasing on $(-2, \infty)$; decreasing on $(-3, -2)$

Begin by finding the derivative of the function. The derivative of $f(x) = x\sqrt{x+3} = x(x+3)^{1/2}$ is

$$f'(x) = 1(x+3)^{1/2} + x\left(\frac{1}{2}(x+3)^{-1/2}\right)$$

$$= \frac{\sqrt{x+3}}{1} + \frac{x}{2\sqrt{x+3}}$$

$$= \frac{2(x+3) + x}{2\sqrt{x+3}}$$

$$= \frac{3x+6}{2\sqrt{x+3}}$$

Next, find any critical numbers that fall inside the interval $(-3, \infty)$ by setting the derivative equal to zero and solving for x. (You only need to set the numerator equal to zero, because the denominator will never equal zero.) This gives you $3x + 6 = 0$, or $x = -2$.

Next, pick a value in $(-3, -2)$ and determine whether the derivative is positive or negative. So if $x = -2.5$, then $f'(-2.5) = \frac{3(-2.5)+6}{2\sqrt{-2.5+3}} < 0$; therefore, the function is decreasing on $(-3, -2)$. Likewise, you can show that $f'(x) > 0$ on $(-2, \infty)$, which means that $f(x)$ is increasing on $(-2, \infty)$.

458. increasing on $\left(\frac{\pi}{2}, \frac{7\pi}{6}\right), \left(\frac{3\pi}{2}, \frac{11\pi}{6}\right)$; decreasing on $\left(0, \frac{\pi}{2}\right), \left(\frac{7\pi}{6}, \frac{3\pi}{2}\right), \left(\frac{11\pi}{6}, 2\pi\right)$

Begin by finding the derivative of the function. The derivative of $f(x) = \cos^2 x - \sin x$ is

$$f'(x) = -2\cos x \sin x - \cos x$$

Next, find any critical numbers that fall inside the specified interval by setting the derivative equal to zero and solving: $-2\cos x \sin x = 0$, or $(-\cos x)(2\sin x + 1) = 0$. Solving $-\cos x = 0$ gives you the solutions $x = \frac{\pi}{2}$ and $x = \frac{3\pi}{2}$, and solving $2\sin + 1 = 0$ gives you $\sin x = -\frac{1}{2}$, which has the solutions $x = \frac{7\pi}{6}$ and $x = \frac{11\pi}{6}$.

To determine where the original function is increasing or decreasing, substitute a test point from inside each interval into the derivative to see whether the derivative is positive or negative. So for the interval $\left(0, \frac{\pi}{2}\right)$, you can use $x = \frac{\pi}{6}$. In that case, $f'\left(\frac{\pi}{6}\right) = -\cos\frac{\pi}{6}\left(2\sin\frac{\pi}{6} + 1\right) < 0$, so the function is decreasing on $\left(0, \frac{\pi}{2}\right)$. Proceeding in a similar manner, you can show that for a point inside the interval $\left(\frac{\pi}{2}, \frac{7\pi}{6}\right)$, $f' > 0$; that

for a point inside the interval $\left(\frac{7\pi}{6}, \frac{3\pi}{2}\right)$, $f' < 0$; that for a point inside $\left(\frac{3\pi}{2}, \frac{11\pi}{6}\right)$, $f' > 0$;

and that for a point inside $\left(\frac{11\pi}{6}, 2\pi\right)$, $f' < 0$. Therefore, $f(x)$ is increasing on $\left(\frac{\pi}{2}, \frac{7\pi}{6}\right)$

and $\left(\frac{3\pi}{2}, \frac{11\pi}{6}\right)$, and $f(x)$ is decreasing on $\left(0, \frac{\pi}{2}\right), \left(\frac{7\pi}{6}, \frac{3\pi}{2}\right)$, and $\left(\frac{11\pi}{6}, 2\pi\right)$.

459. increasing on $\left(0, \frac{\pi}{3}\right), \left(\pi, \frac{5\pi}{3}\right)$; decreasing on $\left(\frac{\pi}{3}, \pi\right), \left(\frac{5\pi}{3}, 2\pi\right)$

Begin by finding the derivative of the function. The derivative of $f(x) = 2 \cos x - \cos 2x$ is

$$f'(x) = -2\sin x + (\sin(2x))(2)$$
$$= -2\sin x + 2(2\sin x \cos x)$$
$$= 2\sin x(2\cos x - 1)$$

Now find the critical numbers by setting each factor equal to zero and solving for x. The equation $\sin x = 0$ gives you the solutions $x = 0$, $x = \pi$, and $x = 2\pi$. Solving $2 \cos x - 1 = 0$ gives you $\cos x = \frac{1}{2}$, which has solutions $x = \frac{\pi}{3}$ and $x = \frac{5\pi}{3}$.

To determine whether the function is increasing or decreasing on $\left(0, \frac{\pi}{3}\right)$, pick a point in the interval and substitute it into the derivative. So if you use $x = \frac{\pi}{4}$, then you have $f'\left(\frac{\pi}{4}\right) = 2\sin\left(\frac{\pi}{4}\right)\left(2\cos\frac{\pi}{4} - 1\right) > 0$, so the function is increasing on $\left(0, \frac{\pi}{3}\right)$. Likewise, if you use the point $x = \frac{2\pi}{3}$ in the interval $\left(\frac{\pi}{3}, \pi\right)$, you have $f'\left(\frac{2\pi}{3}\right) = 2\sin\left(\frac{2\pi}{3}\right)\left(2\cos\frac{2\pi}{3} - 1\right) < 0$. If you use $x = \frac{3\pi}{2}$ in the interval $\left(\pi, \frac{5\pi}{3}\right)$, then $f'\left(\frac{3\pi}{2}\right) = 2\sin\left(\frac{3\pi}{2}\right)\left(2\cos\frac{3\pi}{2} - 1\right) > 0$. And if you use $x = \frac{7\pi}{4}$ in the interval $\left(\frac{5\pi}{3}, 2\pi\right)$, you have $f'\left(\frac{7\pi}{4}\right) = 2\sin\left(\frac{7\pi}{4}\right)\left(2\cos\frac{7\pi}{4} - 1\right) < 0$.

Therefore, $f(x)$ is increasing on the intervals $\left(0, \frac{\pi}{3}\right)$ and $\left(\pi, \frac{5\pi}{3}\right)$, and $f(x)$ is decreasing on the intervals $\left(\frac{\pi}{3}, \pi\right)$ and $\left(\frac{5\pi}{3}, 2\pi\right)$.

460. increasing on $(0, 1)$; decreasing on $(1, \infty)$

Begin by finding the derivative of the function. The derivative of $f(x) = 4\ln x - 2x^2$ is

$$f'(x) = 4\left(\frac{1}{x}\right) - 4x$$
$$= \frac{4}{x} - \frac{4x^2}{x}$$
$$= \frac{4(1 - x^2)}{x}$$
$$= \frac{4(1-x)(1+x)}{x}$$

Now find the critical numbers. Setting the numerator equal to zero gives you $(1 - x)(1 + x) = 0$ so that $x = 1$ or -1. Setting the denominator of the derivative equal to zero gives you $x = 0$. Notice that neither $x = 0$ nor $x = -1$ is in the domain of the original function.

To determine whether the function is increasing or decreasing on (0, 1), take a point in the interval and substitute it into the derivative. If $x = \frac{1}{2}$, then $f'\left(\frac{1}{2}\right) = \dfrac{4\left(1-\left(\frac{1}{2}\right)^2\right)}{\frac{1}{2}} > 0$,

so the function is increasing on (0, 1). Likewise, if you use $x = 2$, then $f'(2) = \dfrac{4(1-4)}{2} < 0$, so the function is decreasing on (1, ∞).

---

461. local maximum at (–1, 7); local minimum at (2, –20)

Begin by finding the derivative of the function $f(x) = 2x^3 - 3x^2 - 12x$:

$$f'(x) = 6x^2 - 6x - 12$$

Then set this derivative equal to zero and solve for x to find the critical numbers:

$$6x^2 - 6x - 12 = 0$$
$$x^2 - x - 2 = 0$$
$$(x-2)(x+1) = 0$$
$$x = 2, -1$$

Next, determine whether the function is increasing or decreasing on the intervals (–∞, –1), (–1, 2), and (2, ∞) by picking a point inside each interval and substituting it into the derivative. Using the values $x = -2$, $x = 0$, and $x = 3$ gives you

$$f'(-2) = 6(-2)^2 - 6(-2) - 12 > 0$$
$$f'(0) = 6(0)^2 - 6(0) - 12 < 0$$
$$f'(3) = 6(3)^3 - 6(3) - 12 > 0$$

Therefore, $f(x)$ is increasing on (–∞, –1), decreasing on (–1, 2), and increasing again on (2, ∞). That means there's a local maximum at $x = -1$ and a local minimum at $x = 2$.

Now enter these values in the original function to find the coordinates of the local maximum and minimum. Because $f(-1) = 2(-1)^3 - 3(-1)^2 - 12(-1) = 7$, the local maximum occurs at (–1, 7). Because $f(2) = 2(2)^3 - 3(2)^2 - 12(2) = -20$, the local minimum occurs at (2, –20).

---

462. no local maxima; local minimum at (16, –16)

Begin by finding the derivative of the function $f(x) = x - 8\sqrt{x} = x - 8x^{1/2}$:

$$f'(x) = 1 - 8\left(\frac{1}{2}x^{-1/2}\right)$$
$$= 1 - \frac{4}{\sqrt{x}}$$
$$= \frac{\sqrt{x} - 4}{\sqrt{x}}$$

Then find the critical numbers. Setting the numerator equal to zero gives you $\sqrt{x} - 4 = 0$ so that $\sqrt{x} = 4$, or $x = 16$. Setting the denominator equal to zero gives you $x = 0$. Next, determine whether the function is increasing or decreasing on the intervals (0, 16) and (16, ∞) by taking a point in each interval and substituting it into the derivative.

So if $x = 1$, you have $f'(1) = 1 - \dfrac{4}{\sqrt{1}} < 0$ so that $f(x)$ is decreasing on $(0, 16)$. And if $x = 25$, you have $f'(25) = 1 - \dfrac{4}{\sqrt{25}} > 0$, so $f(x)$ is increasing on $(16, \infty)$. Therefore, the function has a local minimum at $x = 16$. Finish by finding the coordinates: $f(16) = 16 - 8\sqrt{16} = -16$, so the local minimum is at $(16, -16)$.

463.   local maximum at $(64, 32)$; local minimum at $(0, 0)$

Begin by finding the derivative of the function $f(x) = 6x^{2/3} - x$:

$$f'(x) = 6\left(\frac{2}{3}x^{-1/3}\right) - 1$$

$$= \frac{4}{\sqrt[3]{x}} - \frac{\sqrt[3]{x}}{\sqrt[3]{x}}$$

$$= \frac{4 - \sqrt[3]{x}}{\sqrt[3]{x}}$$

Then find the critical numbers. Setting the numerator equal to zero gives you $4 - \sqrt[3]{x} = 0$ so that $4 = x^{1/3}$, or $64 = x$. Setting the denominator equal to zero gives you $x = 0$ as a solution.

Next, determine whether the function is increasing or decreasing on the intervals $(-\infty, 0)$, $(0, 64)$, and $(64, \infty)$ by taking a point in each interval and substituting it into the derivative. Using the test points $x = -1$, $x = 1$, and $x = 125$ gives you the following:

$$f'(-1) = \frac{4 - \sqrt[3]{-1}}{\sqrt[3]{-1}} = \frac{4 + 1}{-1} < 0$$

$$f'(1) = \frac{4 - \sqrt[3]{1}}{\sqrt[3]{1}} > 0$$

$$f'(125) = \frac{4 - \sqrt[3]{125}}{\sqrt[3]{125}} = \frac{4 - 5}{5} < 0$$

Therefore, $f(x)$ is decreasing on $(-\infty, 0)$, increasing on $(0, 64)$, and decreasing on $(64, \infty)$. That means $f(x)$ has a local minimum at $x = 0$; $f(0) = 0$, so a local minimum occurs at $(0, 0)$. Also, $f(x)$ has a local maximum at $x = 64$; $f(64) = 6(64)^{2/3} - 64 = 32$, so the local maximum occurs at $(64, 32)$.

464.   local maximum at $\left(\dfrac{2\pi}{3}, \dfrac{3\sqrt{3}}{2}\right)$; local minimum at $\left(\dfrac{4\pi}{3}, -\dfrac{3\sqrt{3}}{2}\right)$

Begin by finding the derivative of the function $f(x) = 2\sin x - \sin 2x$:

$$f'(x) = 2\cos x - (\cos 2x)(2)$$

$$= 2\cos x - 2\cos(2x)$$

$$= 2\cos x - 2(2\cos^2 x - 1)$$

$$= -4\cos^2 x + 2\cos x + 2$$

Next, find the critical numbers by setting the derivative equal to zero:

$$-4\cos^2 x + 2\cos x + 2 = 0$$

$$2\cos^2 x - \cos x - 1 = 0$$

$$(2\cos x + 1)(\cos x - 1) = 0$$

Setting the first factor equal to zero gives you $\cos x = -\frac{1}{2}$, which has the solutions $x = \frac{2\pi}{3}$ and $x = \frac{4\pi}{3}$. Setting the second factor equal to zero gives you $\cos x = 1$, which has the solutions $x = 0$ and 2π.

Next, determine whether the function is increasing or decreasing on the intervals $\left(0, \frac{2\pi}{3}\right), \left(\frac{2\pi}{3}, \frac{4\pi}{3}\right),$ and $\left(\frac{4\pi}{3}, 2\pi\right)$ by taking a point inside each interval and substituting it into the derivative to see whether it's positive or negative. Using the points $\frac{\pi}{3}, \pi,$ and $\frac{3\pi}{2}$ gives you

$$f'\left(\frac{\pi}{3}\right) = -4\left(\cos\frac{\pi}{3}\right)^2 + 2\cos\left(\frac{\pi}{3}\right) + 2$$
$$= -1 + 2\left(\frac{1}{2}\right) + 2 > 0$$

$$f'(\pi) = -4(\cos\pi)^2 + 2\cos\pi + 2$$
$$= -4 - 2 + 2$$
$$= -4 < 0$$

$$f'\left(\frac{3\pi}{2}\right) = -4\left(\cos\frac{3\pi}{2}\right)^2 + 2\cos\left(\frac{3\pi}{2}\right) + 2$$
$$= 2 > 0$$

So $f(x)$ has a local maximum when $x = \frac{2\pi}{3}$ because the function changes from increasing to decreasing at this value, and $f(x)$ has a local minimum when $x = \frac{4\pi}{3}$ because the function changes from decreasing to increasing at this value.

To find the points on the original function, substitute in these values. If $x = \frac{2\pi}{3}$, then

$$f\left(\frac{2\pi}{3}\right) = 2\sin\left(\frac{2\pi}{3}\right) - \sin\left(2\left(\frac{2\pi}{3}\right)\right)$$
$$= 2\left(\frac{\sqrt{3}}{2}\right) - \left(-\frac{\sqrt{3}}{2}\right)$$
$$= \frac{3\sqrt{3}}{2}$$

So the local maximum occurs at $\left(\frac{2\pi}{3}, \frac{3\sqrt{3}}{2}\right)$. If $x = \frac{4\pi}{3}$, then

$$f\left(\frac{4\pi}{3}\right) = 2\sin\left(\frac{4\pi}{3}\right) - \sin\left(2\left(\frac{4\pi}{3}\right)\right)$$
$$= 2\left(\frac{-\sqrt{3}}{2}\right) - \frac{\sqrt{3}}{2}$$
$$= -\frac{3\sqrt{3}}{2}$$

Therefore, the local minimum occurs at $\left(\frac{4\pi}{3}, -\frac{3\sqrt{3}}{2}\right)$.

465. local maxima at $\left(\dfrac{-11\pi}{6}, \dfrac{-11\pi}{6}+\sqrt{3}\right), \left(\dfrac{\pi}{6}, \dfrac{\pi}{6}+\sqrt{3}\right)$; local minima at $\left(\dfrac{-7\pi}{6}, \dfrac{-7\pi}{6}-\sqrt{3}\right),$ $\left(\dfrac{5\pi}{6}, \dfrac{5\pi}{6}-\sqrt{3}\right)$

Begin by finding the derivative of the function $f(x) = x + 2\cos x$:

$$f'(x) = 1 - 2\sin x$$

Next, find the critical numbers by setting the function equal to zero and solving for x on the given interval:

$$1 - 2\sin x = 0$$
$$\sin x = \frac{1}{2}$$
$$x = \frac{-11\pi}{6}, \frac{-7\pi}{6}, \frac{\pi}{6}, \frac{5\pi}{6}$$

Then determine whether the function is increasing or decreasing on the intervals $\left(-2\pi, -\dfrac{11\pi}{6}\right), \left(-\dfrac{11\pi}{6}, -\dfrac{7\pi}{6}\right), \left(-\dfrac{7\pi}{6}, \dfrac{\pi}{6}\right), \left(\dfrac{\pi}{6}, \dfrac{5\pi}{6}\right),$ and $\left(\dfrac{5\pi}{6}, 2\pi\right)$ by taking a point inside each interval and substituting it into the derivative. Using the value $x = -6.27$, which is slightly larger than -2π, gives you $f'(-6.27) \approx f'(-2\pi) = 1 > 0$. Likewise, using the points $\dfrac{-3\pi}{2}, 0, \dfrac{\pi}{2},$ and $\dfrac{3\pi}{2}$ from each of the remaining four intervals gives you the following:

$$f'\left(\frac{-3\pi}{2}\right) = 1 - 2\sin\left(\frac{-3\pi}{2}\right) = 1 - 2 < 0$$
$$f'(0) = 1 - 2\sin(0) = 1 > 0$$
$$f'\left(\frac{\pi}{2}\right) = 1 - 2\sin\left(\frac{\pi}{2}\right) = 1 - 2 < 0$$
$$f'\left(\frac{3\pi}{2}\right) = 1 - 2\sin\left(\frac{3\pi}{2}\right) = 3 > 0$$

Therefore, the function has local maxima when $x = -\dfrac{11\pi}{6}$ and $x = \dfrac{\pi}{6}$ and local minima when $x = \dfrac{7\pi}{6}$ and $x = \dfrac{5\pi}{6}$. To find the points on the original function, substitute these x values into the original function:

$$f\left(\frac{-11\pi}{6}\right) = \frac{-11\pi}{6} + 2\cos\left(\frac{-11\pi}{6}\right)$$
$$= \frac{-11\pi}{6} + 2\left(\frac{\sqrt{3}}{2}\right)$$
$$= \frac{-11\pi}{6} + \sqrt{3}$$

$$f\left(\frac{-7\pi}{6}\right) = \frac{-7\pi}{6} + 2\cos\left(\frac{-7\pi}{6}\right)$$
$$= \frac{-7\pi}{6} + 2\left(\frac{-\sqrt{3}}{2}\right)$$
$$= \frac{-7\pi}{6} - \sqrt{3}$$

$$f\left(\frac{\pi}{6}\right) = \frac{\pi}{6} + 2\cos\left(\frac{\pi}{6}\right)$$

$$= \frac{\pi}{6} + 2\left(\frac{\sqrt{3}}{2}\right)$$

$$= \frac{\pi}{6} + \sqrt{3}$$

$$f\left(\frac{5\pi}{6}\right) = \frac{5\pi}{6} + 2\cos\left(\frac{5\pi}{6}\right)$$

$$= \frac{5\pi}{6} + 2\left(\frac{-\sqrt{3}}{2}\right)$$

$$= \frac{5\pi}{6} - \sqrt{3}$$

Therefore, the local maxima are $\left(\frac{-11\pi}{6}, \frac{-11\pi}{6} + \sqrt{3}\right)$ and $\left(\frac{\pi}{6}, \frac{\pi}{6} + \sqrt{3}\right)$, and the local minima are $\left(\frac{-7\pi}{6}, \frac{-7\pi}{6} - \sqrt{3}\right)$ and $\left(\frac{5\pi}{6}, \frac{5\pi}{6} - \sqrt{3}\right)$.

466. concave up on $(1, \infty)$; concave down on $(-\infty, 1)$

To determine concavity, examine the second derivative. Because a derivative measures a "rate of change" and the first derivative of a function gives the slopes of tangent lines, the derivative of the derivative (the second derivative) measures the rate of change of the slopes of the tangent lines. If the second derivative is positive on an interval, the slopes of the tangent lines are increasing, so the function is bending upward, or is concave up. Likewise, if the second derivative is negative on an interval, the slopes of the tangent lines are decreasing, so the function is bending downward, or is concave down.

Begin by finding the first and second derivatives of the function $f(x) = x^3 - 3x^2 + 4$:

$$f'(x) = 3x^2 - 6x$$

$$f''(x) = 6x - 6$$

Setting the second derivative equal to zero gives you $6x - 6 = 0$ so that $x = 1$.

To determine the concavity on the intervals $(-\infty, 1)$ and $(1, \infty)$, pick a point from each interval and substitute it into the second derivative to see whether it's positive or negative. Using the values $x = 0$ and $x = 2$, you have $f''(0) = 6(0) - 6 < 0$ and $f''(2) = 6(2) - 6 > 0$. Therefore, $f(x)$ is concave up on the interval $(1, \infty)$ and concave down on the interval $(-\infty, 1)$.

467. concave up nowhere; concave down on $(-\infty, 0)$, $(0, \infty)$

Begin by finding the first and second derivatives of the function $f(x) = 9x^{2/3} - x$:

$$f'(x) = 9\left(\frac{2}{3}x^{-1/3}\right) - 1$$

$$= 6x^{-1/3} - 1$$

$$f''(x) = 6\left(-\frac{1}{3}x^{-4/3}\right)$$

$$= \frac{-2}{x^{4/3}}$$

Next, set the denominator of the second derivative equal to zero to get $x^{4/3} = 0$ so that $x = 0$.

To determine the concavity on the intervals $(-\infty, 0)$ and $(0, \infty)$, pick a point from each interval and substitute it into the second derivative to see whether it's positive or negative. Using the values $x = -1$ and $x = 1$, you have the following:

$$f''(-1) = \frac{-2}{(-1)^{4/3}} < 0$$

$$f''(1) = \frac{-2}{(1)^{4/3}} < 0$$

Therefore, the $f(x)$ is concave down on the intervals $(-\infty, 0)$ and $(0, \infty)$.

468. concave up on $(-\infty, 0)$, $\left(\frac{1}{2}, \infty\right)$; concave down on $\left(0, \frac{1}{2}\right)$

Begin by finding the first derivative of the function $f(x) = x^{1/3}(x + 1) = x^{4/3} + x^{1/3}$:

$$f'(x) = \frac{4}{3}x^{1/3} + \frac{1}{3}x^{-2/3}$$

Next, use the power rule to find the second derivative:

$$f''(x) = \frac{4}{9}x^{-2/3} - \frac{2}{9}x^{-5/3}$$

$$= \frac{2}{9}x^{-5/3}(2x - 1)$$

$$= \frac{2(2x - 1)}{9x^{5/3}}$$

Then find where the second derivative is equal to zero or undefined. Setting the numerator equal to zero gives you $2x - 1 = 0$ so that $x = \frac{1}{2}$, and setting the denominator equal to zero gives you $9x^{5/3} = 0$ so that $x = 0$.

To determine the concavity on the intervals $(-\infty, 0)$, $\left(0, \frac{1}{2}\right)$, and $\left(\frac{1}{2}, \infty\right)$, pick a point from each interval and substitute it into the second derivative to see whether it's positive or negative. Using the values $x = -1$, $x = \frac{1}{4}$, and $x = 1$ gives you the following:

$$f''(-1) = \frac{4(-1) - 2}{9(-1)^{5/3}} > 0$$

$$f''\left(\frac{1}{4}\right) = \frac{4\left(\frac{1}{4}\right) - 2}{9\left(\frac{1}{4}\right)^{5/3}} < 0$$

$$f''(1) = \frac{4(1) - 2}{9(1)^{5/3}} > 0$$

Therefore, $f(x)$ is concave up on the intervals $(-\infty, 0)$ and $\left(\frac{1}{2}, \infty\right)$, and $f(x)$ is concave down on the interval $\left(0, \frac{1}{2}\right)$.

Answers 401–500

469. concave up on $(-\infty, -2)$, $\left(\dfrac{-2}{\sqrt{5}}, \dfrac{2}{\sqrt{5}}\right)$, $(2, \infty)$; concave down on $\left(-2, \dfrac{-2}{\sqrt{5}}\right)$, $\left(\dfrac{2}{\sqrt{5}}, 2\right)$

Begin by finding the first and second derivatives of the function $f(x) = (x^2 - 4)^3$:

$$f'(x) = 3\left(x^2 - 4\right)^2 (2x)$$

$$= 6x\left(x^2 - 4\right)^2$$

$$f''(x) = 6\left(x^2 - 4\right)^2 + (6x)(2)\left(x^2 - 4\right)(2x)$$

$$= 6\left(x^2 - 4\right)^2 + 24x^2\left(x^2 - 4\right)$$

$$= 6\left(x^2 - 4\right)\left(\left(x^2 - 4\right) + 4x^2\right)$$

$$= 6\left(x^2 - 4\right)\left(5x^2 - 4\right)$$

Next, set each factor in the second derivative equal to zero and solve for x: $x^2 - 4 = 0$ gives you the solutions $x = 2$ and $x = -2$, and $5x^2 - 4 = 0$, or $x^2 = \dfrac{4}{5}$, has the solutions $x = \dfrac{2}{\sqrt{5}}$ and $x = \dfrac{-2}{\sqrt{5}}$.

To determine the concavity on the intervals $(-\infty, -2)$, $\left(-2, -\dfrac{2}{\sqrt{5}}\right)$, $\left(-\dfrac{2}{\sqrt{5}}, \dfrac{2}{\sqrt{5}}\right)$, $\left(\dfrac{2}{\sqrt{5}}, 2\right)$, and $(2, \infty)$, pick a point from each interval and substitute it into the second derivative to see whether it's positive or negative. Using the values $x = -3$, $x = -1$, $x = 0$, $x = 1$, and $x = 3$, you have the following:

$$f''(-3) = 6\left((-3)^2 - 4\right)\left(5(-3)^2 - 4\right) > 0$$

$$f''(-1) = 6\left((-1)^2 - 4\right)\left(5(-1)^2 - 4\right) < 0$$

$$f''(0) = 6(-4)(-4) > 0$$

$$f''(1) = 6\left((1)^2 - 4\right)\left(5(1)^2 - 4\right) < 0$$

$$f''(3) = 6\left((3)^2 - 4\right)\left(5(3)^2 - 4\right) > 0$$

Therefore, $f(x)$ is concave up on the intervals $(-\infty, -2)$, $\left(\dfrac{-2}{\sqrt{5}}, \dfrac{2}{\sqrt{5}}\right)$, and $(2, \infty)$, and $f(x)$ is concave down on the intervals $\left(-2, \dfrac{-2}{\sqrt{5}}\right)$ and $\left(\dfrac{2}{\sqrt{5}}, 2\right)$.

470. concave up on $\left(0.253, \dfrac{\pi}{2}\right)$, $\left(\pi - 0.253, \dfrac{3\pi}{2}\right)$; concave down on the intervals $(0, 0.253)$, $\left(\dfrac{\pi}{2}, \pi - 0.253\right)$, $\left(\dfrac{3\pi}{2}, 2\pi\right)$

Begin by finding the first and second derivatives of the function $f(x) = 2\cos x - \sin(2x)$:

$$f'(x) = -2\sin x - (\cos(2x))(2)$$

$$= -2\sin x - 2\cos(2x)$$

$$f''(x) = -2\cos x + (2\sin(2x))(2)$$
$$= 4\sin(2x) - 2\cos x$$
$$= 4(2\sin x \cos x) - 2\cos x$$
$$= 2\cos x(4\sin x - 1)$$

Next, set each factor of the second derivative equal to zero and solve: $\cos x = 0$ has the solutions $x = \frac{\pi}{2}$ and $x = \frac{3\pi}{2}$, and $4\sin x - 1 = 0$, or $\sin x = \frac{1}{4}$, has the solutions $x = \sin^{-1}\left(\frac{1}{4}\right) \approx 0.253$ and $x = \pi - 0.253$.

To determine the concavity on the intervals $(0, 0.25)$, $\left(0.253, \frac{\pi}{2}\right)$, $\left(\frac{\pi}{2}, \pi - 0.253\right)$, $\left(\pi - 0.253, \frac{3\pi}{2}\right)$, and $\left(\frac{3\pi}{2}, 2\pi\right)$, pick a point from each interval and substitute it into the second derivative to see whether it's positive or negative. Using the values $x = 0.1$, $x = \frac{\pi}{3}$, $x = \frac{2\pi}{3}$, $x = \pi$, and $x = \frac{7\pi}{4}$ gives you the following:

$$f''(0.1) = 2\cos(0.1)(4\sin(0.1) - 1) < 0$$

$$f''\left(\frac{\pi}{3}\right) = 2\cos\left(\frac{\pi}{3}\right)\left(4\sin\left(\frac{\pi}{3}\right) - 1\right) > 0$$

$$f''\left(\frac{2\pi}{3}\right) = 2\cos\left(\frac{2\pi}{3}\right)\left(4\sin\left(\frac{2\pi}{3}\right) - 1\right) < 0$$

$$f''(\pi) = 2\cos(\pi)(4\sin(\pi) - 1) > 0$$

$$f''\left(\frac{7\pi}{4}\right) = 2\cos\left(\frac{7\pi}{4}\right)\left(4\sin\left(\frac{7\pi}{4}\right) - 1\right) < 0$$

Therefore, $f(x)$ is concave up on the intervals $\left(0.253, \frac{\pi}{2}\right)$ and $\left(\pi - 0.253, \frac{3\pi}{2}\right)$, and $f(x)$ is concave down on the intervals $(0, 0.253)$, $\left(\frac{\pi}{2}, \pi - 0.253\right)$, and $\left(\frac{3\pi}{2}, 2\pi\right)$.

471.   no inflection points

Inflection points are points where the function changes concavity. To determine concavity, examine the second derivative. Because a derivative measures a "rate of change" and the first derivative of a function gives the slopes of tangent lines, the derivative of the derivative (the second derivative) measures the rate of change of the slopes of the tangent lines. If the second derivative is positive on an interval, the slopes of the tangent lines are increasing, so the function is bending upward, or is concave up. Likewise, if the second derivative is negative on an interval, the slopes of the tangent lines are decreasing, so the function is bending downward, or is concave down.

Begin by finding the first and second derivatives of the function $f(x) = \frac{1}{x^2 - 9} = (x^2 - 9)^{-1}$. The first derivative is

$$f'(x) = -(x^2 - 9)^{-2}(2x)$$
$$= \frac{-2x}{(x^2 - 9)^2}$$

Now find the second derivative:

$$f''(x) = \frac{(x^2-9)^2(-2)-(-2x)(2(x^2-9)(2x))}{(x^2-9)^4}$$

$$= \frac{-2(x^2-9)\left[(x^2-9)-4x^2\right]}{(x^2-9)^4}$$

$$= \frac{-2(-3x^2-9)}{(x^2-9)^3}$$

$$= \frac{6(x^2+3)}{(x^2-9)^3}$$

Notice that the numerator of the second derivative never equals zero and that the denominator equals zero when $x = \pm 3$; however, $x = \pm 3$ isn't in the domain of the original function. Therefore, there can be no inflection points.

472. $\left(-\frac{1}{6}, \frac{1}{54}\right)$

Begin by finding the first and second derivatives of the function $f(x) = 2x^3 + x^2$. The first derivative is

$$f'(x) = 6x^2 + 2x$$

And the second derivative is

$$f''(x) = 12x + 2$$

Setting the second derivative equal to zero gives you $12x + 2 = 0$, which has the solution $x = \frac{-1}{6}$.

To determine the concavity on the intervals $\left(-\infty, -\frac{1}{6}\right)$ and $\left(-\frac{1}{6}, \infty\right)$, pick a point from each interval and substitute it into the second derivative to see whether it's positive or negative. Using the values $x = -1$ and $x = 0$ gives you the following:

$$f''(-1) = 12(-1) + 2 < 0$$

$$f''(0) = 12(0) + 2 > 0$$

Therefore, the function is concave down on the interval $\left(-\infty, -\frac{1}{6}\right)$ and concave up on the interval $\left(-\frac{1}{6}, \infty\right)$, so $x = -\frac{1}{6}$ is an inflection point. The corresponding y value on $f(x)$ is

$$f\left(-\frac{1}{6}\right) = 2\left(-\frac{1}{6}\right)^3 + \left(-\frac{1}{6}\right)^2$$

$$= \frac{-2}{216} + \frac{6}{216}$$

$$= \frac{4}{216}$$

$$= \frac{1}{54}$$

Therefore, the inflection point is $\left(-\frac{1}{6}, \frac{1}{54}\right)$.

473. no inflection points

Being by finding the first and second derivatives of the function $f(x) = \dfrac{\sin x}{1 + \cos x}$. The first derivative is

$$f'(x) = \frac{(1 + \cos x)\cos x - \sin x(-\sin x)}{(1 + \cos x)^2}$$

$$= \frac{\cos x + \cos^2 x + \sin^2 x}{(1 + \cos x)^2}$$

$$= \frac{1 + \cos x}{(1 + \cos x)^2}$$

$$= \frac{1}{1 + \cos x}$$

$$= (1 + \cos x)^{-1}$$

And the second derivative is

$$f''(x) = -1(1 + \cos x)^{-2}(-\sin x)$$

$$= \frac{\sin x}{(1 + \cos x)^2}$$

Next, find where the second derivative is equal to zero or undefined. Setting the numerator equal to zero gives you $\sin x = 0$, which has solutions $x = 0$, $x = \pi$, and $x = 2\pi$. Setting the denominator equal to zero gives you $1 + \cos x = 0$, or $\cos x = -1$, which has the solution $x = \pi$.

To determine the concavity on the intervals $(0, \pi)$ and $(\pi, 2\pi)$, pick a point from each interval and substitute it into the second derivative to see whether it's positive or negative. Using the values $x = \dfrac{\pi}{2}$ and $x = \dfrac{3\pi}{2}$ gives you the following:

$$f''\left(\frac{\pi}{2}\right) = \frac{\sin\left(\frac{\pi}{2}\right)}{\left(1 + \cos\frac{\pi}{2}\right)^2} > 0$$

$$f''\left(\frac{3\pi}{2}\right) = \frac{\sin\left(\frac{3\pi}{2}\right)}{\left(1 + \cos\frac{3\pi}{2}\right)^2} < 0$$

However, note that at $x = \pi$, the original function is undefined, so there are no inflection points. In fact, if you noticed this at the beginning, there's really no need to determine the concavity on the intervals!

474. $(\pi, 0)$

Begin by finding the first and second derivatives of the function $f(x) = 3\sin x - \sin^3 x$. The first derivative is

$$f'(x) = 3\cos x - 3(\sin x)^2(\cos x)$$

$$= 3\cos x(1 - \sin^2 x)$$

$$= 3\cos x(\cos^2 x)$$

$$= 3\cos^3 x$$

And the second derivative is

$$f''(x) = 3\left(3(\cos x)^2(-\sin x)\right)$$

$$= -9\cos^2 x \sin x$$

Next, find where the second derivative is equal to zero by setting each factor equal to zero: $\cos x$ has the solutions $x = \dfrac{\pi}{2}$ and $x = \dfrac{3\pi}{2}$, and $\sin x = 0$ has the solutions $x = 0$, $x = \pi$, and $x = 2\pi$.

To determine the concavity on the intervals $\left(0, \dfrac{\pi}{2}\right)$, $\left(\dfrac{\pi}{2}, \pi\right)$, $\left(\pi, \dfrac{3\pi}{2}\right)$, and $\left(\dfrac{3\pi}{2}, 2\pi\right)$, pick a point from each interval and substitute it into the second derivative to see whether it's positive or negative. Using the values $x = \dfrac{\pi}{4}$, $x = \dfrac{3\pi}{4}$, $x = \dfrac{5\pi}{4}$, and $x = \dfrac{7\pi}{4}$, you have the following:

$$f''\left(\frac{\pi}{4}\right) = -9\left(\cos\left(\frac{\pi}{4}\right)\right)^2 \sin\left(\frac{\pi}{4}\right) < 0$$

$$f''\left(\frac{3\pi}{4}\right) = -9\left(\cos\left(\frac{3\pi}{4}\right)\right)^2 \sin\left(\frac{3\pi}{4}\right) < 0$$

$$f''\left(\frac{5\pi}{4}\right) = -9\left(\cos\left(\frac{5\pi}{4}\right)\right)^2 \sin\left(\frac{5\pi}{4}\right) > 0$$

$$f''\left(\frac{7\pi}{4}\right) = -9\left(\cos\left(\frac{7\pi}{4}\right)\right)^2 \sin\left(\frac{7\pi}{4}\right) > 0$$

Therefore, the concavity changes when $x = \pi$. The corresponding y value on $f(x)$ is $f(\pi) = 3 \sin \pi - (\sin \pi)^3 = 0$, so the inflection point is $(\pi, 0)$.

475. $(-1, -6)$

Begin by finding the first and second derivatives of the function $f(x) = x^{5/3} - 5x^{2/3}$. The first derivative is

$$f'(x) = \frac{5}{3} x^{2/3} - 5\left(\frac{2}{3} x^{-1/3}\right)$$

$$= \frac{5}{3} x^{2/3} - \frac{10}{3} x^{-1/3}$$

And the second derivative is

$$f''(x) = \frac{5}{3}\left(\frac{2}{3} x^{-1/3}\right) - \frac{10}{3}\left(-\frac{1}{3} x^{-4/3}\right)$$

$$= \frac{10}{9x^{1/3}} + \frac{10}{9x^{4/3}}$$

$$= \frac{10x + 10}{9x^{4/3}}$$

Next, find where the second derivative is equal to zero or undefined by setting the numerator and the denominator equal to zero. For the numerator, you have $10x + 10 = 0$, which has the solution $x = -1$. For the denominator, you have $9x^{4/3} = 0$, which has the solution $x = 0$.

To determine the concavity on the intervals $(-\infty, -1)$, $(1, 0)$, and $(0, \infty)$, pick a point from each interval and substitute it into the second derivative to see whether it's positive or negative. Using the values $x = -2$, $x = -\frac{1}{2}$, and $x = 1$ gives you

$$f''(-2) = \frac{10(-2)+10}{9(-2)^{4/3}} < 0$$

$$f''\left(-\frac{1}{2}\right) = \frac{10\left(-\frac{1}{2}\right)+10}{9\left(-\frac{1}{2}\right)^{4/3}} > 0$$

$$f''(1) = \frac{10(1)+10}{9(1)^{4/3}} > 0$$

Because the concavity changes at $x = -1$, the original function has an inflection point there. The corresponding y value on $f(x)$ is

$$f(-1) = (-1)^{5/3} - 5(-1)^{2/3}$$
$$= -1-5$$
$$= -6$$

Therefore, the inflection point is $(-1, -6)$.

476. no local maxima; local minimum at $(0, 1)$

Begin by finding the first derivative of the function $f(x) = \sqrt[3]{(x^2+1)^2}$:

$$f'(x) = \frac{2}{3}(x^2+1)^{-1/3}(2x)$$
$$= \frac{4x}{3}(x^2+1)^{-1/3}$$
$$= \frac{4x}{3(x^2+1)^{1/3}}$$

Next, find any critical numbers of the function. Setting the numerator of the derivative equal to zero gives you the solution $x = 0$. Note that no real values can make the denominator equal to zero.

Then find the second derivative:

$$f''(x) = \frac{3(x^2+1)^{1/3}(4)-(4x)\left(3\frac{1}{3}(x^2+1)^{-2/3}(2x)\right)}{\left(3(x^2+1)^{1/3}\right)^2}$$

$$= \frac{12(x^2+1)^{1/3}-8x^2(x^2+1)^{-2/3}}{9(x^2+1)^{2/3}}$$

$$= \frac{4(x^2+1)^{-2/3}\left(3(x^2+1)-2x^2\right)}{9(x^2+1)^{2/3}}$$

$$= \frac{4(3x^2+3-2x^2)}{9(x^2+1)^{4/3}}$$

$$= \frac{4x^2+12}{9(x^2+1)^{4/3}}$$

To see whether the original function is concave up or concave down at the critical number, substitute $x = 0$ into the second derivative:

$$f''(0) = \frac{4(0)^2 + 12}{9((0)^2 + 1)^{4/3}} = \frac{4}{3} > 0$$

The second derivative is positive, so the original function is concave up at the critical point; therefore, a local minimum is at $x = 0$. The corresponding y value on the original function is $f(0) = \sqrt[3]{(0^2 + 1)^2} = 1$, so the local minimum is at $(0, 1)$.

477. local maximum at $(0, 1)$; local minima at $\left(-\sqrt{2}, -3\right), \left(\sqrt{2}, -3\right)$

Begin by finding the first derivative of the function $f(x) = x^4 - 4x^2 + 1$:

$$f'(x) = 4x^3 - 8x$$
$$= 4x(x^2 - 2)$$

Set the first derivative equal to zero and solve for x to find the critical numbers of f. The critical numbers are $x = 0$, $x = \sqrt{2}$, and $x = -\sqrt{2}$.

Next, find the second derivative:

$$f''(x) = 12x^2 - 8$$

Substitute each of the critical numbers into the second derivative to see whether the second derivative is positive or negative at those values:

$$f''(0) = 12(0)^2 - 8 = -8 < 0$$
$$f''\left(\sqrt{2}\right) = 12\left(\sqrt{2}\right)^2 - 8 = 16 > 0$$
$$f''\left(-\sqrt{2}\right) = 12\left(-\sqrt{2}\right)^2 - 8 = 16 > 0$$

Therefore, the original function has a local maximum when $x = 0$ and local minima when $x = \sqrt{2}$ and $x = -\sqrt{2}$. The corresponding y values on the original function are $f\left(-\sqrt{2}\right) = \left(-\sqrt{2}\right)^4 - 4\left(-\sqrt{2}\right)^2 + 1 = -3$, $f(0) = 0^4 - 4(0)^2 + 1 = 1$, and $f\left(\sqrt{2}\right) = \left(\sqrt{2}\right)^4 - 4\left(\sqrt{2}\right)^2 + 1 = -3$. Therefore, the local maximum occurs at $(0, 1)$, and the local minima occur at $\left(-\sqrt{2}, -3\right)$ and $\left(\sqrt{2}, -3\right)$.

478. local maxima at $\left(-\frac{1}{\sqrt{2}}, \frac{1}{2}\right), \left(\frac{1}{\sqrt{2}}, \frac{1}{2}\right)$; local minimum at $(0, 0)$

Begin by finding the first derivative of the function $f(x) = 2x^2(1 - x^2) = 2x^2 - 2x^4$:

$$f'(x) = 4x - 8x^3$$
$$= 4x(1 - 2x^2)$$

Set the first derivative equal to zero and solve for x to find the critical numbers.

Solving $4x(1 - 2x^2) = 0$ gives you $x = 0$, $x = \dfrac{1}{\sqrt{2}}$, and $x = -\dfrac{1}{\sqrt{2}}$ as the critical numbers. Next, find the second derivative:

$$f''(x) = 4 - 24x^2$$

Substitute the critical numbers into the second derivative to see whether the second derivative is positive or negative at those values:

$$f''\left(-\frac{1}{\sqrt{2}}\right) = 4 - 24\left(-\frac{1}{\sqrt{2}}\right)^2 = 4 - 12 < 0$$

$$f''(0) = 4 - 24(0)^2 = 4 > 0$$

$$f''\left(\frac{1}{\sqrt{2}}\right) = 4 - 24\left(\frac{1}{\sqrt{2}}\right)^2 = 4 - 12 < 0$$

So the original function has local maxima when $x = \dfrac{1}{\sqrt{2}}$ and $x = -\dfrac{1}{\sqrt{2}}$ and a local minimum when $x = 0$. Finding the corresponding y values on the original function gives you the following:

$$f\left(\frac{1}{\sqrt{2}}\right) = 2\left(\frac{1}{\sqrt{2}}\right)^2 - 2\left(\frac{1}{\sqrt{2}}\right)^4 = \frac{2}{2} - \frac{2}{4} = \frac{1}{2}$$

$$f(0) = 2(0)^2 - 2(0)^4 = 0$$

$$f\left(-\frac{1}{\sqrt{2}}\right) = 2\left(-\frac{1}{\sqrt{2}}\right)^2 - 2\left(-\frac{1}{\sqrt{2}}\right)^4 = \frac{2}{2} - \frac{2}{4} = \frac{1}{2}$$

Therefore, the local maxima are at $\left(-\dfrac{1}{\sqrt{2}}, \dfrac{1}{2}\right)$ and $\left(\dfrac{1}{\sqrt{2}}, \dfrac{1}{2}\right)$, and the local minimum is at $(0, 0)$.

479.   local maximum at $\left(2, \dfrac{1}{4}\right)$; local minimum at $\left(-2, -\dfrac{1}{4}\right)$

Begin by finding the first derivative of the function $f(x) = \dfrac{x}{x^2 + 4}$:

$$f'(x) = \frac{(x^2 + 4)(1) - x(2x)}{(x^2 + 4)^2}$$

$$= \frac{-x^2 + 4}{(x^2 + 4)^2}$$

Then set the numerator and denominator of the first derivative equal to zero to find the critical numbers. Setting the numerator equal to zero gives you $-x^2 + 4 = 0$, which has the solutions $x = 2$ and $x = -2$. Notice that the denominator of the derivative doesn't equal zero for any value of x.

Next, find the second derivative of the function:

$$f''(x) = \frac{\left(x^2+4\right)^2(-2x)-\left(-x^2+4\right)\left(2\left(x^2+4\right)(2x)\right)}{\left(x^2+4\right)^4}$$

$$= \frac{(-2x)\left(x^2+4\right)^2 - 4x\left(x^2+4\right)\left(-x^2+4\right)}{\left(x^2+4\right)^4}$$

$$= \frac{(-2x)\left(x^2+4\right)\left[\left(x^2+4\right)+2\left(-x^2+4\right)\right]}{\left(x^2+4\right)^4}$$

$$= \frac{(-2x)\left(12-x^2\right)}{\left(x^2+4\right)^3}$$

Substitute the critical numbers into the second derivative to see whether the second derivative is positive or negative at those values:

$$f''(-2) = \frac{(4)(8)}{(8)^3} > 0$$

$$f''(2) = \frac{(-4)(8)}{(8)^3} < 0$$

Because $f(x)$ is concave up when $x = -2$ and concave down when $x = 2$, a local minimum occurs at $x = -2$, and a local maximum occurs when $x = 2$. Because $f(-2) = \frac{-2}{(-2)^2+4} = -\frac{1}{4}$ and $f(2) = \frac{2}{2^2+4} = \frac{1}{4}$, the local minimum occurs at $\left(-2, -\frac{1}{4}\right)$, and the local maximum occurs at $\left(2, \frac{1}{4}\right)$.

480. local maximum at $\left(\frac{\pi}{3}, \sqrt{3} - \frac{\pi}{3}\right)$; local minimum at $\left(\frac{5\pi}{3}, -\sqrt{3} - \frac{5\pi}{3}\right)$

Begin by finding the first derivative of the function $f(x) = 2\sin x - x$:

$$f'(x) = 2\cos x - 1$$

Set the first derivative equal to zero to find the critical numbers. From the equation $2\cos x - 1 = 0$, you get $\cos x = \frac{1}{2}$, so the critical numbers are $x = \frac{\pi}{3}$ and $x = \frac{5\pi}{3}$.

Next, find the second derivative:

$$f''(x) = -2\sin x$$

Substitute the critical numbers into the second derivative to see whether the second derivative is positive or negative at these values:

$$f''\left(\frac{\pi}{3}\right) = -2\sin\frac{\pi}{3} = -2\left(\frac{\sqrt{3}}{2}\right) < 0$$

$$f''\left(\frac{5\pi}{3}\right) = -2\sin\frac{5\pi}{3} = -2\left(-\frac{\sqrt{3}}{2}\right) > 0$$

Because the function is concave down at $x = \frac{\pi}{3}$, it has a local maximum there, and because the function is concave up at $x = \frac{5\pi}{3}$, it has a local minimum there. To find the corresponding y value on $f(x)$, substitute the critical numbers into the original function:

$$f\left(\frac{\pi}{3}\right) = 2\sin\left(\frac{\pi}{3}\right) - \frac{\pi}{3}$$

$$= 2\left(\frac{\sqrt{3}}{2}\right) - \frac{\pi}{3}$$

$$= \sqrt{3} - \frac{\pi}{3}$$

$$f\left(\frac{5\pi}{3}\right) = 2\sin\left(\frac{5\pi}{3}\right) - \frac{5\pi}{3}$$

$$= 2\left(-\frac{\sqrt{3}}{2}\right) - \frac{5\pi}{3}$$

$$= -\sqrt{3} - \frac{5\pi}{3}$$

Therefore, the local maximum is at $\left(\frac{\pi}{3}, \sqrt{3} - \frac{\pi}{3}\right)$, and the local minimum is at $\left(\frac{5\pi}{3}, -\sqrt{3} - \frac{5\pi}{3}\right)$.

481. 3

Recall Rolle's theorem: If f is a function that satisfies the following three hypotheses:

✔ f is continuous on the closed interval $[a, b]$

✔ f is differentiable on the open interval (a, b)

✔ $f(a) = f(b)$

then there is a number c in (a, b) such that $f'(c) = 0$.

Notice that the given function is differentiable everywhere (and therefore continuous everywhere) because it's a polynomial. Then verify that $f(0) = f(6)$ so that Rolle's theorem can be applied:

$$f(0) = 0^2 - 0 + 1 = 1$$
$$f(6) = 6^2 - 6(6) + 1 = 1$$

Next, find the derivative of the function, set it equal to zero, and solve for c:

$$f'(c) = 2c - 6$$
$$2c - 6 = 0$$
$$c = 3$$

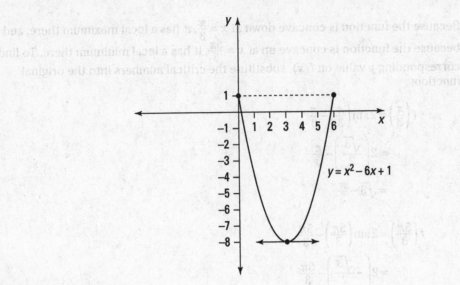

482. $-\dfrac{16}{3}$

Notice that the given function is differentiable on $(-8, 0)$. Then verify that $f(-8) = f(0)$ so that Rolle's theorem can be applied:

$$f(-8) = -8\sqrt{-8+8} = -8\sqrt{0} = 0$$
$$f(0) = 0\sqrt{0+8} = 0$$

Next, find the derivative of the function, set it equal to zero, and solve. Here's the derivative:

$$f'(c) = 1(c+8)^{1/2} + c\left(\frac{1}{2}(c+8)^{-1/2}\right)$$

$$= (c+8)^{-1/2}\left[(c+8) + \frac{1}{2}c\right]$$

$$= \frac{\frac{3}{2}c + 8}{\sqrt{c+8}}$$

Setting the numerator equal to zero and solving for c gives you

$$\frac{3}{2}c + 8 = 0$$

$$\frac{3}{2}c = -8$$

$$c = -\frac{16}{3}$$

483. $0, \pm\dfrac{1}{2}, \pm 1$

Notice that the given function is differentiable everywhere. Then verify that $f(-1) = f(1)$ so that Rolle's theorem can be applied:

$$f(-1) = \cos(-2\pi) = 1$$
$$f(1) = \cos(2\pi) = 1$$

Next, find the derivative of the function, set it equal to zero, and solve for c. Here's the derivative:

$$f'(c) = (-\sin(2\pi c))(2\pi)$$
$$= -2\pi \sin(2\pi c)$$

You need to find the solutions of $\sin(2\pi c) = 0$ on the interval $-1 \le c \le 1$ so that $-2\pi \le 2\pi c \le 2\pi$. Notice that the solutions occur when $2\pi c = -2\pi, -\pi, 0, \pi,$ and 2π. Solving each of these equations for c gives you the solutions $c = -1$, $c = -\dfrac{1}{2}$, $c = 0$, $c = \dfrac{1}{2}$, and $c = 1$.

484. $\dfrac{2\sqrt{3}}{3}$

Recall the mean value theorem: If f is a function that satisfies the following hypotheses:

  ✔ f is continuous on the closed interval $[a, b]$

  ✔ f is differentiable on the open interval (a, b)

then there is a number c in (a, b) such that $f'(c) = \dfrac{f(b) - f(a)}{b - a}$.

Notice that the given function is differentiable everywhere, so you can apply the mean value theorem. Next, find the value of $\dfrac{f(b) - f(a)}{b - a}$ on the given interval:

$$\frac{f(2) - f(0)}{2 - 0} = \frac{\left(2^3 + 3(2) - 1\right) - (-1)}{2}$$
$$= \frac{14}{2}$$
$$= 7$$

Now find the derivative of $f(x) = x^3 + 3x - 1$, the given function, and set it equal to 7. The derivative is $f' = 3x^2 + 3$, so

$$3c^2 + 3 = 7$$
$$3c^2 = 4$$
$$c^2 = \frac{4}{3}$$

The negative root falls outside the given interval, so keeping only the positive root gives you the solution $c = \sqrt{\dfrac{4}{3}} = \dfrac{2}{\sqrt{3}} = \dfrac{2\sqrt{3}}{3}$.

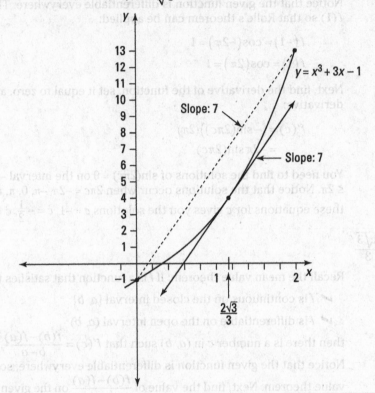

485. $\left(\dfrac{1}{3}\right)^{3/2}$

Notice that the given function is differentiable on $(0, 1)$ and continuous on $[0, 1]$, so you can apply the mean value theorem. Next, find the value of $\dfrac{f(b)-f(a)}{b-a}$ on the given interval:

$$\frac{f(1)-f(0)}{1-0} = \frac{2-0}{1} = 2$$

Now find the derivative of $f(x) = 2\sqrt[3]{x}$, the given function, and set it equal to 2.

The derivative is $f'(x) = 2\left(\dfrac{1}{3}x^{-2/3}\right) = \dfrac{2}{3x^{2/3}}$, so you have

$$\frac{2}{3c^{2/3}} = 2$$

$$\frac{1}{3} = c^{2/3}$$

$$\left(\frac{1}{3}\right)^{3/2} = c$$

486. $3\sqrt{2}-2$

Notice that the given function is differentiable everywhere, so you can apply the mean value theorem. Next, find the value of $\dfrac{f(b)-f(a)}{b-a}$ on the given interval:

$$\frac{f(4)-f(1)}{4-1}=\frac{\dfrac{4}{4+2}-\dfrac{1}{1+2}}{3}$$

$$=\frac{\dfrac{2}{3}-\dfrac{1}{3}}{3}$$

$$=\frac{1}{9}$$

Now find the derivative of $f(x)=\dfrac{x}{x+2}$, the given function, and set it equal to $\dfrac{1}{9}$.

The derivative is $f'(x)=\dfrac{(x+2)(1)-x(1)}{(x+2)^2}=\dfrac{2}{(x+2)^2}$, so you have

$$\frac{2}{(c+2)^2}=\frac{1}{9}$$

$$18=(c+2)^2$$

Keeping only the positive root, you have $\sqrt{18}=c+2$, which has the solution $3\sqrt{2}-2=c$. Note that the negative root would give you $c=-3\sqrt{2}-2$, which is outside of the interval.

487.   24

By the mean value theorem, you have the following equation for some c on the interval [1, 5]:

$$\frac{f(5)-f(1)}{5-1}=f'(c)$$

$$f(5)-f(1)=4f'(c)$$

Next, solve for $f(5)$ using the fact that $f(1) = 12$ and $f'(x) \geq 3$ (and so $f'(c) \geq 3$):

$$f(5)=f(1)+4f'(c)$$

$$=12+4f'(c)$$

$$\geq 12+4(3)=24$$

488. $8 \leq f(8)-f(4) \leq 24$

If $2 \leq f'(x) \leq 6$, then by the mean value theorem, you have the following equation for some c in the interval [4, 8]:

$$\frac{f(8)-f(4)}{8-4}=f'(c)$$

$$f(8)-f(4)=4f'(c)$$

Using $2 \leq f'(x) \leq 6$, you can bound the value of $4f'(c)$ because $2 \leq f'(c) \leq 6$:

$$(4)2 \leq (4)f'(c) \leq (4)6$$

$$8 \leq 4f'(c) \leq 24$$

Because $f(8) - f(4) = 4f'(c)$, it follows that

$$8 \le f(8) - f(4) \le 24$$

489. $2 < \sqrt[3]{9} < \dfrac{25}{12}$

The given function $f(x) = x^{1/3} = \sqrt[3]{x}$ is continuous and differentiable on the given interval. By the mean value theorem, there exists a number c in the interval $[8, 9]$ such that

$$f'(c) = \frac{\sqrt[3]{9} - \sqrt[3]{8}}{9 - 8} = \sqrt[3]{9} - 2$$

But because $f'(x) = \frac{1}{3}x^{-2/3}$, you have $f'(c) = \frac{1}{3}c^{-2/3}$. Replacing $f'(c)$ in the

equation $f'(c) = \dfrac{\sqrt[3]{9} - \sqrt[3]{8}}{9 - 8} = \sqrt[3]{9} - 2$ gives you

$$\frac{1}{3}c^{-2/3} = \sqrt[3]{9} - 2$$

Because $\frac{1}{3}c^{-2/3} > 0$, you have $\sqrt[3]{9} - 2 > 0$, so

$$\sqrt[3]{9} > 2$$

Notice also that f' is decreasing (f'' will be negative on the given interval), so the following is true:

$$f'(c) < f'(8) = \frac{1}{3}(8)^{-2/3}$$
$$= \frac{1}{3(8)^{2/3}}$$
$$= \frac{1}{3(4)}$$
$$= \frac{1}{12}$$

Because $\frac{1}{12} > f'(c) = \sqrt[3]{9} - 2$, you have $\sqrt[3]{9} < 2 + \frac{1}{12} = \frac{25}{12}$. Therefore, it follows that

$$2 < \sqrt[3]{9} < \frac{25}{12}$$

490. velocity: 2; acceleration: 2

Begin by taking the derivative of the position function $s(t)$ to find the velocity function:

$$s'(t) = v(t) = 2t - 8$$

Substituting in the value of $t = 5$, find the velocity:

$$v(5) = 2(5) - 8 = 2$$

Next, take the derivative of the velocity function to find the acceleration function:

$$s''(t) = v'(t) = a(t) = 2$$

Because the acceleration is a constant at $t = 5$, the acceleration is $a(5) = 2$.

491. velocity: 1; acceleration: –2

Begin by taking the derivative of the position function $s(t)$ to find the velocity function:

$$s'(t) = v(t) = 2\cos t + \sin t$$

Substituting in the value $t = \frac{\pi}{2}$ gives you the velocity:

$$v\left(\frac{\pi}{2}\right) = 2\cos\frac{\pi}{2} + \sin\frac{\pi}{2} = 2(0) + 1 = 1$$

Next, take the derivative of the velocity function to find the acceleration function:

$$s''(t) = v'(t) = a(t) = -2\sin t + \cos t$$

Substituting in the value of $t = \frac{\pi}{2}$ gives you the acceleration:

$$s''(t) = v'(t) = a\left(\frac{\pi}{2}\right) = -2\sin\frac{\pi}{2} + \cos\frac{\pi}{2}$$
$$= -2(1) + 0 = -2$$

492. velocity: 0; acceleration: –1

Begin by taking the derivative of the position function $s(t)$ to find the velocity function:

$$s'(t) = v(t) = \frac{(t^2+1)(2) - (2t)(2t)}{(t^2+1)^2}$$

$$= \frac{2t^2 + 2 - 4t^2}{(t^2+1)^2}$$

$$= \frac{2 - 2t^2}{(t^2+1)^2}$$

Substituting in the value $t = 1$ gives you the velocity:

$$v(1) = \frac{2 - 2(1)^2}{((1)^2+1)^2} = 0$$

Next, take the derivative of the velocity function to find the acceleration function:

$$a(t) = \frac{(t^2+1)^2(-4t) - (2 - 2t^2)2(t^2+1)2t}{(t^2+1)^4}$$

$$= \frac{(t^2+1)\left[(-4t)(t^2+1) - 4t(2 - 2t^2)\right]}{(t^2+1)^4}$$

$$= \frac{-4t^3 - 4t - 8t + 8t^3}{(t^2+1)^3}$$

$$= \frac{4t^3 - 12t}{(t^2+1)^3}$$

Substituting in the value of $t = 1$ gives you the acceleration:

$$a(1) = \frac{4(1)^3 - 12(1)}{((1)^2 + 1)^3} = \frac{-8}{8} = -1$$

493. compressing; speed: 8 ft/s

Begin by taking the derivative of the equation of motion to find the velocity function:

$$x(t) = 8\sin(2t)$$
$$x'(t) = v(t) = 8(\cos(2t))(2)$$
$$v(t) = 16\cos(2t)$$

At $t = \frac{\pi}{3}$, you have

$$v\left(\frac{\pi}{3}\right) = 16\cos\left(2\frac{\pi}{3}\right)$$
$$= 16\left(-\frac{1}{2}\right)$$
$$= -8 \text{ ft/s}$$

The velocity is negative, so the spring is compressing. Because the speed is the absolute value of the velocity, the speed is 8 feet per second.

494. maximum height: 25 ft; upward velocity: $8\sqrt{5}$ ft/s; downward velocity: $-8\sqrt{5}$ ft/s

One way to find the maximum height of the stone is to determine when the velocity of the stone is zero, because that's when the stone stops going upward and begins falling. (The stone follows a parabolic path, so you can also use the formula to find the vertex, but try the calculus way instead.)

Finding the derivative of the position function gives you the velocity function:

$$s = 40t - 16t^2$$
$$v = s' = 40 - 32t$$

Set the velocity function equal to zero and solve for time t:

$$40 - 32t = 0$$
$$\frac{40}{32} = t$$
$$t = \frac{5}{4}$$

Substituting this time value into the position equation gives you

$$s = 40\left(\frac{5}{4}\right) - 16\left(\frac{5}{4}\right)^2 = 25$$

Therefore, the maximum height of the stone is 25 feet.

To answer the second part of the question, use algebra to find at what points in time the stone is at a height of 20 feet:

$$40t - 16t^2 = 20$$
$$-16t^2 + 40t - 20 = 0$$
$$4t^2 - 10t + 5 = 0$$

Using the quadratic formula with $a = 4$, $b = -10$, and $c = 5$, you get the solutions $t = \frac{1}{4}\left(5 - \sqrt{5}\right)$ and $t = \frac{1}{4}\left(5 + \sqrt{5}\right)$. Substituting these values into the derivative gives you the velocity at these times:

$$v_1 = 40 - 32\left(\frac{1}{4}\left(5 - \sqrt{5}\right)\right)$$
$$= 40 - 8\left(5 - \sqrt{5}\right)$$
$$= 8\sqrt{5}$$

$$v_2 = 40 - 32\left(\frac{1}{4}\left(5 + \sqrt{5}\right)\right)$$
$$= 40 - 8\left(5 + \sqrt{5}\right)$$
$$= -8\sqrt{5}$$

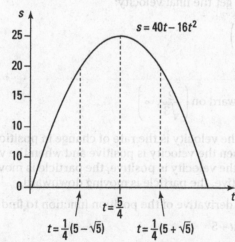

The velocities are approximately equal to 17.89 feet per second and –17.89 feet per second. The signs on the answers reflect that the stone is moving up at the first time and down at the second time.

---

495.   maximum height: 6.25 ft; velocity on impact: –20 ft/s

One way to find the height of the stone is to determine when the velocity of the stone is zero, because that's when the stone stops going upward and begins falling. (The stone follows a parabolic path, so you can also use the formula to find the vertex, but try the calculus way instead.)

Finding the derivative of the position function gives you the velocity function:

$$s = 20t - 16t^2$$
$$v = s' = 20 - 32t$$

Set the velocity function equal to zero and solve for time t:

$$20 - 32t = 0$$
$$t = \frac{20}{32} = \frac{5}{8}$$

Substitute in this value to find the height:

$$s\left(\frac{5}{8}\right) = 20\left(\frac{5}{8}\right) - 16\left(\frac{5}{8}\right)^2 = \frac{25}{4} = 6.25 \text{ ft}$$

To find the velocity of the stone when it hits the ground, first determine when the stone hits the ground by setting the position equation equal to zero and solving for t:

$$20t - 16t^2 = 0$$
$$4t(5 - 4t) = 0$$
$$t = 0, \frac{5}{4}$$

Because $t = 0$ corresponds to when the stone is first released, substitute $t = \frac{5}{4}$ into the velocity function to get the final velocity:

$$v = 20 - 32\left(\frac{5}{4}\right)$$
$$= -20 \text{ ft/s}$$

496. downward on $\left[0, \frac{2}{\sqrt{3}}\right]$; upward on $\left(\frac{2}{\sqrt{3}}, \infty\right)$

Note that because the velocity is the rate of change in position with respect to time, you want to find when the velocity is positive and when the velocity is negative. Assume that when the velocity is positive, the particle is moving upward, and when the velocity is negative, the particle is moving downward.

Begin by taking the derivative of the position function to find the velocity function:

$$y = t^3 - 4t + 5$$
$$v = y' = 3t^2 - 4$$

Set the velocity function equal to zero and solve for time t:

$$3t^2 - 4 = 0$$
$$t = \pm\sqrt{\frac{4}{3}} = \pm\frac{2}{\sqrt{3}}$$

Use only the positive solution based on the given interval for t.

Now take a point from the interval $\left(0, \frac{2}{\sqrt{3}}\right)$ and a point from $\left(\frac{2}{\sqrt{3}}, \infty\right)$ and substitute them into the velocity function to see whether the velocity is positive or negative on those intervals. Using $t = 1$, you have $v(1) = 3(1)^2 - 4 < 0$, and using $t = 10$, you have $v(10) = 3(10)^2 - 4 > 0$; therefore, the particle is moving downward on the interval $\left[0, \frac{2}{\sqrt{3}}\right]$ and upward on the interval $\left(\frac{2}{\sqrt{3}}, \infty\right)$.

497. downward on $\left[0, \frac{3}{4}\right)$; upward on $\left(\frac{3}{4}, \infty\right)$

Note that because the velocity is the rate of change in position with respect to time, you want to find when the velocity is positive and when the velocity is negative. Assume that when the velocity is positive, the particle is moving upward, and when the velocity is negative, the particle is moving downward.

Begin by taking the derivative of the position function to find the velocity function:

$$y = 4t^2 - 6t - 2$$
$$v = y' = 8t - 6$$

Then find when the velocity is equal to zero:

$$8t - 6 = 0$$
$$t = \frac{6}{8} = \frac{3}{4}$$

Now take a point from the interval $\left(0, \frac{3}{4}\right)$ and a point from $\left(\frac{3}{4}, \infty\right)$ and substitute them into the velocity function to see whether the velocity is positive or negative on those intervals. If you use $t = \frac{1}{2}$, you have $v\left(\frac{1}{2}\right) = 8\left(\frac{1}{2}\right) - 6 < 0$, and if $t = 1$, you have $v(1) = 8(1) - 6 > 0$; therefore, the particle is moving downward on the interval $\left[0, \frac{3}{4}\right)$ and upward on the interval $\left(\frac{3}{4}, \infty\right)$.

498. −25, 25

Let x and y be the two numbers so that $x - y = 50$. The function that you want to minimize, the product, is given by

$$P(x, y) = (x)(y)$$

Solve for x so you can write the product in terms of one variable:

$$x - y = 50$$
$$x = 50 + y$$

Then substitute the value of x into P:

$$P = (50 + y)(y)$$
$$= y^2 + 50y$$

Next, find the derivative, set it equal to zero, and solve for y:

$$P' = 2y + 50$$
$$0 = 2y + 50$$
$$-25 = y$$

Using $y = -25$ and $x = 50 + y$ gives you $x = 25$, so the product is

$$P = (25)(-25) = -625$$

You can verify that $y = -25$ gives you a minimum by using the first derivative test.

499. 20, 20

Let x and y be the two positive numbers so that $xy = 400$. The function that you want to minimize, the sum, is given by

$$S(x, y) = x + y$$

You can write the product in terms of one variable:

$$xy = 400$$

$$y = \frac{400}{x}$$

Then substitute the value of y into the sum function:

$$S(x) = x + \frac{400}{x} = x + 400x^{-1}$$

Next, find the derivative of the function:

$$S'(x) = 1 - 400x^{-2}$$

$$= 1 - \frac{400}{x^2}$$

$$= \frac{x^2 - 400}{x^2}$$

Setting the derivative equal to zero gives you $x^2 - 400 = 0$, and keeping only the positive solution, you have $x = 20$. You can verify that this value gives you a minimum by using the first derivative test.

Using $x = 20$ and $xy = 400$ gives you $y = 20$.

500. 15 m × 15 m

If you let x and y be the side lengths of the rectangle, you have the following equation for the perimeter:

$$2x + 2y = 60$$

The function that you want to maximize, the area, is given by

$$A(x, y) = xy$$

You can write this as a function of one variable by using $2x + 2y = 60$ to get $y = 30 - x$. Substituting the value of y into the area equation gives you

$$A(x) = x(30 - x)$$

$$= 30x - x^2$$

Take the derivative and set it equal to zero to find x:

$$A' = 30 - 2x$$
$$0 = 30 - 2x$$
$$2x = 30$$
$$x = 15 \text{ m}$$

You can verify that this gives you a maximum by using the first derivative test.

Using $2x + 2y = 60$ and $x = 15$ meters, you get $y = 15$ meters.

501.   56,250 ft^2

Let x and y be the lengths of the sides of the pen as in the following diagram:

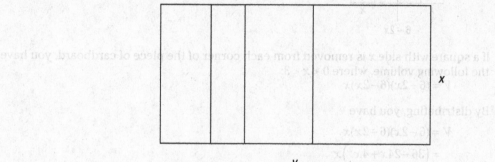

From the diagram, you have the following perimeter formula:

$$5x + 2y = 1,500$$

The function you want to maximize, the area, is given by

$$A(x, y) = xy$$

You can write this as a function of one variable by using the equation $5x + 2y = 1,500$ to get $y = 750 - \frac{5}{2}x$. Substituting the value of y into the area equation gives you

$$A(x) = x\left(750 - \frac{5}{2}x\right)$$
$$= 750x - \frac{5}{2}x^2$$

Take the derivative of this function, set it equal to zero, and solve for x:

$$A' = 750 - 5x$$
$$0 = 750 - 5x$$
$$5x = 750$$
$$x = 150$$

You can verify that this value gives you a maximum by using the first derivative test. Using $5x + 2y = 1,500$ and $x = 150$ gives you $y = 375$; therefore, the area is

$$A = (150)(375) = 56,250 \text{ ft}^2$$

502.   16 ft^3

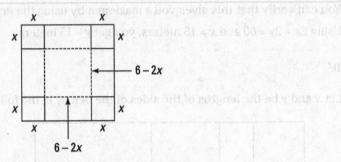

If a square with side x is removed from each corner of the piece of cardboard, you have the following volume, where $0 < x < 3$:
$$V = (6 - 2x)(6 - 2x)x$$

By distributing, you have
$$V = (6 - 2x)(6 - 2x)x$$
$$= \left(36 - 24x + 4x^2\right)x$$
$$= 4x^3 - 24x^2 + 36x$$

You want the maximum volume, so take the derivative of this function:
$$V' = 12x^2 - 48x + 36$$

Then set the derivative equal to zero and solve for x:
$$12x^2 - 48x + 36 = 0$$
$$12\left(x^2 - 4x + 3\right) = 0$$
$$12(x - 3)(x - 1) = 0$$
$$x = 3, 1$$

Because $x = 3$ makes the volume zero, you know that $x = 1$ gives you a maximum, which you can verify by using the first derivative test.

Using $V = (6 - 2x)(6 - 2x)x$ and $x = 1$ gives you the volume:
$$V(1) = (4)(4)(1)$$
$$= 16 \text{ ft}^3$$

503. $20\sqrt[3]{4}$ cm $\times$ $20\sqrt[3]{4}$ cm $\times$ $10\sqrt[3]{4}$ cm

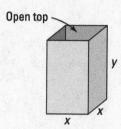

Open top

y

x x

x

If you let x be a length of one side of the base and let y be the height of the box, the volume is

$$V = x(x)(y)$$
$$16{,}000 = x^2 y$$

The square base has an area of x^2, and the box has four sides, each with an area of xy, so the surface area is given by

$$S(x, y) = x^2 + 4xy$$

You can write this as a function of one variable by using $x^2 y = 16{,}000$ to get $y = \dfrac{16{,}000}{x^2}$. Substituting the value of y into the surface area formula gives you

$$S(x) = x^2 + 4x\left(\frac{16{,}000}{x^2}\right)$$
$$= x^2 + 64{,}000 x^{-1}$$

You want the minimum surface area, so take the derivative of this function:

$$S' = 2x - 64{,}000 x^{-2}$$
$$= \frac{2x^3 - 64{,}000}{x^2}$$

Then set the derivative equal to zero and solve for x:

$$2x^3 - 64{,}000 = 0$$
$$2x^3 = 64{,}000$$
$$x = \sqrt[3]{32{,}000}$$
$$x = \sqrt[3]{(8{,}000)(4)}$$
$$x = 20\sqrt[3]{4}$$

Use $x = 20\sqrt[3]{4}$ to find y:

$$y = \frac{16{,}000}{\left((20)4^{1/3}\right)^2}$$

$$= \frac{16{,}000}{(400)\left(4^{2/3}\right)}$$

$$= \frac{40}{4^{2/3}}$$

$$= \frac{(40)4^{1/3}}{4}$$

$$= (10)4^{1/3}$$

$$= 10\sqrt[3]{4}$$

Therefore, the dimensions are $20\sqrt[3]{4}$ cm × $20\sqrt[3]{4}$ cm × $10\sqrt[3]{4}$ cm (approximately 31.75 cm × 31.75 cm × 15.87 cm).

Note that you can easily use the second derivative test to verify that $x = 20\sqrt[3]{4}$ gives you a minimum because $S'' > 0$ for all $x > 0$.

504. $\left(-\frac{1}{7}, \frac{8\sqrt{6}}{7}\right), \left(-\frac{1}{7}, -\frac{8\sqrt{6}}{7}\right)$

You want to find the maximum distance. The distance from a point (x, y) to the point $(1, 0)$ is given by

$$D(x,y) = \sqrt{(x-1)^2 + (y-0)^2}$$

$$= \sqrt{x^2 - 2x + 1 + y^2}$$

Rewrite the equation of the ellipse as $y^2 = 8 - 8x^2$ and substitute the value of y^2 into the distance equation:

$$D(x) = \sqrt{x^2 - 2x + 1 + 8 - 8x^2}$$

$$= \sqrt{9 - 2x - 7x^2}$$

Tip: You can take the derivative of this function and use the first derivative test to find a maximum, but it's easier to use the square of the distance, which gets rid of the radical. For a function that satisfies $f(x) \geq 0$, its local maxima and minima occur at the same x values as the local maxima and minima of its square, $[f(x)]^2$. Obviously, the corresponding y values would change, but that doesn't matter here!

The square of the distance is

$$S = 9 - 2x - 7x^2$$

And the derivative of this function is

$$S' = -2 - 14x$$

Setting the derivative equal to zero and solving gives you $-2 - 14x = 0$, or $-2 = 14x$, which has the solution $x = -\frac{1}{7}$. You can verify that $x = -\frac{1}{7}$ gives you a maximum by using the first derivative test.

Using $x = -\frac{1}{7}$ and $y^2 = 8 - 8x^2$, find the y coordinate:

$$y^2 = 8 - 8\left(\frac{-1}{7}\right)^2$$

$$y^2 = 8 - \frac{8}{49}$$

$$y^2 = \frac{384}{49}$$

$$y = \pm\sqrt{\frac{384}{49}} = \pm\frac{8\sqrt{6}}{7}$$

Therefore, the two points that are farthest from (1, 0) are the points $\left(\frac{-1}{7}, \frac{8\sqrt{6}}{7}\right)$ and $\left(\frac{-1}{7}, -\frac{8\sqrt{6}}{7}\right)$.

505. $\left(\frac{-24}{17}, \frac{6}{17}\right)$

You want to find the minimum distance. The distance from a point (x, y) to the origin is

$$D(x, y) = \sqrt{(x-0)^2 + (y-0)^2}$$
$$= \sqrt{x^2 + y^2}$$

Using $y = 4x + 6$, the distance is

$$D(x) = \sqrt{x^2 + (4x+6)^2}$$
$$= \sqrt{17x^2 + 48x + 36}$$

Tip: You can take the derivative of this function and use the first derivative test to find a minimum, but it's easier to use the square of the distance, which gets rid of the radical. For a function that satisfies $f(x) \geq 0$, its local maxima and minima occur at the same x values as the local maxima and minima of its square, $[f(x)]^2$. Obviously, the corresponding y values would change, but that doesn't matter here!

Using the square of the distance, you have

$$S = 17x^2 + 48x + 36$$

The derivative is

$$S' = 34x + 48$$

Setting the derivative equal to zero gives you $34x + 48 = 0$, which has the solution $x = \frac{-24}{17}$. You can verify that this value gives you a minimum by using the first derivative test.

Using $x = \frac{-24}{17}$ and $y = 4x + 6$, find the y coordinate:

$$y = 4\left(\frac{-24}{17}\right) + 6$$

$$= \frac{-96}{17} + \frac{102}{17}$$

$$= \frac{6}{17}$$

Therefore, the point closest to the origin is $\left(\frac{-24}{17}, \frac{6}{17}\right)$.

506. $3\sqrt{5}$ in. × $6\sqrt{5}$ in.

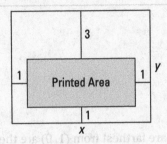

You want to maximize the printed area. Let x and y be the length and width of the poster. Because the total area must be 90 square inches, you have $A = xy = 90$.

The poster has a 1-inch margin at the bottom and sides and a 3-inch margin at the top, so the printed area of the poster is

$$A(x,y) = (x-2)(y-4)$$

Using $y = \dfrac{90}{x}$, the printed area is

$$A(x) = (x-2)\left(\frac{90}{x}-4\right)$$

$$= 90 - 4x - \frac{180}{x} + 8$$

$$= 98 - 4x - \frac{180}{x}$$

Find the derivative of the function for the printed area:

$$A'(x) = -4 + \frac{180}{x^2} = \frac{-4x^2 + 180}{x^2}$$

Set the derivative equal to zero and solve for x: $-4x^2 + 180 = 0$ so that $x^2 = 45$. Keeping the positive solution, you have $x = \sqrt{45} = \sqrt{(9)(5)} = 3\sqrt{5}$.

Using the first derivative test, you can verify that $x = 3\sqrt{5}$ gives you a maximum. From the equation $y = \dfrac{90}{x}$, you get the y value:

$$y = \frac{90}{3\sqrt{5}} = \frac{30}{\sqrt{5}} = \frac{30\sqrt{5}}{5} = 6\sqrt{5}$$

Therefore, the dimensions are $3\sqrt{5}$ inches × $6\sqrt{5}$ inches (approximately 6.7 inches × 13.4 inches).

507. $-\sqrt{\dfrac{3}{2}}$ and $\sqrt{\dfrac{3}{2}}$

You want to locate the maximum slope of $f(x) = 2 + 20x^3 - 4x^5$. The slope of the tangent line is given by the derivative

$$f'(x) = 60x^2 - 20x^4$$

At a point p, the slope is

$$s(p) = 60p^2 - 20p^4$$

You want to maximize this function, so take another derivative:

$$s'(p) = 120p - 80p^3$$

To find the critical numbers, set this function equal to zero and solve for p:

$$120p - 80p^3 = 0$$
$$40p(3 - 2p^2) = 0$$
$$p = 0, \pm\sqrt{\frac{3}{2}}$$

Take a point from each interval and test it in $s'(p)$ to see whether the answer is positive or negative. By taking a point from the interval $\left(-\infty, -\sqrt{\frac{3}{2}}\right)$, you have $s'(p) > 0$; by using a point from $\left(-\sqrt{\frac{3}{2}}, 0\right)$, you have $s'(p) < 0$; in $\left(0, \sqrt{\frac{3}{2}}\right)$, you have $s'(p) > 0$; and in $\left(\sqrt{\frac{3}{2}}, \infty\right)$, you have $s'(p) < 0$. Because $s(p)$ approaches $-\infty$ as x approaches $\pm\infty$, the maximum value must occur at one (or both) of $p = \pm\sqrt{\frac{3}{2}}$.

Substituting these values into the slope equation gives you

$$s\left(\sqrt{\frac{3}{2}}\right) = 60\left(\sqrt{\frac{3}{2}}\right)^2 - 20\left(\sqrt{\frac{3}{2}}\right)^4 = 45$$

$$s\left(-\sqrt{\frac{3}{2}}\right) = 60\left(-\sqrt{\frac{3}{2}}\right)^2 - 20\left(-\sqrt{\frac{3}{2}}\right)^4 = 45$$

Therefore, the maximum slope occurs when $x = \sqrt{\frac{3}{2}}$ and when $x = -\sqrt{\frac{3}{2}}$.

508. $396.23

If you let x be the length of the base and let y be the height, the volume is

$$V = (2x)(x)y$$
$$20 = (2x)(x)y$$
$$20 = 2x^2 y$$

The area of the base is $2x^2$, and the box has four sides, each with an area of xy. With the base material at $20 per square meter and the side material at $12 per square meter, the total cost is

$$C(x, y) = 20(2x^2) + 12(4xy)$$
$$= 40x^2 + 48xy$$

You can write this as a function of one variable by using the volume equation to get $y = \frac{20}{2x^2} = \frac{10}{x^2}$. The cost becomes

$$C(x) = 40x^2 + 48x\left(\frac{10}{x^2}\right)$$
$$= 40x^2 + 480x^{-1}$$

You want the minimum cost, so find the derivative of this function:

$$C' = 80x - 480x^{-2}$$

$$= \frac{80x^3 - 480}{x^2}$$

Setting this equal to zero, you get $80x^3 - 480 = 0$, which has the solution $x = \sqrt[3]{6} \approx 1.82$ meters. Substituting this value into the cost function $C(x) = 40x^2 + 480x^{-1}$ gives you the minimum cost:

$$C\left(\sqrt[3]{6}\right) = 40\left(\sqrt[3]{6}\right)^2 + 480\left(\sqrt[3]{6}\right)^{-1}$$

$$\approx \$396.23$$

509.   20 m

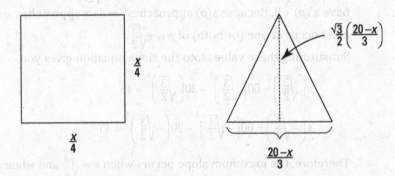

You want to maximize the total area. If x is the length of wire used for the square, each side of the square has a length of $\frac{x}{4}$ meters, so the area of the square is $\left(\frac{x}{4}\right)^2$. You'll have $(20 - x)$ meters of wire left for the triangle, so each side of the triangle is $\left(\frac{20-x}{3}\right)$ meters. Because the triangle is equilateral, its height is $\frac{\sqrt{3}}{2}\left(\frac{20-x}{3}\right)$ meters, so the area of the triangle is $\frac{1}{2}bh = \frac{1}{2}\left(\frac{20-x}{3}\right)\left(\frac{\sqrt{3}}{2}\left(\frac{20-x}{3}\right)\right)$. Therefore, the total area of the square and the triangle together is

$$A(x) = \left(\frac{x}{4}\right)^2 + \frac{1}{2}\left(\frac{20-x}{3}\right)\frac{\sqrt{3}}{2}\left(\frac{20-x}{3}\right)$$

$$= \frac{x^2}{16} + \frac{\sqrt{3}}{36}(20-x)^2$$

where $0 \le x \le 20$.

Find the derivative of this function:

$$A'(x) = \frac{2x}{16} + \frac{\sqrt{3}}{36}2(20-x)(-1)$$

$$= \frac{1}{8}x - \frac{\sqrt{3}}{18}(20-x)$$

Then set the derivative equal to zero and solve for x:

$$\frac{1}{8}x - \frac{\sqrt{3}}{18}(20-x) = 0$$

$$\frac{1}{8}x - \frac{10\sqrt{3}}{9} + \frac{\sqrt{3}}{18}x = 0$$

$$\frac{9}{72}x - \frac{80\sqrt{3}}{72} + \frac{4\sqrt{3}}{72}x = 0$$

$$\frac{9}{72}x + \frac{4\sqrt{3}}{72}x = \frac{80\sqrt{3}}{72}$$

$$x\left(\frac{9+4\sqrt{3}}{72}\right) = \frac{80\sqrt{3}}{72}$$

$$x = \frac{80\sqrt{3}}{9+4\sqrt{3}}$$

It's reasonable to let $x = 0$ (the wire is used entirely to make the triangle) or $x = 20$ (the wire is used entirely to make the square), so check those values as well. Here are the areas:

$$A(0) = \frac{0^2}{16} + \frac{\sqrt{3}}{36}(20-0)^2 = \frac{\sqrt{3}}{36}(400) \approx 19.25$$

$$A(20) = \frac{20^2}{16} + \frac{\sqrt{3}}{36}(20-20)^2 = \frac{400}{16} = 25$$

$$A\left(\frac{80\sqrt{3}}{9+4\sqrt{3}}\right) = \frac{\left(\frac{80\sqrt{3}}{9+4\sqrt{3}}\right)^2}{16} + \frac{\sqrt{3}}{36}\left(20 - \left(\frac{80\sqrt{3}}{9+4\sqrt{3}}\right)\right)^2 \approx 10.87$$

The maximum occurs when $x = 20$.

510. $\dfrac{80\sqrt{3}}{9+4\sqrt{3}}$ meters

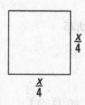 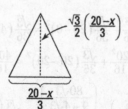

You want to minimize the total area. If x is the length of wire used for the square, each side of the square has a length of $\frac{x}{4}$ meters, so the area of the square is $\left(\frac{x}{4}\right)^2$. You'll have $(20-x)$ meters of wire left for the triangle, so each side of the triangle is $\left(\frac{20-x}{3}\right)$ meters. Because the triangle is equilateral, its height is $\frac{\sqrt{3}}{2}\left(\frac{20-x}{3}\right)$ meters, so the area of the triangle is $\frac{1}{2}bh = \frac{1}{2}\left(\frac{20-x}{3}\right)\left(\frac{\sqrt{3}}{2}\left(\frac{20-x}{3}\right)\right)$. Therefore, the total area of the square and the triangle together is

$$A(x) = \left(\frac{x}{4}\right)^2 + \frac{1}{2}\left(\frac{20-x}{3}\right)\frac{\sqrt{3}}{2}\left(\frac{20-x}{3}\right)$$

$$= \frac{x^2}{16} + \frac{\sqrt{3}}{36}(20-x)^2$$

where $0 \le x \le 20$.

Find the derivative of this function:

$$A'(x) = \frac{2x}{16} + \frac{\sqrt{3}}{36}2(20-x)(-1)$$

$$= \frac{1}{8}x - \frac{\sqrt{3}}{18}(20-x)$$

Set the derivative equal to zero and solve for x:

$$\frac{1}{8}x - \frac{\sqrt{3}}{18}(20-x) = 0$$

$$\frac{1}{8}x - \frac{10\sqrt{3}}{9} + \frac{\sqrt{3}}{18}x = 0$$

$$\frac{9}{72}x - \frac{80\sqrt{3}}{72} + \frac{4\sqrt{3}}{72}x = 0$$

$$\frac{9}{72}x + \frac{4\sqrt{3}}{72}x = \frac{80\sqrt{3}}{72}$$

$$x\left(\frac{9+4\sqrt{3}}{72}\right) = \frac{80\sqrt{3}}{72}$$

$$x = \frac{80\sqrt{3}}{9+4\sqrt{3}}$$

It's reasonable to let $x = 0$ (the wire is used entirely to make the triangle) or $x = 20$ (the wire is used entirely to make the square), so check those values as well. Here are the areas:

$$A(0) = \frac{0^2}{16} + \frac{\sqrt{3}}{36}(20-0)^2 = \frac{\sqrt{3}}{36}(400) \approx 19.25$$

$$A(20) = \frac{20^2}{16} + \frac{\sqrt{3}}{36}(20-20)^2 = \frac{400}{16} = 25$$

$$A\left(\frac{80\sqrt{3}}{9+4\sqrt{3}}\right) = \frac{\left(\frac{80\sqrt{3}}{9+4\sqrt{3}}\right)^2}{16} + \frac{\sqrt{3}}{36}\left(20-\left(\frac{80\sqrt{3}}{9+4\sqrt{3}}\right)\right)^2 \approx 10.87$$

Therefore, the minimum occurs when

$$x = \frac{80\sqrt{3}}{9+4\sqrt{3}} \approx 8.70 \text{ meters}$$

511. $\dfrac{20\sqrt[3]{5}}{1+\sqrt[3]{5}}$ feet from the bright light source

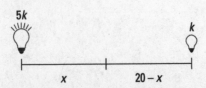

Let x be the distance of the object from the brighter light source and k be the strength of the weaker light source. The illumination of each light source is directly proportional to the strength of the light source ($5k$ and k) and inversely proportional to the square of the distance from the source (x and $20 - x$), so the total illumination is

$$I(x) = \frac{5k}{x^2} + \frac{k}{(20-x)^2}$$

where $0 < x < 20$.

Taking the derivative of this function gives you

$$I'(x) = \frac{-10k}{x^3} + \frac{2k}{(20-x)^3}$$

Set this derivative equal to zero and simplify:

$$\frac{-10k(20-x)^3}{x^3(20-x)^3} + \frac{2kx^3}{x^3(20-x)^3} = 0$$

$$2kx^3 = 10k(20-x)^3$$

$$x^3 = 5(20-x)^3$$

Then solve for x. Taking the cube root of both sides and solving gives you

$$x^3 = 5(20-x)^3$$

$$x = \sqrt[3]{5}(20-x)$$

$$x = 20\sqrt[3]{5} - \sqrt[3]{5}x$$

$$x + \sqrt[3]{5}x = 20\sqrt[3]{5}$$

$$x(1+\sqrt[3]{5}) = 20\sqrt[3]{5}$$

$$x = \frac{20\sqrt[3]{5}}{1+\sqrt[3]{5}} \approx 12.62 \text{ ft}$$

Note that you can use the first derivative test to verify that this value does in fact give a minimum.

512.   12

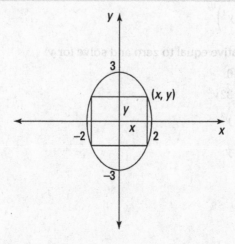

You want to maximize the area of the rectangle. From the diagram, the area of the rectangle is

$$A(x,y) = (2x)(2y) = 4xy$$

Using the equation of the ellipse, solve for x:

$$\frac{x^2}{4} + \frac{y^2}{9} = 1$$

$$\frac{x^2}{4} = 1 - \frac{y^2}{9}$$

$$x^2 = 4 - \frac{4y^2}{9}$$

$$x = \pm\sqrt{4 - \frac{4y^2}{9}}$$

Keeping only the positive solution, you have $x = \sqrt{4 - \frac{4y^2}{9}}$ (although you can just as easily work with the negative solution). Therefore, the area in terms of y becomes

$$A(y) = 4\left(4 - \frac{4}{9}y^2\right)^{1/2}(y)$$

$$= 4y\left(4 - \frac{4}{9}y^2\right)^{1/2}$$

Take the derivative by using the product rule and the chain rule:

$$A' = 4\left(4 - \frac{4}{9}y^2\right)^{1/2} + 4y\left[\frac{1}{2}\left(4 - \frac{4}{9}y^2\right)^{-1/2}\left(-\frac{8}{9}y\right)\right]$$

$$= 4\left(4 - \frac{4}{9}y^2\right)^{1/2} - \frac{16y^2}{9\left(4 - \frac{4}{9}y^2\right)^{1/2}}$$

$$= \frac{36\left(4 - \frac{4}{9}y^2\right) - 16y^2}{9\left(4 - \frac{4}{9}y^2\right)^{1/2}}$$

$$= \frac{144 - 32y^2}{9\left(4 - \frac{4}{9}y^2\right)^{1/2}}$$

Then set the derivative equal to zero and solve for y:

$$144 - 32y^2 = 0$$

$$144 = 32y^2$$

$$\frac{144}{32} = y^2$$

$$\pm\sqrt{\frac{9}{2}} = y$$

Keeping the positive solution (either works), you have $y = \dfrac{3}{\sqrt{2}} = \dfrac{3\sqrt{2}}{2}$, which gives you the following x value:

$$x = \sqrt{4 - \frac{4}{9}\left(\frac{3\sqrt{2}}{2}\right)^2}$$

$$= \sqrt{4 - \frac{4}{9}\left(\frac{9(2)}{4}\right)}$$

$$= \sqrt{4 - 2}$$

$$= \sqrt{2}$$

Therefore, the area of the rectangle is

$$A = 4\sqrt{2}\left(\frac{3\sqrt{2}}{2}\right) = 12$$

Note that you can verify that the value $y = \dfrac{3}{\sqrt{2}} = \dfrac{3\sqrt{2}}{2}$ does indeed give a maximum by using the first derivative test.

---

513.   0.84771

Use the formula $x_{n+1} = x_n - \dfrac{f(x_n)}{f'(x_n)}$ with $f(x) = x^3 + 4x - 4$, $f'(x) = 3x^2 + 4$, and $x_1 = 1$. Note that the formula gives you $x_2 = x_1 - \dfrac{f(x_1)}{f'(x_1)}$, $x_3 = x_2 - \dfrac{f(x_2)}{f'(x_2)}$, and so on. Therefore, you have

$$x_2 = 1 - \frac{(1)^3 + 4(1) - 4}{3(1)^2 + 4} = \frac{6}{7}$$

$$x_3 = \frac{6}{7} - \frac{\left(\frac{6}{7}\right)^3 + 4\left(\frac{6}{7}\right) - 4}{3\left(\frac{6}{7}\right)^2 + 4} = \frac{451}{532} \approx 0.84774$$

$$x_4 = 0.84774 - \frac{(0.84774)^3 + 4(0.84774) - 4}{3(0.84774)^2 + 4} \approx 0.84771$$

$$x_5 = 0.84771 - \frac{(0.84771)^3 + 4(0.84771) - 4}{3(0.84771)^2 + 4} \approx 0.84771$$

---

514.   2.0597671

Use the formula $x_{n+1} = x_n - \dfrac{f(x_n)}{f'(x_n)}$ with $f(x) = x^4 - 18$, $f'(x) = 4x^3$, and $x_1 = 2$. Note that the formula gives you $x_2 = x_1 - \dfrac{f(x_1)}{f'(x_1)}$, $x_3 = x_2 - \dfrac{f(x_2)}{f'(x_2)}$, and so on. Therefore, you have

$$x_2 = 2 - \frac{(2)^4 - 18}{4(2)^3} = \frac{33}{16} = 2.0625$$

$$x_3 = 2.0625 - \frac{(2.0625)^4 - 18}{4(2.0625)^3} \approx 2.0597725$$

$$x_4 = 2.0597725 - \frac{(2.0597725)^4 - 18}{4(2.0597725)^3} \approx 2.0597671$$

$$x_5 = 2.0597671 - \frac{(2.0597671)^4 - 18}{4(2.0597671)^3} \approx 2.0597671$$

515.   −1.2457342

Use the formula $x_{n+1} = x_n - \frac{f(x_n)}{f'(x_n)}$ with $f(x) = x^5 + 3$, $f'(x) = 5x^4$, and $x_1 = -1$. Note that

the formula gives you $x_2 = x_1 - \frac{f(x_1)}{f'(x_1)}$, $x_3 = x_2 - \frac{f(x_2)}{f'(x_2)}$, and so on. Therefore, you have

$$x_2 = -1 - \frac{(-1)^5 + 3}{5(-1)^4} = -1.4$$

$$x_3 = -1.4 - \frac{(-1.4)^5 + 3}{5(-1.4)^4} \approx -1.2761849$$

$$x_4 = -1.2761849 - \frac{(-1.2761849)^5 + 3}{5(-1.2761849)^4} \approx -1.2471501$$

$$x_5 = -1.2471501 - \frac{(-1.2471501)^5 + 3}{5(-1.2471501)^4} \approx -1.2457342$$

516.   0.73909

Newton's method begins by making one side of the equation zero, labeling the other side $f(x)$, and then picking a value of x_1 that's "close" to a root of $f(x)$. There's certainly a bit of trial and error involved; in this case, you could graph both $y = \cos x$ and $y = x$ and look for a point of intersection to get a rough idea of what the root may be.

Use the formula $x_{n+1} = x_n - \frac{f(x_n)}{f'(x_n)}$ with $f(x) = \cos x - x$ and $f'(x) = -\sin x - 1$. Note that

the formula gives you $x_2 = x_1 - \frac{f(x_1)}{f'(x_1)}$, $x_3 = x_2 - \frac{f(x_2)}{f'(x_2)}$, and so on. You also have to

decide on a value for x_1. Notice that $f(0) = 1 - 0 = 1$, which is close to the desired value of 0, so you can start with the value $x_1 = 0$. Therefore, you have

$$x_2 = 0 - \frac{\cos(0) - (0)}{-\sin(0) - 1} = 1$$

$$x_3 = 1 - \frac{\cos(1) - (1)}{-\sin(1) - 1} \approx 0.75036$$

$$x_4 = 0.75036 - \frac{\cos(0.75036) - (0.75036)}{-\sin(0.75036) - 1} \approx 0.73911$$

$$x_5 = 0.73911 - \frac{\cos(0.73911) - (0.73911)}{-\sin(0.73911) - 1} \approx 0.73909$$

$$x_6 = 0.73909 - \frac{\cos(0.73909) - (0.73909)}{-\sin(0.73909) - 1} \approx 0.73909$$

The approximation is no longer changing, so the solution is 0.73909.

517. 1.69562

Newton's method begins by making one side of the equation zero, labeling the other side $f(x)$, and then picking a value of x_1 that's "close" to a root of $f(x)$. In general, a bit of trial and error is involved, but in this case, you're given a closed interval to work with, which considerably narrows down the choices!

Use the formula $x_{n+1} = x_n - \dfrac{f(x_n)}{f'(x_n)}$ with $f(x) = x^3 - x^2 - 2$ and $f'(x) = 3x^2 - 2x$. Note that

the formula gives you $x_2 = x_1 - \dfrac{f(x_1)}{f'(x_1)}$, $x_3 = x_2 - \dfrac{f(x_2)}{f'(x_2)}$, and so on. You can choose any

value in the interval $[1, 2]$ for x_1, but $x_1 = 1$ makes the computations a bit easier for the first step, so use that value:

$$x_2 = 1 - \frac{(1)^3 - (1)^2 - 2}{3(1)^2 - 2(1)} = 3$$

$$x_3 = 3 - \frac{(3)^3 - (3)^2 - 2}{3(3)^2 - 2(3)} = \frac{47}{21} \approx 2.23810$$

$$x_4 = 2.23810 - \frac{(2.23810)^3 - (2.23810)^2 - 2}{3(2.23810)^2 - 2(2.23810)} \approx 1.83987$$

$$x_5 = 1.83987 - \frac{(1.83987)^3 - (1.83987)^2 - 2}{3(1.83987)^2 - 2(1.83987)} \approx 1.70968$$

$$x_6 = 1.70968 - \frac{(1.70968)^3 - (1.70968)^2 - 2}{3(1.70968)^2 - 2(1.70968)} \approx 1.69577$$

$$x_7 = 1.69577 - \frac{(1.69577)^3 - (1.69577)^2 - 2}{3(1.69577)^2 - 2(1.69577)} \approx 1.69562$$

$$x_8 = 1.69562 - \frac{(1.69562)^3 - (1.69562)^2 - 2}{3(1.69562)^2 - 2(1.69562)} \approx 1.69562$$

The approximation is no longer changing, so the solution is 1.69562.

518. 1.22074

Newton's method begins by making one side of the equation zero, labeling the other side $f(x)$, and then picking a value of x_1 that's "close" to a root of $f(x)$. There's certainly a bit of trial and error involved; in this case, you could graph both $y = \sqrt{x+1}$ and $y = x^2$ and look for a point of intersection to get a rough idea of what the root may be.

Use the formula $x_{n+1} = x_n - \dfrac{f(x_n)}{f'(x_n)}$ with $f(x) = \sqrt{x+1} - x^2$ and $f'(x) = \dfrac{1}{2\sqrt{x+1}} - 2x$. Note

that the formula gives you $x_2 = x_1 - \dfrac{f(x_1)}{f'(x_1)}$, $x_3 = x_2 - \dfrac{f(x_2)}{f'(x_2)}$, and so on. Notice that

$f(1) = \sqrt{1+1} - 1^2 = \sqrt{2} - 1$, which is close to the desired value of 0, so you can start with $x_1 = 1$:

$$x_2 = 1 - \frac{\sqrt{(1)+1}-(1)^2}{\frac{1}{2\sqrt{(1)+1}}-2(1)} \approx 1.25158$$

$$x_3 = 1.25158 - \frac{\sqrt{(1.25158)+1}-(1.25158)^2}{\frac{1}{2\sqrt{(1.25158)+1}}-2(1.25158)} \approx 1.22120$$

$$x_4 = 1.22120 - \frac{\sqrt{(1.22120)+1}-(1.22120)^2}{\frac{1}{2\sqrt{(1.22120)+1}}-2(1.22120)} \approx 1.22074$$

$$x_5 = 1.22074 - \frac{\sqrt{(1.22074)+1}-(1.22074)^2}{\frac{1}{2\sqrt{(1.22074)+1}}-2(1.22074)} \approx 1.22074$$

The approximation is no longer changing, so the solution is 1.22074.

519. $\frac{23}{4}$

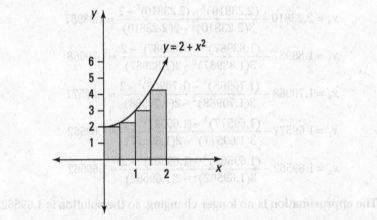

Because the given function has values that are strictly greater than or equal to zero on the given interval, you can interpret the Riemann sum as approximating the area that's underneath the curve and bounded below by the x-axis.

You want to use four rectangles of equal width to estimate the area under $f(x) = 2 + x^2$ over the interval $0 \le x \le 2$. Begin by dividing the length of the interval by 4 to find the width of each rectangle:

$$\Delta x = \frac{2-0}{4} = \frac{1}{2}$$

Then divide the interval $[0, 2]$ into 4 equal pieces, each with a width of $\frac{1}{2}$, to get the intervals $\left[0, \frac{1}{2}\right], \left[\frac{1}{2}, 1\right], \left[1, \frac{3}{2}\right]$, and $\left[\frac{3}{2}, 2\right]$. Using the left endpoint of each interval to calculate the heights of the rectangles gives you the values $x = 0$, $x = \frac{1}{2}$, $x = 1$, and $x = \frac{3}{2}$.

The height of each rectangle is $f(x)$. To approximate the area under the curve, multiply the width of each rectangle by $f(x)$ and add the areas:

$$\frac{1}{2}f(0) + \frac{1}{2}f\left(\frac{1}{2}\right) + \frac{1}{2}f(1) + \frac{1}{2}f\left(\frac{3}{2}\right)$$

$$= \frac{1}{2}\left[2 + 0^2\right] + \frac{1}{2}\left[2 + \left(\frac{1}{2}\right)^2\right] + \frac{1}{2}\left[2 + (1)^2\right] + \frac{1}{2}\left[2 + \left(\frac{3}{2}\right)^2\right]$$

$$= \frac{23}{4}$$

520. 10.43

Because the given function has values that are strictly greater than or equal to zero on the given interval, you can interpret the Riemann sum as approximating the area that's underneath the curve and bounded below by the x-axis.

You want to use five rectangles of equal width to estimate the area under $f(x) = \sqrt[3]{x} + x$ over the interval $1 \le x \le 4$. Begin by dividing the length of the interval by 5 to find the width of each rectangle:

$$\Delta x = \frac{4-1}{5} = \frac{3}{5}$$

Then divide the interval [1, 4] into 5 equal pieces, each with a width of $\frac{3}{5}$, to get the intervals $\left[1, \frac{8}{5}\right]$, $\left[\frac{8}{5}, \frac{11}{5}\right]$, $\left[\frac{11}{5}, \frac{14}{5}\right]$, $\left[\frac{14}{5}, \frac{17}{5}\right]$, and $\left[\frac{17}{5}, 4\right]$. Using the left endpoint of each interval to calculate the heights of the rectangles gives you the values $x = 1$, $x = \frac{8}{5}$, $x = \frac{11}{5}$, $x = \frac{14}{5}$, and $x = \frac{17}{5}$. The height of each rectangle is $f(x)$. To approximate the area under the curve, multiply the width of each rectangle by $f(x)$ and add the areas:

$$\frac{3}{5}f(1) + \frac{3}{5}f\left(\frac{8}{5}\right) + \frac{3}{5}f\left(\frac{11}{5}\right) + \frac{3}{5}f\left(\frac{14}{5}\right) + \frac{3}{5}f\left(\frac{17}{5}\right)$$

$$= \frac{3}{5}\left[\sqrt[3]{1} + 1\right] + \frac{3}{5}\left[\sqrt[3]{\frac{8}{5}} + \frac{8}{5}\right] + \frac{3}{5}\left[\sqrt[3]{\frac{11}{5}} + \frac{11}{5}\right] + \frac{3}{5}\left[\sqrt[3]{\frac{14}{5}} + \frac{14}{5}\right] + \frac{3}{5}\left[\sqrt[3]{\frac{17}{5}} + \frac{17}{5}\right]$$

$$\approx 10.43$$

521. 22.66

Because the given function has values that are strictly greater than or equal to zero on the given interval, you can interpret the Riemann sum as approximating the area that's underneath the curve and bounded below by the x-axis.

You want to use seven rectangles of equal width to estimate the area under $f(x) =$ $4 \ln x + 2x$ over the interval $1 \le x \le 4$. Begin by dividing the length of the interval by 7 to find the width of each rectangle:

$$\Delta x = \frac{4-1}{7} = \frac{3}{7}$$

Then divide the interval [1, 4] into 7 equal pieces, each with a width of $\frac{3}{7}$, to get the intervals $\left[1, \frac{10}{7}\right], \left[\frac{10}{7}, \frac{13}{7}\right], \left[\frac{13}{7}, \frac{16}{7}\right], \left[\frac{16}{7}, \frac{19}{7}\right], \left[\frac{19}{7}, \frac{22}{7}\right], \left[\frac{22}{7}, \frac{25}{7}\right]$, and $\left[\frac{25}{7}, 4\right]$. Using the left endpoint of each interval to calculate the heights of the rectangles gives you the values $x = 1$, $x = \frac{10}{7}$, $x = \frac{13}{7}$, $x = \frac{16}{7}$, $x = \frac{19}{7}$, $x = \frac{22}{7}$, and $x = \frac{25}{7}$. The height of each rectangle is $f(x)$. To approximate the area under the curve, multiply the width of each rectangle by $f(x)$ and add the areas:

$$\frac{3}{7}f(1) + \frac{3}{7}f\left(\frac{10}{7}\right) + \frac{3}{7}f\left(\frac{13}{7}\right) + \frac{3}{7}f\left(\frac{16}{7}\right) + \frac{3}{7}f\left(\frac{19}{7}\right) + \frac{3}{7}f\left(\frac{22}{7}\right) + \frac{3}{7}f\left(\frac{25}{7}\right)$$

$$= \frac{3}{7}\left[4\ln 1 + 2(1)\right] + \frac{3}{7}\left[4\ln\frac{10}{7} + 2\left(\frac{10}{7}\right)\right] + \frac{3}{7}\left[4\ln\frac{13}{7} + 2\left(\frac{13}{7}\right)\right] + \frac{3}{7}\left[4\ln\frac{16}{7} + 2\left(\frac{16}{7}\right)\right]$$

$$+ \frac{3}{7}\left[4\ln\frac{19}{7} + 2\left(\frac{19}{7}\right)\right] + \frac{3}{7}\left[4\ln\frac{22}{7} + 2\left(\frac{22}{7}\right)\right] + \frac{3}{7}\left[4\ln\frac{25}{7} + 2\left(\frac{25}{7}\right)\right]$$

$$\approx 22.66$$

---

522. 2.788×10^{10}

Because the given function has values that are strictly greater than or equal to zero on the given interval, you can interpret the Riemann sum as approximating the area that's underneath the curve and bounded below by the x-axis.

You want to use eight rectangles of equal width to estimate the area under $f(x) = e^{3x} + 4$ over the interval $1 \le x \le 9$. Begin by dividing the length of the interval by 8 to find the width of each rectangle:

$$\Delta x = \frac{9-1}{8} = 1$$

Then divide the interval [1, 9] into 8 equal pieces, each with a width of 1, to get the intervals [1, 2], [2, 3], [3, 4], [4, 5], [5, 6], [6, 7], [7, 8], and [8, 9]. Using the left endpoint of each interval to calculate the heights of the rectangles gives you the values $x = 1$, $x = 2$, $x = 3$, $x = 4$, $x = 5$, $x = 6$, $x = 7$, and $x = 8$. The height of each rectangle is $f(x)$. To approximate the area under the curve, multiply the width of each rectangle by $f(x)$ and add the areas:

$$1f(1) + 1f(2) + 1f(3) + 1f(4) + 1f(5) + 1f(6) + 1f(7) + 1f(8)$$

$$= \left[e^{3(1)} + 4\right] + \left[e^{3(2)} + 4\right] + \left[e^{3(3)} + 4\right] + \left[e^{3(4)} + 4\right] + \left[e^{3(5)} + 4\right] + \left[e^{3(6)} + 4\right]$$

$$+ \left[e^{3(7)} + 4\right] + \left[e^{3(8)} + 4\right]$$

$$\approx 2.788 \times 10^{10}$$

---

523. 24

Because the given function has values that are strictly greater than or equal to zero on the given interval, you can interpret the Riemann sum as approximating the area that's underneath the curve and bounded below by the x-axis.

You want to use four rectangles of equal width to estimate the area under $f(x) = 1 + 2x$ over the interval $0 \le x \le 4$. Begin by dividing the length of the interval by 4 to find the width of each rectangle:

$$\Delta x = \frac{4-0}{4} = 1$$

Then divide the interval $[0, 4]$ into 4 equal pieces, each with a width of 1, to get the intervals $[0, 1]$, $[1, 2]$, $[2, 3]$, and $[3, 4]$. Using the right endpoint of each interval to calculate the heights of the rectangles gives you the values $x = 1$, $x = 2$, $x = 3$, and $x = 4$. The height of each rectangle is $f(x)$. To approximate the area under the curve, multiply the width of each rectangle by $f(x)$ and add the areas:

$$1f(1) + 1f(2) + 1f(3) + 1f(4)$$
$$= [1 + 2(1)] + [1 + 2(2)] + [1 + 2(3)] + [1 + 2(4)]$$
$$= 24$$

---

524.   −8.89

In this example, the given function has values that are both positive and negative on the given interval, so don't make the mistake of thinking the Riemann sum approximates the area between the curve and the x-axis. In this case, the Riemann sum approximates the value of

$$\left(\begin{array}{c} \text{the area of the region above the } x\text{-axis} \\ \text{that's bounded above by the function} \end{array} \right) - \left(\begin{array}{c} \text{the area of the region below the } x\text{-axis} \\ \text{that's bounded below by the function} \end{array} \right)$$

You want to use five rectangles of equal width over the interval $2 \le x \le 6$. Begin by dividing the length of the interval by 5 to find the width of each rectangle:

$$\Delta x = \frac{6-2}{5} = \frac{4}{5}$$

Then divide the interval $[2, 6]$ into 5 equal pieces, each with a width of $\frac{4}{5}$, to get the intervals $\left[2, \frac{14}{5}\right]$, $\left[\frac{14}{5}, \frac{18}{5}\right]$, $\left[\frac{18}{5}, \frac{22}{5}\right]$, $\left[\frac{22}{5}, \frac{26}{5}\right]$, and $\left[\frac{26}{5}, 6\right]$. Using the right endpoint of each interval to calculate the heights of the rectangles gives you the values $x = \frac{14}{5}$, $x = \frac{18}{5}$, $x = \frac{22}{5}$, $x = \frac{26}{5}$, and $x = 6$. The height of each rectangle is $f(x)$. (Note that a "negative height" indicates that the rectangle is below the x-axis.) To find the Riemann sum, multiply the width of each rectangle by $f(x)$ and add the areas:

$$\frac{4}{5}f\left(\frac{14}{5}\right) + \frac{4}{5}f\left(\frac{18}{5}\right) + \frac{4}{5}f\left(\frac{22}{5}\right) + \frac{4}{5}f\left(\frac{26}{5}\right) + \frac{4}{5}f(6)$$

$$= \frac{4}{5}\left[\frac{14}{5}\sin\frac{14}{5}\right] + \frac{4}{5}\left[\frac{18}{5}\sin\frac{18}{5}\right] + \frac{4}{5}\left[\frac{22}{5}\sin\frac{22}{5}\right] + \frac{4}{5}\left[\frac{26}{5}\sin\frac{26}{5}\right] + \frac{4}{5}[6\sin6]$$

$$\approx -8.89$$

Answers
501–600

525.   3.24

In this example, the given function has values that are both positive and negative on the given interval, so the Riemann sum approximates the value of

$$\left(\begin{array}{c}\text{the area of the region above the } x\text{-axis}\\ \text{that's bounded above by the function}\end{array}\right) - \left(\begin{array}{c}\text{the area of the region below the } x\text{-axis}\\ \text{that's bounded below by the function}\end{array}\right)$$

You want to use six rectangles of equal width over the interval $0 \le x \le 5$. Begin by dividing the length of the interval by 6 to find the width of each rectangle:

$$\Delta x = \frac{5-0}{6} = \frac{5}{6}$$

Then divide the interval [0, 5] into 6 equal pieces, each with a width of $\frac{5}{6}$, to get the intervals $\left[0, \frac{5}{6}\right], \left[\frac{5}{6}, \frac{10}{6}\right], \left[\frac{10}{6}, \frac{15}{6}\right], \left[\frac{15}{6}, \frac{20}{6}\right], \left[\frac{20}{6}, \frac{25}{6}\right]$, and $\left[\frac{25}{6}, 5\right]$. Using the right endpoint of each interval to calculate the height of the rectangles gives you the values $x = \frac{5}{6}, x = \frac{10}{6}, x = \frac{15}{6}, x = \frac{20}{6}, x = \frac{25}{6}$, and $x = 5$. The height of each rectangle is $f(x)$. (Note that a "negative height" indicates that the rectangle is below the x-axis.) To find the Riemann sum, multiply the width of each rectangle by $f(x)$ and add the areas:

$$\frac{5}{6}f\left(\frac{5}{6}\right) + \frac{5}{6}f\left(\frac{10}{6}\right) + \frac{5}{6}f\left(\frac{15}{6}\right) + \frac{5}{6}f\left(\frac{20}{6}\right) + \frac{5}{6}f\left(\frac{25}{6}\right) + \frac{5}{6}f(5)$$

$$= \frac{5}{6}\left[\sqrt{\frac{5}{6}} - 1\right] + \frac{5}{6}\left[\sqrt{\frac{10}{6}} - 1\right] + \frac{5}{6}\left[\sqrt{\frac{15}{6}} - 1\right] + \frac{5}{6}\left[\sqrt{\frac{20}{6}} - 1\right] + \frac{5}{6}\left[\sqrt{\frac{25}{6}} - 1\right] + \frac{5}{6}\left[\sqrt{5} - 1\right]$$

$$\approx 3.24$$

526.   1.34

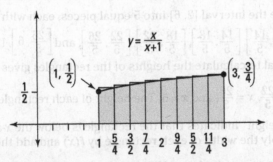

Because the given function has values that are strictly greater than or equal to zero on the given interval, you can interpret the Riemann sum as approximating the area that's underneath the curve and bounded below by the x-axis.

You want to use eight rectangles of equal width to estimate the area under $f(x) = \frac{x}{x+1}$ over the interval $1 \le x \le 3$. Begin by dividing the length of the interval by 8 to find the width of each rectangle:

$$\Delta x = \frac{3-1}{8} = \frac{1}{4}$$

Then divide the interval $[1, 3]$ into eight equal pieces, each with a width of $\frac{1}{4}$, to get the intervals $\left[1, \frac{5}{4}\right], \left[\frac{5}{4}, \frac{6}{4}\right], \left[\frac{6}{4}, \frac{7}{4}\right], \left[\frac{7}{4}, 2\right], \left[2, \frac{9}{4}\right], \left[\frac{9}{4}, \frac{10}{4}\right], \left[\frac{10}{4}, \frac{11}{4}\right]$, and $\left[\frac{11}{4}, 3\right]$. Using the right endpoint of each interval to calculate the height of the rectangles gives you the values $x = \frac{5}{4}$, $x = \frac{6}{4}$, $x = \frac{7}{4}$, $x = 2$, $x = \frac{9}{4}$, $x = \frac{10}{4}$, $x = \frac{11}{4}$, and $x = 3$. The height of each rectangle is $f(x)$. To approximate the area under the curve, multiply the width of each rectangle by $f(x)$ and add the areas:

$$\frac{1}{4}f\left(\frac{5}{4}\right) + \frac{1}{4}f\left(\frac{6}{4}\right) + \frac{1}{4}f\left(\frac{7}{4}\right) + \frac{1}{4}f(2) + \frac{1}{4}f\left(\frac{9}{4}\right) + \frac{1}{4}f\left(\frac{10}{4}\right) + \frac{1}{4}f\left(\frac{11}{4}\right) + \frac{1}{4}f(3)$$

$$= \frac{1}{4}\left[\frac{5/4}{5/4+1}\right] + \frac{1}{4}\left[\frac{6/4}{6/4+1}\right] + \frac{1}{4}\left[\frac{7/4}{7/4+1}\right] + \frac{1}{4}\left[\frac{2}{2+1}\right] + \frac{1}{4}\left[\frac{9/4}{9/4+1}\right] + \frac{1}{4}\left[\frac{10/4}{10/4+1}\right]$$

$$+ \frac{1}{4}\left[\frac{11/4}{11/4+1}\right] + \frac{1}{4}\left[\frac{3}{3+1}\right]$$

$$\approx 1.34$$

---

527.   0.29

In this example, the given function has values that are both positive and negative on the given interval, so don't make the mistake of thinking the Riemann sum approximates the area between the curve and the x-axis. In this case, the Riemann sum approximates the value of

$$\left(\begin{array}{c}\text{the area of the region above the } x\text{-axis}\\ \text{that's bounded above by the function}\end{array}\right) - \left(\begin{array}{c}\text{the area of the region below the } x\text{-axis}\\ \text{that's bounded below by the function}\end{array}\right)$$

You want to use four rectangles of equal width over the interval $0 \le x \le 3$. Begin by dividing the length of the interval by 4 to find the width of each rectangle:

$$\Delta x = \frac{3-0}{4} = \frac{3}{4}$$

Then divide the interval $[0, 3]$ into 4 equal pieces, each with a width of $\frac{3}{4}$, to get the intervals $\left[0, \frac{3}{4}\right], \left[\frac{3}{4}, \frac{6}{4}\right], \left[\frac{6}{4}, \frac{9}{4}\right]$, and $\left[\frac{9}{4}, 3\right]$. Recall that to find the midpoint of an interval, you simply add the left and the right endpoints together and then divide by 2 (that is, you average the two values). In this case, the midpoints are $x = \frac{3}{8}$, $x = \frac{9}{8}$, $x = \frac{15}{8}$, and $x = \frac{21}{8}$. The height of each rectangle is $f(x)$. (Note that a "negative height" indicates that the rectangle is below the x-axis.) To find the Riemann sum, multiply the width of each rectangle by $f(x)$ and add the areas:

$$\frac{3}{4}f\left(\frac{3}{8}\right) + \frac{3}{4}f\left(\frac{9}{8}\right) + \frac{3}{4}f\left(\frac{15}{8}\right) + \frac{3}{4}f\left(\frac{21}{8}\right)$$

$$= \frac{3}{4}\left[2\cos\left(\frac{3}{8}\right)\right] + \frac{3}{4}\left[2\cos\left(\frac{9}{8}\right)\right] + \frac{3}{4}\left[2\cos\left(\frac{15}{8}\right)\right] + \frac{3}{4}\left[2\cos\left(\frac{21}{8}\right)\right]$$

$$\approx 0.29$$

528.   0.32

In this example, the given function has values that are both positive and negative on the given interval, so the Riemann sum approximates the value of

$$\left(\begin{array}{c}\text{the area of the region above the } x\text{-axis}\\ \text{that's bounded above by the function}\end{array}\right) - \left(\begin{array}{c}\text{the area of the region below the } x\text{-axis}\\ \text{that's bounded below by the function}\end{array}\right)$$

You want to use five rectangles of equal width over the interval $1 \le x \le 5$. Begin by dividing the length of the interval by 5 to find the width of each rectangle:

$$\Delta x = \frac{5-1}{5} = \frac{4}{5}$$

Divide the interval [1, 5] into 5 equal pieces, each with a width of $\frac{4}{5}$, to get the intervals $\left[1, \frac{9}{5}\right], \left[\frac{9}{5}, \frac{13}{5}\right], \left[\frac{13}{5}, \frac{17}{5}\right], \left[\frac{17}{5}, \frac{21}{5}\right]$, and $\left[\frac{21}{5}, 5\right]$. Recall that to find the midpoint of an interval, you simply add the left and the right endpoints together and then divide by 2 (that is, you average the two values). In this case, the midpoints are $x = \frac{7}{5}$, $x = \frac{11}{5}$, $x = \frac{15}{5} = 3$, $x = \frac{19}{5}$, and $x = \frac{23}{5}$. The height of each rectangle is $f(x)$. (Note that a "negative height" indicates that the rectangle is below the x-axis.) To find the Riemann sum, multiply the width of each rectangle by $f(x)$ and add the areas:

$$\frac{4}{5}f\left(\frac{7}{5}\right) + \frac{4}{5}f\left(\frac{11}{5}\right) + \frac{4}{5}f(3) + \frac{4}{5}f\left(\frac{19}{5}\right) + \frac{4}{5}f\left(\frac{23}{5}\right)$$

$$= \frac{4}{5}\left[\frac{\sin\frac{7}{5}}{\frac{7}{5}+1}\right] + \frac{4}{5}\left[\frac{\sin\frac{11}{5}}{\frac{11}{5}+1}\right] + \frac{4}{5}\left[\frac{\sin 3}{3+1}\right] + \frac{4}{5}\left[\frac{\sin\frac{19}{5}}{\frac{19}{5}+1}\right] + \frac{4}{5}\left[\frac{\sin\frac{23}{5}}{\frac{23}{5}+1}\right]$$

$$\approx 0.32$$

529.   160.03

Because the given function has values that are strictly greater than or equal to zero on the given interval, you can interpret the Riemann sum as approximating the area that's underneath the curve and bounded below by the x-axis.

You want to use six rectangles of equal width over the interval $1 \le x \le 3$. Begin by dividing the length of the interval by 6 to find the width of each rectangle:

$$\Delta x = \frac{4-1}{6} = \frac{1}{2}$$

Then divide the interval [1, 4] into 6 equal pieces, each with a width of $\frac{1}{2}$, to get the intervals $\left[1, \frac{3}{2}\right], \left[\frac{3}{2}, 2\right], \left[2, \frac{5}{2}\right], \left[\frac{5}{2}, 3\right], \left[3, \frac{7}{2}\right]$, and $\left[\frac{7}{2}, 4\right]$. Recall that to find the midpoint of an interval, you simply add the left and the right endpoints together and then divide by 2 (that is, you average the two values). In this case, the midpoints are $x = \frac{5}{4}$, $x = \frac{7}{4}$,

$x = \dfrac{9}{4}$, $x = \dfrac{11}{4}$, $x = \dfrac{13}{4}$, and $x = \dfrac{15}{4}$. The height of each rectangle is $f(x)$. To find the Riemann sum, multiply the width of each rectangle by $f(x)$ and add the areas:

$$\tfrac{1}{2}f\left(\tfrac{5}{4}\right) + \tfrac{1}{2}f\left(\tfrac{7}{4}\right) + \tfrac{1}{2}f\left(\tfrac{9}{4}\right) + \tfrac{1}{2}f\left(\tfrac{11}{4}\right) + \tfrac{1}{2}f\left(\tfrac{13}{4}\right) + \tfrac{1}{2}f\left(\tfrac{15}{4}\right)$$

$$= \tfrac{1}{2}\left[3e^{5/4} + 2\right] + \tfrac{1}{2}\left[3e^{7/4} + 2\right] + \tfrac{1}{2}\left[3e^{9/4} + 2\right] + \tfrac{1}{2}\left[3e^{11/4} + 2\right] + \tfrac{1}{2}\left[3e^{13/4} + 2\right] + \tfrac{1}{2}\left[3e^{15/4} + 2\right]$$

$$\approx 160.03$$

530. 18.79

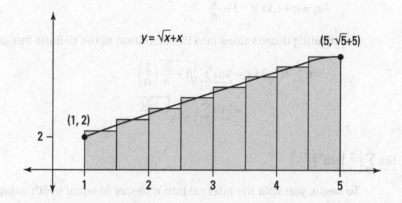

Because the given function has values that are strictly greater than or equal to zero on the given interval, you can interpret the Riemann sum as approximating the area that's underneath the curve and bounded below by the x-axis.

You want to use eight rectangles of equal width over the interval $1 \le x \le 5$. Begin by dividing the length of the interval by 8 to find the width of each rectangle:

$$\Delta x = \frac{5-1}{8} = \frac{1}{2}$$

Then divide the interval $[1, 5]$ into 8 equal pieces, each with a width of $\dfrac{1}{2}$, to get the intervals $\left[1, \tfrac{3}{2}\right]$, $\left[\tfrac{3}{2}, 2\right]$, $\left[2, \tfrac{5}{2}\right]$, $\left[\tfrac{5}{2}, 3\right]$, $\left[3, \tfrac{7}{2}\right]$, $\left[\tfrac{7}{2}, 4\right]$, $\left[4, \tfrac{9}{2}\right]$, and $\left[\tfrac{9}{2}, 5\right]$. Recall that to find the midpoint of an interval, you simply add the left and the right endpoints together and then divide by 2 (that is, you average the two values). In this case, the midpoints are $x = \dfrac{5}{4}$, $x = \dfrac{7}{4}$, $x = \dfrac{9}{4}$, $x = \dfrac{11}{4}$, $x = \dfrac{13}{4}$, $x = \dfrac{15}{4}$, $x = \dfrac{17}{4}$, and $x = \dfrac{19}{4}$. The height of each rectangle is $f(x)$. To approximate the area under the curve, multiply the width of each rectangle by $f(x)$ and add the areas:

$$\tfrac{1}{2}f\left(\tfrac{5}{4}\right) + \tfrac{1}{2}f\left(\tfrac{7}{4}\right) + \tfrac{1}{2}f\left(\tfrac{9}{4}\right) + \tfrac{1}{2}f\left(\tfrac{11}{4}\right) + \tfrac{1}{2}f\left(\tfrac{13}{4}\right) + \tfrac{1}{2}f\left(\tfrac{15}{4}\right) + \tfrac{1}{2}f\left(\tfrac{17}{4}\right) + \tfrac{1}{2}f\left(\tfrac{19}{4}\right)$$

$$= \tfrac{1}{2}\left[\sqrt{\tfrac{5}{4}} + \tfrac{5}{4}\right] + \tfrac{1}{2}\left[\sqrt{\tfrac{7}{4}} + \tfrac{7}{4}\right] + \tfrac{1}{2}\left[\sqrt{\tfrac{9}{4}} + \tfrac{9}{4}\right] + \tfrac{1}{2}\left[\sqrt{\tfrac{11}{4}} + \tfrac{11}{4}\right] + \tfrac{1}{2}\left[\sqrt{\tfrac{13}{4}} + \tfrac{13}{4}\right] + \tfrac{1}{2}\left[\sqrt{\tfrac{15}{4}} + \tfrac{15}{4}\right]$$

$$+ \tfrac{1}{2}\left[\sqrt{\tfrac{17}{4}} + \tfrac{17}{4}\right] + \tfrac{1}{2}\left[\sqrt{\tfrac{19}{4}} + \tfrac{19}{4}\right]$$

$$\approx 18.79$$

531. $\displaystyle\lim_{n\to\infty}\sum_{i=1}^{n}\left(\frac{3}{n}\right)\sqrt{1+\frac{3i}{n}}$

To begin, you split the interval into n pieces of equal width using the formula $\Delta x = \dfrac{b-a}{n}$, where a is the lower limit of integration and b is the upper limit of integration. In this case, you have

$$\Delta x = \frac{4-1}{n} = \frac{3}{n}$$

You also want to select a point from each interval. The formula $x_i = a + (\Delta x)i$ gives you the right endpoint from each interval. Here, you have

$$x_i = a + (\Delta x)i = 1 + \frac{3i}{n}$$

Substituting those values into the definition of the definite integral gives you

$$\lim_{n\to\infty}\sum_{i=1}^{n} f(x_i)\Delta x = \lim_{n\to\infty}\sum_{i=1}^{n}\sqrt{1+\frac{3i}{n}}\left(\frac{3}{n}\right)$$

$$= \lim_{n\to\infty}\sum_{i=1}^{n}\left(\frac{3}{n}\right)\sqrt{1+\frac{3i}{n}}$$

532. $\displaystyle\lim_{n\to\infty}\sum_{i=1}^{n}\left(\frac{\pi}{n}\right)\sin^2\left(\frac{\pi i}{n}\right)$

To begin, you split the interval into n pieces of equal width using the formula $\Delta x = \dfrac{b-a}{n}$, where a is the lower limit of integration and b is the upper limit of integration. In this case, you have

$$\Delta x = \frac{\pi-0}{n} = \frac{\pi}{n}$$

You also want to select a point from each interval. The formula $x_i = a + (\Delta x)i$ gives you the right endpoint from each interval. Here, you have

$$x_i = a + (\Delta x)i = 0 + \frac{\pi i}{n} = \frac{\pi i}{n}$$

Substituting those values into the definition of the definite integral gives you

$$\lim_{n\to\infty}\sum_{i=1}^{n} f(x_i)\Delta x = \lim_{n\to\infty}\sum_{i=1}^{n}\left[\sin^2\left(\frac{\pi i}{n}\right)\right]\left(\frac{\pi}{n}\right)$$

$$= \lim_{n\to\infty}\sum_{i=1}^{n}\left(\frac{\pi}{n}\right)\sin^2\left(\frac{\pi i}{n}\right)$$

533. $\displaystyle\lim_{n\to\infty}\sum_{i=1}^{n}\left[\left(1+\frac{4i}{n}\right)^2+\left(1+\frac{4i}{n}\right)\right]\frac{4}{n}$

To begin, you split the interval into n pieces of equal width using the formula $\Delta x = \dfrac{b-a}{n}$, where a is the lower limit of integration and b is the upper limit of integration. In this case, you have

$$\Delta x = \frac{5-1}{n} = \frac{4}{n}$$

You also want to select a point from each interval. The formula $x_i = a + (\Delta x)i$ gives you the right endpoint from each interval. Here, you have

$$x_i = a + (\Delta x)i = 1 + \frac{4i}{n}$$

Substituting those values into the definition of the definite integral gives you

$$\lim_{n \to \infty} \sum_{i=1}^{n} f(x_i)\Delta x = \lim_{n \to \infty} \sum_{i=1}^{n} \left[\left(1 + \frac{4i}{n}\right)^2 + \left(1 + \frac{4i}{n}\right) \right] \frac{4}{n}$$

534. $\lim_{n \to \infty} \sum_{i=1}^{n} \frac{\pi}{4n} \left[\tan\left(\frac{\pi i}{4n}\right) + \sec\left(\frac{\pi i}{4n}\right) \right]$

To begin, you split the interval into n pieces of equal width using the formula $\Delta x = \frac{b-a}{n}$, where a is the lower limit of integration and b is the upper limit of integration. In this case, you have

$$\Delta x = \frac{\pi/4 - 0}{n} = \frac{\pi}{4n}$$

You also want to select a point from each interval. The formula $x_i = a + (\Delta x)i$ gives you the right endpoint from each interval. Here, you have

$$x_i = a + (\Delta x)i = 0 + \frac{\pi i}{4n}$$

Substituting those values into the definition of the definite integral gives you

$$\lim_{n \to \infty} \sum_{i=1}^{n} f(x_i)\Delta x = \lim_{n \to \infty} \sum_{i=1}^{n} \left[\tan\left(\frac{\pi i}{4n}\right) + \sec\left(\frac{\pi i}{4n}\right) \right] \frac{\pi}{4n}$$

$$= \lim_{n \to \infty} \sum_{i=1}^{n} \frac{\pi}{4n} \left[\tan\left(\frac{\pi i}{4n}\right) + \sec\left(\frac{\pi i}{4n}\right) \right]$$

535. $\lim_{n \to \infty} \sum_{i=1}^{n} \left(\frac{2}{n}\right) \left[3\left(4 + \frac{2i}{n}\right)^3 + \left(4 + \frac{2i}{n}\right)^2 - \left(4 + \frac{2i}{n}\right) + 5 \right]$

To begin, you split the interval into n pieces of equal width using the formula $\Delta x = \frac{b-a}{n}$, where a is the lower limit of integration and b is the upper limit of integration. In this case, you have

$$\Delta x = \frac{6-4}{n} = \frac{2}{n}$$

You also want to select a point from each interval. The formula $x_i = a + (\Delta x)i$ gives you the right endpoint from each interval. Here, you have

$$x_i = a + (\Delta x)i = 4 + \frac{2i}{n}$$

Substituting those values into the definition of the definite integral gives you

$$\lim_{n\to\infty}\sum_{i=1}^{n}f(x_i)\Delta x$$

$$=\lim_{n\to\infty}\sum_{i=1}^{n}\left[3\left(4+\frac{2i}{n}\right)^3+\left(4+\frac{2i}{n}\right)^2-\left(4+\frac{2i}{n}\right)+5\right]\left(\frac{2}{n}\right)$$

$$=\lim_{n\to\infty}\sum_{i=1}^{n}\left(\frac{2}{n}\right)\left[3\left(4+\frac{2i}{n}\right)^3+\left(4+\frac{2i}{n}\right)^2-\left(4+\frac{2i}{n}\right)+5\right]$$

536. $\int_4^7 x^6\,dx$

Recall that in the limit representation of a definite integral, you divide the interval over which you're integrating into n pieces of equal width. You also have to select a point from each interval; the formula $a+(\Delta x)i$ lets you select the right endpoint of each interval.

Begin by looking for the factor that would represent Δx. In this example, $\Delta x=\frac{3}{n}$; if you ignore the n in this expression, you're left with the length of the interval over which the definite integral is being evaluated. In this case, the length of the interval is 3. Notice that if you consider $4+\frac{3i}{n}$, this is of the form $a+(\Delta x)i$, where $a=4$.

You now have a value for a and know the length of the interval, so you can conclude that you're integrating over the interval [4, 7]. To produce the function, replace each expression of the form $4+\frac{3i}{n}$ that appears in the summation with the variable x. In this

example, you replace $\left(4+\frac{3i}{n}\right)^6$ with x^6 so that $f(x)=x^6$. Therefore, the Riemann sum could represent the definite integral $\int_4^7 x^6\,dx$.

537. $\int_0^{\pi/3}\sec x\,dx$

Recall that in the limit representation of a definite integral, you divide the interval over which you're integrating into n pieces of equal width. You also have to select a point from each interval; the formula $a+(\Delta x)i$ lets you select the right endpoint of each interval.

Begin by looking for the factor that would represent Δx. In this example, $\Delta x=\frac{\pi}{3n}$; if you ignore the n in this expression, you're left with the length of the interval over which the definite integral is being evaluated. In this case, the length of the interval is $\frac{\pi}{3}$. Notice that if you consider $\frac{i\pi}{3n}$, this is of the form $a+(\Delta x)i$, where $a=0$.

You now have a value for a and know the length of the interval, so you can conclude that you're integrating over the interval $\left[0,\frac{\pi}{3}\right]$. To produce the function, replace each expression of the form $\frac{i\pi}{3n}$ that appears in the summation with the variable x. In this

example, you replace $\sec\left(\frac{i\pi}{3n}\right)$ with $\sec x$ so that $f(x)=\sec x$. Therefore, the Riemann sum could represent the definite integral $\int_0^{\pi/3}\sec x\,dx$.

538. $\int_6^7 x\, dx$

Recall that in the limit representation of a definite integral, you divide the interval over which you're integrating into n pieces of equal width. You also have to select a point from each interval; the formula $a + (\Delta x)i$ lets you select the right endpoint of each interval.

Begin by looking for the factor that would represent Δx. In this example, $\Delta x = \frac{1}{n}$; if you ignore the n in this expression, you're left with the length of the interval over which the definite integral is being evaluated. In this case, the length of the interval is 1. Notice that if you consider $6 + \frac{i}{n}$, this is of the form $a + (\Delta x)i$, where $a = 6$.

You now have a value for a and know the length of the interval, so you can conclude that you're integrating over the interval $[6, 7]$. To produce the function, replace each expression of the form $6 + \frac{i}{n}$ that appears in the summation with the variable x. In this example, you replace $6 + \frac{i}{n}$ with x so that $f(x) = x$. Therefore, the Riemann sum could represent the definite integral $\int_6^7 x\, dx$.

539. $\int_0^{\pi/2} (\cos x + \sin x)\, dx$

Recall that in the limit representation of a definite integral, you divide the interval over which you're integrating into n pieces of equal width. You also have to select a point from each interval; the formula $a + (\Delta x)i$ lets you select the right endpoint of each interval.

Begin by looking for the factor that would represent Δx. In this example, $\Delta x = \frac{\pi}{2n}$; if you ignore the n in this expression, you're left with the length of the interval over which the definite integral is being evaluated. In this case, the length of the interval is $\frac{\pi}{2}$. Notice that if you consider $\left[0, \frac{\pi}{2}\right]$, this is of the form $a + (\Delta x)i$, where $a = 0$.

You now have a value for a and know the length of the interval, so you can conclude that you're integrating over the interval $\left[0, \frac{\pi}{2}\right]$. To produce the function, replace each expression of the form $\frac{i\pi}{2n}$ that appears in the summation with the variable x. In this example, you replace $\cos\left(\frac{i\pi}{2n}\right) + \sin\left(\frac{i\pi}{2n}\right)$ with $\cos x + \sin x$ so that $f(x) = \cos x + \sin x$. Therefore, the Riemann sum could represent the definite integral $\int_0^{\pi/2} (\cos x + \sin x)\, dx$.

540. $\int_0^5 \sqrt{x + x^3}\, dx$

Recall that in the limit representation of a definite integral, you divide the interval over which you're integrating into n pieces of equal width. You also have to select a point from each interval; the formula $a + (\Delta x)i$ lets you select the right endpoint of each interval.

Begin by looking for the factor that would represent Δx. In this example, $\Delta x = \frac{5}{n}$; if you ignore the n in this expression, you're left with the length of the interval over which the definite integral is being evaluated. In this case, the length of the interval is 5.

Notice that if you consider $\dfrac{5i}{n}$, this is of the form $a + (\Delta x)i$, where $a = 0$.

You now have a value for a and know the length of the interval, so you can conclude that you're integrating over the interval $[0, 5]$. To produce the function, replace each expression of the form $\dfrac{5i}{n}$ that appears in the summation with the variable x. Notice that you can rewrite the summation as $\displaystyle\lim_{n\to\infty}\sum_{i=1}^{n}\frac{5}{n}\sqrt{\frac{5i}{n}+\frac{125i^3}{n^3}} = \lim_{n\to\infty}\sum_{i=1}^{n}\frac{5}{n}\sqrt{\frac{5i}{n}+\left(\frac{5i}{n}\right)^3}$; that

means you can replace $\sqrt{\dfrac{5i}{n}+\left(\dfrac{5i}{n}\right)^3}$ with $\sqrt{x+x^3}$ so that $f(x) = \sqrt{x+x^3}$. Therefore, the Riemann sum could represent the definite integral $\displaystyle\int_0^5 \sqrt{x+x^3}\,dx$.

---

541.   6

To begin, you split the interval into n pieces of equal width using the formula $\Delta x = \dfrac{b-a}{n}$, where a is the lower limit of integration and b is the upper limit of integration. In this case, you have

$$\Delta x = \frac{2-0}{n} = \frac{2}{n}$$

You also want to select a point from each interval. The formula $x_i = a + (\Delta x)i$ gives you the right endpoint from each interval. Here, you have

$$x_i = a+(\Delta x)i = 0+\frac{2i}{n} = \frac{2i}{n}$$

Substituting those values into the definition of the integral gives you the following:

$$\int_0^2 (1+2x)\,dx = \lim_{n\to\infty}\sum_{i=1}^{n} f(x_i)\Delta x$$

$$= \lim_{n\to\infty}\sum_{i=1}^{n}\left[1+2\left(\frac{2i}{n}\right)\right]\left(\frac{2}{n}\right)$$

$$= \lim_{n\to\infty}\sum_{i=1}^{n}\frac{2}{n}\left[1+\frac{4i}{n}\right]$$

$$= \lim_{n\to\infty}\frac{2}{n}\left[\sum_{i=1}^{n}1+\frac{4}{n}\sum_{i=1}^{n}i\right]$$

$$= \lim_{n\to\infty}\frac{2}{n}\left[n+\frac{4}{n}\frac{n(n+1)}{2}\right]$$

$$= \lim_{n\to\infty}\left[2+4\frac{n^2+n}{n^2}\right]$$

$$= 2+4(1)$$

$$= 6$$

Note that formulas $\displaystyle\sum_{i=1}^{n}k = kn$ and $\displaystyle\sum_{i=1}^{n}i = \frac{n(n+1)}{2}$ were used to simplify the summations.

542. 196

To begin, you split the interval into n pieces of equal width using the formula $\Delta x = \dfrac{b-a}{n}$, where a is the lower limit of integration and b is the upper limit of integration. In this case, you have

$$\Delta x = \frac{4-0}{n} = \frac{4}{n}$$

You also want to select a point from each interval. The formula $x_i = a + (\Delta x)i$ gives you the right endpoint from each interval. Here, you have

$$x_i = a + (\Delta x)i = 0 + \frac{4i}{n} = \frac{4i}{n}$$

Substituting those values into the definition of the integral gives you the following:

$$\int_0^4 (1 + 3x^3)\,dx = \lim_{n \to \infty} \sum_{i=1}^{n} f(x_i)\Delta x$$

$$= \lim_{n \to \infty} \sum_{i=1}^{n} \left[1 + 3\left(\frac{4i}{n}\right)^3 \right]\frac{4}{n}$$

$$= \lim_{n \to \infty} \frac{4}{n}\left[\frac{192}{n^3}\sum_{i=1}^{n} i^3 + \sum_{i=1}^{n} 1 \right]$$

$$= \lim_{n \to \infty} \frac{4}{n}\left[\frac{192}{n^3}\left(\frac{n(n+1)}{2}\right)^2 + n \right]$$

$$= \lim_{n \to \infty} \left[192\frac{n^2(n+1)^2}{n^4} + 4 \right]$$

$$= 192(1) + 4$$

$$= 196$$

Note that the formulas $\displaystyle\sum_{i=1}^{n} k = kn$ and $\displaystyle\sum_{i=1}^{n} i^3 = \left(\frac{n(n+1)}{2}\right)^2$ were used to simplify the summations.

543. $\dfrac{9}{2}$

To begin, you split the interval into n pieces of equal width using the formula $\Delta x = \dfrac{b-a}{n}$, where a is the lower limit of integration and b is the upper limit of integration. In this case, you have

$$\Delta x = \frac{4-1}{n} = \frac{3}{n}$$

You also want to select a point from each interval. The formula $x_i = a + (\Delta x)i$ gives you the right endpoint from each interval. Here, you have

$$x_i = a + (\Delta x)i = 1 + \frac{3i}{n}$$

Substituting those values into the definition of the integral gives you the following:

$$\int_1^4 (4-x)dx = \lim_{n\to\infty}\sum_{i=1}^n f(x_i)\Delta x$$

$$= \lim_{n\to\infty}\sum_{i=1}^n\left[4-\left(1+\frac{3i}{n}\right)\right]\frac{3}{n}$$

$$= \lim_{n\to\infty}\frac{3}{n}\left[\sum_{i=1}^n 3 - \frac{3}{n}\sum_{i=1}^n i\right]$$

$$= \lim_{n\to\infty}\frac{3}{n}\left[3n - \frac{3}{n}\left(\frac{n(n+1)}{2}\right)\right]$$

$$= \lim_{n\to\infty}\left[9 - \frac{9}{2}\frac{n(n+1)}{n^2}\right]$$

$$= 9 - \frac{9}{2}$$

$$= \frac{9}{2}$$

Note that the formulas $\sum_{i=1}^n k = kn$ and $\sum_{i=1}^n i = \frac{n(n+1)}{2}$ were used to simplify the summations.

544. $\frac{3}{2}$

To begin, you split the interval into n pieces of equal width using the formula $\Delta x = \frac{b-a}{n}$, where a is the lower limit of integration and b is the upper limit of integration. In this case, you have

$$\Delta x = \frac{3-0}{n} = \frac{3}{n}$$

You also want to select a point from each interval. The formula $x_i = a + (\Delta x)i$ gives you the right endpoint from each interval. Here, you have

$$x_i = a + (\Delta x)i = 0 + \frac{3i}{n} = \frac{3i}{n}$$

Substituting those values into the definition of the integral gives you the following:

$$\int_0^3 (2x^2 - x - 4)dx = \lim_{n\to\infty}\sum_{i=1}^n f(x_i)\Delta x$$

$$= \lim_{n\to\infty}\sum_{i=1}^n\left[2\left(\frac{3i}{n}\right)^2 - \left(\frac{3i}{n}\right) - 4\right]\left(\frac{3}{n}\right)$$

$$= \lim_{n\to\infty}\frac{3}{n}\left[\frac{18}{n^2}\sum_{i=1}^n i^2 - \frac{3}{n}\sum_{i=1}^n i - \sum_{i=1}^n 4\right]$$

$$= \lim_{n\to\infty}\frac{3}{n}\left[\frac{18}{n^2}\left(\frac{n(n+1)(2n+1)}{6}\right) - \frac{3}{n}\left(\frac{n(n+1)}{2}\right) - 4n\right]$$

$$= \lim_{n\to\infty}\left[9\frac{n(n+1)(2n+1)}{n^3} - \frac{9}{2}\frac{n(n+1)}{n^2} - 12\right]$$

$$= 9(2) - \frac{9}{2} - 12$$

$$= \frac{3}{2}$$

Note that the formulas $\sum_{i=1}^{n} k = kn$, $\sum_{i=1}^{n} i = \dfrac{n(n+1)}{2}$, and $\sum_{i=1}^{n} i^2 = \dfrac{n(n+1)(2n+1)}{6}$ were used to simplify the summations.

---

545. $\dfrac{8}{3}$

To begin, you split the interval into n pieces of equal width using the formula $\Delta x = \dfrac{b-a}{n}$, where a is the lower limit of integration and b is the upper limit of integration. In this case, you have

$$\Delta x = \dfrac{3-1}{n} = \dfrac{2}{n}$$

You also want to select a point from each interval. The formula $x_i = a + (\Delta x)i$ gives you the right endpoint from each interval. Here, you have

$$x_i = a + (\Delta x)i = 1 + \dfrac{2i}{n}$$

Substituting those values into the definition of the integral gives you the following:

$$\int_{1}^{3} (x^2 + x - 5)\,dx = \lim_{n\to\infty} \sum_{i=1}^{n} f(x_i)\Delta x$$

$$= \lim_{n\to\infty} \sum_{i=1}^{n} \left[\left(1 + \dfrac{2i}{n}\right)^2 + \left(1 + \dfrac{2i}{n}\right) - 5 \right]\dfrac{2}{n}$$

$$= \lim_{n\to\infty} \sum_{i=1}^{n} \dfrac{2}{n}\left[1 + \dfrac{4i}{n} + \dfrac{4i^2}{n^2} + 1 + \dfrac{2i}{n} - 5 \right]$$

$$= \lim_{n\to\infty} \sum_{i=1}^{n} \dfrac{2}{n}\left[\dfrac{4i^2}{n^2} + \dfrac{6i}{n} - 3 \right]$$

$$= \lim_{n\to\infty} \dfrac{2}{n}\left[\dfrac{4}{n^2}\sum_{i=1}^{n} i^2 + \dfrac{6}{n}\sum_{i=1}^{n} i - \sum_{i=1}^{n} 3 \right]$$

$$= \lim_{n\to\infty} \dfrac{2}{n}\left[\dfrac{4}{n^2}\left(\dfrac{n(n+1)(2n+1)}{6} \right) + \dfrac{6}{n}\left(\dfrac{n(n+1)}{2} \right) - 3n \right]$$

$$= \lim_{n\to\infty} \left[\dfrac{4}{3}\dfrac{n(n+1)(2n+1)}{n^3} + 6\dfrac{n(n+1)}{n^2} - 6 \right]$$

$$= \dfrac{4}{3}(2) + 6 - 6$$

$$= \dfrac{8}{3}$$

Note that the formulas $\sum_{i=1}^{n} k = kn$, $\sum_{i=1}^{n} i = \dfrac{n(n+1)}{2}$, and $\sum_{i=1}^{n} i^2 = \dfrac{n(n+1)(2n+1)}{6}$ were used to simplify the summations.

546. $\sqrt{1+4x}$

Part of the fundamental theorem of calculus states that if the function g is continuous on $[a, b]$, then the function f defined by

$$f(x) = \int_a^x g(t)dt \quad (\text{where } a \le x \le b)$$

is continuous on $[a, b]$ and is differentiable on (a, b). Furthermore, $f'(x) = g(x)$.

To find the derivative of the function $f(x) = \int_0^x \sqrt{1+4t}\, dt$, simply substitute the upper limit of integration, x, into the integrand:

$$f'(x) = \sqrt{1+4x}$$

547. $(2+x^6)^4$

To find the derivative of the function $f(x) = \int_3^x (2+t^6)^4\, dt$, simply substitute the upper limit of integration, x, into the integrand:

$$f'(x) = (2+x^6)^4$$

548. $x^3 \cos x$

To find the derivative of the function $f(x) = \int_0^x t^3 \cos(t)dt$, simply substitute the upper limit of integration, x, into the integrand:

$$f'(x) = x^3 \cos x$$

549. $-e^{x^2}$

Part of the fundamental theorem of calculus states that if the function g is continuous on $[a, b]$, then the function f defined by

$$f(x) = \int_a^x g(t)dt \quad (\text{where } a \le x \le b)$$

is continuous on $[a, b]$ and is differentiable on (a, b). Furthermore, $f'(x) = g(x)$.

To use the fundamental theorem of calculus, you need to have the variable in the upper limit of integration. Therefore, to find the derivative of the function $f(x) = \int_x^4 e^{t^2}\, dt$, first flip the limits of integration and change the sign:

$$f(x) = \int_x^4 e^{t^2}\, dt$$
$$= -\int_4^x e^{t^2}\, dt$$

Then substitute the upper limit of integration into the integrand to find the derivative:

$$f'(x) = -(e^{x^2})$$
$$= -e^{x^2}$$

550. $-\cos^3 x$

Part of the fundamental theorem of calculus states that if the function g is continuous on $[a, b]$, then the function f defined by

$$f(x) = \int_a^x g(t)dt \quad (\text{where } a \le x \le b)$$

is continuous on $[a, b]$ and is differentiable on (a, b). Furthermore, $f'(x) = g(x)$.

To use the fundamental theorem of calculus, you need to have the variable in the upper limit of integration. Therefore, to find the derivative of the function $f(x) = \int_{\sin x}^0 (1-t^2)dt$, first flip the limits of integration and change the sign:

$$f(x) = \int_{\sin x}^0 (1-t^2)dt$$
$$= -\int_0^{\sin x} (1-t^2)dt$$

Note also that to find $\dfrac{d}{dx}\int_a^{h(x)} g(t)dt$, you can use the substitution $u = h(x)$ and then apply the chain rule as follows:

$$\frac{d}{dx}\int_a^{h(x)} g(t)dt = \frac{d}{dx}\int_a^u g(t)dt$$
$$= \frac{d}{du}\left(\int_a^u g(t)dt\right)\frac{du}{dx}$$
$$= \left(g'(u)\right)\frac{du}{dx}$$
$$= \left(g'(h(x))\right)\frac{du}{dx}$$

All this tells you to substitute the upper limit of integration into the integrand and multiply by the derivative of the upper limit of integration. Therefore, the derivative of $f(x) = -\int_0^{\sin x}(1-t^2)dt$ is

$$f'(x) = -\left(1-(\sin x)^2\right)(\cos x)$$
$$= -\cos x(1-\sin^2 x)$$
$$= -\cos x(\cos^2 x)$$
$$= -\cos^3 x$$

Note that the identity $1 - \sin^2 x = \cos^2 x$ was used to simplify the derivative.

551. $-2x$

Part of the fundamental theorem of calculus states that if the function g is continuous on $[a, b]$, then the function f defined by

$$f(x) = \int_a^x g(t)dt \quad (\text{where } a \le x \le b)$$

is continuous on $[a, b]$ and is differentiable on (a, b). Furthermore, $f'(x) = g(x)$.

To use the fundamental theorem of calculus, you need to have the variable in the upper limit of integration. Therefore, to find the derivative of the function $f(x) = \int_{\ln(x^2+1)}^{4} e^t \, dt$, flip the limits of integration and change the sign of the integral:

$$f(x) = \int_{\ln(x^2+1)}^{4} e^t \, dt$$

$$= -\int_{4}^{\ln(x^2+1)} e^t \, dt$$

Note also that to find $\dfrac{d}{dx}\int_{a}^{h(x)} g(t)\,dt$, you can use the substitution $u = h(x)$ and then apply the chain rule as follows:

$$\frac{d}{dx}\int_{a}^{h(x)} g(t)\,dt = \frac{d}{dx}\int_{a}^{u} g(t)\,dt$$

$$= \frac{d}{du}\left(\int_{a}^{u} g(t)\,dt\right)\frac{du}{dx}$$

$$= \left(g'(u)\right)\frac{du}{dx}$$

$$= \left(g'(h(x))\right)\frac{du}{dx}$$

All this tells you to substitute the upper limit of integration into the integrand and multiply by the derivative of the upper limit of integration. Therefore, the derivative of $f(x) = -\int_{4}^{\ln(x^2+1)} e^t \, dt$ is

$$f'(x) = \left(-e^{\ln(x^2+1)}\right)\left(\frac{1}{x^2+1}\right)(2x)$$

$$= -(x^2+1)\left(\frac{2x}{x^2+1}\right)$$

$$= -2x$$

552. $\sin x\left(\cos^2 x + \sin(\cos x)\right)$

Part of the fundamental theorem of calculus states that if the function g is continuous on $[a, b]$, then the function f defined by

$$f(x) = \int_{a}^{x} g(t)\,dt \quad (\text{where } a \le x \le b)$$

is continuous on $[a, b]$ and is differentiable on (a, b). Furthermore, $f'(x) = g(x)$.

To use the fundamental theorem of calculus, you need to have the variable in the upper limit of integration. Therefore, to find the derivative of the function $f(x) = \int_{\cos x}^{1}\left(t^2 + \sin t\right)dt$, first flip the limits of integration and change the sign:

$$f(x) = \int_{\cos x}^{1}\left(t^2 + \sin t\right)dt$$

$$= -\int_{1}^{\cos x}\left(t^2 + \sin t\right)dt$$

Note also that to find $\dfrac{d}{dx}\int_{a}^{h(x)} g(t)\,dt$, you can use the substitution

$u = h(x)$ and then apply the chain rule as follows:

$$\frac{d}{dx}\int_a^{h(x)} g(t)dt = \frac{d}{dx}\int_a^u g(t)dt$$

$$= \frac{d}{du}\left(\int_a^u g(t)dt\right)\frac{du}{dx}$$

$$= \left(g'(u)\right)\frac{du}{dx}$$

$$= \left(g'(h(x))\right)\frac{du}{dx}$$

All this tells you to substitute the upper limit of integration into the integrand and multiply by the derivative of the upper limit of integration. Therefore, the derivative of $f(x) = -\int_1^{\cos x}\left(t^2 + \sin t\right)dt$ is

$$f'(x) = -\left(\cos^2 x + \sin(\cos x)\right)\left(-\sin x\right)$$

$$= \sin x\left(\cos^2 x + \sin(\cos x)\right)$$

---

553. $\dfrac{\sin^2\left(\dfrac{1}{x}\right)}{x^2}$

Part of the fundamental theorem of calculus states that if the function g is continuous on $[a, b]$, then the function f defined by

$$f(x) = \int_a^x g(t)dt \quad (\text{where } a \le x \le b)$$

is continuous on $[a, b]$ and is differentiable on (a, b). Furthermore, $f'(x) = g(x)$.

To use the fundamental theorem of calculus, you need to have the variable in the upper limit of integration. Therefore, to find the derivative of the function $f(x) = \int_{1/x}^1 \sin^2(t)dt$, first flip the limits of integration and change the sign:

$$f(x) = \int_{1/x}^1 \sin^2(t)dt$$

$$= -\int_1^{1/x} \sin^2(t)dt$$

Note also that to find $\dfrac{d}{dx}\int_a^{h(x)} g(t)dt$, you can use the substitution $u = h(x)$ and then apply the chain rule as follows:

$$\frac{d}{dx}\int_a^{h(x)} g(t)dt = \frac{d}{dx}\int_a^u g(t)dt$$

$$= \frac{d}{du}\left(\int_a^u g(t)dt\right)\frac{du}{dx}$$

$$= \left(g'(u)\right)\frac{du}{dx}$$

$$= \left(g'(h(x))\right)\frac{du}{dx}$$

All this tells you to substitute the upper limit of integration into the integrand and multiply by the derivative of the upper limit of integration. Therefore, the derivative of $f(x) = -\int_1^{1/x} \sin^2(t)dt$ is

$$f'(x) = \left(-\sin^2\left(\frac{1}{x}\right)\right)\left(\frac{-1}{x^2}\right)$$

$$= \frac{\sin^2\left(\frac{1}{x}\right)}{x^2}$$

554. $\dfrac{3}{x + x^{10}}$

Part of the fundamental theorem of calculus states that if the function g is continuous on $[a, b]$, then the function f defined by

$$f(x) = \int_a^x g(t)\,dt \quad (\text{where } a \le x \le b)$$

is continuous on $[a, b]$ and is differentiable on (a, b). Furthermore, $f'(x) = g(x)$.

Note also that to find $\dfrac{d}{dx}\int_a^{h(x)} g(t)\,dt$, you can use the substitution $u = h(x)$ and then apply the chain rule as follows:

$$\frac{d}{dx}\int_a^{h(x)} g(t)\,dt = \frac{d}{dx}\int_a^u g(t)\,dt$$

$$= \frac{d}{du}\left(\int_a^u g(t)\,dt\right)\frac{du}{dx}$$

$$= \left(g'(u)\right)\frac{du}{dx}$$

$$= \left(g'(h(x))\right)\frac{du}{dx}$$

All this tells you to substitute the upper limit of integration into the integrand and multiply by the derivative of the upper limit of integration. Therefore, the derivative of $f(x) = \int_1^{x^3} \dfrac{1}{t + t^4}\,dt$ is

$$f'(x) = \frac{1}{x^3 + \left(x^3\right)^4}\left(3x^2\right)$$

$$= \frac{3x^2}{x^3 + x^{12}}$$

$$= \frac{3}{x + x^{10}}$$

555. $\dfrac{3x^2}{\sqrt{3 + x^3}} - \dfrac{\sec^2 x}{\sqrt{3 + \tan x}}$

Part of the fundamental theorem of calculus states that if the function g is continuous on $[a, b]$, then the function f defined by

$$f(x) = \int_a^x g(t)\,dt \quad (\text{where } a \le x \le b)$$

is continuous on $[a, b]$ and is differentiable on (a, b). Furthermore, $f'(x) = g(x)$.

To use the fundamental theorem of calculus, you need to have the variable in the upper limit of integration. Therefore, to find the derivative of the function

$f(x) = \int_{\tan x}^{x^3} \dfrac{1}{\sqrt{3+t}}\, dt$, first split the integral into two separate integrals:

$$f(x) = \int_{\tan x}^{0} \frac{1}{\sqrt{3+t}}\, dt + \int_{0}^{x^3} \frac{1}{\sqrt{3+t}}\, dt$$

Then flip the limits of integration and change the sign of the first integral:

$$f(x) = \int_{\tan x}^{0} \frac{1}{\sqrt{3+t}}\, dt + \int_{0}^{x^3} \frac{1}{\sqrt{3+t}}\, dt$$

$$= -\int_{0}^{\tan x} \frac{1}{\sqrt{3+t}}\, dt + \int_{0}^{x^3} \frac{1}{\sqrt{3+t}}\, dt$$

Note also that to find $\dfrac{d}{dx}\displaystyle\int_{a}^{h(x)} g(t)\, dt$, you can use the substitution $u = h(x)$ and then apply the chain rule as follows:

$$\frac{d}{dx}\int_{a}^{h(x)} g(t)\, dt = \frac{d}{dx}\int_{a}^{u} g(t)\, dt$$

$$= \frac{d}{du}\left(\int_{a}^{u} g(t)\, dt\right)\frac{du}{dx}$$

$$= \left(g'(u)\right)\frac{du}{dx}$$

$$= \left(g'(h(x))\right)\frac{du}{dx}$$

All this tells you to substitute the upper limit of integration into the integrand and multiply by the derivative of the upper limit of integration. Therefore, the derivative of $f(x) = -\displaystyle\int_{0}^{\tan x} \frac{1}{\sqrt{3+t}}\, dt + \int_{0}^{x^3} \frac{1}{\sqrt{3+t}}\, dt$ is

$$f'(x) = \left(-\frac{1}{\sqrt{3+\tan x}}\right)\left(\sec^2 x\right) + \frac{1}{\sqrt{3+x^3}}\left(3x^2\right)$$

$$= \frac{3x^2}{\sqrt{3+x^3}} - \frac{\sec^2 x}{\sqrt{3+\tan x}}$$

---

556. $-2\left(\dfrac{4x^2-1}{16x^4+1}\right) + 6\left(\dfrac{36x^2-1}{1{,}296x^4+1}\right)$

Part of the fundamental theorem of calculus states that if the function g is continuous on $[a, b]$, then the function f defined by

$$f(x) = \int_{a}^{x} g(t)\, dt \quad (\text{where } a \le x \le b)$$

is continuous on $[a, b]$ and is differentiable on (a, b). Furthermore, $f'(x) = g(x)$.

To use the fundamental theorem of calculus, you need to have the variable in the upper limit of integration. Therefore, to find the derivative of the function $f(x) = \displaystyle\int_{2x}^{6x} \frac{t^2-1}{t^4+1}\, dt$, first split the integral into two separate integrals:

$$f(x) = \int_{2x}^{6x} \frac{t^2-1}{t^4+1}\, dt$$

$$= \int_{2x}^{0} \frac{t^2-1}{t^4+1}\, dt + \int_{0}^{6x} \frac{t^2-1}{t^4+1}\, dt$$

Then flip the limits of integration and change the sign of the first integral:

$$f(x) = \int_{2x}^{0} \frac{t^2-1}{t^4+1}dt + \int_{0}^{6x} \frac{t^2-1}{t^4+1}dt$$

$$= -\int_{0}^{2x} \frac{t^2-1}{t^4+1}dt + \int_{0}^{6x} \frac{t^2-1}{t^4+1}dt$$

Note also that to find $\frac{d}{dx}\int_{a}^{h(x)} g(t)dt$, you can use the substitution $u = h(x)$ and then apply the chain rule as follows:

$$\frac{d}{dx}\int_{a}^{h(x)} g(t)dt = \frac{d}{dx}\int_{a}^{u} g(t)dt$$

$$= \frac{d}{du}\left(\int_{a}^{u} g(t)dt\right)\frac{du}{dx}$$

$$= \left(g'(u)\right)\frac{du}{dx}$$

$$= \left(g'(h(x))\right)\frac{du}{dx}$$

All this tells you to substitute the upper limit of integration into the integrand and multiply by the derivative of the upper limit of integration. Therefore, the derivative of $f(x) = -\int_{0}^{2x} \frac{t^2-1}{t^4+1}dt + \int_{0}^{6x} \frac{t^2-1}{t^4+1}dt$ is

$$f'(x) = -\left(\frac{(2x)^2-1}{(2x)^4+1}\right)(2) + \left(\frac{(6x)^2-1}{(6x)^4+1}\right)(6)$$

$$= -2\left(\frac{4x^2-1}{16x^4+1}\right) + 6\left(\frac{36x^2-1}{1,296x^4+1}\right)$$

557. $2x\left(5^{x^2}\right) - \frac{1}{\ln 5}$

Part of the fundamental theorem of calculus states that if the function g is continuous on $[a, b]$, then the function f defined by

$$f(x) = \int_{a}^{x} g(t)dt \quad (\text{where } a \le x \le b)$$

is continuous on $[a, b]$ and is differentiable on (a, b). Furthermore, $f'(x) = g(x)$.

To use the fundamental theorem of calculus, you need to have the variable in the upper limit of integration. Therefore, to find the derivative of the function $f(x) = \int_{\log_5 x}^{x^2} 5^t dt$, first split the integral into two separate integrals:

$$f(x) = \int_{\log_5 x}^{x^2} 5^t dt$$

$$= \int_{\log_5 x}^{0} 5^t dt + \int_{0}^{x^2} 5^t dt$$

Then flip the limits of integration and change the sign of the first integral:

$$f(x) = \int_{\log_5 x}^{0} 5^t dt + \int_{0}^{x^2} 5^t dt$$

$$= -\int_{0}^{\log_5 x} 5^t dt + \int_{0}^{x^2} 5^t dt$$

Note also that to find $\dfrac{d}{dx}\displaystyle\int_a^{h(x)} g(t)\,dt$, you can use the substitution $u = h(x)$ and then apply the chain rule as follows:

$$\frac{d}{dx}\int_a^{h(x)} g(t)\,dt = \frac{d}{dx}\int_a^{u} g(t)\,dt$$

$$= \frac{d}{du}\left(\int_a^{u} g(t)\,dt\right)\frac{du}{dx}$$

$$= \left(g'(u)\right)\frac{du}{dx}$$

$$= \left(g'(h(x))\right)\frac{du}{dx}$$

All this tells you to substitute the upper limit of integration into the integrand and multiply by the derivative of the upper limit of integration. Therefore, the derivative of $f(x) = -\displaystyle\int_0^{\log_5 x} 5^t\,dt + \int_0^{x^2} 5^t\,dt$ is

$$f'(x) = -5^{\log_5 x}\left(\frac{1}{x\ln 5}\right) + 5^{x^2}(2x)$$

$$= \frac{-x}{x\ln 5} + 2x\left(5^{x^2}\right)$$

$$= 2x\left(5^{x^2}\right) - \frac{1}{\ln 5}$$

---

558. 10

Using elementary antiderivative formulas gives you

$$\int_1^3 5\,dx = 5x\Big|_1^3$$

$$= 5(3) - 5(1)$$

$$= 10$$

---

559. $\dfrac{\sqrt{2}}{2}$

Using elementary antiderivative formulas gives you

$$\int_0^{\pi/4} \cos x\,dx = \sin x\Big|_0^{\pi/4}$$

$$= \sin\frac{\pi}{4} - \sin 0$$

$$= \frac{\sqrt{2}}{2} - 0$$

$$= \frac{\sqrt{2}}{2}$$

---

560. 1

Using elementary antiderivative formulas gives you

$$\int_0^{\pi/4} \sec^2 t\,dt = \tan t\Big|_0^{\pi/4}$$

$$= \tan\left(\frac{\pi}{4}\right) - \tan(0)$$

$$= 1$$

561. 1

Using elementary antiderivative formulas gives you

$$\int_0^{\pi/3} \sec x \tan x \, dx = \sec x \Big|_0^{\pi/3}$$

$$= \sec \frac{\pi}{3} - \sec 0$$

$$= \frac{1}{\cos \frac{\pi}{3}} - \frac{1}{\cos 0}$$

$$= 2 - 1$$

$$= 1$$

562. $\dfrac{12}{\ln 4}$

Using the basic antiderivative formula for an exponential function gives you

$$\int_1^2 4^x \, dx = \frac{4^x}{\ln 4} \Big|_1^2$$

$$= \frac{4^2}{\ln 4} - \frac{4^1}{\ln 4}$$

$$= \frac{12}{\ln 4}$$

563. $\dfrac{\pi^2}{2} - 2$

Using elementary antiderivative formulas gives you

$$\int_0^\pi (x - \sin x) \, dx = \left(\frac{x^2}{2} + \cos x \right) \Big|_0^\pi$$

$$= \left(\frac{\pi^2}{2} + \cos \pi \right) - (0 + \cos 0)$$

$$= \left(\frac{\pi^2}{2} - 1 \right) - (0 + 1)$$

$$= \frac{\pi^2}{2} - 2$$

564. $\dfrac{285}{4}$

Using elementary antiderivative formulas gives you

$$\int_1^4 (x + x^3) \, dx = \left(\frac{x^2}{2} + \frac{x^4}{4} \right) \Big|_1^4$$

$$= \frac{4^2}{2} + \frac{4^4}{4} - \left(\frac{1^2}{2} + \frac{1^4}{4} \right)$$

$$= \frac{285}{4}$$

565. $\dfrac{45}{2}$

Rewrite the radical using exponential notation:

$$\int_1^8 2\sqrt[3]{x}\,dx = \int_1^8 2x^{1/3}\,dx$$

Then use elementary antiderivative formulas:

$$\int_1^8 2x^{1/3}\,dx = 2\left(\frac{x^{4/3}}{\frac{4}{3}}\right)\Bigg|_1^8$$

$$= 2\left(\frac{3x^{4/3}}{4}\right)\Bigg|_1^8$$

$$= \frac{3}{2}\left(x^{4/3}\right)\Bigg|_1^8$$

$$= \frac{3}{2}\left(8^{4/3} - 1^{4/3}\right)$$

$$= \frac{3}{2}(16 - 1)$$

$$= \frac{45}{2}$$

566. $1 - \dfrac{\pi}{4}$

Using elementary antiderivative formulas gives you

$$\int_{\pi/4}^{\pi/2}\left(\csc^2 x - 1\right)dx = (-\cot x - x)\big|_{\pi/4}^{\pi/2}$$

$$= \left(-\cot\frac{\pi}{2} - \frac{\pi}{2}\right) - \left(-\cot\frac{\pi}{4} - \frac{\pi}{4}\right)$$

$$= 0 - \frac{\pi}{2} - \left(-1 - \frac{\pi}{4}\right)$$

$$= 1 - \frac{\pi}{4}$$

567. -10

Using elementary antiderivative formulas gives you

$$\int_0^2\left(1 + 2y - 4y^3\right)dy = \left(y + \frac{2y^2}{2} - 4\frac{y^4}{4}\right)\Bigg|_0^2$$

$$= \left(y + y^2 - y^4\right)\Big|_0^2$$

$$= \left(2 + 2^2 - 2^4\right) - \left(0 + 0^2 - 0^4\right)$$

$$= 2 + 4 - 16$$

$$= -10$$

568. $\dfrac{7}{6}$

Begin by rewriting the square root using an exponent:

$$\int_0^1 \left(\sqrt{x} + x\right) dx = \int_0^1 \left(x^{1/2} + x\right) dx$$

Then use elementary antiderivative formulas:

$$\int_0^1 \left(x^{1/2} + x\right) dx = \left(\frac{x^{3/2}}{3/2} + \frac{x^2}{2}\right)\Bigg|_0^1$$

$$= \left(\frac{2}{3}x^{3/2} + \frac{x^2}{2}\right)\Bigg|_0^1$$

$$= \left(\frac{2}{3}1^{3/2} + \frac{1^2}{2}\right) - \left(\frac{2}{3}0^{3/2} + \frac{0^2}{2}\right)$$

$$= \frac{7}{6}$$

569. $\dfrac{-7}{2}$

Integrate each piece of the function over the given integral. Here's the piecewise function:

$$f(x) = \begin{cases} x, & -3 \le x \le 0 \\ \cos x, & 0 < x \le \dfrac{\pi}{2} \end{cases}$$

And here are the calculations:

$$\int_{-3}^0 x\,dx + \int_0^{\pi/2} \cos x\,dx = \frac{x^2}{2}\Bigg|_{-3}^0 + \sin x\Big|_0^{\pi/2}$$

$$= \left(\frac{0^2}{2} - \frac{(-3)^2}{2}\right) + \left(\sin\frac{\pi}{2} - \sin(0)\right)$$

$$= \frac{-9}{2} + 1$$

$$= \frac{-7}{2}$$

570. $\dfrac{29}{2}$

The given integral is

$$\int_{-3}^4 |x - 2|\,dx$$

Begin by splitting up the integral by using the definition of the absolute value function:

$$|x - 2| = \begin{cases} (x - 2), & x \ge 2 \\ -(x - 2), & x < 2 \end{cases}$$

So to evaluate $\int_{-3}^4 |x - 2|\,dx$, use the function $-(x - 2)$ over the interval $[-3, 2]$ and use the function $(x - 2)$ over the interval $[2, 4]$.

Therefore, the integral becomes

$$\int_{-3}^{4} |x-2| \, dx = \int_{-3}^{2} -(x-2) \, dx + \int_{2}^{4} (x-2) \, dx$$

$$= \int_{-3}^{2} (2-x) \, dx + \int_{2}^{4} (x-2) \, dx$$

$$= \left(2x - \frac{x^2}{2} \right) \Big|_{-3}^{2} + \left(\frac{x^2}{2} - 2x \right) \Big|_{2}^{4}$$

$$= \left[\left(2(2) - \frac{2^2}{2} \right) - \left(2(-3) - \frac{(-3)^2}{2} \right) \right] + \left[\left(\frac{4^2}{2} - 2(4) \right) - \left(\frac{2^2}{2} - 2(2) \right) \right]$$

$$= (4-2) - \left(-6 - \frac{9}{2} \right) + (8-8) - (2-4)$$

$$= \frac{29}{2}$$

---

571. $\frac{4}{7} x^{7/4} + C$

Add 1 to the exponent and divide by that new exponent to get the antiderivative:

$$\int x^{3/4} \, dx = \frac{x^{7/4}}{7/4} + C$$

$$= \frac{4}{7} x^{7/4} + C$$

---

572. $-\frac{5}{6x^6} + C$

First move the variable to the numerator:

$$\int \frac{5}{x^7} \, dx = \int 5x^{-7} \, dx$$

$$= 5 \int x^{-7} \, dx$$

Then integrate by adding 1 to the exponent while dividing by that new exponent:

$$5 \int x^{-7} \, dx = 5 \left(\frac{x^{-6}}{-6} \right) + C$$

$$= -\frac{5}{6x^6} + C$$

---

573. $\frac{x^4}{4} + x^3 - x + C$

Find the antiderivative of each term:

$$\int \left(x^3 + 3x^2 - 1 \right) dx = \frac{x^4}{4} + 3\frac{x^3}{3} - x + C$$

$$= \frac{x^4}{4} + x^3 - x + C$$

---

574. $3 \sin x + 4 \cos x + C$

Use elementary antiderivative formulas:

$$\int (3 \cos x - 4 \sin x) \, dx = 3 \sin x + 4 \cos x + C$$

575. $4x + C$

This problem can be a bit difficult, unless you remember the trigonometric identities! In this problem, simply factor out the 4 and use a trigonometric identity:

$$\int \left(4\cos^2 x + 4\sin^2 x\right) dx = \int 4\left(\cos^2 x + \sin^2 x\right) dx$$
$$= \int 4(1) dx$$
$$= \int 4 dx$$
$$= 4x + C$$

576. $\tan x - x + C$

Use a trigonometric identity followed by elementary antiderivatives:

$$\int \tan^2 x\, dx = \int \left(\sec^2 x - 1\right) dx$$
$$= \tan x - x + C$$

577. $x^3 + x^2 + x + C$

Find the antiderivative of each term:

$$\int \left(3x^2 + 2x + 1\right) dx = 3\frac{x^3}{3} + 2\frac{x^2}{2} + x + C$$
$$= x^3 + x^2 + x + C$$

578. $\frac{2}{7} x^{7/3} + x^5 + C$

Use elementary antiderivative formulas:

$$\int \left(\frac{2}{3} x^{4/3} + 5x^4\right) dx = \frac{2}{3}\left(\frac{x^{7/3}}{7/3}\right) + 5\left(\frac{x^5}{5}\right) + C$$
$$= \frac{2}{7} x^{7/3} + x^5 + C$$

579. $4x + \frac{x^2}{2} + \tan x + C$

Begin by using a trigonometric identity to replace $\tan^2 x$. Then simplify:

$$\int \left(5 + x + \tan^2 x\right) dx = \int \left(5 + x + \sec^2 x - 1\right) dx$$
$$= \int \left(4 + x + \sec^2 x\right) dx$$
$$= 4x + \frac{x^2}{2} + \tan x + C$$

580. $\dfrac{36}{11}x^{11/6}+C$

Begin by rewriting the radicals using exponential notation. Then use an elementary antiderivative formula and simplify:

$$\int 6\sqrt{x}\sqrt[3]{x}\,dx = \int 6x^{1/2}x^{1/3}\,dx$$
$$= \int 6x^{5/6}\,dx$$
$$= 6\dfrac{x^{11/6}}{11\big/6}$$
$$= \dfrac{36}{11}x^{11/6}+C$$

581. $\dfrac{2\sqrt{5}}{3}x^{3/2}+C$

Rewrite the integrand using exponential notation and factor out the constant. Then use elementary antiderivative formulas:

$$\int \sqrt{5x}\,dx = \int \sqrt{5}\sqrt{x}\,dx$$
$$= \sqrt{5}\int x^{1/2}\,dx$$
$$= \sqrt{5}\,\dfrac{x^{3/2}}{3\big/2}+C$$
$$= \dfrac{2\sqrt{5}}{3}x^{3/2}+C$$

582. $\dfrac{6\sqrt[6]{4}}{5}x^{5/6}+C$

Begin by writing the radical using exponential notation:

$$\int \sqrt[6]{\dfrac{4}{x}}\,dx = \int \dfrac{\sqrt[6]{4}}{\sqrt[6]{x}}\,dx$$
$$= \sqrt[6]{4}\int \dfrac{1}{x^{1/6}}\,dx$$
$$= \sqrt[6]{4}\int x^{-1/6}\,dx$$

Then use an elementary antiderivative formula:

$$\sqrt[6]{4}\int x^{-1/6}\,dx = \sqrt[6]{4}\,\dfrac{x^{5/6}}{5\big/6}+C$$
$$= \dfrac{6\sqrt[6]{4}}{5}x^{5/6}+C$$

583. $\frac{1}{3}x + C$

Use a trigonometric identity in the numerator of the integrand and simplify:

$$\int \frac{1 - \sin^2 x}{3 \cos^2 x}\, dx = \frac{1}{3} \int \frac{\cos^2 x}{\cos^2 x}\, dx$$
$$= \frac{1}{3} \int 1\, dx$$
$$= \frac{1}{3}x + C$$

584. $\frac{x^5}{5} + \frac{3x^4}{4} + C$

Begin by multiplying out the integrand. Then integrate each term:

$$\int x^2 \left(x^2 + 3x \right) dx = \int \left(x^4 + 3x^3 \right) dx$$
$$= \frac{x^5}{5} + 3\frac{x^4}{4} + C$$
$$= \frac{x^5}{5} + \frac{3x^4}{4} + C$$

585. $-\frac{1}{x} - \frac{1}{2x^2} - \frac{1}{3x^3} + C$

Begin by breaking up the fraction and simplifying each term:

$$\int \frac{x^2 + x + 1}{x^4}\, dx = \int \left(\frac{x^2}{x^4} + \frac{x}{x^4} + \frac{1}{x^4} \right) dx$$
$$= \int \left(x^{-2} + x^{-3} + x^{-4} \right) dx$$

Then find the antiderivative of each term:

$$\int \left(x^{-2} + x^{-3} + x^{-4} \right) dx = \frac{x^{-1}}{-1} + \frac{x^{-2}}{-2} + \frac{x^{-3}}{-3} + C$$
$$= -\frac{1}{x} - \frac{1}{2x^2} - \frac{1}{3x^3} + C$$

586. $x + \frac{2}{3}x^3 + \frac{1}{5}x^5 + C$

Expand the integrand and use elementary antiderivative formulas:

$$\int \left(1 + x^2 \right)^2 dx = \int \left(1 + x^2 \right)\left(1 + x^2 \right) dx$$
$$= \int \left(1 + 2x^2 + x^4 \right) dx$$
$$= x + \frac{2}{3}x^3 + \frac{1}{5}x^5 + C$$

587. $3x + \cot x + C$

Use a trigonometric identity on the integrand and simplify:

$$\int\left(2-\cot^2 x\right)dx = \int\left(2-\left(\csc^2 x-1\right)\right)dx$$
$$= \int\left(3-\csc^2 x\right)dx$$
$$= 3x+\cot x+C$$

---

588. $\quad 2\sqrt{x}-2x^2+C$

Begin by splitting up the fraction and writing the radical using exponential notation:

$$\int\frac{\sqrt{x}-4x^2}{x}dx = \int\left(\frac{x^{1/2}}{x}-\frac{4x^2}{x}\right)dx$$
$$= \int\left(x^{-1/2}-4x\right)dx$$

Then use elementary antiderivative formulas and simplify:

$$\int\left(x^{-1/2}-4x\right)dx = \frac{x^{1/2}}{\frac{1}{2}}-4\frac{x^2}{2}+C$$
$$= 2\sqrt{x}-2x^2+C$$

---

589. $\quad \frac{3}{5}x^{5/3}+\frac{3}{2}x^{4/3}+x+C$

Begin by writing the cube root using exponential notation and multiplying out the integrand:

$$\int\left(\sqrt[3]{x}+1\right)^2 dx = \int\left(x^{1/3}+1\right)\left(x^{1/3}+1\right)dx$$
$$= \int\left(x^{2/3}+2x^{1/3}+1\right)dx$$

Then use elementary antiderivative formulas:

$$\int\left(x^{2/3}+2x^{1/3}+1\right)dx = \frac{x^{5/3}}{\frac{5}{3}}+2\frac{x^{4/3}}{\frac{4}{3}}+x+C$$
$$= \frac{3}{5}x^{5/3}+\frac{3}{2}x^{4/3}+x+C$$

---

590. $\quad \frac{2}{3}x^{3/2}+\frac{6}{7}x^{7/6}+C$

Begin by rewriting the radicals using exponential notation and then distribute:

$$\int\sqrt{x}\left(1+\frac{1}{\sqrt[3]{x}}\right)dx = \int x^{1/2}\left(1+\frac{1}{x^{1/3}}\right)dx$$
$$= \int x^{1/2}\left(1+x^{-1/3}\right)dx$$
$$= \int\left(x^{1/2}+x^{1/6}\right)dx$$

Next, use elementary antiderivative formulas for each term:

$$\int\left(x^{1/2}+x^{1/6}\right)dx = \frac{x^{3/2}}{\frac{3}{2}}+\frac{x^{7/6}}{\frac{7}{6}}+C$$
$$= \frac{2}{3}x^{3/2}+\frac{6}{7}x^{7/6}+C$$

591. $-\csc x + C$

Rewrite the integrand. Then use trigonometric identities followed by using a basic antiderivative formula:

$$\int \frac{\cos x}{\sin^2 x}\, dx = \int \frac{\cos x}{\sin x}\left(\frac{1}{\sin x}\right) dx$$

$$= \int \cot x \csc x\, dx$$

$$= -\csc x + C$$

592. $-\dfrac{1}{x} + \dfrac{5}{4}x^4 + C$

Begin by splitting up the fraction and simplifying the integrand:

$$\int \frac{x + 5x^6}{x^3}\, dx = \int \left(\frac{x}{x^3} + \frac{5x^6}{x^3}\right) dx$$

$$= \int \left(x^{-2} + 5x^3\right) dx$$

Then use elementary antiderivative formulas:

$$\int \left(x^{-2} + 5x^3\right) dx = \frac{x^{-1}}{-1} + 5\frac{x^4}{4} + C$$

$$= -\frac{1}{x} + \frac{5}{4}x^4 + C$$

593. $\dfrac{7}{13}x^{13/7} + \dfrac{7}{6}x^{6/7} + C$

Begin by splitting up the fraction and simplifying the integrand:

$$\int \frac{x+1}{\sqrt[7]{x}}\, dx = \int \left(\frac{x}{x^{1/7}} + \frac{1}{x^{1/7}}\right) dx$$

$$= \int \left(x^{6/7} + x^{-1/7}\right) dx$$

Then use elementary antiderivative formulas:

$$\int \left(x^{6/7} + x^{-1/7}\right) dx = \frac{x^{13/7}}{13/7} + \frac{x^{6/7}}{6/7} + C$$

$$= \frac{7}{13}x^{13/7} + \frac{7}{6}x^{6/7} + C$$

594. $-\cot x + x + C$

Begin by splitting up the fraction and using trigonometric identities to rewrite the integrand:

$$\int \frac{1 + \sin^2 x}{\sin^2 x}\, dx = \int \left(\frac{1}{\sin^2 x} + \frac{\sin^2 x}{\sin^2 x}\right) dx$$

$$= \int \left(\csc^2 x + 1\right) dx$$

Now apply elementary antiderivative formulas:

$$\int \left(\csc^2 x + 1\right) dx$$
$$= -\cot x + x + C$$

595.

$$2x - \frac{x^2}{2} - \frac{x^3}{3} + C$$

Multiply out the integrand and then use elementary antiderivative formulas on each term:

$$\int (1-x)(2+x)dx = \int \left(2 - x - x^2\right) dx$$
$$= 2x - \frac{x^2}{2} - \frac{x^3}{3} + C$$

596.

$$\tan x - x + C$$

Begin by multiplying out the integrand. Then simplify and use elementary antiderivative formulas:

$$\int \sec x(\sec x - \cos x)dx = \int \left(\sec^2 x - \sec x \cos x\right) dx$$
$$= \int \left(\sec^2 x - \frac{1}{\cos x}\cos x\right) dx$$
$$= \int \left(\sec^2 x - 1\right) dx$$
$$= \tan x - x + C$$

597.

$$\frac{x^2}{2} - 3x + C$$

Factor the numerator out of the integrand. Then simplify and use elementary antiderivative formulas:

$$\int \frac{x^2 - 5x + 6}{x-2} dx = \int \frac{(x-2)(x-3)}{(x-2)} dx$$
$$= \int (x-3)dx$$
$$= \frac{x^2}{2} - 3x + C$$

598.

$$\frac{x^3}{3} - \frac{x^2}{2} + x + C$$

Begin by factoring the numerator. Then simplify and use elementary antiderivative formulas:

$$\int \frac{x^3 + 1}{x+1} dx = \int \frac{(x+1)\left(x^2 - x + 1\right)}{(x+1)} dx$$
$$= \int \left(x^2 - x + 1\right) dx$$
$$= \frac{x^3}{3} - \frac{x^2}{2} + x + C$$

599. $\dfrac{20}{13}x^{2.6} - \dfrac{5}{19}x^{3.8} + C$

When finding the antiderivative, don't be thrown off by the decimal exponents; simply add 1 to the exponent of each term and divide by that new exponent for each term to get the solution:

$$\int\left(4x^{1.6} - x^{2.8}\right)dx = 4\frac{x^{2.6}}{2.6} - \frac{x^{3.8}}{3.8} + C$$

$$= 4\frac{x^{2.6}}{26/10} - \frac{x^{3.8}}{38/10} + C$$

$$= \frac{20}{13}x^{2.6} - \frac{5}{19}x^{3.8} + C$$

600. $-\dfrac{50}{11x^{1.1}} - \dfrac{10}{x^{0.1}} + C$

Begin by splitting up the fraction and simplifying:

$$\int\frac{5+x}{x^{2.1}}dx = \int\left(\frac{5}{x^{2.1}} + \frac{x}{x^{2.1}}\right)dx$$

$$= \int\left(5x^{-2.1} + x^{-1.1}\right)dx$$

Then use elementary antiderivative formulas. To get the final answer, turn the decimals in the denominators into fractions and simplify:

$$\int\left(5x^{-2.1} + x^{-1.1}\right)dx = 5\frac{x^{-1.1}}{-1.1} + \frac{x^{-0.1}}{-0.1} + C$$

$$= -\frac{50}{11x^{1.1}} - \frac{10}{x^{0.1}} + C$$

601. $\dfrac{2}{9}x^{9/2} - \dfrac{2}{3}x^{3/2} + \dfrac{5}{4}x^{4/5} + C$

Start by bringing all variables to the numerator. Then use elementary antiderivative formulas:

$$\int\left(x^{7/2} - x^{1/2} + \frac{1}{x^{1/5}}\right)dx = \int\left(x^{7/2} - x^{1/2} + x^{-1/5}\right)dx$$

$$= \frac{x^{9/2}}{9/2} - \frac{x^{3/2}}{3/2} + \frac{x^{4/5}}{4/5} + C$$

$$= \frac{2}{9}x^{9/2} - \frac{2}{3}x^{3/2} + \frac{5}{4}x^{4/5} + C$$

602. $-\dfrac{1}{2}\cos x + C$

Use a trigonometric identity and simplify:

$$\int\frac{\sin 2x}{4\cos x}dx = \int\frac{2\sin x \cos x}{4\cos x}dx$$

$$= \frac{1}{2}\int\sin x\, dx$$

$$= -\frac{1}{2}\cos x + C$$

603.

$x + C$

Start by using a trigonometric identity for $1 + \cot^2 x$, followed by rewriting $\csc^2 x$ as $\dfrac{1}{\sin^2 x}$. Then simplify:

$$\int \sin^2 x \left(1 + \cot^2 x\right) dx = \int \sin^2 x \left(\csc^2 x\right) dx$$

$$= \int \sin^2 x \left(\frac{1}{\sin^2 x}\right) dx$$

$$= \int 1 dx$$

$$= x + C$$

604.

$4x - \dfrac{5}{x} + \dfrac{1}{x^2} + C$

Begin by splitting the fraction into three terms:

$$\int \left(\frac{4x^3 + 5x - 2}{x^3}\right) dx = \int \left(\frac{4x^3}{x^3} + \frac{5x}{x^3} - \frac{2}{x^3}\right) dx$$

$$= \int \left(4 + 5x^{-2} - 2x^{-3}\right) dx$$

Then use elementary antiderivative formulas and simplify:

$$\int \left(4 + 5x^{-2} - 2x^{-3}\right) dx = 4x + \frac{5x^{-1}}{-1} - 2\frac{x^{-2}}{-2} + C$$

$$= 4x - \frac{5}{x} + \frac{1}{x^2} + C$$

605.

$7x + C$

Begin by factoring the 3 out of the first two terms of the integrand. Then use a trigonometric identity:

$$\int \left(3\sin^2 x + 3\cos^2 x + 4\right) dx$$

$$= \int \left(3\left(\sin^2 x + \cos^2 x\right) + 4\right) dx$$

$$= \int \left(3(1) + 4\right) dx$$

$$= \int 7 dx = 7x + C$$

606.

$\dfrac{x^3}{3} - \dfrac{5}{2}x^2 + C$

Begin by factoring the numerator. Then simplify and use elementary antiderivative formulas:

$$\int \frac{x^3-25x}{x+5}\,dx = \int \frac{x\left(x^2-25\right)}{(x+5)}\,dx$$
$$= \int \frac{x(x+5)(x-5)}{(x+5)}\,dx$$
$$= \int x(x-5)\,dx$$
$$= \int \left(x^2-5x\right)\,dx$$
$$= \frac{x^3}{3}-\frac{5}{2}x^2+C$$

607. $\tan x + C$

Begin by writing $\tan^2 x$ as $\dfrac{\sin^2 x}{\cos^2 x}$. Then simplify:

$$\int \frac{\tan^2 x}{\sin^2 x}\,dx = \int \left(\frac{\sin^2 x}{\cos^2 x}\frac{1}{\sin^2 x}\right)dx$$
$$= \int \frac{1}{\cos^2 x}\,dx$$
$$= \int \sec^2 x\,dx$$
$$= \tan x + C$$

608. $\sin x + \cos x + C$

Use a trigonometric identity on the numerator of the integrand, factor, and simplify:

$$\int \frac{\cos 2x}{\cos x+\sin x}\,dx = \int \frac{\cos^2 x-\sin^2 x}{\cos x+\sin x}\,dx$$
$$= \int \frac{(\cos x-\sin x)(\cos x+\sin x)}{\cos x+\sin x}\,dx$$
$$= \int (\cos x-\sin x)\,dx$$
$$= \sin x + \cos x + C$$

609. $-\cot x - 2x + C$

Begin by using a trigonometric identity on the numerator of the integrand. Then split up the fraction and rewrite both fractions using trigonometric identities:

$$\int \frac{\cos 2x}{\sin^2 x}\,dx = \int \frac{2\cos^2 x-1}{\sin^2 x}\,dx$$
$$= \int \left(2\frac{\cos^2 x}{\sin^2 x}-\frac{1}{\sin^2 x}\right)dx$$
$$= \int \left(2\cot^2 x-\csc^2 x\right)dx$$

Notice that the first term of the integrand does not have an elementary antiderivative, so you can use a trigonometric identity again to simplify and then integrate:

$$\int\left(2\cot^2 x - \csc^2 x\right)dx$$
$$= \int\left(2\left(\csc^2 x - 1\right) - \csc^2 x\right)dx$$
$$= \int\left(\csc^2 x - 2\right)dx$$
$$= -\cot x - 2x + C$$

610. $\sin x + C$

Begin by factoring $\cos x$ from the numerator and use a trigonometric identity to simplify:

$$\int\frac{\cos x + \cos x \tan^2 x}{\sec^2 x}dx = \int\frac{\cos x\left(1 + \tan^2 x\right)}{\sec^2 x}dx$$
$$= \int\frac{\cos x\left(\sec^2 x\right)}{\sec^2 x}dx$$
$$= \int\cos x\,dx$$
$$= \sin x + C$$

611.   the net change in the baby's weight in pounds during the first 2 weeks of life

The *net change theorem* states that the integral of a rate of change is the net change:

$$\int_a^b F'(x)dx = F(b) - F(a)$$

Because $w'(t)$ is the rate of the baby's growth in pounds per week, $w(t)$ represents the child's weight in pounds at age t. From the net change theorem, you have

$$\int_0^2 w'(t)dt = w(2) - w(0)$$

Therefore, the integral represents the increase in the child's weight (in pounds) from birth to the start of the second week.

612.   the total amount of oil, in gallons, that leaks from the tanker for the first 180 minutes after the leak begins.

Because $r(t)$ is the rate at which oil leaks from a tanker in gallons per minute, you can write $r(t) = -V'(t)$, where $V(t)$ is the volume of the oil at time t. Notice that you need the minus sign because the volume is decreasing, so $V'(t)$ is negative but $r(t)$ is positive. By the net change theorem, you have

$$\int_0^{180} r(t)dt = -\int_0^{180} V'(t)dt$$
$$= -[V(180) - V(0)]$$
$$= V(0) - V(180)$$

This is the number of gallons of oil that leaks from the tank in the first 180 minutes after the leak begins.

613. the total bird population 6 months after they're placed in the refuge

Because $p'(t)$ is the rate of growth of the bird population per month, $p(t)$ represents the bird population after t months. From the net change theorem, you have

$$\int_0^6 p'(t)\,dt = p(6) - p(0)$$

This represents the change in the bird population in 6 months. Therefore, $100 + \int_0^6 p'(t)\,dt$ represents the total bird population after 6 months.

614. the net change of the particle's position, or displacement, during the first 10 seconds

Because $s'(t) = v(t)$, where $s(t)$ is the position function of the particle after t seconds, you have

$$\int_0^{10} v(t)\,dt = s(10) - s(0)$$

This represents the net change of the particle's position, or *displacement*, during the first 10 seconds. Notice that because the velocity is in meters per second, the displacement units are in meters.

615. the net change in the car's velocity, in meters per second, from the end of the third second to the end of the fifth second

Because $v'(t) = a(t)$, where $v(t)$ is the velocity function of the car after t seconds, you have

$$\int_3^5 a(t)\,dt = v(5) - v(3)$$

This represents the net change in the car's velocity, in meters per second, from the end of the third second to the end of the fifth second.

616. the total number of solar panels produced from the end of the second week to the end of the fourth week

Because $P'(t)$ is the rate of production per week, $P(t)$ represents the total number of solar panels produced after t weeks. By the net change theorem, you have

$$\int_2^4 P'(t)\,dt = P(4) - P(2)$$

This represents the total number of solar panels produced from the end of the second week to the end of the fourth week.

617. the net change in the charge from time t_1 to time t_2

$I(t)$ is defined as the derivative of the charge, $Q'(t) = I(t)$, so by the net change theorem, you have

$$\int_{t_1}^{t_2} I(t)\,dt = Q(t_2) - Q(t_1)$$

This represents the net change in the charge from time t_1 to time t_2.

618. the net change in income in dollars during the first 10 years at the job

$I(t) = I'(t)$, where $I(t)$ represents your total income after t years, so by the net change theorem, you have

$$\int_0^{10} I'(t)dt = I(10) - I(0)$$

This represents the net change in income in dollars during the first 10 years at the job.

619. the net change in the amount of water in the pool from the end of the 60th minute to the end of the 120th minute

Because $w'(t)$ is the rate at which water enters the pool in gallons per minute, $w(t)$ represents the amount of water in the pool at time t. By the net change theorem, you have

$$\int_{60}^{120} w'(t)dt = w(120) - w(60)$$

This represents the net change in the amount of water in the pool from the end of the 60th minute to the end of the 120th minute.

620. 5

Note that *displacement,* or change in position, can be positive or negative or zero. You can think of the particle moving to the left if the displacement is negative and moving to the right if the displacement is positive.

Velocity is the rate of change in displacement with respect to time, so if you integrate the velocity function over an interval where the velocity is negative, you're finding how far the particle travels to the left over that time interval (the value of the integral is negative to indicate that the displacement is to the left). Likewise, if you integrate the velocity function over an interval where the velocity is positive, you're finding the distance that the particle travels to the right over that time interval (in this case, the value of the integral is positive). By combining these two values — that is, by integrating the velocity function over the given time interval — you find the net displacement.

To find the displacement, simply integrate the velocity function over the given interval:

$$s(5) - s(0) = \int_0^5 (2t - 4)dt$$

$$= \left(\frac{2t^2}{2} - 4t \right) \Big|_0^5$$

$$= \left(5^2 - 4(5) \right) - (0 - 0)$$

$$= 5$$

621. $\dfrac{2}{3}$

To find the displacement, simply integrate the velocity function over the given interval:

$$s(4) - s(2) = \int_2^4 \left(t^2 - t - 6\right) dt$$

$$= \left(\frac{t^3}{3} - \frac{t^2}{2} - 6t\right)\Bigg|_2^4$$

$$= \left(\frac{4^3}{3} - \frac{4^2}{2} - 6(4)\right) - \left(\frac{2^3}{3} - \frac{2^2}{2} - 6(2)\right)$$

$$= \frac{64}{3} - 32 - \frac{8}{3} + 14$$

$$= \frac{56}{3} - \frac{54}{3}$$

$$= \frac{2}{3}$$

622. 0

To find the displacement, simply integrate the velocity function over the given interval:

$$s(\pi) - s(0) = \int_0^\pi 2\cos t\, dt = 2\sin t\Big|_0^\pi$$

$$= 2\sin\pi - 2\sin 0$$

$$= 2(0) - 2(0)$$

$$= 0$$

623. $\dfrac{\sqrt{3}-3}{2}$

To find the displacement, simply integrate the velocity function over the given interval:

$$s\left(\frac{\pi}{2}\right) - s\left(\frac{-\pi}{6}\right) = \int_{-\pi/6}^{\pi/2} \left(\sin t - \cos t\right) dt$$

$$= \left(-\cos t - \sin t\right)\Big|_{-\pi/6}^{\pi/2}$$

$$= \left(-\cos\frac{\pi}{2} - \sin\frac{\pi}{2}\right) - \left(-\cos\left(\frac{-\pi}{6}\right) - \sin\left(\frac{-\pi}{6}\right)\right)$$

$$= (0-1) - \left(\frac{-\sqrt{3}}{2} + \frac{1}{2}\right)$$

$$= \frac{\sqrt{3}-3}{2}$$

624. $\dfrac{-40}{3}$

To find the displacement, simply integrate the velocity function over the given interval:

$$s(25) - s(1) = \int_1^{25}\left(t^{1/2} - 4\right)dt = \left(\frac{2t^{3/2}}{3} - 4t\right)\Big|_1^{25}$$

$$= \frac{2}{3}\Bigl(\sqrt{25}\Bigr)^3 - 4(25) - \left(\frac{2}{3} - 4\right)$$

$$= \frac{250}{3} - 100 - \frac{2}{3} + 4$$

$$= \frac{248}{3} - 96$$

$$= \frac{-40}{3}$$

625. 13

Note that *displacement,* or change in position, can be positive or negative or zero. You can think of the particle moving to the left if the displacement is negative and moving to the right if the displacement is positive.

Velocity is the rate of change in displacement with respect to time, so if you integrate the velocity function over an interval where the velocity is negative, you're finding how far the particle travels to the left over that time interval (the value of the integral is negative to indicate that the displacement is to the left). Likewise, if you integrate the velocity function over an interval where the velocity is positive, you're finding the distance that the particle travels to the right over that time interval (in this case, the value of the integral is positive). By combining these two values, you find the net displacement.

Unlike displacement, distance can't be negative. In order to find the *distance* traveled, you can integrate the absolute value of the velocity function because you're then integrating a function that's greater than or equal to zero on the given interval.

$$\int_0^5 |v(t)|\,dt$$

Find any zeros of the function on the given interval so you can determine where the velocity function is positive or negative. In this case, you have $2t - 4 = 0$ so that $t = 2$. Because the velocity function is negative on the interval $(0, 2)$ and positive on the interval $(2, 5)$, the distance traveled is

$$\int_0^2 -(2t - 4)\,dt + \int_2^5 (2t - 4)\,dt$$

$$= \left(-t^2 + 4t\right)\Big|_0^2 + \left(t^2 - 4t\right)\Big|_2^5$$

$$= \left[-\left(2^2\right) + 4(2) - (0 + 0)\right] + \left[\left(5^2 - 4(5)\right) - \left(2^2 - 4(2)\right)\right]$$

$$= 4 + 9$$

$$= 13$$

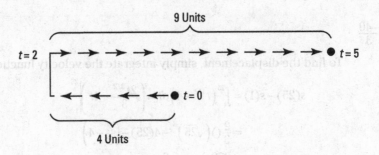

---

626.   5

Unlike displacement, distance can't be negative. In order to find the distance traveled, you can integrate the absolute value of the velocity function because you're then integrating a function that's greater than or equal to zero on the given interval.

$$\int_2^4 \left| t^2 - t - 6 \right| dt$$

Find any zeros of the function on the given interval so you can determine where the velocity function is positive or negative. In this case, you have $t^2 - t - 6 = 0$, or $(t - 3)(t + 2) = 0$, which has solutions $t = 3$ and $t = -2$. Because the velocity function is negative on the interval $(2, 3)$ and positive on the interval $(3, 4)$, the distance traveled is

$$\int_2^3 -\left(t^2 - t - 6\right) dt + \int_3^4 \left(t^2 - t - 6\right) dt$$

$$= \left(-\frac{t^3}{3} + \frac{t^2}{2} + 6t \right)\Big|_2^3 + \left(\frac{t^3}{3} - \frac{t^2}{2} - 6t \right)\Big|_3^4$$

$$= \left(-\frac{3^3}{3} + \frac{3^2}{2} + 6(3) \right) - \left(-\frac{2^3}{3} + \frac{2^2}{2} + 6(2) \right) + \left(\frac{4^3}{3} - \frac{4^2}{2} - 6(4) \right) - \left(\frac{3^3}{3} - \frac{3^2}{2} - 6(3) \right)$$

$$= 5$$

---

627.   4

Unlike displacement, distance can't be negative. In order to find the distance traveled, you can integrate the absolute value of the velocity function because you're then integrating a function that's greater than or equal to zero on the given interval.

$$\int_0^\pi \left| 2\cos t \right| dt$$

Find any zeros of the function on the given interval so you can determine where the velocity function is positive or negative. In this case, you have $2\cos t \geq 0$ on $\left(0, \frac{\pi}{2} \right)$ and $2\cos t \leq 0$ on $\left(\frac{\pi}{2}, \pi \right)$. Therefore, the distance traveled is

$$\int_0^{\pi/2} 2\cos t + \int_{\pi/2}^\pi -2\cos t \, dt$$

$$= 2\sin t \Big|_0^{\pi/2} - 2\sin t \Big|_{\pi/2}^\pi$$

$$= \left(2\sin\frac{\pi}{2} - 2\sin 0 \right) - \left(2\sin\pi - 2\sin\left(\frac{\pi}{2}\right) \right)$$

$$= 2 + 2$$

$$= 4$$

628. $\dfrac{4\sqrt{2}-\sqrt{3}-1}{2}$

Unlike displacement, distance can't be negative. In order to find the distance traveled, you can integrate the absolute value of the velocity function because you're then integrating a function that's greater than or equal to zero on the given interval.

$$\int_{-\pi/6}^{\pi/2}\left|\sin t-\cos t\right|dt$$

Find any zeros of the function on the given interval so you can determine where the velocity function is positive or negative. In this case, you have $\sin t-\cos t=0$ so that $\sin t=\cos t$, which has a solution $t=\dfrac{\pi}{4}$ on the given interval. Note that on the interval $\left(\dfrac{-\pi}{6},\dfrac{\pi}{4}\right)$, you have $\sin t-\cos t<0$ and that on the interval $\left(\dfrac{\pi}{4},\dfrac{\pi}{2}\right)$, $\sin t-\cos t>0$. Therefore, the distance traveled is

$$\int_{-\pi/6}^{\pi/4}-(\sin t-\cos t)dt+\int_{\pi/4}^{\pi/2}(\sin t-\cos t)dt$$

$$=(\cos t+\sin t)\Big|_{-\pi/6}^{\pi/4}+(-\cos t-\sin t)\Big|_{\pi/4}^{\pi/2}$$

$$=\left(\cos\tfrac{\pi}{4}+\sin\tfrac{\pi}{4}\right)-\left(\cos\tfrac{-\pi}{6}+\sin\tfrac{-\pi}{6}\right)-\left[\left(\cos\tfrac{\pi}{2}+\sin\tfrac{\pi}{2}\right)-\left(\cos\tfrac{\pi}{4}+\sin\tfrac{\pi}{4}\right)\right]$$

$$=\frac{\sqrt{2}}{2}+\frac{\sqrt{2}}{2}-\frac{\sqrt{3}}{2}+\frac{1}{2}-\left[(0+1)-\left(\frac{\sqrt{2}}{2}+\frac{\sqrt{2}}{2}\right)\right]$$

$$=\frac{\sqrt{2}}{2}+\frac{\sqrt{2}}{2}-\frac{\sqrt{3}}{2}+\frac{1}{2}-1+\frac{\sqrt{2}}{2}+\frac{\sqrt{2}}{2}$$

$$=\frac{4\sqrt{2}-\sqrt{3}-1}{2}$$

629. $\dfrac{68}{3}$

Unlike displacement, distance can't be negative. In order to find the distance traveled, you can integrate the absolute value of the velocity function because you're then integrating a function that's greater than or equal to zero on the given interval.

$$\int_{1}^{25}\left|\sqrt{t}-4\right|dt$$

Find the any zeros of the function on the given interval so you can determine where the velocity function is positive or negative. In this case, you have $\sqrt{t}-4=0$ so that $\sqrt{t}=4$, or $t=16$. Notice that on the interval $(1,16)$, you have $\sqrt{t}-4<0$, whereas on the interval $(16,25)$, you have $\sqrt{t}-4>0$. Therefore, the distance traveled is

$$\int_{1}^{16}-\left(t^{1/2}-4\right)dt+\int_{16}^{25}\left(t^{1/2}-4\right)dt$$

$$=\left(\frac{-2t^{3/2}}{3}+4t\right)\Bigg|_{1}^{16}+\left(\frac{2}{3}t^{3/2}-4t\right)\Bigg|_{16}^{25}$$

$$=\left[\frac{-2}{3}(16)^{3/2}+4(16)-\left(-\frac{2}{3}+4\right)\right]+\left[\frac{2}{3}(25)^{3/2}-4(25)-\left(\frac{2}{3}(16)^{3/2}-4(16)\right)\right]$$

$$=\frac{-128}{3}+64-\frac{10}{3}+\frac{250}{3}-100-\frac{128}{3}+64$$

$$=\frac{68}{3}$$

630. $\dfrac{304}{3}$

Because the derivative of the position function is the rate of change in position with respect to time, the derivative of the position function gives you the *velocity* function. Likewise, the derivative of the velocity function is the rate of change in velocity with respect to time, so the derivative of the velocity function gives you the *acceleration* function. It follows that the antiderivative of the acceleration function is the velocity function and that the antiderivative of the velocity function is the position function.

First find the velocity function by evaluating the antiderivative of the acceleration function, $a(t) = t + 2$:

$$v(t) = \int (t+2)dt$$
$$= \frac{t^2}{2} + 2t + C$$

Next, use the initial condition, $v(0) = -6$, to solve for the arbitrary constant of integration:

$$v(0) = \frac{0^2}{2} + 20 + C$$
$$-6 = \frac{0^2}{2} + 2(0) + C$$
$$-6 = C$$

Therefore, the velocity function is

$$v(t) = \frac{t^2}{2} + 2t - 6$$

To find the displacement, simply integrate the velocity function over the given interval, $0 \le t \le 8$:

$$s(8) - s(0) = \int_0^8 \left(\frac{1}{2}t^2 + 2t - 6\right)dt$$
$$= \left(\frac{1}{6}t^3 + t^2 - 6t\right)\Big|_0^8$$
$$= \frac{1}{6}(8)^3 + 8^2 - 6(8) - (0 + 0 - 0)$$
$$= \frac{304}{3}$$

631. $\dfrac{-35}{6}$

First find the velocity function by evaluating the antiderivative of the acceleration function, $a(t) = 2t + 1$:

$$v(t) = \int (2t+1)dt$$
$$= t^2 + t + C$$

Next, use the initial condition, $v(0) = -12$, to solve for the arbitrary constant of integration:

$$v(0) = 0^2 + 0 + C$$
$$-12 = 0^2 + 0 + C$$
$$-12 = C$$

Therefore, the velocity function is

$$v(t) = t^2 + t - 12$$

Next, find the displacement by integrating the velocity function over the given interval, $0 \le t \le 5$:

$$s(5) - s(0) = \int_0^5 \left(t^2 + t - 12\right) dt$$

$$= \left(\frac{t^3}{3} + \frac{t^2}{2} - 12t\right)\Big|_0^5$$

$$= \frac{5^3}{3} + \frac{5^2}{2} - 12(5) - (0+0-0)$$

$$= \frac{-35}{6}$$

---

632. $\quad \dfrac{3+\sqrt{3}}{2}$

First find the velocity function by evaluating the antiderivative of the acceleration function, $a(t) = \sin t + \cos t$:

$$v(t) = \int (\sin t + \cos t) dt$$

$$= -\cos t + \sin t + C$$

Next, use the initial condition, $v\left(\frac{\pi}{4}\right) = 0$, to solve for the arbitrary constant of integration:

$$v\left(\frac{\pi}{4}\right) = -\cos\frac{\pi}{4} + \sin\frac{\pi}{4} + C$$

$$0 = -\cos\frac{\pi}{4} + \sin\frac{\pi}{4} + C$$

$$0 = \frac{-\sqrt{2}}{2} + \frac{\sqrt{2}}{2} + C$$

$$0 = C$$

Therefore, the velocity function is

$$v(t) = -\cos t + \sin t$$

Next, find the displacement by integrating the velocity function over the given interval, $\frac{\pi}{6} \le t \le \pi$:

$$s(\pi) - s\left(\frac{\pi}{6}\right) = \int_{\pi/6}^{\pi} (-\cos t + \sin t) dt$$

$$= (-\sin t - \cos t)\Big|_{\pi/6}^{\pi}$$

$$= -\sin\pi - \cos\pi - \left(-\sin\frac{\pi}{6} - \cos\frac{\pi}{6}\right)$$

$$= -0 + 1 - \left(-\frac{1}{2} - \frac{\sqrt{3}}{2}\right)$$

$$= \frac{3+\sqrt{3}}{2}$$

633. $\dfrac{344}{3}$

Because the derivative of the position function is the rate of change in position with respect to time, the derivative of the position function gives you the *velocity* function. Likewise, the derivative of the velocity function is the rate of change in velocity with respect to time, so the derivative of the velocity function gives you the *acceleration* function. It follows that the antiderivative of the acceleration function is the velocity function and that the antiderivative of the velocity function is the position function.

Note that *displacement,* or change in position, can be positive or negative or zero. You can think of the particle moving to the left if the displacement is negative and moving to the right if the displacement is positive. But unlike displacement, distance can't be negative. To find the distance traveled, you can integrate the absolute value of the velocity function because you're then integrating a function that's greater than or equal to zero on the given interval.

Begin by finding the velocity function by evaluating the antiderivative of the acceleration function, $a(t) = t + 2$:

$$v(t) = \int (t+2)\,dt$$

$$= \frac{t^2}{2} + 2t + C$$

Next, use the initial condition, $v(0) = -6$, to solve for the arbitrary constant of integration:

$$v(0) = \frac{0^2}{2} + 20 + C$$

$$-6 = \frac{0^2}{2} + 2(0) + C$$

$$-6 = C$$

Therefore, the velocity function is

$$v(t) = \frac{t^2}{2} + 2t - 6$$

To find the distance traveled, integrate the absolute value of the velocity function over the given interval:

$$\int_0^8 \left| \frac{t^2}{2} + 2t - 6 \right| dt$$

Find any zeros of the function on the given interval so that you can determine where the velocity function is positive or negative: $\frac{t^2}{2} + 2t - 6 = 0$, or $t^2 + 4t - 12 = 0$, which factors as $(t + 6)(t - 2) = 0$ and has the solutions of $t = -6$, $t = 2$. Notice that on the interval $(0, 2)$, the velocity function is negative and that on the interval $(2, 8)$, the velocity function is positive. Therefore, the distance traveled is

$$\int_0^2 -\left(\frac{t^2}{2}+2t-6\right)dt + \int_2^8\left(\frac{t^2}{2}+2t-6\right)dt$$

$$=\left(\frac{-t^3}{6}-t^2+6t\right)\Big|_0^2 + \left(\frac{t^3}{6}+t^2-6t\right)\Big|_2^8$$

$$=\left[\left(\frac{-2^3}{6}-2^2+6(2)\right)-(0)\right]+\left[\frac{8^3}{6}+8^2-6(8)-\left(\frac{2^3}{6}+2^2-6(2)\right)\right]$$

$$=\frac{344}{3}$$

634. $\dfrac{235}{6}$

Begin by finding the velocity function by evaluating the antiderivative of the acceleration function, $a(t) = 2t + 1$:

$$v(t) = \int(2t+1)dt$$
$$= t^2 + t + C$$

Next, use the initial condition, $v(0) = -12$, to solve for the arbitrary constant of integration:

$$v(0) = 0^2 + 0 + C$$
$$-12 = 0^2 + 0 + C$$
$$-12 = C$$

Therefore, the velocity function is

$$v(t) = t^2 + t - 12$$

To find the distance traveled, integrate the absolute value of the velocity function over the given interval:

$$\int_0^5 \left| t^2 + t - 12 \right| dt$$

Find any zeros of the function on the given interval so you can determine where the velocity function is positive or negative: $t^2 + t - 12 = 0$ factors as $(t + 4)(t - 3) = 0$ and has the solutions $t = -4$, $t = 3$. The velocity function is negative on the interval $(0, 3)$ and positive on the interval $(3, 5)$, so the distance traveled is

$$\int_0^3 -\left(t^2+t-12\right)dt + \int_3^5\left(t^2+t-12\right)dt$$

$$=\left(-\frac{t^3}{3}-\frac{t^2}{2}+12t\right)\Big|_0^3 + \left(\frac{t^3}{3}+\frac{t^2}{2}-12t\right)\Big|_3^5$$

$$=\left(-\frac{3^3}{3}-\frac{3^2}{2}+12(3)\right)-(0)+\left[\left(\frac{5^3}{3}+\frac{5^2}{2}-12(5)\right)-\left(\frac{3^3}{3}+\frac{3^2}{2}-12(3)\right)\right]$$

$$=\frac{235}{6}$$

635. $\dfrac{4\sqrt{2}-\sqrt{3}+1}{2}$

Begin by finding the velocity function by evaluating the antiderivative of the acceleration function, $a(t) = \sin t + \cos t$:

$$v(t) = \int (\sin t + \cos t)\, dt$$
$$= -\cos t + \sin t + C$$

Next, use the initial condition, $v\left(\dfrac{\pi}{4}\right) = 0$, to solve for the arbitrary constant of integration:

$$v\left(\frac{\pi}{4}\right) = -\cos\frac{\pi}{4} + \sin\frac{\pi}{4} + C$$
$$0 = -\cos\frac{\pi}{4} + \sin\frac{\pi}{4} + C$$
$$0 = \frac{-\sqrt{2}}{2} + \frac{\sqrt{2}}{2} + C$$
$$0 = C$$

Therefore, the velocity function is

$$v(t) = -\cos t + \sin t$$

To find the distance traveled, integrate the absolute value of the velocity function over the given interval, $\dfrac{\pi}{6} \le t \le \pi$:

$$\int_{\pi/6}^{\pi} \left| -\cos t + \sin t \right| dt$$

Find any zeros of the velocity function on the given interval so you can determine where the velocity function is positive or negative:

$$-\cos t + \sin t = 0$$
$$\sin t = \cos t$$
$$t = \frac{\pi}{4}$$

The velocity function is negative on the interval $\left(\dfrac{\pi}{6}, \dfrac{\pi}{4}\right)$ and positive on the interval $\left(\dfrac{\pi}{4}, \pi\right)$, so the total distance traveled is

$$\int_{\pi/6}^{\pi/4} -(-\cos t + \sin t)\, dt + \int_{\pi/4}^{\pi} (-\cos t + \sin t)\, dt$$
$$= \left(\sin t + \cos t\right)\Big|_{\pi/6}^{\pi/4} + \left(-\sin t - \cos t\right)\Big|_{\pi/4}^{\pi}$$
$$= \left(\sin\frac{\pi}{4} + \cos\frac{\pi}{4}\right) - \left(\sin\frac{\pi}{6} + \cos\frac{\pi}{6}\right) + \left[(-\sin\pi - \cos\pi) - \left(-\sin\frac{\pi}{4} - \cos\frac{\pi}{4}\right)\right]$$
$$= \frac{4\sqrt{2} - \sqrt{3} + 1}{2}$$

636. $\frac{4}{15}$

Recall that the area A of the region bounded by the curves $y = f(x)$ and $y = g(x)$ and by the lines $x = a$ and $x = b$, where f and g are continuous and $f(x) \geq g(x)$ for all x in $[a, b]$, is

$$A = \int_a^b (f(x) - g(x)) \, dx$$

More generically, in terms of a graph, you can think of the formula as

$$A = \int_a^b ((\text{top function}) - (\text{bottom function})) \, dx$$

Note that lines $x = a$ and $x = b$ may not be given, so the limits of integration often correspond to points of intersection of the functions.

Begin by finding the points of intersection by setting the functions equal to each other and solving for x:

$$x^4 = x^2$$
$$x^4 - x^2 = 0$$
$$x^2(x^2 - 1) = 0$$
$$x = 0, \pm 1$$

Because $x^2 \geq x^4$ on the interval $[-1, 1]$, the integral to find the area is

$$\int_{-1}^1 (x^2 - x^4) \, dx$$

The integrand is an even function, so it's symmetric about the y-axis; therefore, you can instead integrate on the interval $[0, 1]$ and multiply by 2:

$$\int_{-1}^1 (x^2 - x^4) \, dx = 2 \int_0^1 (x^2 - x^4) \, dx$$

This gives you the following:

$$2 \int_0^1 (x^2 - x^4) \, dx = 2 \left[\left(\frac{x^3}{3} - \frac{x^5}{5} \right) \right]_0^1 = 2 \left(\frac{1}{3} - \frac{1}{5} \right) = \frac{4}{15}$$

The following figure shows the region bounded by the given curves:

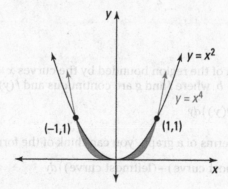

637. $\frac{1}{6}$

Begin by finding the points of intersection by setting the functions equal to each other and solving for x:

$$x = \sqrt{x}$$
$$x^2 = x$$
$$x^2 - x = 0$$
$$x(x-1) = 0$$
$$x = 0, 1$$

To determine which function is larger on the interval $(0, 1)$, take a point inside the interval and substitute it into each function. If you use $x = \frac{1}{4}$, then the first curve gives you $y = \frac{1}{4}$, and the second curve gives you $y = \sqrt{\frac{1}{4}} = \frac{1}{2}$. Therefore, $\sqrt{x} > x$ on $(0, 1)$. That means the integral for the area of the bounded region is

$$\int_0^1 \left(\sqrt{x} - x\right) dx = \int_0^1 \left(x^{1/2} - x\right) dx$$
$$= \left(\frac{2x^{3/2}}{3} - \frac{x^2}{2}\right)\Big|_0^1$$
$$= \frac{2}{3} - \frac{1}{2} - (0-0)$$
$$= \frac{1}{6}$$

638. $\sin 1 + \frac{1}{2}$

Because $\cos x + 1 \geq x$ on $[0, 1]$, the integral to find the area is

$$\int_0^1 (\cos x + 1 - x) dx = \left(\sin x + x - \frac{x^2}{2}\right)\Big|_0^1$$
$$= \sin 1 + 1 - \frac{1}{2} - (0 + 0 - 0)$$
$$= \sin 1 + \frac{1}{2}$$

639. $\frac{8}{3}$

Recall that the area A of the region bounded by the curves $x = f(y)$ and $x = g(y)$ and by the lines $y = a$ and $y = b$, where f and g are continuous and $f(y) \geq g(y)$ for all y in $[a, b]$, is

$$A = \int_a^b \left(f(y) - g(y)\right) dy$$

More generically, in terms of a graph, you can think of the formula as

$$A = \int_a^b \left((\text{rightmost curve}) - (\text{leftmost curve})\right) dy$$

Note that lines $y = a$ and $y = b$ may not be given, so the limits of integration often correspond to points of intersection of the curves.

In this case, you could integrate with respect to x, but you'd have to solve each equation for y by completing the square, which would needlessly complicate the problem.

Begin by finding the points of intersection by setting the expressions equal to each other and solving for y:

$$y^2 - y = 3y - y^2$$
$$2y^2 - 4y = 0$$
$$2y(y-2) = 0$$
$$y = 0, 2$$

To determine which function has larger x values for y in the interval $(0, 2)$, pick a point in the interval and substitute it into each function. So if $y = 1$, then $x = 1^2 - 1 = 0$ and $x = 3(1) - 1^2 = 2$. Therefore, the integral to find the area is

$$\int_0^2 \left[(3y - y^2) - (y^2 - y) \right] dy$$
$$= \int_0^2 (4y - 2y^2) dy$$
$$= \left(2y^2 - \frac{2y^3}{3} \right) \Bigg|_0^2$$
$$= 2(2)^2 - \frac{2(2)^3}{3} - (0 - 0)$$
$$= 8 - \frac{16}{3}$$
$$= \frac{8}{3}$$

640. $\quad \dfrac{4}{3}$

In this case, you can easily solve the two equations for x, so integrating with respect to y makes sense. If you instead solved the equations for y, you'd have to use more than one integral when integrating with respect to x to set up the area.

Begin by solving the first equation for x to get $x = y^2 - 1$. Because $\sqrt{y} \geq y^2 - 1$ on the interval $[0, 1]$, the integral to find the area is

$$\int_0^1 \left(\sqrt{y} - (y^2 - 1) \right) dy = \int_0^1 \left(y^{1/2} - y^2 + 1 \right) dy$$
$$= \left(\frac{2}{3} y^{3/2} - \frac{1}{3} y^3 + y \right) \Bigg|_0^1$$
$$= \frac{2}{3} - \frac{1}{3} + 1$$
$$= \frac{4}{3}$$

641. $\dfrac{125}{6}$

Notice that you can easily solve the two equations for x, so integrating with respect to y makes sense. If you were to solve the first equation for y in order to integrate with respect to x, you'd have to use more than one integral to set up the area.

Begin by solving the second equation for x to get $x = y + 7$. Then find the points of intersection by setting the expressions equal to each other and solving for y:

$$y + 7 = 1 + y^2$$
$$0 = y^2 - y - 6$$
$$0 = (y - 3)(y + 2)$$
$$y = 3, -2$$

To determine which curve has the larger x values for y on the interval $(-2, 3)$, pick a point in the interval and substitute it into each equation. If $y = 0$, then $x = 1 + 0^2 = 1$ and $x = 0 + 7 = 7$. Therefore, the integral to find the area is

$$\int_{-2}^{3} \left[(y + 7) - (1 + y^2) \right] dy$$
$$= \int_{-2}^{3} \left[y + 6 - y^2 \right] dy$$
$$= \left(\frac{y^2}{2} + 6y - \frac{y^3}{3} \right) \Bigg|_{-2}^{3}$$
$$= \frac{(3)^2}{2} + 6(3) - \frac{(3)^3}{3} - \left(\frac{(-2)^2}{2} + 6(-2) - \frac{(-2)^3}{3} \right)$$
$$= \frac{125}{6}$$

The following figure shows the region bounded by the given curves:

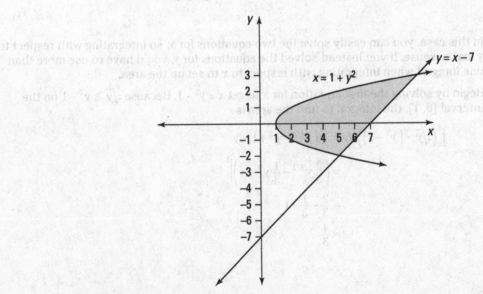

642. $\frac{1}{6}$

Begin by finding the point of intersection of the two curves by setting them equal to each other and solving for y:

$$y^2 = 3y - 2$$
$$y^2 - 3y + 2 = 0$$
$$(y-1)(y-2) = 0$$
$$y = 1, 2$$

Notice that $3y - 2 > y^2$ for y in the interval $(1, 2)$. Therefore, the integral to find the area of the region is

$$\int_1^2 \left[(3y-2)-\left(y^2\right)\right] dy$$
$$= \left(3\frac{y^2}{2} - 2y - \frac{y^3}{3}\right)\Big|_1^2$$
$$= \left(3\frac{(2)^2}{2} - 2(2) - \frac{(2)^3}{3}\right) - \left(3\frac{(1)^2}{2} - 2(1) - \frac{(1)^3}{3}\right)$$
$$= \frac{1}{6}$$

643. $\frac{9}{8}$

In this case, integrating with respect to y makes sense. You could integrate with respect to x, but you'd have to solve $x = 2y^2$ for y and then use two integrals to compute the area, because the "top function" isn't the same for the entire region.

Begin by isolating x in the second equation to get $x = 1 - y$. Then find the points of intersection by setting the expressions equal to each other and solving for y:

$$1 - y = 2y^2$$
$$0 = 2y^2 + y - 1$$
$$0 = (2y-1)(y+1)$$
$$y = \frac{1}{2}, -1$$

To determine which curve has the larger x values for y in the interval $\left(-1, \frac{1}{2}\right)$, pick a point in the interval and substitute it into each equation. So if $y = 0$, then $x = 2(0)^2 = 0$ and $x + 0 = 1$. Therefore, $x = 1 - y$ is the rightmost curve and $x = 2y^2$ is the leftmost curve, which means the integral to find the area is

$$\int_{-1}^{1/2} \left(1 - y - 2y^2\right) dy$$

$$= \left(y - \frac{y^2}{2} - \frac{2y^3}{3}\right)\Bigg|_{-1}^{1/2}$$

$$= \frac{1}{2} - \frac{\left(\frac{1}{2}\right)^2}{2} - \frac{2\left(\frac{1}{2}\right)^3}{3} - \left(-1 - \frac{(-1)^2}{2} - \frac{2(-1)^3}{3}\right)$$

$$= \frac{9}{8}$$

The following figure shows the region bounded by the given curves:

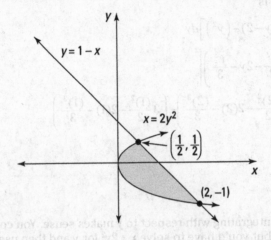

644. 36

Begin by finding the points of intersection by setting the functions equal to each other and solving for x:

$$2x = 8 - x^2$$

$$x^2 + 2x - 8 = 0$$

$$(x + 4)(x - 2) = 0$$

$$x = -4, 2$$

To determine which function is larger on the interval $(-4, 2)$, take a point inside the interval and substitute it into each function. If you let $x = 0$, then $y = 2(0) = 0$ and $y = 8 - 0^2 = 8$. Therefore, $8 - x^2 > 2x$ on $(-4, 2)$, so the integral for the area of the bounded region is

$$\int_{-4}^{2}\left[\left(8-x^2\right)-(2x)\right]dx$$

$$=\int_{-4}^{2}\left(-x^2-2x+8\right)dx$$

$$=\left(\frac{-x^3}{3}-x^2+8x\right)\Bigg|_{-4}^{2}$$

$$=\frac{-(2)^3}{3}-2^2+8(2)-\left(-\frac{(-4)^3}{3}-(-4)^2+8(-4)\right)$$

$$=-\frac{8}{3}-4+16-\frac{64}{3}+16+32$$

$$=60-24$$

$$=36$$

The following figure shows the region bounded by the given curves:

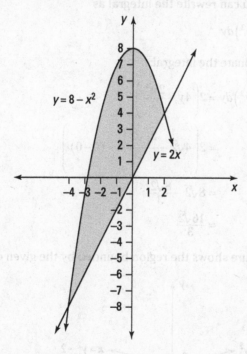

645. $\dfrac{16\sqrt{2}}{3}$

In this example, integrating with respect to *y* makes sense. You could integrate with respect to *x*, but you'd have to solve both $x = 2 - y^2$ and $x = y^2 - 2$ for *y* and then use two integrals to compute the area, because the "top function" and the "bottom function" aren't the same for the entire region. (You could actually reduce the region to a single integral by using symmetry, but that isn't possible in general.)

Begin by finding the points of intersection by setting the expressions equal to each other and solving for y:

$$2 - y^2 = y^2 - 2$$
$$4 = 2y^2$$
$$2 = y^2$$
$$\pm\sqrt{2} = y$$

To determine which curve has the larger x values for y in the interval $\left(-\sqrt{2}, \sqrt{2}\right)$, take a point inside the interval and substitute it into each function. So if $y = 0$, then $x = 2 - 0^2 = 2$ and $x = 0^2 - 2 = -2$; therefore, $2 - y^2 > y^2 - 2$ on $\left(-\sqrt{2}, \sqrt{2}\right)$. That means the integral to find the area is

$$\int_{-\sqrt{2}}^{\sqrt{2}} \left[\left(2 - y^2\right) - \left(y^2 - 2\right)\right] dy$$

By symmetry, you can rewrite the integral as

$$2\int_0^{\sqrt{2}} \left(4 - 2y^2\right) dy$$

Now you can evaluate the integral:

$$2\int_0^{\sqrt{2}} \left(4 - 2y^2\right) dy = 2\left(4y - \frac{2y^3}{3}\right)\Bigg|_0^{\sqrt{2}}$$

$$= 2\left(4\sqrt{2} - \frac{2\left(\sqrt{2}\right)^3}{3} - (0 - 0)\right)$$

$$= 8\sqrt{2} - \frac{8\sqrt{2}}{3}$$

$$= \frac{16\sqrt{2}}{3}$$

The following figure shows the region bounded by the given curves:

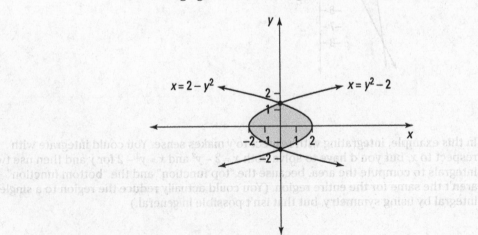

646. 72

Begin by finding the points of intersection by setting the functions equal to each other and solving for x:

$$14 - x^2 = x^2 - 4$$
$$18 = 2x^2$$
$$9 = x^2$$
$$\pm 3 = x$$

To determine which function is larger on the interval $(-3, 3)$, take a point inside the interval and substitute into each function. If $x = 0$, then $y = 14 - 0^2 = 14$ and $y = 0^2 - 4 = -4$. Therefore, $14 - x^2 > x^2$ on the interval $(-3, 3)$, so the integral for the area of the bounded region becomes

$$\int_{-3}^{3} \left[\left(14 - x^2\right) - \left(x^2 - 4\right) \right] dx$$

The integrand is an even function, so it's symmetric about the y-axis; therefore, you can instead integrate on the interval $[0, 3]$ and multiply by 2:

$$\int_{-3}^{3} \left[\left(14 - x^2\right) - \left(x^2 - 4\right) \right] dx$$
$$= 2\int_{0}^{3} \left(-2x^2 + 18\right) dx$$
$$= 2\left(\frac{-2x^3}{3} + 18x \right)\Bigg|_{0}^{3}$$
$$= 2\left(\frac{-2(3)^3}{3} + 18(3) - (0 + 0) \right)$$
$$= 2(-18 + 54)$$
$$= 72$$

647. $\frac{1}{3}$

In this example, integrating with respect to y makes sense. You could integrate with respect to x, but you'd have to solve $4x + y^2 = -3$ for y and then use two integrals to compute the area, because the "top function" isn't the same for the entire region. Begin by solving the second equation for x to get $x = \dfrac{-3 - y^2}{4}$. To find the points of intersection, set the expressions equal to each other and solve for y:

$$4y + y^2 = -3$$
$$y^2 + 4y + 3 = 0$$
$$(y + 1)(y + 3) = 0$$
$$y = -1, -3$$

To determine which curve has larger x values for y in the interval $(-3, 1)$, pick a point inside the interval and substitute it into each function. So if $y = -2$, then $x = -2$ and $x = \dfrac{-3 - (-2)^2}{4} = \dfrac{-7}{4}$. Therefore, the integral to find the area is

$$\int_{-3}^{-1}\left(\frac{-3-y^2}{4}-y\right)dy$$

$$=\int_{-3}^{-1}\left[\frac{-3}{4}-\frac{1}{4}y^2-y\right]dy$$

$$=\int_{-3}^{-1}-\left[\frac{3}{4}+\frac{1}{4}y^2+y\right]dy$$

$$=-\left(\frac{3}{4}y+\frac{1}{12}y^3+\frac{y^2}{2}\right)\Bigg|_{-3}^{-1}$$

$$=-\left[\left(\frac{3}{4}(-1)+\frac{1}{12}(-1)^3+\frac{(-1)^2}{2}\right)-\left(\frac{3}{4}(-3)+\frac{1}{12}(-3)^3+\frac{(-3)^2}{2}\right)\right]$$

$$=-\left[-\frac{3}{4}-\frac{1}{12}+\frac{1}{2}+\frac{9}{4}+\frac{27}{12}-\frac{9}{2}\right]$$

$$=-\left[\frac{6}{4}+\frac{26}{12}-\frac{8}{2}\right]$$

$$=-\left[\frac{9}{6}+\frac{13}{6}-\frac{24}{6}\right]$$

$$=-\left[-\frac{1}{3}\right]=\frac{1}{3}$$

---

648. $\quad \dfrac{9}{2}$

Begin by finding the points of intersection of the two curves by setting them equal to each other and solving for y:

$$\frac{3+y}{3}=1+\sqrt{y}$$

$$3+y=3+3\sqrt{y}$$

$$y=3\sqrt{y}$$

$$y^2=9y$$

$$y^2-9y=0$$

$$y(y-9)=0$$

$$y=0,9$$

To determine which curve has the larger x values for y on the interval $(0, 9)$, pick a value in the interval and substitute it into each equation. So if $y = 1$, then $x = 1+\sqrt{1}=2$ and $x=\dfrac{3+1}{3}=\dfrac{4}{3}$. Therefore, the integral to find the area is

$$\int_0^9\left(1+\sqrt{y}-\left(\frac{3+y}{3}\right)\right)dy=\int_0^9\left(1+y^{1/2}-\left(1+\frac{1}{3}y\right)\right)dy$$

$$=\int_0^9\left(y^{1/2}-\frac{1}{3}y\right)dy$$

$$=\left(\frac{2}{3}y^{3/2}-\frac{1}{6}y^2\right)\Bigg|_0^9$$

$$=\frac{2}{3}(9)^{3/2}-\frac{1}{6}(9)^2$$

$$=\frac{9}{2}$$

649. $2 + \dfrac{\pi^2}{4}$

Notice that the functions $y = x - \frac{\pi}{2}$ and $y = \cos x$ intersect when $x = \frac{\pi}{2}$. On the interval $\left(0, \frac{\pi}{2}\right)$, you have $\cos x \geq \left(x - \frac{\pi}{2}\right)$, and on the interval $\left(\frac{\pi}{2}, \pi\right)$, you have $x - \frac{\pi}{2} \geq \cos x$. Therefore, the integrals to find the area of the region are

$$\int_0^{\pi/2}\left[\cos x - \left(x - \frac{\pi}{2}\right)\right]dx + \int_{\pi/2}^{\pi}\left[\left(x - \frac{\pi}{2}\right) - \cos x\right]dx$$

$$= \left[\sin x - \frac{x^2}{2} + \frac{\pi}{2}x\right]\Bigg|_0^{\pi/2} + \left[\left(\frac{x^2}{2} - \frac{\pi}{2}x\right) - \sin x\right]\Bigg|_{\pi/2}^{\pi}$$

$$= 1 - \frac{1}{2}\left(\frac{\pi}{2}\right)^2 + \frac{\pi}{2}\left(\frac{\pi}{2}\right) + \left[\left(\frac{\pi^2}{2} - \frac{\pi^2}{2} - 0\right) - \left(\frac{1}{2}\left(\frac{\pi}{2}\right)^2 - \left(\frac{\pi}{2}\right)^2 - 1\right)\right]$$

$$= 2 + \frac{\pi^2}{4}$$

650. $\dfrac{9}{2}$

Begin by finding the points of intersection by setting the functions equal to each other and solving for x:

$$x^3 - x = 2x$$
$$x^3 - 3x = 0$$
$$x\left(x^2 - 3\right) = 0$$
$$x = 0, \pm\sqrt{3}$$

To determine which function is larger on the interval $\left(-\sqrt{3}, 0\right)$, take a point in the interval and substitute it into each function to determine which is larger. If $x = -1$, then $y = (-1)^3 - (-1) = 0$ and $y = 2(-1) = -2$; therefore, $x^3 - x > 2x$ on $\left(-\sqrt{3}, 0\right)$. In a similar manner, check which function is larger on the interval $\left(0, \sqrt{3}\right)$. By letting $x = 1$, you can show that $2x > x^3 - x$ on $\left(0, \sqrt{3}\right)$. Therefore, the integral to find the area of the region is

$$\int_{-\sqrt{3}}^{0}\left[\left(x^3 - x\right) - (2x)\right]dx + \int_0^{\sqrt{3}}\left[(2x) - \left(x^3 - x\right)\right]dx$$

$$= \int_{-\sqrt{3}}^{0}\left[x^3 - 3x\right]dx + \int_0^{\sqrt{3}}\left(3x - x^3\right)dx$$

$$= \left(\frac{x^4}{4} - \frac{3x^2}{2}\right)\Bigg|_{-\sqrt{3}}^{0} + \left(\frac{3x^2}{2} - \frac{x^4}{4}\right)\Bigg|_0^{\sqrt{3}}$$

$$= (0 - 0) - \left(\frac{9}{4} - \frac{9}{2}\right) + \left(\frac{9}{2} - \frac{9}{4}\right)$$

$$= \frac{18}{2} - \frac{18}{4} = \frac{18}{4} = \frac{9}{2}$$

The following figure shows the region bounded by the given curves:

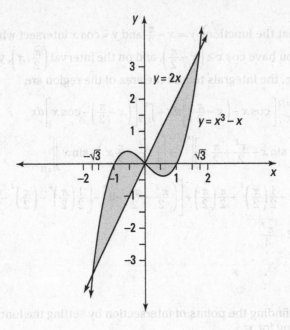

651. $\dfrac{125}{6}$

Notice that you can easily solve the two equations for x, so integrating with respect to y makes sense. If you were to solve the second equation for y in order to integrate with respect to x, you'd have to use more than one integral to set up the area.

Solve the first equation for x to get $x = -y$. Then find the points of intersection by setting the equations equal to each other and solving for y:

$$-y = y^2 + 4y$$
$$0 = y^2 + 5y$$
$$0 = y(y+5)$$
$$y = 0, -5$$

To determine which curve has larger x values on the interval $(-5, 0)$, pick a point in the interval and substitute it into each equation. If $y = -1$, then $x = -(-1) = 1$ and $x = (-1)^2 + 4(-1) = -3$. Therefore, the integral to find the area is

$$\int_{-5}^{0}\left(-y-\left(y^2+4y\right)\right)dy = \int_{-5}^{0}\left(-y-y^2-4y\right)dy$$
$$= \int_{-5}^{0}\left(-y^2-5y\right)dy$$
$$= \left(-\frac{y^3}{3}-\frac{5y^2}{2}\right)\bigg|_{-5}^{0}$$
$$= 0-\left(-\frac{(-5)^3}{3}-\frac{5(-5)^2}{2}\right)$$
$$= \frac{125}{6}$$

652. $\dfrac{4}{3}$

Begin by setting the expressions equal to each other and solving for y to find the points of intersection:

$$\sqrt{y+3} = \frac{y+3}{2}$$
$$2\sqrt{y+3} = y+3$$
$$4(y+3) = y^2+6y+9$$
$$0 = y^2+2y-3$$
$$0 = (y+3)(y-1)$$
$$y = -3, 1$$

To determine which curve has larger x values for y on the interval $(-3, 1)$, take a point in the interval and substitute it into each equation. So if $y = -2$, then $x = \sqrt{-2+3} = 1$ and $x = \dfrac{-2+3}{2} = \dfrac{1}{2}$. Therefore, the integral to find the area of the region is

$$\int_{-3}^{1}\left[\sqrt{(y+3)} - \left(\frac{y+3}{2}\right)\right]dy$$

$$= \int_{-3}^{1}\left[(y+3)^{1/2} - \left(\frac{y}{2}+\frac{3}{2}\right)\right]dy$$

$$= \left(\frac{2(y+3)^{3/2}}{3} - \frac{1}{4}y^2 - \frac{3}{2}y\right)\Bigg|_{-3}^{1}$$

$$= \left(\frac{2}{3}(4)^{3/2} - \frac{1}{4} - \frac{3}{2}\right) - \left(0 - \frac{1}{4}(9) + \frac{3}{2}(3)\right)$$

$$= \frac{4}{3}$$

653. $2\sqrt{2} - 1$

Begin by finding the point of intersection on the interval $\left[-\dfrac{\pi}{4}, \dfrac{\pi}{2}\right]$ by setting the functions equal to each other and solving for x: $\sin x = \cos x$ has a solution when $x = \dfrac{\pi}{4}$.

To determine which function is larger on each interval, pick a point in the interval and substitute it into each function. On the interval $\left[-\dfrac{\pi}{4}, \dfrac{\pi}{4}\right]$, you have $\cos x \ge \sin x$, and on the interval $\left[\dfrac{\pi}{4}, \dfrac{\pi}{2}\right]$, you have $\sin x \ge \cos x$. Therefore, the integrals required to find the area of the region are

$$\int_{-\pi/4}^{\pi/4}(\cos x - \sin x)dx + \int_{\pi/4}^{\pi/2}(\sin x - \cos x)dx$$

$$= (\sin x + \cos x)\Big|_{-\pi/4}^{\pi/4} - (\cos x + \sin x)\Big|_{\pi/4}^{\pi/2}$$

$$= \left[\left(\frac{\sqrt{2}}{2}+\frac{\sqrt{2}}{2}\right) - \left(-\frac{\sqrt{2}}{2}+\frac{\sqrt{2}}{2}\right)\right] - \left[(0+1) - \left(\frac{\sqrt{2}}{2}+\frac{\sqrt{2}}{2}\right)\right]$$

$$= 2\sqrt{2} - 1$$

The following figure shows the region bounded by the functions.

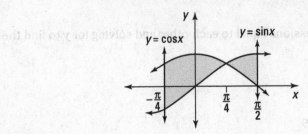

654. $\dfrac{10-2^{5/2}}{3}$

Begin by finding the points of intersection of the two curves by setting the expressions equal to each other and solving for y:

$$y^2 = \sqrt{y}$$
$$y^4 = y$$
$$y^4 - y = 0$$
$$y\left(y^3 - 1\right) = 0$$
$$y = 0, 1$$

To determine which expression has larger x values for y on the interval $(0, 1)$, pick a point in the interval and substitute it into each equation. So if $y = \frac{1}{4}$, then $x = \left(\frac{1}{4}\right)^2 = \frac{1}{16}$ and $x = \sqrt{\frac{1}{4}} = \frac{1}{2}$. For y on the interval $(0, 1)$, you have $\sqrt{y} > y^2$. Similarly, you can show that $y^2 > \sqrt{y}$ for y on the interval $(1, 2)$. Therefore, the integrals to find the area of the region are

$$\int_0^1 \left(\sqrt{y} - y^2\right) dy + \int_1^2 \left(y^2 - \sqrt{y}\right) dy$$

$$= \left(\frac{2}{3} y^{3/2} - \frac{y^3}{3}\right)\Bigg|_0^1 + \left(\frac{y^3}{3} - \frac{2}{3} y^{3/2}\right)\Bigg|_1^2$$

$$= \frac{2}{3} - \frac{1}{3} + \left[\left(\frac{8}{3} - \frac{2(2)^{3/2}}{3}\right) - \left(\frac{1}{3} - \frac{2}{3}\right)\right]$$

$$= \frac{10}{3} - \frac{2^{5/2}}{3}$$

$$= \frac{10 - 2^{5/2}}{3}$$

655. $\dfrac{125}{6}$

Begin by finding the points of intersection of the two curves by setting the expressions equal to each other:

$$y^2 - y = 4y$$
$$y^2 - 5y = 0$$
$$y(y - 5) = 0$$
$$y = 0, 5$$

To determine which curve has the larger x values for y on the interval $(0, 5)$, take a point in the interval and substitute it into each equation. So if $y = 1$, then $x = 1^2 - 1 = 0$ and $x = 4(1) = 4$. Therefore, the integral to find the area of the region is

$$\int_0^5 \left[4y - \left(y^2 - y \right) \right] dy = \int_0^5 5y - y^2 dy$$

$$= \left(\frac{5y^2}{2} - \frac{y^3}{3} \right) \Big|_0^5$$

$$= \frac{5(5)^2}{2} - \frac{(5)^3}{3}$$

$$= \frac{125}{6}$$

---

656.   18

Notice that you can easily solve the two equations for x, so integrating with respect to y makes sense. If you were to solve the second equation for y in order to integrate with respect to x, you'd have to complete the square to solve for y, which would needlessly complicate the problem.

Begin by solving both equations for x to get $x = \frac{y^2 - 6}{2}$ and $x = y + 1$. Then set the expressions equal to each other and solve for y to find the points of intersection:

$$y + 1 = \frac{y^2 - 6}{2}$$
$$2y + 2 = y^2 - 6$$
$$y^2 - 2y - 8 = 0$$
$$(y - 4)(y + 2) = 0$$
$$y = 4, -2$$

To find which curve has the larger x values for y on the interval $(-2, 4)$, take a point in the interval and substitute it into each equation. So if $y = 0$, then $x = \frac{0 - 6}{2} = -3$ and $x = 0 + 1 = 1$. Therefore, the integral to find the area of the region is

$$\int_{-2}^4 \left[(y + 1) - \left(\frac{y^2 - 6}{2} \right) \right] dy$$

$$= \int_{-2}^4 \left[(y + 1) - \left(\frac{y^2}{2} - 3 \right) \right] dy$$

$$= \int_{-2}^4 \left(y + 4 - \frac{1}{2} y^2 \right) dy$$

$$= \left(\frac{y^2}{2} + 4y - \frac{1}{6} y^3 \right) \Big|_{-2}^4$$

$$= \left(\frac{16}{2} + 16 - \frac{1}{6}(64) \right) - \left(2 - 8 + \frac{1}{6}(8) \right)$$

$$= 18$$

657. $\frac{3}{2}$

Begin by finding all the points of intersection of the three functions. To do so, solve each equation for y and then set the functions equal to each other. Solving the second equation for y gives you $y = -\frac{x}{2}$, and solving the third equation for y gives you $y = 3 - 2x$. Set these two functions equal to each other and solve for x:

$$-\frac{x}{2} = 3 - 2x$$

$$\frac{3x}{2} = 3$$

$$x = 2$$

Likewise, finding the other points of intersection gives you $x = -\frac{x}{2}$ so that $x = 0$ is a solution and gives you $x = 3 - 2x$ so that $x = 1$ is a solution.

Notice that on the interval $(0, 1)$, the region is bounded above by the function $y = x$ and below by the function $y = -\frac{x}{2}$. On the interval $(1, 2)$, the region is bounded above by the function $y = 3 - 2x$ and below by the function $y = -\frac{x}{2}$. Therefore, the integrals to find the area of the region are

$$\int_0^1 \left(x - \left(-\frac{x}{2} \right) \right) dx + \int_1^2 \left[(3 - 2x) - \left(-\frac{x}{2} \right) \right] dx$$

$$= \int_0^1 \frac{3}{2} x \, dx + \int_1^2 \left(3 - \frac{3}{2} x \right) dx$$

$$= \left(\frac{3}{4} x^2 \right) \Big|_0^1 + \left(3x - \frac{3}{4} x^2 \right) \Big|_1^2$$

$$= \frac{3}{4} + \left((6 - 3) - \left(3 - \frac{3}{4} \right) \right)$$

$$= \frac{3}{2}$$

658. $\frac{4}{3}$

Begin by finding the points of intersection by setting the functions equal to each other. Square both sides of the equation and factor to solve for x:

$$\sqrt{x + 4} = \frac{x + 4}{2}$$

$$x + 4 = \frac{x^2 + 8x + 16}{4}$$

$$4x + 16 = x^2 + 8x + 16$$

$$0 = x^2 + 4x$$

$$0 = x(x + 4)$$

$$x = 0, -4$$

To determine which function is larger on the interval $(-4, 0)$, pick a point inside the interval and substitute it into each equation. So if $x = -3$, then $y = \sqrt{-3+4} = 1$ and $y = \frac{-3+4}{2} = \frac{1}{2}$. Because $\sqrt{x+4} > \frac{x+4}{2}$ on $(-4, 0)$, the integral to find the area is

$$\int_{-4}^{0}\left[(x+4)^{1/2} - \frac{x+4}{2}\right]dx$$

$$= \int_{-4}^{0}\left[(x+4)^{1/2} - \frac{x}{2} - 2\right]dx$$

$$= \left(\frac{2(x+4)^{3/2}}{3} - \frac{x^2}{4} - 2x\right)\Bigg|_{-4}^{0}$$

$$= \frac{2}{3}(4)^{3/2} - \frac{0^2}{4} - 2(0) - \left(0 - \frac{(-4)^2}{4} - 2(-4)\right)$$

$$= \frac{4}{3}$$

---

659. 18

To find the points of intersection, begin by noting that $y = |2x| = 2x$ on the interval $[0, \infty)$. To find the point of intersection on $[0, 8)$, set the functions equal to each other and solve for x:

$$2x = x^2 - 3$$

$$0 = x^2 - 2x - 3$$

$$0 = (x-3)(x+1)$$

$$x = 3, -1$$

Because the interval under consideration is $[0, 8)$, use the solution $x = 3$.

Likewise, on $(-\infty, 0)$, you have $y = |2x| = -2x$. Again, find the point of intersection by setting the functions equal to each other and solving for x:

$$-2x = x^2 - 3$$

$$0 = x^2 + 2x - 3$$

$$0 = (x+3)(x-1)$$

$$x = -3, 1$$

Because the interval under consideration is $(-\infty, 0)$, keep only the solution $x = -3$. (You could have also noted that because both $y = |2x|$ and $y = x^2 - 3$ are even functions, then if there's a point of intersection at $x = 3$, there must also be a point of intersection at $x = -3$.)

On the interval $(-3, 3)$, you have $|2x| > x^2 - 3$. Therefore, the integral to find the area is

$$\int_{-3}^{3}\left(|2x| - \left(x^2 - 3\right)\right)dx$$

By symmetry, you can rewrite the integral as

$$2\int_{0}^{3}\left(2x - \left(x^2 - 3\right)\right)dx$$

Finally, simplify and evaluate this integral:

$$2\int_0^3 \left(3+2x-x^2\right)dx$$

$$= 2\left(3x+x^2-\frac{x^3}{3}\right)\Bigg|_0^3$$

$$= 2\left[3(3)+3^2-\frac{3^3}{3}-(0+0-0)\right]$$

$$= 2[9+9-9]$$

$$= 18$$

The following figure shows the region bounded by the given curves:

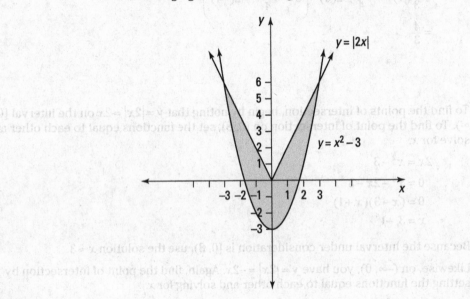

660. $\frac{1}{2}$

Begin by finding the points of intersection on the interval $\left[0, \frac{\pi}{2}\right]$ by setting the functions equal to each other:

$$\cos x = \sin 2x$$

Use an identity on the right-hand side of the equation and factor:

$$\cos x = 2\sin x\cos x$$

$$0 = 2\sin x\cos x - \cos x$$

$$0 = \cos x(2\sin x - 1)$$

Now set each factor equal to zero and solve for x: $\cos x = 0$ and $2\sin x - 1 = 0$, or $\sin x = \frac{1}{2}$. On the interval $\left[0, \frac{\pi}{2}\right]$, you have $\cos x = 0$ if $x = \frac{\pi}{2}$ and $\sin x = \frac{1}{2}$ if $x = \frac{\pi}{6}$. On the interval $\left(0, \frac{\pi}{6}\right)$, you have $\cos x > \sin(2x)$, and on $\left(\frac{\pi}{6}, \frac{\pi}{2}\right)$, you have $\sin(2x) > \cos x$. Therefore, the integrals to find the area are

$$\int_0^{\pi/6}\left[\cos x-\sin(2x)\right]dx+\int_{\pi/6}^{\pi/2}\left[\sin(2x)-\cos x\right]dx$$

$$=\left(\sin x+\frac{\cos(2x)}{2}\right)\Bigg|_0^{\pi/6}+\left(\frac{-\cos(2x)}{2}-\sin x\right)\Bigg|_{\pi/6}^{\pi/2}$$

$$=\left[\sin\frac{\pi}{6}+\frac{\cos\left(\frac{\pi}{3}\right)}{2}-\left(\sin 0+\frac{\cos(0)}{2}\right)\right]+\left[\frac{-\cos\pi}{2}-\sin\frac{\pi}{2}-\left(\frac{-\cos\left(\frac{\pi}{3}\right)}{2}-\sin\frac{\pi}{6}\right)\right]$$

$$=\left[\frac{1}{2}+\frac{1}{4}-0-\frac{1}{2}\right]+\left[\frac{1}{2}-1+\frac{1}{4}+\frac{1}{2}\right]$$

$$=\frac{1}{4}+\frac{1}{4}=\frac{1}{2}$$

The following figure shows the region bounded by the given curves:

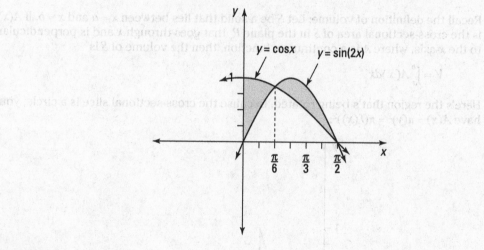

---

661. $\dfrac{13}{4}+\ln\left(\dfrac{1}{8}\right)$

Begin by finding the point of intersection by setting the functions $y=2e^{2x}$ and $y=3-5e^x$ equal to each other and solving for x:

$$2e^{2x}=3-5e^x$$

$$2e^{2x}+5e^x-3=0$$

$$\left(2e^x-1\right)\left(e^x+3\right)=0$$

$$e^x=\frac{1}{2},-3$$

Because $e^x=-3$ has no solution, the only solution is $x=\ln\left(\frac{1}{2}\right)$. Note that $\ln\left(\frac{1}{2}\right)<0$, so the line $x=0$ bounds the region on the right so that the upper limit of integration is $b=0$.

To determine which function is larger on the interval $\left(\ln\left(\frac{1}{2}\right),0\right)$, pick a point inside the interval and substitute it into each equation. On the interval $\left(\ln\left(\frac{1}{2}\right),0\right)$, notice that $2e^{2x}>5e^x$. Therefore, the integral to find the area is

$$\int_{\ln(1/2)}^{0}\left[2e^{2x}-\left(3-5e^{x}\right)\right]dx$$

$$=\left(e^{2x}-3x+5e^{x}\right)\Big|_{\ln(1/2)}^{0}$$

$$=e^{0}-3(0)+5e^{0}-\left(\left(e^{\ln(1/2)}\right)^{2}-3\ln\left(\frac{1}{2}\right)+5e^{\ln(1/2)}\right)$$

$$=1+5-\left(\frac{1}{2}\right)^{2}+3\ln\left(\frac{1}{2}\right)-5\left(\frac{1}{2}\right)$$

$$=6-\frac{1}{4}+\ln\left(\frac{1}{2}\right)^{3}-\frac{5}{2}$$

$$=\frac{13}{4}+\ln\left(\frac{1}{8}\right)$$

662. $\quad \dfrac{\pi}{9}$

Recall the definition of volume: Let S be a solid that lies between $x = a$ and $x = b$. If $A(x)$ is the cross-sectional area of S in the plane P_x that goes through x and is perpendicular to the x-axis, where A is a continuous function, then the volume of S is

$$V = \int_{a}^{b} A(x)\,dx$$

Here's the region that's being rotated. Because the cross-sectional slice is a circle, you have $A(x) = \pi(y)^2 = \pi(f(x))^2$.

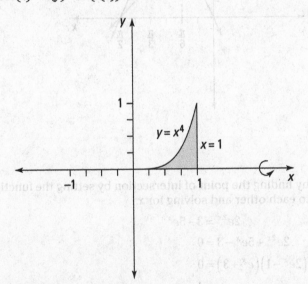

Note that when $y = 0$, you have $0 = x^4$ so that $x = 0$ is the lower limit of integration.

The integral to find the volume is

$$\pi\int_0^1\left(x^4\right)^2 dx = \pi\int_0^1 x^8 dx$$

$$= \pi\left(\frac{x^9}{9}\right)\Bigg|_0^1$$

$$= \pi\left(\frac{1}{9}-0\right)$$

$$= \frac{\pi}{9}$$

663. 2π

Recall the definition of volume: Let S be a solid that lies between $y = c$ and $y = d$. If $A(y)$ is the cross-sectional area of S in the plane P_y that goes through y and is perpendicular to the y-axis, where A is a continuous function, then the volume of S is

$$V = \int_c^d A(y)dx$$

Here's the region that's being rotated. Because the cross-sectional slice is a circle, you have $A(y) = \pi(x)^2 = \pi(g(y))^2$.

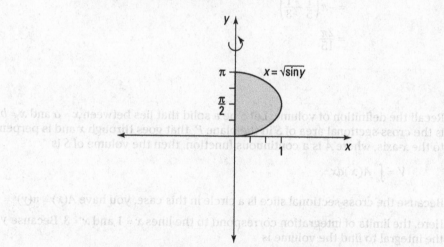

In this case, the limits of integration correspond to $y = 0$ and $y = \pi$. Because $g(y) = \sqrt{\sin y}$, the integral to find the area becomes

$$\pi\int_0^\pi \left(\sqrt{\sin y}\right)^2 dy = \pi\int_0^\pi \sin y\, dy$$

$$= \pi\left(-\cos y\right)\Big|_0^\pi$$

$$= -\pi\left(\cos\pi - \cos 0\right)$$

$$= -\pi(-1-1)$$

$$= 2\pi$$

664. $\dfrac{2\pi}{15}$

Recall the definition of volume: Let S be a solid that lies between $x = a$ and $x = b$. If $A(x)$ is the cross-sectional area of S in the plane P_x that goes through x and is perpendicular to the x-axis, where A is a continuous function, then the volume of S is

$$V = \int_a^b A(x)\,dx$$

Because the cross-sectional slice is a circle in this case, you have $A(x) = \pi(y)^2 = \pi(f(x))^2$. Here, the limits of integration correspond to the lines $x = 3$ and $x = 5$. Because $y = \dfrac{1}{x}$, the integral to find the volume is

$$\pi \int_3^5 \left(\frac{1}{x}\right)^2 dx = \pi \int_3^5 x^{-2}\,dx$$

$$= \pi\left(-x^{-1}\right)\Big|_3^5$$

$$= -\pi\left(\frac{1}{x}\right)\Big|_3^5$$

$$= -\pi\left(\frac{1}{5} - \frac{1}{3}\right)$$

$$= \frac{2\pi}{15}$$

665. $\pi \ln 3$

Recall the definition of volume: Let S be a solid that lies between $x = a$ and $x = b$. If $A(x)$ is the cross-sectional area of S in the plane P_x that goes through x and is perpendicular to the x-axis, where A is a continuous function, then the volume of S is

$$V = \int_a^b A(x)\,dx$$

Because the cross-sectional slice is a circle in this case, you have $A(x) = \pi(y)^2 = \pi(f(x))^2$. Here, the limits of integration correspond to the lines $x = 1$ and $x = 3$. Because $y = \dfrac{1}{\sqrt{x}}$, the integral to find the volume is

$$\pi \int_1^3 \left(\frac{1}{\sqrt{x}}\right)^2 dx = \pi \int_1^3 \frac{1}{x}\,dx$$

$$= \pi \ln|x|\Big|_1^3$$

$$= \pi(\ln 3 - \ln 1)$$

$$= \pi \ln 3$$

666. π

Recall the definition of volume: Let S be a solid that lies between $x = a$ and $x = b$. If $A(x)$ is the cross-sectional area of S in the plane P_x that goes through x and is perpendicular to the x-axis, where A is a continuous function, then the volume of S is

$$V = \int_a^b A(x)\,dx$$

Here's the region that's being rotated. Because the cross-sectional slice is a circle, you have $A(x) = \pi(y)^2 = \pi(f(x))^2$.

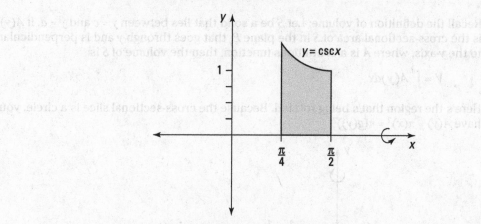

In this case, the limits of integration correspond to the lines $x = \frac{\pi}{4}$ and $x = \frac{\pi}{2}$. Because $y = \csc x$, the integral to find the volume is

$$\pi\int_{\pi/4}^{\pi/2}(\csc x)^2\,dx = \pi\left(-\cot x\right)\Big|_{\pi/4}^{\pi/2}$$

$$= -\pi\left(\cot\frac{\pi}{2} - \cot\frac{\pi}{4}\right)$$

$$= -\pi(0-1)$$

$$= \pi$$

---

667. $\dfrac{4\pi}{3}$

Recall the definition of volume: Let S be a solid that lies between $x = a$ and $x = b$. If $A(x)$ is the cross-sectional area of S in the plane P_x that goes through x and is perpendicular to the x-axis, where A is a continuous function, then the volume of S is

$$V = \int_a^b A(x)\,dx$$

Because the cross-sectional slice is a circle in this case, you have $A(x) = \pi(y)^2 = \pi(f(x))^2$.

Begin by isolating y in the first equation:

$$y = \frac{-x+4}{4}$$

$$= -\frac{1}{4}x + 1$$

If you let $y = 0$ in this equation, you get $x = 4$, which corresponds to the upper limit of integration. Therefore, you have the following integral:

$$\pi\int_0^4\left(-\frac{1}{4}x+1\right)^2 dx = \pi\int_0^4\left(\frac{1}{16}x^2 - \frac{1}{2}x + 1\right)dx$$

$$= \pi\left(\frac{x^3}{48} - \frac{x^2}{4} + x\right)\Big|_0^4$$

$$= \pi\left(\frac{4^3}{48} - \frac{4^2}{4} + 4 - (0 - 0 + 0)\right)$$

$$= \frac{4\pi}{3}$$

668. $\dfrac{\pi}{105}$

Recall the definition of volume: Let S be a solid that lies between $y = c$ and $y = d$. If $A(y)$ is the cross-sectional area of S in the plane P_y that goes through y and is perpendicular to the y-axis, where A is a continuous function, then the volume of S is

$$V = \int_c^d A(y)\,dx$$

Here's the region that's being rotated. Because the cross-sectional slice is a circle, you have $A(y) = \pi(x)^2 = \pi(g(y))^2$.

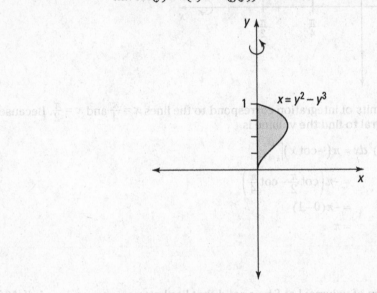

Begin by setting the expressions equal to each other and solving for y in order to find the limits of integration:

$$0 = y^2 - y^3$$
$$0 = y^2(1 - y)$$
$$y = 0, 1$$

Therefore, the integral to find the volume is

$$\pi \int_0^1 \left(y^2 - y^3\right)^2 dy = \pi \int_0^1 \left(y^4 - 2y^5 + y^6\right) dy$$

$$= \pi \left(\frac{y^5}{5} - \frac{y^6}{3} + \frac{y^7}{7} \right) \Bigg|_0^1$$

$$= \pi \left(\frac{1^5}{5} - \frac{1^6}{3} + \frac{1^7}{7} \right)$$

$$= \frac{\pi}{105}$$

669.

8π

Recall the definition of volume: Let S be a solid that lies between $x = a$ and $x = b$. If $A(x)$ is the cross-sectional area of S in the plane P_x that goes through x and is perpendicular to the x-axis, where A is a continuous function, then the volume of S is

$$V = \int_a^b A(x)\,dx$$

Because the cross-sectional slice is a circle in this case, you have $A(x) = \pi(y)^2 = \pi(f(x))^2$.

To get the lower limit of integration, find where the function $y = \sqrt{x-1}$ intersects the line $y = 0$. Set the equations equal to each other and solve for x:

$$0 = \sqrt{x-1}$$
$$0 = x - 1$$
$$1 = x$$

Therefore, $x = 1$ and $x = 5$ are the limits of integration. Because $y = \sqrt{x-1}$, the integral becomes

$$\pi\int_1^5 \left(\sqrt{x-1}\right)^2 dx = \pi\int_1^5 (x-1)\,dx$$

$$= \pi\left(\frac{x^2}{2} - x\right)\Big|_1^5$$

$$= \pi\left[\left(\frac{5^2}{2} - 5\right) - \left(\frac{1^2}{2} - 1\right)\right]$$

$$= 8\pi$$

670.

$\dfrac{448\sqrt{2}}{15}\pi$

Recall the definition of volume: Let S be a solid that lies between $x = a$ and $x = b$. If $A(x)$ is the cross-sectional area of S in the plane P_x that goes through x and is perpendicular to the x-axis, where A is a continuous function, then the volume of S is

$$V = \int_a^b A(x)\,dx$$

In this case, the cross-sectional slice has the shape of a washer. When the cross-sectional slice is a washer, you can find $A(x)$ by using

$$A = \pi(\text{outer radius})^2 - \pi(\text{inner radius})^2$$

$$= \pi\left(r_{\text{out}}\right)^2 - \pi\left(r_{\text{in}}\right)^2$$

where the outer radius, r_{out}, is the distance of the function farther away from the line of rotation and the inner radius, r_{in}, is the distance of the function closer to the line of rotation at a particular value of x.

First find the limits of integration by finding the points of intersection of the two curves. Set the functions equal to each other and solve for x:

$$2 = 4 - \frac{x^2}{4}$$

$$\frac{x^2}{4} = 2$$

$$x^2 = 8$$

$$x = \pm 2\sqrt{2}$$

On the interval $\left(-2\sqrt{2}, 2\sqrt{2}\right)$, you have $4 - \frac{x^2}{4} > 2$ so that $r_{\text{out}} = 4 - \frac{x^2}{4}$ and $r_{\text{in}} = 2$.

Therefore, the integral to find the volume is

$$\pi \int_{-2\sqrt{2}}^{2\sqrt{2}} \left[\left(4 - \frac{x^2}{4} \right)^2 - 2^2 \right] dx$$

$$= \pi \int_{-2\sqrt{2}}^{2\sqrt{2}} \left(12 - 2x^2 + \frac{1}{16} x^4 \right) dx$$

Notice that $y = 12 - 2x^2 + \frac{1}{16} x^4$ is an even function, so it's symmetric about the y-axis.

Therefore, you can instead make zero the lower limit of integration and double the value of the integral:

$$\pi \int_{-2\sqrt{2}}^{2\sqrt{2}} \left(12 - 2x^2 + \frac{1}{16} x^4 \right) dx$$

$$= 2\pi \int_{0}^{2\sqrt{2}} \left(12 - 2x^2 + \frac{1}{16} x^4 \right) dx$$

$$= 2\pi \left(12x - \frac{2}{3} x^3 + \frac{1}{80} x^5 \right) \Big|_{0}^{2\sqrt{2}}$$

$$= 2\pi \left(12 \left(2\sqrt{2} \right) - \frac{2}{3} \left(2\sqrt{2} \right)^3 + \frac{1}{80} \left(2\sqrt{2} \right)^5 \right)$$

$$= \frac{448\sqrt{2}}{15} \pi$$

671. $\dfrac{384\pi}{7}$

Recall the definition of volume: Let S be a solid that lies between $y = c$ and $y = d$. If $A(y)$ is the cross-sectional area of S in the plane P_y that goes through y and is perpendicular to the y-axis, where A is a continuous function, then the volume of S is

$$V = \int_{c}^{d} A(y) \, dx$$

Because the cross-sectional slice is a circle in this case, you have $A(y) = \pi(x)^2 = \pi(g(y))^2$.

Note that $y = 8$ gives you one of the limits of integration; find the other limit of integration by setting the functions equal to each other and solving for y:

$$0 = y^{2/3}$$

$$0 = y$$

With the limits of integration $y = 0$ and $y = 8$ and with $g(y) = y^{2/3}$, the volume of the region is

$$\pi\int_0^8 \left(y^{2/3}\right)^2 dy = \pi\int_0^8 y^{4/3}dy$$

$$= \pi\frac{3}{7}y^{7/3}\Big|_0^8$$

$$= \frac{3\pi}{7}\left(8^{7/3}-0\right)$$

$$= \frac{384\pi}{7}$$

672. $\frac{4\pi}{3}r^3$

Recall the definition of volume: Let S be a solid that lies between $x = a$ and $x = b$. If $A(x)$ is the cross-sectional area of S in the plane P_x that goes through x and is perpendicular to the x-axis, where A is a continuous function, then the volume of S is

$$V = \int_a^b A(x)dx$$

Because the cross-sectional slice is a circle in this case, you have $A(x) = \pi(y)^2 = \pi(f(x))^2$.

Notice that $y = \sqrt{r^2 - x^2}$ intersects $y = 0$ when $x = r$ and when $x = -r$, which corresponds to the limits of integration. Therefore, the integral to find the volume becomes

$$\pi\int_{-r}^r \left(\sqrt{r^2-x^2}\right)^2 dx = \pi\int_{-r}^r \left(r^2-x^2\right)dx$$

Because $y = r^2 - x^2$ is an even function, you can change the lower limit of integration to zero and multiply by 2 to evaluate the integral:

$$\pi\int_{-r}^r \left(r^2-x^2\right)dx$$

$$= 2\pi\int_0^r \left(r^2-x^2\right)dx$$

$$= 2\pi\left(r^2 x - \frac{x^3}{3}\right)\Big|_0^r$$

$$= 2\pi\left(r^2(r) - \frac{r^3}{3}\right)$$

$$= \frac{4\pi}{3}r^3$$

By rotating the semicircular region, you've found the formula for the volume of a sphere!

673. π

Recall the definition of volume: Let S be a solid that lies between $x = a$ and $x = b$. If $A(x)$ is the cross-sectional area of S in the plane P_x that goes through x and is perpendicular to the x-axis, where A is a continuous function, then the volume of S is

$$V = \int_a^b A(x)dx$$

Here's the region that's being rotated:

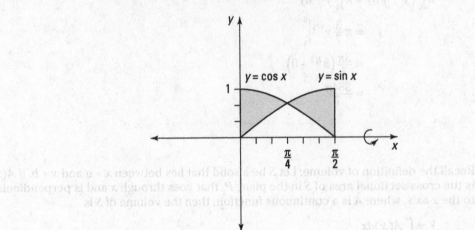

In this case, the cross-sectional slice has the shape of a washer. When the cross-sectional slice is a washer, you can find $A(x)$ by using

$$A = \pi(\text{outer radius})^2 - \pi(\text{inner radius})^2$$

$$= \pi\left(r_{out}\right)^2 - \pi\left(r_{in}\right)^2$$

where the outer radius, r_{out}, is the distance of the function farther away from the line of rotation and the inner radius, r_{in}, is the distance of the function closer to the line of rotation at a particular value of x.

Begin by finding the point of intersection of the two functions by setting them equal to each other: $\sin x = \cos x$, which has the solution $x = \frac{\pi}{4}$ on the interval $\left(0, \frac{\pi}{2}\right)$. Notice that on the interval $\left(0, \frac{\pi}{4}\right)$, you have $\cos x > \sin x$ so that $r_{out} = \cos x$ and $r_{in} = \sin x$, and on the interval $\left(\frac{\pi}{4}, \frac{\pi}{2}\right)$, you have $\sin x > \cos x$ so that $r_{out} = \sin x$ and $r_{in} = \sin x$.

Therefore, the integrals to find the volume are

$$\pi\int_0^{\pi/4} (\cos x)^2 - (\sin x)^2 \, dx + \pi\int_{\pi/4}^{\pi/2} (\sin x)^2 - (\cos x)^2 \, dx$$

Because $\pi\int_0^{\pi/4} (\cos x)^2 - (\sin x)^2 \, dx$ and $\pi\int_{\pi/4}^{\pi/2} (\sin x)^2 - (\cos x)^2 \, dx$ are equal, you can compute the volume by evaluating the integral on the interval $\left(0, \frac{\pi}{4}\right)$ and doubling the answer:

$$2\pi\int_0^{\pi/4} (\cos x)^2 - (\sin x)^2 \, dx = 2\pi\int_0^{\pi/4} \cos 2x \, dx$$

$$= 2\pi\left(\frac{\sin 2x}{2}\right)\Big|_0^{\pi/4}$$

$$= 2\pi\left[\frac{\sin\frac{\pi}{2}}{2} - \frac{\sin 0}{2}\right]$$

$$= \pi$$

674. $\dfrac{\pi^2}{8}+\dfrac{\pi}{4}$

Recall the definition of volume: Let S be a solid that lies between $x = a$ and $x = b$. If $A(x)$ is the cross-sectional area of S in the plane P_x that goes through x and is perpendicular to the x-axis, where A is a continuous function, then the volume of S is

$$V = \int_a^b A(x)\,dx$$

Here's the region that's being rotated. Because the cross-sectional slice is a circle, you have $A(x) = \pi(y)^2 = \pi(f(x))^2$.

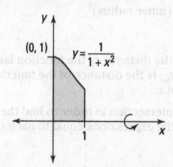

Because the function is bounded by $x = 0$ and $x = 1$, these are the limits of integration. With $y = \dfrac{1}{1+x^2}$, the integral to find the volume becomes

$$\pi\int_0^1 \left(\frac{1}{1+x^2}\right)^2 dx = \pi\int_0^1 \frac{1}{\left(1+x^2\right)^2}\,dx$$

Now use the trigonometric substitution $x = \tan\theta$ so that $dx = \sec^2\theta\,d\theta$. Find the new limits of integration by noting that if $x = 1$, then $1 = \tan\theta$ so that $\dfrac{\pi}{4} = \theta$. Likewise, if $x = 0$, then $0 = \tan\theta$ so that $0 = \theta$. Using these new values, you produce the following integral:

$$\pi\int_0^{\pi/4} \frac{\sec^2\theta}{\left(1+\tan^2\theta\right)^2}\,d\theta = \pi\int_0^{\pi/4} \frac{\sec^2\theta}{\left(\sec^2\theta\right)^2}\,d\theta$$

$$= \pi\int_0^{\pi/4} \cos^2\theta\,d\theta$$

Now use the identity $\cos^2\theta = \dfrac{1}{2}(1+\cos(2\theta))$:

$$\pi\int_0^{\pi/4} \frac{1}{2}(1+\cos(2\theta))\,d\theta = \frac{\pi}{2}\left(\theta + \frac{\sin 2\theta}{2}\right)\Bigg|_0^{\pi/4}$$

$$= \frac{\pi}{2}\left(\frac{\pi}{4} + \frac{\sin\frac{\pi}{2}}{2} - (0+0)\right)$$

$$= \frac{\pi^2}{8} + \frac{\pi}{4}$$

675. $\dfrac{108\pi}{5}$

Recall the definition of volume: Let S be a solid that lies between $x = a$ and $x = b$. If $A(x)$ is the cross-sectional area of S in the plane P_x that goes through x and is perpendicular to the x-axis, where A is a continuous function, then the volume of S is

$$V = \int_a^b A(x)\,dx$$

In this case, the cross-sectional slice has the shape of a washer. When the cross-sectional slice is a washer, you can find $A(x)$ by using

$$A = \pi(\text{outer radius})^2 - \pi(\text{inner radius})^2$$
$$= \pi\left(r_{\text{out}}\right)^2 - \pi\left(r_{\text{in}}\right)^2$$

where the outer radius, r_{out}, is the distance of the function farther away from the line of rotation and the inner radius, r_{in}, is the distance of the function closer to the line of rotation at a particular value of x.

Begin by finding the points of intersection in order to find the limits of integration. Solve the second equation for y, set the expressions equal to each other, and solve for x:

$$3 - x = 3 + 2x - x^2$$
$$0 = 3x - x^2$$
$$0 = x(3 - x)$$
$$x = 0, 3$$

To find out which function is greater on the interval $(0, 3)$ without graphing (and is therefore farther away from the line of rotation), take a point in this interval and substitute that value into each function. So if $x = 1$, you have $y = 3 + 2 - 1^2 = 4$ and $y = 3 - 1 = 2$. Therefore, $r_{\text{out}} = 3 + 2x - x^2$ and $r_{\text{in}} = 3 - x$, so the integral becomes

$$\pi \int_0^3 \left[\left(3 + 2x - x^2\right)^2 - (3 - x)^2\right]dx$$
$$= \pi \int_0^3 \left[\left(x^4 - 4x^3 - 2x^2 + 12x + 9\right) - \left(9 - 6x + x^2\right)\right]dx$$
$$= \pi \int_0^3 \left(x^4 - 4x^3 - 3x^2 + 18x\right)dx$$
$$= \pi\left(\frac{x^5}{5} - x^4 - x^3 + 9x^2\right)\Bigg|_0^3$$
$$= \pi\left(\frac{3^5}{5} - 3^4 - 3^3 + 9(3)^2\right)$$
$$= \frac{108\pi}{5}$$

676. $\dfrac{3\pi}{10}$

Recall the definition of volume: Let S be a solid that lies between $y = c$ and $y = d$. If $A(y)$ is the cross-sectional area of S in the plane P_y that goes through y and is perpendicular to the y-axis, where A is a continuous function, then the volume of S is

$$V = \int_c^d A(y)\,dx$$

In this case, the cross-sectional slice has the shape of a washer. When the cross-sectional slice is a washer, you can find $A(y)$ by using

$$A = \pi(\text{outer radius})^2 - \pi(\text{inner radius})^2$$
$$= \pi\left(r_{\text{out}}\right)^2 - \pi\left(r_{\text{in}}\right)^2$$

where the outer radius, r_{out}, is the distance of the expression farther away from the line of rotation and the inner radius, r_{in}, is the distance of the expression closer to the line of rotation at a particular value of y.

The region is being rotated about the line the y-axis, so solve the first equation for x:

$$\sqrt{y} = x$$

Note that you keep the positive root because the region being rotated is in the first quadrant.

Find the limits of integration by setting the functions equal to each other and solving for y:

$$\sqrt{y} = y^2$$
$$y = y^4$$
$$0 = y^4 - y$$
$$0 = y\left(y^3 - 1\right)$$
$$y = 0, 1$$

For a value in the interval $(0, 1)$, you have $\sqrt{y} > y^2$ so that $r_{\text{out}} = \sqrt{y}$ and $r_{\text{in}} = y^2$. Therefore, the integral to find the volume becomes

$$\pi\int_0^1\left[\left(\sqrt{y}\right)^2 - \left(y^2\right)^2\right]dy = \pi\int_0^1\left(y - y^4\right)dy$$
$$= \pi\left(\frac{y^2}{2} - \frac{y^5}{5}\right)\Bigg|_0^1$$
$$= \pi\left(\frac{1}{2} - \frac{1}{5}\right)$$
$$= \frac{3\pi}{10}$$

677. $\quad \dfrac{36}{35}\pi$

Recall the definition of volume: Let S be a solid that lies between $x = a$ and $x = b$. If $A(x)$ is the cross-sectional area of S in the plane P_x that goes through x and is perpendicular to the x-axis, where A is a continuous function, then the volume of S is

$$V = \int_a^b A(x)dx$$

Here's the region being rotated:

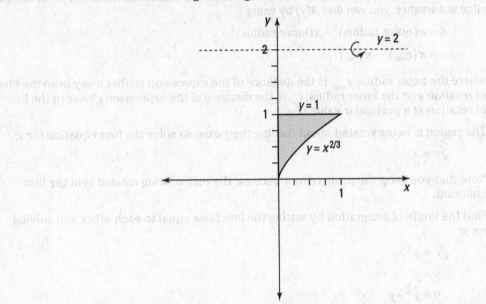

In this case, the cross-sectional slice has the shape of a washer. When the cross-sectional slice is a washer, you can find $A(x)$ by using

$$A = \pi(\text{outer radius})^2 - \pi(\text{inner radius})^2$$

$$= \pi\left(r_{\text{out}}\right)^2 - \pi\left(r_{\text{in}}\right)^2$$

where the outer radius, r_{out}, is the distance of the function farther away from the line of rotation and the inner radius, r_{in}, is the distance of the function closer to the line of rotation at a particular value of x.

Begin by finding the point of intersection by setting the functions equal to each other and solving for x:

$$x^{2/3} = 1$$

$$x = 1$$

The limits of integration are therefore $x = 0$ (given) and $x = 1$. Note that $r_{\text{out}} = 2 - x^{2/3}$ and $r_{\text{in}} = 2 - 1 = 1$ because you're rotating the region about the line $y = 2$. Therefore, the integral to find the volume is

$$\pi\int_0^1\left[\left(2 - x^{2/3}\right)^2 - (2-1)^2\right]dx = \pi\int_0^1\left(3 - 4x^{2/3} + x^{4/3}\right)dx$$

$$= \pi\left(3x - \frac{12x^{5/3}}{5} + \frac{3x^{7/3}}{7}\right)\Bigg|_0^1$$

$$= \pi\left(3 - \frac{12}{5} + \frac{3}{7}\right)$$

$$= \frac{36}{35}\pi$$

678. $\dfrac{21}{20}\pi$

Recall the definition of volume: Let S be a solid that lies between $y = c$ and $y = d$. If $A(y)$ is the cross-sectional area of S in the plane P_y that goes through y and is perpendicular to the y-axis, where A is a continuous function, then the volume of S is

$$V = \int_c^d A(y)\,dx$$

In this case, the cross-sectional slice has the shape of a washer. When the cross-sectional slice is a washer, you can find $A(y)$ by using

$$A = \pi(\text{outer radius})^2 - \pi(\text{inner radius})^2$$
$$= \pi\left(r_{\text{out}}\right)^2 - \pi\left(r_{\text{in}}\right)^2$$

where the outer radius, r_{out}, is the distance of the expression farther away from the line of rotation and the inner radius, r_{in}, is the distance of the expression closer to the line of rotation at a particular value of y.

At $x = 0$, you have $y = 0^{2/3} = 0$, which will be the lower limit of integration; $y = 1$ corresponds to the upper limit of integration.

Solving the equation $y = x^{2/3}$ for x gives you $x = y^{3/2}$. Note that because you're rotating the region about the line $x = -1$, you have $r_{\text{out}} = y^{3/2} + 1$ and $r_{\text{in}} = 0 + 1 = 1$. Therefore, the integral to find the volume is

$$\pi\int_0^1\left[\left(y^{3/2}+1\right)^2 - (0+1)^2\right]dy = \pi\int_0^1\left(2y^{3/2}+y^3\right)dy$$
$$= \pi\left(\frac{4}{5}y^{5/2}+\frac{y^4}{4}\right)\Bigg|_0^1$$
$$= \pi\left(\frac{4}{5}+\frac{1}{4}\right)$$
$$= \frac{21}{20}\pi$$

679. $\left(8\ln\left|2+\sqrt{3}\right|-\sqrt{3}\right)\pi$

Recall the definition of volume: Let S be a solid that lies between $x = a$ and $x = b$. If $A(x)$ is the cross-sectional area of S in the plane P_x that goes through x and is perpendicular to the x-axis, where A is a continuous function, then the volume of S is

$$V = \int_a^b A(x)\,dx$$

Here's the region being rotated:

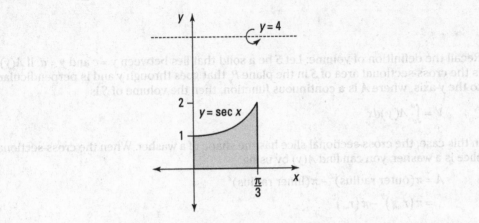

In this case, the cross-sectional slice has the shape of a washer. When the cross-sectional slice is a washer, you can find $A(x)$ by using

$$A = \pi(\text{outer radius})^2 - \pi(\text{inner radius})^2$$

$$= \pi\left(r_{\text{out}}\right)^2 - \pi\left(r_{\text{in}}\right)^2$$

where the outer radius, r_{out}, is the distance of the function farther away from the line of rotation and the inner radius, r_{in}, is the distance of the function closer to the line of rotation at a particular value of x.

On the interval $\left(0, \dfrac{\pi}{3}\right)$, you have $\sec x < 4$. Because you're rotating the region about the line $y = 4$, you have $r_{\text{out}} = 4$ and $r_{\text{in}} = 4 - \sec x$. Therefore, the integral to find the volume is

$$\pi \int_0^{\pi/3} \left[4^2 - (4 - \sec x)^2\right] dx$$

$$= \pi \int_0^{\pi/3} \left(8 \sec x - \sec^2 x\right) dx$$

$$= \pi \left[8 \ln|\sec x + \tan x| - \tan x\right]\Big|_0^{\pi/3}$$

$$= \pi \left[8 \ln\left|\sec \frac{\pi}{3} + \tan \frac{\pi}{3}\right| - \tan \frac{\pi}{3} - \left(8 \ln|\sec 0 + \tan 0| - \tan 0\right)\right]$$

$$= \left(8 \ln\left|2 + \sqrt{3}\right| - \sqrt{3}\right)\pi$$

680. $\dfrac{832}{15}\pi$

Recall the definition of volume: Let S be a solid that lies between $y = c$ and $y = d$. If $A(y)$ is the cross-sectional area of S in the plane P_y that goes through y and is perpendicular to the y-axis, where A is a continuous function, then the volume of S is

$$V = \int_c^d A(y)\,dx$$

In this case, the cross-sectional slice has the shape of a washer. When the cross-sectional slice is a washer, you can find $A(y)$ by using

$$A = \pi(\text{outer radius})^2 - \pi(\text{inner radius})^2$$

$$= \pi\left(r_{\text{out}}\right)^2 - \pi\left(r_{\text{in}}\right)^2$$

where the outer radius, r_{out}, is the distance of the expression farther away from the line of rotation and the inner radius, r_{in}, is the distance of the expression closer to the line of rotation at a particular value of y.

Begin by setting the expressions equal to each other to find the points of intersection, which correspond to the limits of integration:

$$y^2 = 4$$
$$y = -2, 2$$

Because you're rotating the region about the line $x = 5$, you have $r_{out} = 5 - y^2$ and $r_{in} = 5 - 4 = 1$. Therefore, the integral to find the volume is

$$\pi \int_{-2}^{2} \left[\left(5 - y^2\right)^2 - (5-4)^2 \right] dy$$

$$= \pi \int_{-2}^{2} \left(24 - 10y^2 + y^4 \right) dy$$

$$= 2\pi \int_{0}^{2} \left(24 - 10y^2 + y^4 \right) dy$$

$$= 2\pi \left(24y - \frac{10y^3}{3} + \frac{y^5}{5} \right) \Bigg|_{0}^{2}$$

$$= 2\pi \left(24(2) - \frac{10(2)^3}{3} + \frac{(2)^5}{5} \right)$$

$$= \frac{832}{15}\pi$$

681. $\left(\dfrac{5}{2} - \dfrac{1}{2e^2} - \dfrac{2}{e} \right)\pi$

Recall the definition of volume: Let S be a solid that lies between $x = a$ and $x = b$. If $A(x)$ is the cross-sectional area of S in the plane P_x that goes through x and is perpendicular to the x-axis, where A is a continuous function, then the volume of S is

$$V = \int_{a}^{b} A(x)\,dx$$

Here's the region that's being rotated:

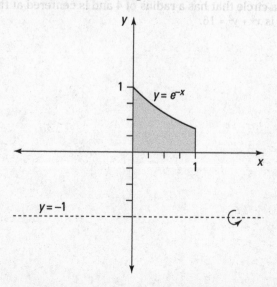

In this case, the cross-sectional slice has the shape of a washer. When the cross-sectional slice is a washer, you can find $A(x)$ by using

$$A = \pi(\text{outer radius})^2 - \pi(\text{inner radius})^2$$
$$= \pi\left(r_{\text{out}}\right)^2 - \pi\left(r_{\text{in}}\right)^2$$

where the outer radius, r_{out}, is the distance of the function farther away from the line of rotation and the inner radius, r_{in}, is the distance of the function closer to the line of rotation at a particular value of x.

Note that the lines $x = 0$ and $x = 1$ correspond to the limits of integration. Because you're rotating the region about the line $y = -1$, you have $r_{\text{out}} = e^{-x} + 1$ and $r_{\text{in}} = 0 + 1 = 1$. Therefore, the integral to find the volume becomes

$$\pi\int_0^1\left[\left(e^{-x}+1\right)^2 - (0+1)^2\right]dx$$
$$= \pi\int_0^1\left(e^{-2x}+2e^{-x}\right)dx$$
$$= \pi\left(-\frac{e^{-2x}}{2}-2e^{-x}\right)\Bigg|_0^1$$
$$= \pi\left[\left(-\frac{e^{-2}}{2}-2e^{-1}\right)-\left(-\frac{1}{2}-2\right)\right]$$
$$= \left(\frac{5}{2}-\frac{1}{2e^2}-\frac{2}{e}\right)\pi$$

---

682. $\dfrac{1,024}{3}$

Recall the definition of volume: Let S be a solid that lies between $x = a$ and $x = b$. If $A(x)$ is the cross-sectional area of S in the plane P_x that goes through x and is perpendicular to the x-axis, where A is a continuous function, then the volume of S is

$$V = \int_a^b A(x)\,dx$$

Here, you want to find an expression for $A(x)$.

The base is a circle that has a radius of 4 and is centered at the origin, so the equation of the circle is $x^2 + y^2 = 16$.

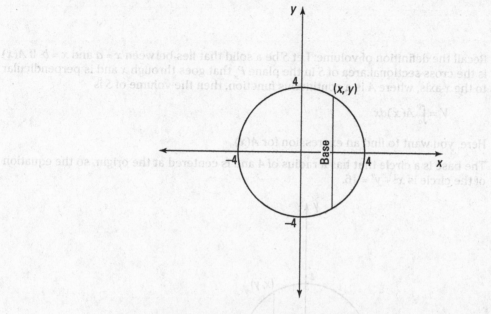

If you consider a cross-sectional slice at point (x, y) on the circle, where $y > 0$, the base of the square equals $2y$; therefore, the area of the cross-sectional slice is $(2y)^2 = 4y^2$.

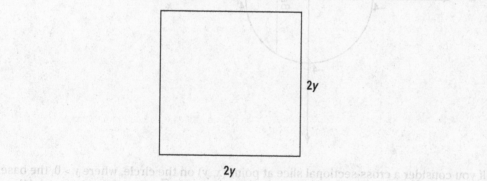

In order to integrate with respect to x, you can solve the equation of the circle for y^2 and write the area as $4y^2 = 4(16 - x^2) = A(x)$. Because the base of the solid varies for x over the interval $[-4, 4]$, the limits of integration are -4 and 4. Therefore, the integral to find the volume is

$$\int_{-4}^{4} 4\left(16 - x^2\right) dx = \int_{-4}^{4} \left(64 - 4x^2\right) dx$$

$$= 2\int_{0}^{4} \left(64 - 4x^2\right) dx$$

$$= 2\left(64x - \frac{4x^3}{3} \right)\Bigg|_{0}^{4}$$

$$= 2\left(64(4) - \frac{4(4)^3}{3} \right)$$

$$= \frac{1,024}{3}$$

683. $256\sqrt{3}$

Recall the definition of volume: Let S be a solid that lies between $x = a$ and $x = b$. If $A(x)$ is the cross-sectional area of S in the plane P_x that goes through x and is perpendicular to the x-axis, where A is a continuous function, then the volume of S is

$$V = \int_a^b A(x)\,dx$$

Here, you want to find an expression for $A(x)$.

The base is a circle that has a radius of 4 and is centered at the origin, so the equation of the circle is $x^2 + y^2 = 16$.

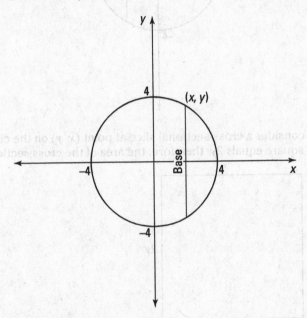

If you consider a cross-sectional slice at point (x, y) on the circle, where $y > 0$, the base of the equilateral triangle equals $2y$ and has a height of $\sqrt{3}y$; therefore, the area of the triangle is $\frac{1}{2}bh = \frac{1}{2}(2y)\left(\sqrt{3}y\right) = \sqrt{3}y^2$.

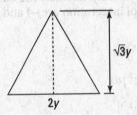

In order to integrate with respect to x, you can solve the equation of the circle for y^2 and write the area as $\sqrt{3}y^2 = \sqrt{3}\left(16 - x^2\right) = A(x)$. Because the base of the solid varies for x over the interval $[-4, 4]$, the limits of integration are -4 and 4. Therefore, the integral to find the volume is

$$\int_{-4}^{4}\left(\sqrt{3}\left(16-x^{2}\right)\right)dx = 2\sqrt{3}\int_{0}^{4}\left(16-x^{2}\right)dx$$

$$= 2\sqrt{3}\left(16x-\frac{x^{3}}{3}\right)\Big|_{0}^{4}$$

$$= 2\sqrt{3}\left(16(4)-\frac{(4)^{3}}{3}\right)$$

$$= 256\sqrt{3}$$

684.   96

Recall the definition of volume: Let S be a solid that lies between $y = c$ and $y = d$. If $A(y)$ is the cross-sectional area of S in the plane P_y that goes through y and is perpendicular to the y-axis, where A is a continuous function, then the volume of S is

$$V = \int_{c}^{d} A(y)dx$$

Here, you want to find an expression for $A(y)$.

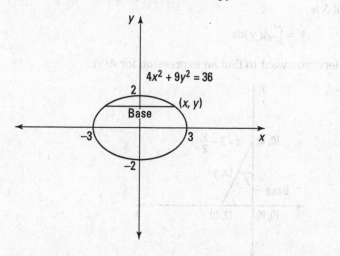

If you consider a cross-sectional slice that goes through a point (x, y) on the ellipse, where $x > 0$, then one side of the square has length $2x$. The area of the square is

$$(2x)^{2} = 4x^{2} = 4\left(\frac{1}{4}\left(36-9y^{2}\right)\right) = 36-9y^{2} = A(y)$$

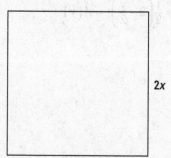

Because the base of the solid varies for y over the interval $[-2, 2]$ the limits of integration are -2 and 2. Therefore, the integral to find the volume is

$$\int_{-2}^{2}\left(36-9y^{2}\right)dy = 2\int_{0}^{2}\left(36-9y^{2}\right)dy$$
$$= 2\left(36y-3y^{3}\right)\Big|_{0}^{2}$$
$$= 2\left(36(2)-3(2)^{3}\right)$$
$$= 96$$

Note that because $36 - 9y^2$ is an even function of y, the lower limit of integration was changed to zero and the resulting integral was multiplied by 2 to make the integral a bit easier to compute.

685. $\dfrac{8}{3}$

Recall the definition of volume: Let S be a solid that lies between $y = c$ and $y = d$. If $A(y)$ is the cross-sectional area of S in the plane P_y that goes through y and is perpendicular to the y-axis, where A is a continuous function, then the volume of S is

$$V = \int_{c}^{d} A(y)dx$$

Here, you want to find an expression for $A(y)$.

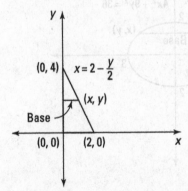

The base of S is the region bounded by the curves $x = 0$, $y = 0$, and $x = 2 - \dfrac{y}{2}$. If you consider a cross-sectional slice that goes through a point (x, y) on the line $x = 2 - \dfrac{y}{2}$, the cross-sectional slice is an isosceles triangle with height equal to the base, so the area is

$$\tfrac{1}{2}bh = \tfrac{1}{2}(x)(x) = \tfrac{1}{2}x^{2} = \tfrac{1}{2}\left(2-\tfrac{y}{2}\right)^{2} = A(y)$$

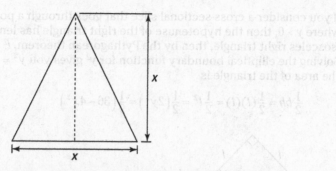

Because the base of the solid varies for y over the interval $[0, 4]$, the limits of integration are 0 and 4. Therefore, the integral to find the volume is

$$\int_0^4 \frac{1}{2}\left(2-\frac{y}{2}\right)^2 dy = \int_0^4 \left(2-y+\frac{1}{8}y^2\right)dy$$

$$= \left(2y-\frac{1}{2}y^2+\frac{1}{24}y^3\right)\Big|_0^4$$

$$= 2(4)-\frac{1}{2}(4)^2+\frac{1}{24}(4)^3$$

$$= \frac{8}{3}$$

686.   16

Recall the definition of volume: Let S be a solid that lies between $x = a$ and $x = b$. If $A(x)$ is the cross-sectional area of S in the plane P_x that goes through x and is perpendicular to the x-axis, where A is a continuous function, then the volume of S is

$$V = \int_a^b A(x)dx$$

Here, you want to find an expression for $A(x)$.

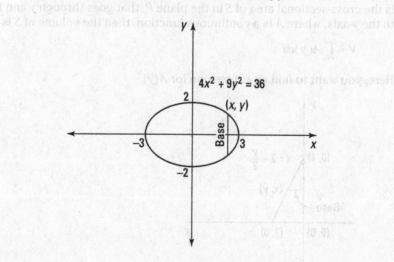

If you consider a cross-sectional slice that goes through a point (x, y) on the ellipse, where $y > 0$, then the hypotenuse of the right triangle has length $2y$. If l is a leg of the isosceles right triangle, then by the Pythagorean theorem, $l^2 + l^2 = (2y)^2$ so that $l^2 = 2y^2$. Solving the elliptical boundary function for y^2 gives you $y^2 = \frac{1}{9}\left(36 - 4x^2\right)$. Therefore, the area of the triangle is

$$\frac{1}{2}bh = \frac{1}{2}(l)(l) = \frac{1}{2}l^2 = \frac{1}{2}\left(2y^2\right) = \frac{1}{9}\left(36 - 4x^2\right)$$

Because the base of the solid varies for x over the interval $[-3, 3]$, the limits of integration are -3 and 3. Therefore, the integral to find the volume is

$$\int_{-3}^{3} \frac{1}{9}\left(36 - 4x^2\right)dx = 2\int_{0}^{3} \frac{1}{9}\left(36 - 4x^2\right)dx$$

$$= \frac{2}{9}\left(36x - \frac{4x^3}{3}\right)\Bigg|_0^3$$

$$= \frac{2}{9}\left(36(3) - \frac{4(3)^3}{3}\right)$$

$$= 16$$

Note that because $y = 36 - 4x^2$ is an even function, the lower limit of integration was changed to zero and the resulting integral was multiplied by 2 to make the integral a bit easier to compute.

687. $\quad \dfrac{2\pi}{3}$

Recall the definition of volume: Let S be a solid that lies between $y = c$ and $y = d$. If $A(y)$ is the cross-sectional area of S in the plane P_y that goes through y and is perpendicular to the y-axis, where A is a continuous function, then the volume of S is

$$V = \int_c^d A(y)dx$$

Here, you want to find an expression for $A(y)$.

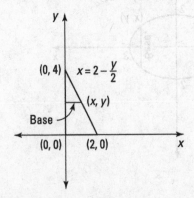

The base of S is the region bounded by the curves $x = 0$, $y = 0$, and $x = 2 - \frac{y}{2}$. If you consider a cross-sectional slice that goes through a point (x, y) on the line $x = 2 - \frac{y}{2}$, that cross-sectional slice is a semicircle with a radius of $\frac{x}{2}$. That means the area of the slice is

$$\frac{1}{2}\pi r^2 = \frac{\pi\left(\frac{x}{2}\right)^2}{2} = \frac{\pi x^2}{8} = \frac{\pi\left(2 - \frac{y}{2}\right)^2}{8} = A(y)$$

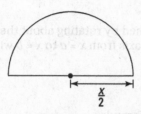

Because the base of the solid varies for y over the interval $[0, 4]$, the limits of integration are 0 and 4. Therefore, the integral to find the volume is

$$\frac{\pi}{8}\int_0^4 \left(2 - \frac{y}{2}\right)^2 dy = \frac{\pi}{8}\int_0^4 \left(4 - 2y + \frac{1}{4}y^2\right) dy$$

$$= \frac{\pi}{8}\left(4y - y^2 + \frac{1}{12}y^3\right)\Big|_0^4$$

$$= \frac{\pi}{8}\left(4(4) - (4)^2 + \frac{1}{12}(4)^3\right)$$

$$= \frac{2\pi}{3}$$

688. 4π

To find the volume of a solid obtained by rotating about the y-axis a region that's under the curve $y = f(x)$ and above the x-axis from $x = a$ to $x = b$ with cylindrical shells, use the integral

$$V = \int_a^b 2\pi x f(x)\, dx$$

More generically, you can use the formula

$$V = 2\pi \int_a^b (\text{shell radius})(\text{shell height})dx$$

To find the shell radius, let x be in the interval $[a, b]$; the shell radius is the distance from x to the line of rotation. To find the shell height, let $f(x)$ be the function that bounds the region above and let $g(x)$ be the function that bounds the region below; the shell height is given by $f(x) - g(x)$. Note that the limits of integration often correspond to points of intersection.

Here, the limits of integration are given as $x = 1$ and $x = 3$ and the shell height is $\frac{1}{x} - 0$, so the integral becomes

$$2\pi \int_1^3 x\left(\frac{1}{x}\right)dx = 2\pi \int_1^3 1\,dx$$

$$= 2\pi\, x\Big|_1^3$$

$$= 2\pi(3-1)$$

$$= 4\pi$$

689. 8π

To find the volume of a solid obtained by rotating about the y-axis a region that's under the curve $y = f(x)$ and above the x-axis from $x = a$ to $x = b$ with cylindrical shells, use the integral

$$V = \int_a^b 2\pi x f(x)\,dx$$

More generically, you can use the formula

$$V = 2\pi \int_a^b (\text{shell radius})(\text{shell height})\,dx$$

To find the shell radius, let x be in the interval $[a, b]$; the shell radius is the distance from x to the line of rotation. To find the shell height, let $f(x)$ be the function that bounds the region above and let $g(x)$ be the function that bounds the region below; the shell height is given by $f(x) - g(x)$. Note that the limits of integration often correspond to points of intersection.

Find the other limit of integration by determining where $y = x^2$ and $y = 0$ intersect: $x^2 = 0$ gives you $x = 0$. The shell height is $x^2 - 0$, so the integral becomes

$$2\pi \int_0^2 x\left(x^2\right)dx = 2\pi \int_0^2 x^3\,dx$$

$$= 2\pi \left.\frac{x^4}{4}\right|_0^2$$

$$= \frac{\pi}{2}\left(2^4 - 0\right)$$

$$= 8\pi$$

690. $\dfrac{6\pi}{7}$

To find the volume of a solid obtained by rotating about the x-axis a region that's to the left of the curve $x = f(y)$ and to the right of the y-axis from $y = c$ to $y = d$ with cylindrical shells, use the integral

$$V = \int_c^d 2\pi y f(y)\,dy$$

More generically, you can use the formula

$$V = 2\pi \int_c^d (\text{shell radius})(\text{shell height})\,dy$$

To find the shell radius, let y be in the interval $[c, d]$; the shell radius is the distance from y to the line of rotation. To find the shell height, let $f(y)$ be the curve that bounds the region on the right and let $g(y)$ be the curve that bounds the region on the left; the shell height is given by $f(y) - g(y)$. Note that the limits of integration often correspond to points of intersection.

Because $y = 1$ corresponds to one of the limits of integration, find the other limit of integration by solving the equation $0 = y^{1/3}$, which gives you $0 = y$. Then find the volume using cylindrical shells:

$$2\pi \int_0^1 y\left(y^{1/3}\right) dy = 2\pi \int_0^1 y^{4/3} dy$$

$$= 2\pi \left(\frac{3}{7} y^{7/3}\right)\Big|_0^1$$

$$= 2\pi \left(\frac{3}{7} - 0\right)$$

$$= \frac{6\pi}{7}$$

691. $\dfrac{40\pi}{3}$

The region is being rotated about the line $x = -1$, which is parallel to the y-axis. To find the volume of a solid obtained by rotating about the y-axis a region that's under the curve $y = f(x)$ and above the x-axis from $x = a$ to $x = b$ using cylindrical shells, use the following integral:

$$V = \int_a^b 2\pi x f(x) dx$$

More generically, you can use the formula

$$V = 2\pi \int_a^b (\text{shell radius})(\text{shell height}) dx$$

To find the shell radius, let x be in the interval $[a, b]$; the shell radius is the distance from x to the line of rotation. To find the shell height, let $f(x)$ be the function that bounds the region above and let $g(x)$ be the function that bounds the region below; the shell height is given by $f(x) - g(x)$. Note that the limits of integration often correspond to points of intersection.

The line $x = 2$ gives you one of the limits of integration, so begin by finding the other limit of integration by determining where the function intersects the x-axis; $0 = x^2$ gives you $0 = x$. Notice that the shell radius is $(x + 1)$, so here's the integral to find the volume using cylindrical shells:

$$2\pi \int_0^2 (x+1)x^2 dx = 2\pi \int_0^2 \left(x^3 + x^2\right) dx$$

$$= 2\pi \left(\frac{x^4}{4} + \frac{x^3}{3}\right)\Big|_0^2$$

$$= 2\pi \left(\frac{2^4}{4} + \frac{2^3}{3} - (0 + 0)\right)$$

$$= \frac{40\pi}{3}$$

692. 216π

To find the volume of a solid obtained by rotating about the y-axis a region that's under the curve $y = f(x)$ and above the x-axis from $x = a$ to $x = b$ with cylindrical shells, use the integral

$$V = \int_a^b 2\pi x\, f(x)\, dx$$

More generically, you can use the formula

$$V = 2\pi \int_a^b (\text{shell radius})(\text{shell height})\, dx$$

To find the shell radius, let x be in the interval $[a, b]$; the shell radius is the distance from x to the line of rotation. To find the shell height, let $f(x)$ be the function that bounds the region above and let $g(x)$ be the function that bounds the region below; the shell height is given by $f(x) - g(x)$. Note that the limits of integration often correspond to points of intersection.

Here's the region being rotated about the y-axis:

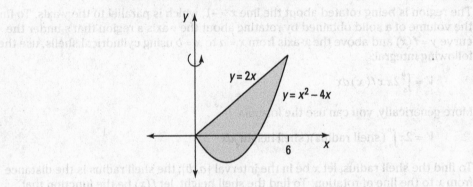

Begin by finding the limits of integration by setting the functions equal to each other and solving for x:

$$2x = x^2 - 4x$$
$$0 = x^2 - 6x$$
$$0 = x(x - 6)$$
$$x = 0, 6$$

Because $2x \geq x^2 - 4x$ on the interval $[0, 6]$, the shell height is given by $2x - (x^2 - 4x)$. Therefore, the integral to find the volume is

$$2\pi \int_0^6 x\left[2x - \left(x^2 - 4x\right)\right] dx$$

$$= 2\pi \int_0^6 \left(6x^2 - x^3\right) dx$$

$$= 2\pi \left(2x^3 - \frac{x^4}{4}\right)\Big|_0^6$$

$$= 2\pi \left(2(6)^3 - \frac{(6)^4}{4}\right)$$

$$= 216\pi$$

693. $\dfrac{4,096\pi}{9}$

To find the volume of a solid obtained by rotating about the *x*-axis a region to the left of the curve $x = f(y)$ and to the right of the *y*-axis from $y = c$ to $y = d$ with cylindrical shells, use the integral

$$V = \int_c^d 2\pi y\, f(y)\, dy$$

More generically, you can use the formula

$$V = 2\pi \int_c^d (\text{shell radius})(\text{shell height})\, dy$$

To find the shell radius, let *y* be in the interval [*c*, *d*]; the shell radius is the distance from *y* to the line of rotation. To find the shell height, let $f(y)$ be the curve that bounds the region on the right and let $g(y)$ be the curve that bounds the region on the left; the shell height is given by $f(y) - g(y)$. Note that the limits of integration often correspond to points of intersection.

Begin by isolating the *y* in the first equation to get $y^{1/4} = x$. Find the lower limit of integration by solving $y^{1/4} = 0$ to get $y = 0$. Then find the volume using cylindrical shells:

$$2\pi \int_0^{16} y\left(y^{1/4}\right) dy = 2\pi \int_0^{16} y^{5/4}\, dy$$

$$= 2\pi\, \frac{4}{9}\, y^{9/4}\Big|_0^{16}$$

$$= 2\pi \left(\frac{4}{9}(16)^{9/4} - 0\right)$$

$$= \frac{4,096\pi}{9}$$

694. $\dfrac{625\pi}{2}$

To find the volume of a solid obtained by rotating about the *x*-axis a region to the left of the curve $x = f(y)$ and to the right of the *y*-axis from $y = c$ to $y = d$ with cylindrical shells, use the integral

$$V = \int_c^d 2\pi y\, f(y)\, dy$$

More generically, you can use the formula

$$V = 2\pi \int_c^d (\text{shell radius})(\text{shell height})\, dy$$

To find the shell radius, let *y* be in the interval [*c*, *d*]; the shell radius is the distance from *y* to the line of rotation. To find the shell height, let $f(y)$ be the curve that bounds the region on the right and let $g(y)$ be the curve that bounds the region on the left; the shell height is given by $f(y) - g(y)$. Note that the limits of integration often correspond to points of intersection.

Here's the region being rotated about the *x*-axis:

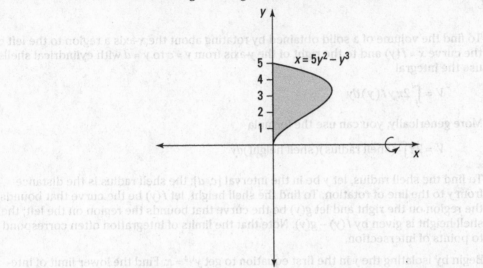

Begin by finding the limits of integration. To do so, find the points of intersection of the two curves by setting the expressions equal to each other and solving for *y*:

$$0 = 5y^2 - y^3$$
$$0 = y^2(5-y)$$
$$y = 0, 5$$

Note that $5y^2 - y^3 \geq 0$ on the interval $[0, 5]$ so that the shell height is $5y^2 - y^3 - 0$, or simply $5y^2 - y^3$. Therefore, the integral to find the volume is

$$2\pi \int_0^5 y\left(5y^2 - y^3\right) dy = 2\pi \int_0^5 \left(5y^3 - y^4\right) dy$$

$$= 2\pi \left(\frac{5y^4}{4} - \frac{y^5}{5}\right)\Bigg|_0^5$$

$$= 2\pi \left(\frac{5(5)^4}{4} - \frac{(5)^5}{5}\right)$$

$$= \frac{625\pi}{2}$$

695. 16π

The region is being rotated about the line $x = 4$, which is parallel to the *y*-axis. To find the volume of a solid obtained by rotating about the *y*-axis a region that's under the curve $y = f(x)$ and above the *x*-axis from $x = a$ to $x = b$ with cylindrical shells, use the integral

$$V = \int_a^b 2\pi x f(x) dx$$

More generically, you can use the formula

$$V = 2\pi \int_a^b (\text{shell radius})(\text{shell height}) dx$$

To find the shell radius, let x be in the interval $[a, b]$; the shell radius is the distance from x to the line of rotation. To find the shell height, let $f(x)$ be the function that bounds the region above and let $g(x)$ be the function that bounds the region below; the shell height is given by $f(x) - g(x)$. Note that the limits of integration often correspond to points of intersection.

Begin by finding the points of intersection by setting the functions equal to each other and solving for x in order to find the limits of integration:

$$4x - x^2 = x^2$$
$$0 = 2x^2 - 4x$$
$$0 = 2x(x - 2)$$
$$x = 0, 2$$

Note that on the interval $[0, 2]$, you have $4x - x^2 \geq x^2$ so that the shell height is $(4x - x^2) - (x^2)$. Also note that the region is to the left of the line of rotation $x = 4$, so the shell radius is $(4 - x)$. Therefore, the integral to find the volume is

$$2\pi \int_0^2 (4 - x)\left[\left(4x - x^2\right) - x^2\right] dx$$
$$= 2\pi \int_0^2 \left(16x - 12x^2 + 2x^3\right) dx$$
$$= 2\pi \left(8x^2 - 4x^3 + \frac{1}{2}x^4\right)\Big|_0^2$$
$$= 2\pi(32 - 32 + 8)$$
$$= 16\pi$$

---

696. $\dfrac{13\pi}{6}$

To find the volume of a solid obtained by rotating about the y-axis a region that's under the curve $y = f(x)$ and above the x-axis from $x = a$ to $x = b$ with cylindrical shells, use the integral

$$V = \int_a^b 2\pi x\, f(x)\, dx$$

More generically, you can use the formula

$$V = 2\pi \int_a^b (\text{shell radius})(\text{shell height})\, dx$$

To find the shell radius, let x be in the interval $[a, b]$; the shell radius is the distance from x to the line of rotation. To find the shell height, let $f(x)$ be the function that bounds the region above and let $g(x)$ be the function that bounds the region below; the shell height is given by $f(x) - g(x)$. Note that the limits of integration often correspond to points of intersection.

Here, the limits of integration correspond to the lines $x = 0$ and $x = 1$, the shell height is $(1 + x + x^2)$, and the shell radius is x. Therefore, the integral to find the volume is

$$2\pi \int_0^1 x\left(1 + x + x^2\right) dx = 2\pi \int_0^1 \left(x + x^2 + x^3\right) dx$$
$$= 2\pi \left(\frac{x^2}{2} + \frac{x^3}{3} + \frac{x^4}{4}\right)\Big|_0^1$$
$$= 2\pi \left(\frac{1^2}{2} + \frac{1^3}{3} + \frac{1^4}{4} - (0)\right)$$
$$= \frac{13\pi}{6}$$

697. $\dfrac{8\pi}{3}$

To find the volume of a solid obtained by rotating about the y-axis a region that's under the curve $y = f(x)$ and above the x-axis from $x = a$ to $x = b$ with cylindrical shells, use the integral

$$V = \int_a^b 2\pi x\, f(x)\, dx$$

More generically, you can use the formula

$$V = 2\pi \int_a^b (\text{shell radius})(\text{shell height})\, dx$$

To find the shell radius, let x be in the interval $[a, b]$; the shell radius is the distance from x to the line of rotation. To find the shell height, let $f(x)$ be the function that bounds the region above and let $g(x)$ be the function that bounds the region below; the shell height is given by $f(x) - g(x)$. Note that the limits of integration often correspond to points of intersection.

Set the functions equal to each other and solve for x in order to find the other limit of integration:

$$4x - x^2 = 4$$
$$0 = x^2 - 4x + 4$$
$$0 = (x-2)(x-2)$$
$$x = 2$$

Here, the shell height is given by $4 - (4x - x^2)$ and the shell radius is x, so the integral to find the volume is

$$2\pi \int_0^2 x\left(4 - \left(4x - x^2\right)\right) dx$$
$$= 2\pi \int_0^2 \left(4x - 4x^2 + x^3\right) dx$$
$$= 2\pi \left(2x^2 - \frac{4}{3}x^3 + \frac{1}{4}x^4\right)\Big|_0^2$$
$$= 2\pi \left(2(2)^2 - \frac{4}{3}(2)^3 + \frac{1}{4}(2)^4 - (0)\right)$$
$$= \frac{8\pi}{3}$$

698. $\dfrac{81}{2}\pi$

To find the volume of a solid obtained by rotating about the x-axis a region to the left of the curve $x = f(y)$ and to the right of the y-axis from $y = c$ to $y = d$ with cylindrical shells, use the integral

$$V = \int_c^d 2\pi y\, f(y)\, dy$$

More generically, you can use the formula

$$V = 2\pi \int_c^d (\text{shell radius})(\text{shell height})\, dy$$

To find the shell radius, let y be in the interval $[c, d]$; the shell radius is the distance from y to the line of rotation. To find the shell height, let $f(y)$ be the curve that bounds the region on the right and let $g(y)$ be the curve that bounds the region on the left; the shell height is given by $f(y) - g(y)$. Note that the limits of integration often correspond to points of intersection.

Notice that if $x = 0$, then $y = \sqrt{9-0} = 3$, so the limits of integration are $y = 0$ (given) and $y = 3$. Solving for x in the equation $y = \sqrt{9-x}$ gives you $x = 9 - y^2$ so that the shell height is $(9 - y^2)$. Therefore, the integral to find the volume is

$$2\pi \int_0^3 y\left(9 - y^2\right) dy = 2\pi \left(\frac{9}{2} y^2 - \frac{1}{4} y^4\right)\Big|_0^3$$

$$= 2\pi \left(\frac{9}{2} 3^2 - \frac{1}{4} 3^4\right)$$

$$= \frac{81}{2}\pi$$

---

699. $\dfrac{16}{3}\pi$

The region is being rotated about the line $x = 2$, which is parallel to the y-axis. To find the volume of a solid obtained by rotating about the y-axis a region that's under the curve $y = f(x)$ and above the x-axis from $x = a$ to $x = b$ with cylindrical shells, use the integral

$$V = \int_a^b 2\pi x f(x)\, dx$$

More generically, you can use the formula

$$V = 2\pi \int_a^b (\text{shell radius})(\text{shell height}) dx$$

To find the shell radius, let x be in the interval $[a, b]$; the shell radius is the distance from x to the line of rotation. To find the shell height, let $f(x)$ be the function that bounds the region above and let $g(x)$ be the function that bounds the region below; the shell height is given by $f(x) - g(x)$. Note that the limits of integration often correspond to points of intersection.

Find the limits of integration by setting the functions equal to each other and solving for x:

$$1 - x^2 = 0$$
$$(1-x)(1+x) = 0$$
$$x = 1, -1$$

Because the line of rotation $x = 2$ is to the right of the region being rotated, the shell radius is $(2 - x)$. Therefore, to find the volume of revolution using cylindrical shells, you use the following integral:

$$2\pi \int_{-1}^1 (2-x)\left(1 - x^2\right) dx$$

$$= 2\pi \int_{-1}^1 \left(x^3 - 2x^2 - x + 2\right) dx$$

$$= 2\pi \left(\frac{1}{4} x^4 - \frac{2}{3} x^3 - \frac{1}{2} x^2 + 2x\right)\Big|_{-1}^1$$

$$= 2\pi \left[\left(\frac{1}{4} - \frac{2}{3} - \frac{1}{2} + 2\right) - \left(\frac{1}{4} + \frac{2}{3} - \frac{1}{2} - 2\right)\right]$$

$$= \frac{16}{3}\pi$$

700. $\dfrac{625\pi}{6}$

To find the volume of a solid obtained by rotating about the y-axis a region that's under the curve $y = f(x)$ and above the x-axis from $x = a$ to $x = b$ with cylindrical shells, use the integral

$$V = \int_a^b 2\pi x f(x)\,dx$$

More generically, you can use the formula

$$V = 2\pi \int_a^b (\text{shell radius})(\text{shell height})\,dx$$

To find the shell radius, let x be in the interval $[a, b]$; the shell radius is the distance from x to the line of rotation. To find the shell height, let $f(x)$ be the function that bounds the region above and let $g(x)$ be the function that bounds the region below; the shell height is given by $f(x) - g(x)$. Note that the limits of integration often correspond to points of intersection.

Here's the region being rotated about the y-axis:

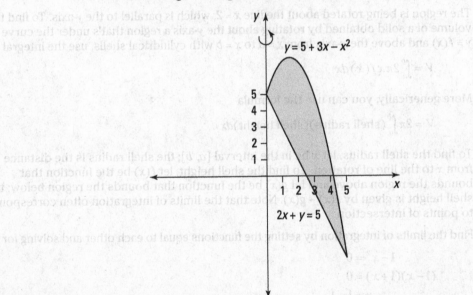

Begin by isolating y in the second equation to get $y = 5 - 2x$. Then set the expressions equal to each other and solve for x to find the x coordinates of the points of intersection:

$$5 + 3x - x^2 = 5 - 2x$$
$$0 = 5x - x^2$$
$$0 = x(5 - x)$$
$$x = 0, 5$$

Because $5 + 3x - x^2 \geq 5 - 2x$ on the interval $[0, 5]$, the shell height is given by $(5 + 3x - x^2) - (5 - 2x)$. Therefore, the integral to find the volume is

$$2\pi \int_0^5 x\left(\left(5 + 3x - x^2\right) - (5 - 2x)\right)dx$$

$$= 2\pi \int_0^5 \left(5x^2 - x^3\right)dx$$

$$= 2\pi \left(\frac{5x^3}{3} - \frac{x^4}{4}\right)\Bigg|_0^5$$

$$= 2\pi \left(\frac{5^4}{3} - \frac{5^4}{4}\right)$$

$$= \frac{625\pi}{6}$$

701. $\quad \dfrac{175}{4}\pi$

The region is being rotated about the line $x = -1$, which is parallel to the y-axis. To find the volume of a solid obtained by rotating about the y-axis a region that's under the curve $y = f(x)$ and above the x-axis from $x = a$ to $x = b$ with cylindrical shells, use the integral

$$V = \int_a^b 2\pi x f(x)\, dx$$

More generically, you can use the formula

$$V = 2\pi \int_a^b (\text{shell radius})(\text{shell height})dx$$

To find the shell radius, let x be in the interval $[a, b]$; the shell radius is the distance from x to the line of rotation. To find the shell height, let $f(x)$ be the function that bounds the region above and let $g(x)$ be the function that bounds the region below; the shell height is given by $f(x) - g(x)$. Note that the limits of integration often correspond to points of intersection.

Here's the region being rotated about the line $x = -1$:

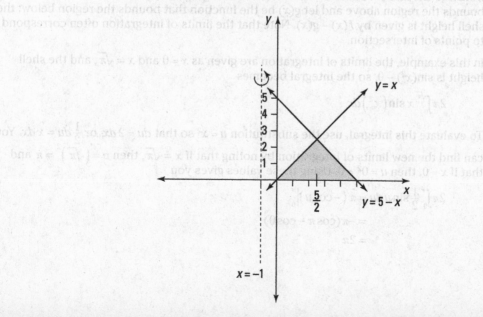

Find the point of intersection using $y = x$, $x + y = 5$, and substitution to get $x + x = 5$ so that $x = \frac{5}{2}$. Notice that on the interval $\left(0, \frac{5}{2}\right)$, the upper boundary of the region is x and the lower boundary is $y = 0$; on $\left(\frac{5}{2}, 5\right)$ the upper boundary of the region is $5 - x$ and the lower boundary is $y = 0$. Because the line of rotation is $x = -1$, the shell radius is $(x + 1)$. Therefore, to find the volume using cylindrical shells, you use two integrals:

$$2\pi \int_0^{5/2} (x+1)(x)\,dx + 2\pi \int_{5/2}^5 (x+1)(5-x)\,dx$$

$$= 2\pi \int_0^{5/2} \left(x^2 + x\right)dx + 2\pi \int_{5/2}^5 \left(5 + 4x - x^2\right)dx$$

$$= 2\pi \left(\frac{x^3}{3} + \frac{x^2}{2}\right)\Bigg|_0^{5/2} + 2\pi \left(5x + 2x^2 - \frac{x^3}{3}\right)\Bigg|_{5/2}^5$$

$$= 2\pi \left(\frac{1}{3}\left(\frac{5}{2}\right)^3 + \frac{1}{2}\left(\frac{5}{2}\right)^2\right) + 2\pi \left[\left(5(5) + 2(5)^2 - \frac{5^3}{3}\right) - \left(5\left(\frac{5}{2}\right) + 2\left(\frac{5}{2}\right)^2 - \frac{1}{3}\left(\frac{5}{2}\right)^3\right)\right]$$

$$= \frac{175}{4}\pi$$

702. 2π

To find the volume of a solid obtained by rotating about the y-axis a region that's under the curve $y = f(x)$ and above the x-axis from $x = a$ to $x = b$ with cylindrical shells, use the integral

$$V = \int_a^b 2\pi x f(x)\,dx$$

More generically, you can use the formula

$$V = 2\pi \int_a^b (\text{shell radius})(\text{shell height})dx$$

To find the shell radius, let x be in the interval $[a, b]$; the shell radius is the distance from x to the line of rotation. To find the shell height, let $f(x)$ be the function that bounds the region above and let $g(x)$ be the function that bounds the region below; the shell height is given by $f(x) - g(x)$. Note that the limits of integration often correspond to points of intersection.

In this example, the limits of integration are given as $x = 0$ and $x = \sqrt{\pi}$, and the shell height is $\sin(x^2) - 0$, so the integral becomes

$$2\pi \int_0^{\sqrt{\pi}} x \sin\left(x^2\right)dx$$

To evaluate this integral, use the substitution $u = x^2$ so that $du = 2\,dx$, or $\frac{1}{2}du = x\,dx$. You can find the new limits of integration by noting that if $x = \sqrt{\pi}$, then $u = \left(\sqrt{\pi}\right)^2 = \pi$ and that if $x = 0$, then $u = 0^2 = 0$. Using these values gives you

$$2\pi \int_0^\pi \frac{1}{2}\sin u\,du = \pi(-\cos u)\Big|_0^\pi$$

$$= -\pi(\cos \pi - \cos 0)$$

$$= 2\pi$$

703. $2\pi(e^2+1)$

To find the volume of a solid obtained by rotating about the x-axis a region to the left of the curve $x = f(y)$ and to the right of the y-axis from $y = c$ to $y = d$ with cylindrical shells, use the integral

$$V = \int_c^d 2\pi y f(y) dy$$

More generically, you can use the formula

$$V = 2\pi \int_c^d (\text{shell radius})(\text{shell height}) dy$$

To find the shell radius, let y be in the interval $[c, d]$; the shell radius is the distance from y to the line of rotation. To find the shell height, let $f(y)$ be the curve that bounds the region on the right and let $g(y)$ be the curve that bounds the region on the left; the shell height is given by $f(y) - g(y)$. Note that the limits of integration often correspond to points of intersection.

In this example, the limits of integration are given as $y = 0$ and $y = 2$, and the shell height is $e^y - 0$. Therefore, to find the volume using cylindrical shells, you use

$$2\pi \int_0^2 y e^y dy$$

To evaluate this integral, use integration by parts with $u = y$ so that $du = dy$, and let $dv = e^y dy$ so that $v = e^y$:

$$2\pi \int_0^2 y e^y dy = 2\pi \left(y e^y \Big|_0^2 - \int_0^2 e^y dy \right)$$

$$= 2\pi \left(y e^y - e^y \right) \Big|_0^2$$

$$= 2\pi \left(2e^2 - e^2 - (0-1) \right)$$

$$= 2\pi \left(e^2 + 1 \right)$$

704. $\pi \left(1 - \dfrac{1}{e^9} \right)$

To find the volume of a solid obtained by rotating about the y-axis a region that's under the curve $y = f(x)$ and above the x-axis from $x = a$ to $x = b$ with cylindrical shells, use the integral

$$V = \int_a^b 2\pi x f(x) dx$$

More generically, you can use the formula

$$V = 2\pi \int_a^b (\text{shell radius})(\text{shell height}) dx$$

To find the shell radius, let x be in the interval $[a, b]$; the shell radius is the distance from x to the line of rotation. To find the shell height, let $f(x)$ be the function that bounds the region above and let $g(x)$ be the function that bounds the region below; the shell height is given by $f(x) - g(x)$. Note that the limits of integration often correspond to points of intersection.

In this example, the limits of integration correspond to the lines $x = 0$ and $x = 3$ and the shell height is given by e^{-x^2}. Therefore, to find the volume of revolution using cylindrical shells, use the following integral:

$$2\pi \int_0^3 x\left(e^{-x^2}\right)dx$$

To evaluate this integral, use the substitution $u = -x^2$ so that $du = -2x\,dx$, or $-\frac{1}{2}du = x\,dx$. Find the new limits of integration by noting that if $x = 3$, then $u = -9$, and if $x = 0$, then $u = 0$. With these new values, the integral becomes

$$-\pi\int_0^{-9} e^u du = \pi\int_{-9}^0 e^u du$$
$$= \pi\left(e^u\right)\Big|_{-9}^0$$
$$= \pi\left(e^0 - e^{-9}\right)$$
$$= \pi\left(1 - \frac{1}{e^9}\right)$$

705. $\frac{25}{21}\pi$

The region is being rotated about the line $x = -1$, which is parallel to the y-axis. To find the volume of a solid obtained by rotating about the y-axis a region that's under the curve $y = f(x)$ and above the x-axis from $x = a$ to $x = b$ with cylindrical shells, use the integral

$$V = \int_a^b 2\pi x\, f(x)\,dx$$

More generically, you can use the formula

$$V = 2\pi\int_a^b (\text{shell radius})(\text{shell height})dx$$

To find the shell radius, let x be in the interval $[a, b]$; the shell radius is the distance from x to the line of rotation. To find the shell height, let $f(x)$ be the function that bounds the region above and let $g(x)$ be the function that bounds the region below; the shell height is given by $f(x) - g(x)$. Note that the limits of integration often correspond to points of intersection.

Here's the region being rotated about the line $x = -1$:

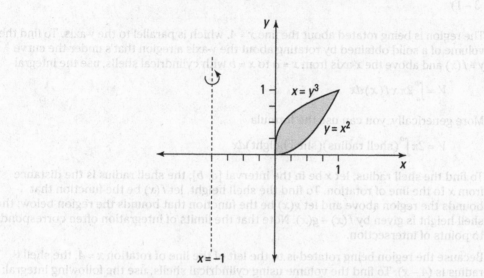

Begin by writing $x = y^3$ as $x^{1/3} = y$. Then set the functions equal to each other to find the points of intersection, which give you the limits of integration. Cube both sides of the equation and factor to solve for x:

$$x^{1/3} = x^2$$
$$x = x^6$$
$$0 = x^6 - x$$
$$0 = x\left(x^5 - 1\right)$$
$$x = 0, 1$$

Notice that $x^{1/3} > x^2$ on the interval $[0, 1]$, so the shell height is $(x^{1/3} - x^2)$. For a value in the interval $[0, 1]$, the distance from x to the line of rotation $x = -1$ is $(x + 1)$, so the shell radius is $(x + 1)$. Therefore, the integral to find the volume of revolution using cylindrical shells is

$$2\pi \int_0^1 (x+1)\left(x^{1/3} - x^2\right) dx$$
$$= 2\pi \int_0^1 \left(x^{4/3} - x^3 + x^{1/3} - x^2\right) dx$$
$$= 2\pi \left(\frac{3}{7} x^{7/3} - \frac{1}{4} x^4 + \frac{3}{4} x^{4/3} - \frac{1}{3} x^3\right)\Bigg|_0^1$$
$$= 2\pi \left(\frac{3}{7} - \frac{1}{4} + \frac{3}{4} - \frac{1}{3}\right)$$
$$= \frac{25}{21}\pi$$

706.

$4\pi(2\ln 3 - 1)$

The region is being rotated about the line $x = 4$, which is parallel to the y-axis. To find the volume of a solid obtained by rotating about the y-axis a region that's under the curve $y = f(x)$ and above the x-axis from $x = a$ to $x = b$ with cylindrical shells, use the integral

$$V = \int_a^b 2\pi x f(x)\,dx$$

More generically, you can use the formula

$$V = 2\pi \int_a^b (\text{shell radius})(\text{shell height})\,dx$$

To find the shell radius, let x be in the interval $[a, b]$; the shell radius is the distance from x to the line of rotation. To find the shell height, let $f(x)$ be the function that bounds the region above and let $g(x)$ be the function that bounds the region below; the shell height is given by $f(x) - g(x)$. Note that the limits of integration often correspond to points of intersection.

Because the region being rotated is to the left of the line of rotation $x = 4$, the shell radius is $(4 - x)$. To find the volume using cylindrical shells, use the following integral:

$$2\pi \int_1^3 (4 - x)\left(\frac{1}{x}\right)dx = 2\pi \int_1^3 \left(\frac{4}{x} - 1\right)dx$$
$$= 2\pi \left(4\ln|x| - x\right)\Big|_1^3$$
$$= 2\pi \left(4\ln 3 - 3 - (4\ln 1 - 1)\right)$$
$$= 2\pi (4\ln 3 - 2)$$
$$= 4\pi (2\ln 3 - 1)$$

707.

$\dfrac{10\pi}{21}$

The region is being rotated about the line $y = 1$, which is parallel to the x-axis. To find the volume of a solid obtained by rotating about the x-axis a region to the left of the curve $x = f(y)$ and to the right of the y-axis from $y = c$ to $y = d$ with cylindrical shells, use the integral

$$V = \int_c^d 2\pi y f(y)\,dy$$

More generically, you can use the formula

$$V = 2\pi \int_c^d (\text{shell radius})(\text{shell height})\,dy$$

To find the shell radius, let y be in the interval $[c, d]$; the shell radius is the distance from y to the line of rotation. To find the shell height, let $f(y)$ be the curve that bounds the region on the right and let $g(y)$ be the curve that bounds the region on the left; the shell height is given by $f(y) - g(y)$. Note that the limits of integration often correspond to points of intersection.

Here's the region being rotated about the line $y = 1$:

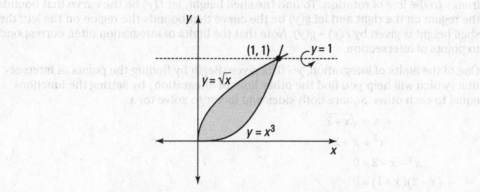

Begin by isolating x in each equation to get $y^2 = x$ and $y^{1/3} = x$. Set these expressions equal to each other to find the points of intersection, which give you the limits of integration. Cube both sides of the equation and factor to solve for y:

$$y^2 = y^{1/3}$$
$$y^6 = y$$
$$y^6 - y = 0$$
$$y(y^5 - 1) = 0$$
$$y = 0, 1$$

Note that $y^{1/3} > y^2$ for a point in the interval $(0, 1)$, so the shell height is $y^{1/3} - y^2$. Also note that the region being rotated is below the line of rotation $y = 1$, so in the interval $(0, 1)$, the distance from y to the line of rotation is $(1 - y)$, making the shell radius $(1 - y)$. Therefore, the integral to find the volume using cylindrical shells is

$$2\pi \int_0^1 (1-y)\left(y^{1/3} - y^2\right)dy$$
$$= 2\pi \int_0^1 \left(y^{1/3} - y^2 - y^{4/3} + y^3\right)dy$$
$$= 2\pi \left(\frac{3}{4}y^{4/3} - \frac{y^3}{3} - \frac{3}{7}y^{7/3} + \frac{y^4}{4}\right)\Bigg|_0^1$$
$$= 2\pi \left(\frac{3}{4}1^{4/3} - \frac{1^3}{3} - \frac{3}{7}1^{7/3} + \frac{1^4}{4}\right)$$
$$= \frac{10\pi}{21}$$

---

708. $\dfrac{16}{3}\pi$

To find the volume of a solid obtained by rotating about the x-axis a region to the left of the curve $x = f(y)$ and to the right of the y-axis from $y = c$ to $y = d$ with cylindrical shells, use the integral

$$V = \int_c^d 2\pi y\, f(y)\, dy$$

More generically, you can use the formula

$$V = 2\pi \int_c^d (\text{shell radius})(\text{shell height})dy$$

To find the shell radius, let y be in the interval $[c, d]$; the shell radius is the distance from y to the line of rotation. To find the shell height, let $f(y)$ be the curve that bounds the region on the right and let $g(y)$ be the curve that bounds the region on the left; the shell height is given by $f(y) - g(y)$. Note that the limits of integration often correspond to points of intersection.

One of the limits of integration, $y = 0$, is given. Begin by finding the points of intersection (which will help you find the other limit of integration) by setting the functions equal to each other. Square both sides and factor to solve for x:

$$x = \sqrt{x+2}$$
$$x^2 = x+2$$
$$x^2 - x - 2 = 0$$
$$(x-2)(x+1) = 0$$
$$x = 2, -1$$

Notice that $x = -1$ is an extraneous solution. If $x = 2$, then $y = \sqrt{2+2} = 2$, which gives you the other limit of integration.

Solving $y = \sqrt{x+2}$ for x gives you $x = y^2 - 2$. Note that $y > y^2 - 2$ in the interval $[0, 2]$, so the shell height is $y - (y^2 - 2)$. Therefore, the integral to find the volume using cylindrical shells is

$$2\pi \int_0^2 y\left[y - \left(y^2 - 2\right)\right] dy$$
$$= 2\pi \int_0^2 \left(y^2 + 2y - y^3\right) dy$$
$$= 2\pi \left(\frac{1}{3}y^3 + y^2 - \frac{1}{4}y^4\right)\Big|_0^2$$
$$= 2\pi \left(\frac{1}{3}(8) + 4 - \frac{1}{4}(16)\right)$$
$$= \frac{16}{3}\pi$$

709. $\quad \sqrt{2\pi}\left(1 - \dfrac{1}{\sqrt{e}}\right)$

To find the volume of a solid obtained by rotating about the y-axis a region that's under the curve $y = f(x)$ and above the x-axis from $x = a$ to $x = b$ with cylindrical shells, use the integral

$$V = \int_a^b 2\pi x f(x)\, dx$$

More generically, you can use the formula

$$V = 2\pi \int_a^b (\text{shell radius})(\text{shell height}) dx$$

To find the shell radius, let x be in the interval $[a, b]$; the shell radius is the distance from x to the line of rotation. To find the shell height, let $f(x)$ be the function that bounds the region above and let $g(x)$ be the function that bounds the region below; the shell height is given by $f(x) - g(x)$. Note that the limits of integration often correspond to points of intersection.

Note that the lines $x = 0$ and $x = 1$ correspond to the limits of integration and that the shell height is $\left(\dfrac{1}{\sqrt{2\pi}} e^{-x^2/2} - 0 \right)$, or simply $\dfrac{1}{\sqrt{2\pi}} e^{-x^2/2}$. To find the volume using cylindrical shells, use the following integral:

$$2\pi \int_0^1 x \left(\frac{1}{\sqrt{2\pi}} e^{-x^2/2} \right) dx = \frac{2\pi}{\sqrt{2\pi}} \int_0^1 x e^{-x^2/2} dx$$

Now use the substitution $u = -\dfrac{1}{2} x^2$ so that $du = -x\,dx$, or $-du = x\,dx$. Find the new limits of integration by noting that if $x = 1$, then $u = -\dfrac{1}{2}$, and if $x = 0$, then $u = 0$. Using these new values gives you the following:

$$-\frac{2\pi}{\sqrt{2\pi}} \int_0^{-1/2} e^u du = \sqrt{2\pi} \int_{-1/2}^0 e^u du$$

$$= \sqrt{2\pi} \left(e^u \right) \Big|_{-1/2}^0$$

$$= \sqrt{2\pi} \left(1 - e^{-1/2} \right)$$

$$= \sqrt{2\pi} \left(1 - \frac{1}{\sqrt{e}} \right)$$

710. $2\pi \left(\dfrac{2^{11/2} + 7}{20} - 2\ln 2 \right)$

To find the volume of a solid obtained by rotating about the y-axis a region that's under the curve $y = f(x)$ and above the x-axis from $x = a$ to $x = b$ with cylindrical shells, use the integral

$$V = \int_a^b 2\pi x f(x)\, dx$$

More generically, you can use the formula

$$V = 2\pi \int_a^b (\text{shell radius})(\text{shell height}) dx$$

To find the shell radius, let x be in the interval $[a, b]$; the shell radius is the distance from x to the line of rotation. To find the shell height, let $f(x)$ be the function that bounds the region above and let $g(x)$ be the function that bounds the region below; the shell height is given by $f(x) - g(x)$. Note that the limits of integration often correspond to points of intersection.

Notice that on the interval $[1, 2]$, you have $\sqrt{x} \geq \ln x$, so the shell height is $\sqrt{x} - \ln x$. Therefore, the integral to find the volume is

$$2\pi \int_1^2 x \left(\sqrt{x} - \ln x \right) dx = 2\pi \int_1^2 \left(x^{3/2} - x \ln x \right) dx$$

To evaluate $\int x \ln x\, dx$, use integration by parts with $u = \ln x$ so that $du = \dfrac{1}{x} dx$, and use $dv = x\,dx$ so that $v = \dfrac{x^2}{2}$. These substitutions give you

$$\int x \ln x\, dx = \frac{x^2}{2} \ln x - \int \frac{x^2}{2} \frac{1}{x} dx$$

$$= \frac{x^2}{2} \ln x - \int \frac{x}{2} dx$$

$$= \frac{x^2}{2} \ln x - \frac{x^2}{4} + C$$

Evaluating the integral $2\pi\int_1^2 \left(x^{3/2} - x\ln x\right)dx$ gives you

$$2\pi\int_1^2 \left(x^{3/2} - x\ln x\right)dx$$

$$= 2\pi\left(\frac{2}{5}x^{5/2} - \left(\frac{x^2}{2}\ln x - \frac{x^2}{4}\right)\right)\Bigg|_1^2$$

$$= 2\pi\left(\frac{2}{5}x^{5/2} - \frac{x^2}{2}\ln x + \frac{x^2}{4}\right)\Bigg|_1^2$$

$$= 2\pi\left(\left(\frac{2}{5}2^{5/2} - \frac{2^2}{2}\ln 2 + \frac{2^2}{4}\right) - \left(\frac{2}{5}1^{5/2} - \frac{1^2}{2}\ln 1 + \frac{1}{4}\right)\right)$$

$$= 2\pi\left(\frac{2^{7/2}}{5} - 2\ln 2 + 1 - \frac{2}{5} - \frac{1}{4}\right)$$

$$= 2\pi\left(\frac{2^{11/2} + 7}{20} - 2\ln 2\right)$$

---

711. $\quad \pi^2 - 2\pi$

To find the volume of a solid obtained by rotating about the x-axis a region to the left of the curve $x = f(y)$ and to the right of the y-axis from $y = c$ to $y = d$ with cylindrical shells, use the integral

$$V = \int_c^d 2\pi y\, f(y)\, dy$$

More generically, you can use the formula

$$V = 2\pi\int_c^d (\text{shell radius})(\text{shell height})\, dy$$

To find the shell radius, let y be in the interval $[c, d]$; the shell radius is the distance from y to the line of rotation. To find the shell height, let $f(y)$ be the curve that bounds the region on the right and let $g(y)$ be the curve that bounds the region on the left; the shell height is given by $f(y) - g(y)$. Note that the limits of integration often correspond to points of intersection.

Here, the limits of integration correspond to the lines $y = 0$ and $y = \frac{\pi}{2}$, and the shell height is $\cos y$. To find the volume using cylindrical shells, use the integral

$$2\pi\int_0^{\pi/2} y\cos y\, dy$$

To evaluate the integral, use integration by parts with $u = y$ so that $du = dy$, and let $dv = \cos y\, dy$ so that $v = \sin y$:

$$2\pi\int_0^{\pi/2} y\cos y\, dy$$

$$= 2\pi\left(y\sin y\Big|_0^{\pi/2} - \int_0^{\pi/2}\sin y\, dy\right)$$

$$= 2\pi\left(y\sin y\Big|_0^{\pi/2} + \cos y\Big|_0^{\pi/2}\right)$$

$$= 2\pi\left(\left(\frac{\pi}{2}\sin\frac{\pi}{2} - 0\sin 0\right) + \left(\cos\frac{\pi}{2} - \cos 0\right)\right)$$

$$= 2\pi\left(\frac{\pi}{2} - 1\right)$$

$$= \pi^2 - 2\pi$$

712. 1,470 J

First find the force by multiplying mass by acceleration:

$$F = mg = (50 \text{ kg})(9.8 \text{ m/s}^2) = 490 \text{ N}$$

The force is constant, so no integral is required to find the work:

$$W = Fd = (490 \text{ N})(3 \text{ m}) = 1,470 \text{ J}$$

713. 9,600 J

The force is constant, so no integral is required to find the work. Enter the numbers in the work equation and solve:

$$W = Fd = (800 \text{ N})(12 \text{ m}) = 9,600 \text{ J}$$

714. 337.5 ft·lb

Let n be the number of subintervals of length Δx, and let x_i^* be a sample point in the ith subinterval $[x_{i-1}, x_i]$.

The portion of the rope that is from x_{i-1} feet to x_i feet below the top of the cliff weighs $(0.75)\Delta x$ pounds and must be lifted approximately x_i^* feet. Therefore, its contribution to the total work is approximately $(0.75)\left(x_i^*\right)\Delta x$ ft·lb. The total work is

$$W = \lim_{n \to \infty} \sum_{i=1}^{n} (0.75)\left(x_i^*\right)\Delta x$$

$$= \int_0^{30} 0.75x \, dx$$

$$= \frac{0.75}{2} x^2 \Big|_0^{30}$$

$$= 337.5 \text{ ft·lb}$$

715. 253.125 ft·lb

Let n be the number of subintervals of length Δx, and let x_i^* be a sample point in the ith subinterval $[x_{i-1}, x_i]$.

First consider the work required to pull the top half of the rope that is from x_{i-1} feet to x_i feet below the top of the cliff, where x is in the interval $[0, 15]$. This part of the rope weighs $(0.75)\Delta x$ pounds and must be lifted approximately x_i^* feet, so its contribution to the total work is approximately $(0.75)\left(x_i^*\right)\Delta x$ foot-pounds. Therefore, the total work for the top half of the rope is

$$W_{\text{top}} = \lim_{n \to \infty} \sum_{i=1}^{n} (0.75)\left(x_i^*\right)\Delta x$$

$$= \int_0^{15} 0.75x \, dx$$

$$= \frac{0.75}{2} x^2 \Big|_0^{15}$$

$$= 84.375 \text{ ft·lb}$$

The bottom half of the rope must be lifted 15 feet, so the work required to lift the bottom half is

$$W_{\text{bottom}} = \lim_{n \to \infty} \sum_{i=1}^{n} (0.75)(15)\Delta x$$

$$= \int_{15}^{30} 0.75(15) dx$$

$$= 11.25x \Big|_{0}^{30}$$

$$= 168.75 \text{ ft·lb}$$

To find the total work required, add the two values: $84.375 + 168.75 = 253.125$ ft·lb.

Note that you can find W_{bottom} without an integral because there's a constant force on this section of the rope!

716.   630,000 ft·lb

Let n be the number of subintervals of length Δx, and let x_i^* be a sample point in the ith subinterval $[x_{i-1}, x_i]$.

A section of cable that is from x_{i-1} feet to x_i feet below the top of the building weighs $4\Delta x$ pounds and must be lifted approximately x_i^* feet, so its contribution to the total work is approximately $(4)(x_i^*)\Delta x$ foot-pounds. Therefore, the work required to lift the cable is

$$W = \lim_{n \to \infty} \sum_{i=1}^{n} (4)(x_i^*)\Delta x$$

$$= \int_{0}^{300} 4x \, dx$$

$$= 2x^2 \Big|_{0}^{300}$$

$$= 180,000 \text{ ft·lb}$$

The work required to lift the piece of metal is $(1,500 \text{ lb})(300 \text{ ft}) = 450,000$ ft·lb. Therefore, the total work required is $180,000 + 450,000 = 630,000$ ft·lb.

717.   22,500 ft·lb

Let n be the number of subintervals of length Δx, and let x_i^* be a sample point in the ith subinterval $[x_{i-1}, x_i]$.

The cable weighs $\dfrac{300 \text{ lb}}{150 \text{ ft}} = 2$ lb/ft. The portion of the cable that is from x_{i-1} feet to x_i feet below the top of the building weighs $2\Delta x$ pounds and must be lifted approximately x_i^* feet, so its contribution to the total work is approximately $2x_i^* \Delta x$ foot-pounds. The total work is

$$W = \lim_{n \to \infty} \sum_{i=1}^{n} 2x_i^* \Delta x$$

$$= \int_{0}^{150} 2x \, dx$$

$$= x^2 \Big|_{0}^{150}$$

$$= (150)^2$$

$$= 22,500 \text{ ft·lb}$$

718.

39,200 J

Let n be the number of subintervals of length Δx, and let x_i^* be a sample point in the ith subinterval $[x_{i-1}, x_i]$.

A horizontal slice of water that is Δx meters thick and is at a distance of x_i^* meters from the top of the tank has a volume of $((4)(2)(\Delta x))$ cubic meters and has a mass of

$$\text{mass} = (\text{density})(\text{volume})$$
$$= \left(1{,}000 \text{ kg/m}^3\right)((4 \text{ m})(2 \text{ m})(\Delta x \text{ m}))$$
$$= 8{,}000\Delta x \text{ kg}$$

Because mass is not a force, you must multiply it by the acceleration due to gravity, 9.8 meters per second squared, in order to find the weight of the slice in newtons:

$$\left(9.8 \text{ m/s}^2\right)(8{,}000\Delta x \text{ kg}) = 78{,}400\Delta x \text{ N}$$

The work required to pump out this slice of water is approximately $78{,}400\left(x_i^*\right)\Delta x$ joules. Therefore, the total work is

$$W = \lim_{n \to \infty} \sum_{i=1}^{n} 78{,}400\left(x_i^*\right)\Delta x$$
$$= \int_0^1 (78{,}400x)dx$$
$$= 39{,}200x^2 \Big|_0^1$$
$$= 39{,}200 \text{ J}$$

719.

7.875 ft·lb

Begin by finding the spring constant, k, using the information about the work required to stretch the spring 2 feet beyond its natural length:

$$W = \int_0^2 kx\,dx$$
$$14 = \frac{kx^2}{2} \Big|_0^2$$
$$14 = \frac{k}{2}\left((2)^2 - (0)^2\right)$$
$$14 = 2k$$
$$7 \text{ lb/ft} = k$$

Next, find the work required to stretch the spring 18 inches, or 1.5 feet, beyond its natural length:

$$W = \int_0^{1.5} 7x\,dx$$
$$= \frac{7}{2}x^2 \Big|_0^{1.5}$$
$$= \frac{7}{2}\left((1.5)^2 - (0)^2\right)$$
$$= 7.875 \text{ ft·lb}$$

720. 1.51 J

Begin by using the equation $F(x) = kx$ to find the value of the spring constant. Make sure you convert the centimeters to meters so you can use the units newton-meters, or joules.

$$8\text{ N} = k\left(\frac{9}{100}\text{ m}\right)$$

$$\frac{800}{9}\text{ N}\cdot\text{m} = k$$

Now compute the work, again making sure to convert the lengths into meters. Because you're stretching the spring from 12 centimeters beyond its natural length to 22 centimeters beyond its natural length, the limits of integration are from $\frac{12}{100}$ to $\frac{22}{100}$; therefore, the integral to compute the work is

$$W = \int_{12/100}^{22/100} \frac{800}{9}x\,dx$$

$$= \frac{800}{18}x^2\bigg|_{0.12}^{0.22}$$

$$= \frac{400}{9}\left((0.22)^2 - (0.12)^2\right)$$

$$= 1.51\text{ J}$$

721. $\frac{6}{\pi}\left(\sqrt{3}-1\right)$ J

Use an integral to find the work:

$$W = \int_1^2 2\sin\left(\frac{\pi x}{6}\right)dx$$

$$= 2\left[-\frac{6}{\pi}\cos\left(\frac{\pi x}{6}\right)\right]\bigg|_1^2$$

$$= -\frac{12}{\pi}\left(\cos\left(\frac{\pi}{3}\right) - \cos\left(\frac{\pi}{6}\right)\right)$$

$$= -\frac{12}{\pi}\left(\frac{1}{2} - \frac{\sqrt{3}}{2}\right)$$

$$= \frac{6}{\pi}\left(\sqrt{3} - 1\right)\text{ J}$$

722. 25.2 J

Begin by finding the spring constant, k, using the information about the initial work required to stretch the spring 10 centimeters, or 0.10 meters, from its natural length:

$$W = \int_0^{0.10} kx \, dx$$

$$5 = \left. \frac{kx^2}{2} \right|_0^{0.10}$$

$$5 = \frac{k}{2}\left((0.10)^2\right)$$

$$5 = \frac{k}{2}\left((0.10)^2\right)$$

$$5 = \frac{k}{200}$$

$$1{,}000 \text{ N·m} = k$$

Now find the work required to stretch the spring from 30 centimeters to 42 centimeters, or from 0.15 meters from its natural length to 0.27 meters from its natural length:

$$W = \int_{0.15}^{0.27} 1{,}000x \, dx$$

$$= \left. 500x^2 \right|_{0.15}^{0.27}$$

$$= 500\left((0.27)^2 - (0.15)^2\right)$$

$$= 25.2 \text{ J}$$

---

723. $\frac{20}{3}$ ft·lb

Begin by using the equation $F(x) = kx$ to find the value of k, the spring constant. The force is measured in foot-pounds, so make sure the length is in feet rather than inches.

$$15 \text{ lb} = k\left(\frac{6}{12} \text{ ft}\right)$$

$$30 \text{ ft·lb} = k$$

Therefore, the force equation is $F(x) = 30x$.

Because the spring starts at its natural length, $x = 0$ is the lower limit of integration. Because the spring is stretched 8 inches, or $\frac{8}{12}$ feet, beyond its natural length, the upper limit of integration is $\frac{8}{12} = \frac{2}{3}$. Therefore, the integral to find the work is

$$W = \int_0^{8/12} 30x \, dx$$

$$= \left. 15x^2 \right|_0^{2/3}$$

$$= 15\left(\left(\frac{2}{3}\right)^2 - 0^2\right)$$

$$= \frac{20}{3} \text{ ft·lb}$$

---

724. 2.25 J

Begin by using the equation $F(x) = kx$ to find the value of the spring constant. Make sure you convert the centimeters to meters so you can use the units newton-meters, or joules. If x is the distance the spring is stretched beyond its natural length, then $x = \frac{5}{100}$ m. Therefore, the spring constant is

$$30 \text{ N} = k\left(\frac{5}{100} \text{ m}\right)$$

$$600 \text{ N} \cdot \text{m} = k$$

Now compute the work, making sure to convert the distances into meters. Because you're computing the work while moving the spring from 5 centimeters beyond its natural length to 10 centimeters beyond its natural length, the limits of integration are from $\frac{5}{100}$ to $\frac{10}{100}$.

$$W = \int_{5/100}^{10/100} 600x \, dx$$

$$= 300x^2 \Big|_{0.05}^{0.10}$$

$$= 300\left((0.1)^2 - (0.05)^2\right)$$

$$= 2.25 \text{ J}$$

---

725. 400 ft·lb

Let n be the number of subintervals of length Δx, and let x_i^* be a sample point in the ith subinterval $[x_{i-1}, x_i]$. Let x_i^* be the distance from the middle of the chain.

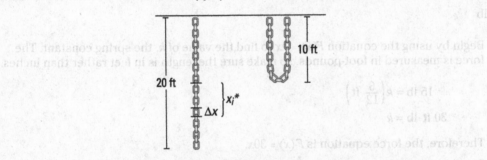

Notice that when you're lifting the end of the chain to the top, only the bottom half of the chain moves. A section of chain that is Δx in length and that is x_i^* feet from the middle of the chain weighs $4\Delta x$ pounds and will move approximately $2x_i^*$ feet. Therefore, the work required is

$$W = \lim_{n \to \infty} \sum_{i=1}^{n} \left(2x_i^*\right)(4\Delta x)$$

$$= \int_0^{10} 8x \, dx$$

$$= 4x^2 \Big|_0^{10}$$

$$= 400 \text{ ft} \cdot \text{lb}$$

726. 613 J

Let n be the number of subintervals of length Δx, and let x_i^* be a sample point in the ith subinterval $[x_{i-1}, x_i]$.

The part of the chain x meters from the lifted end is raised $5 - x$ meters if $0 \le x \le 5$ and is lifted 0 meters otherwise. Therefore, the work required is

$$W = \lim_{n \to \infty} \sum_{i=1}^{n} \left(5 - x_i^*\right)(49)\Delta x$$

$$= 49 \int_0^5 (5 - x)\,dx$$

$$= 49\left(5x - \frac{x^2}{2}\right)\Big|_0^5$$

$$= 49\left(5(5) - \frac{(5)^2}{2}\right)$$

$$= 612.5 \text{ J} \approx 613 \text{ J}$$

727. 3,397,333 J

Let n be the number of subintervals of length Δx, and let x_i^* be a sample point in the ith subinterval $[x_{i-1}, x_i]$.

Consider a horizontal slice of water that is Δx meters thick and is at a height of x_i^* meters from the bottom of the tank. The trough has a triangular face with a width and height of 4 meters, so by similar triangles, the width of the slice of water is the same as the height x_i^*; therefore, the volume of the slice of water is $10\left(x_i^*\right)(\Delta x)$ cubic meters. To find the weight of the water, multiply the acceleration due to gravity ($g = 9.8$ m/s^2) by the *mass*, which equals the density of water (1,000 kg/m^3) multiplied by the volume:

$$\left(9.8 \text{ m/s}^2\right)\left(1{,}000 \text{ kg/m}^3\right)\left((10 \text{ m})\left(x_i^* \text{ m}\right)(\Delta x \text{ m})\right) = (98{,}000)\left(x_i^*\right)(\Delta x) \text{ N}$$

The water must travel a distance of $\left(7 - x_i^*\right)$ meters to exit the tank. Therefore, the total work required is

$$W = \lim_{n \to \infty} \sum_{i=1}^{n} (98{,}000)\left(x_i^*\right)\left(7 - x_i^*\right)(\Delta x)$$

$$= \int_0^4 98{,}000(x)(7 - x)\,dx$$

$$= 98{,}000 \int_0^4 \left(7x - x^2\right)dx$$

$$= 98{,}000\left(\frac{7}{2}x^2 - \frac{1}{3}x^3\right)\Big|_0^4$$

$$= 98{,}000\left(\frac{7}{2}4^2 - \frac{1}{3}4^3\right)$$

$$= \frac{10{,}192{,}000}{3} \text{ J}$$

$$\approx 3{,}397{,}333 \text{ J}$$

728. 169,646 ft·lb

Let n be the number of subintervals of length Δx, and let x_i^* be a sample point in the ith subinterval $[x_{i-1}, x_i]$.

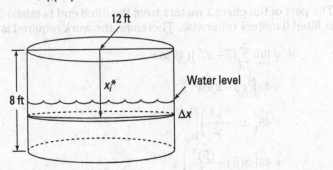

A horizontal slice of water that is Δx feet thick and is at a depth of x_i^* feet from the top of the tank is cylindrical, so it has a volume of $\left(\pi(6)^2(\Delta x)\right)$ cubic feet. Because water weighs 62.5 pounds per cubic foot, this slice of water weighs $62.5(36\pi)(\Delta x)$ pounds. The work required to pump out this slice of water is

$$62.5(36\pi)\left(x_i^*\right)(\Delta x) \text{ ft·lb} = 2{,}250\pi\left(x_i^*\right)(\Delta x) \text{ ft·lb}$$

Therefore, the total work is

$$W = \lim_{n\to\infty} \sum_{i=1}^{n} 2{,}250\pi\left(x_i^*\right)\Delta x$$

$$= \int_4^8 2{,}250\pi(x)\,dx$$

$$= 1{,}125\pi x^2 \Big|_4^8$$

$$= 54{,}000\pi \text{ ft·lb}$$

$$\approx 169{,}646 \text{ ft·lb}$$

729. 3,000 ft·lb

Let n be the number of subintervals of length Δx, and let x_i^* be a sample point in the ith subinterval $[x_{i-1}, x_i]$.

The work required to lift only the bucket is

$$(5 \text{ lb})(100 \text{ ft}) = 500 \text{ ft·lb}$$

Lifting the bucket takes 50 seconds, so the bucket is losing 1 pound of water per second. At time t (in seconds), the bucket is $x_i^* = 2t$ feet above the original 100-foot depth, but it now holds only $(50 - t)$ pounds of water. In terms of distance, the bucket holds $\left(50 - \frac{1}{2}x_i^*\right)$ pounds of water when it's x_i^* feet above the original 100-foot depth. Moving this amount of water a distance of Δx requires $\left(50 - \frac{1}{2}x_i^*\right)\Delta x$ foot-pounds of work. Therefore, the work required to lift the only the water is

$$W_{water} = \lim_{n \to \infty} \sum_{i=1}^{n} \left(50 - \frac{1}{2}x_i^*\right)\Delta x$$

$$= \int_0^{100} \left(50 - \frac{x}{2}\right) dx$$

$$= \left(50x - \frac{x^2}{4}\right)\Big|_0^{100}$$

$$= \left(50(100) - \frac{(100)^2}{4}\right)$$

$$= 5,000 - 2,500$$

$$= 2,500 \text{ ft·lb}$$

Adding the work to lift the bucket gives you the total work required: 500 + 2,500 = 3,000 foot-pounds.

---

730.   4 cm

Begin by using the information about how much work is required to stretch the spring from 8 to 10 centimeters to solve for the spring constant and the natural length of the spring; you do so by setting up a system of equations.

You don't know the natural length L for the integral involved in stretching the spring from 8 centimeters to 10 centimeters beyond its natural length, so the limits of integration for one of the integrals involving work are from $0.08 - L$ to $0.10 - L$. Likewise, the other integral will have limits of integration from $0.10 - L$ to $0.12 - L$. Therefore, you have the following two equations for work:

$$10 \text{ J} = \int_{0.08-L}^{0.10-L} kx\, dx$$

$$14 \text{ J} = \int_{0.10-L}^{0.12-L} kx\, dx$$

Expanding the first of the two integrals gives you

$$10 = \int_{0.08-L}^{0.10-L} kx\, dx$$

$$10 = \frac{kx^2}{2}\Big|_{0.08-L}^{0.10-L}$$

$$10 = \frac{k}{2}\left[(0.10-L)^2 - (0.08-L)^2\right]$$

$$20 = k\left[0.0036 - 0.04L\right]$$

Likewise, expanding the second integral gives you

$$14 = \int_{0.10-L}^{0.12-L} kx\, dx$$

$$14 = \frac{kx^2}{2}\Big|_{0.10-L}^{0.12-L}$$

$$14 = \frac{k}{2}\left[(0.12-L)^2 - (0.10-L)^2\right]$$

$$28 = k\left[0.0044 - 0.04L\right]$$

Now solve for the spring constant k. By subtracting $20 = k[0.0036 - 0.04L]$ from $28 = k[0.0044 - 0.04L]$, you're left with $8 = k(0.0008)$ so that $k = 10,000$.

Finally, find the natural length of the spring. Substituting the value of k into the equation $20 = k[0.0036 - 0.04L]$ gives you $20 = 10,000[0.0036 - 0.04L]$, and solving for L yields 0.04 m $= L$. Therefore, the natural length of the spring is 4 centimeters.

731.   5.29 ft

Let n be the number of subintervals of length Δx, and let x_i^* be a sample point in the ith subinterval $[x_{i-1}, x_i]$.

A horizontal slice of water that is Δx feet thick and is at a distance of x_i^* feet from the top of the tank is cylindrical and therefore has a volume of $\left(\pi(6)^2(\Delta x)\right)$ cubic feet. Because water weighs 62.5 pounds per cubic foot, this slice of water weighs $62.5(36\pi)(\Delta x)$ pounds. The work required to pump out this slice of water is

$$62.5(36\pi)(x_i^*)(\Delta x) \text{ ft·lb} = 2,250\pi(x_i^*)(\Delta x) \text{ ft·lb}$$

You don't know how many feet of water are pumped out, so you need to solve for the upper limit of integration in the equation:

$$13,500\pi = \int_4^d (2,250\pi x)\,dx$$
$$13,500\pi = 1,125\pi x^2 \Big|_4^d$$
$$13,500\pi = 1,125\pi\left(d^2 - 16\right)$$
$$12 = d^2 - 16$$
$$28 = d^2$$
$$2\sqrt{7} = d$$
$$5.29 \text{ ft} \approx d$$

Because the water level in the tank was reduced, the water is now 5.29 feet from the top of the tank.

732.   16.67 cm

Begin by using the information about how much work is required to stretch the spring from 40 to 60 centimeters (0.40 to 0.60 meters) beyond its natural length to solve for the spring constant and the natural length of the spring; to do so, set up a system of equations.

Because you don't know the natural length L for the integral involved in stretching the spring from 0.40 meters to 0.60 meters beyond its natural length, the limits of integration for one of the integrals involving work are from $0.40 - L$ to $0.60 - L$. Likewise, the other integral will have limits of integration from $0.60 - L$ to $0.80 - L$. Therefore, you have the following two equations for work:

$$25 \text{ J} = \int_{0.40-L}^{0.60-L} kx\,dx$$

$$40 \text{ J} = \int_{0.60-L}^{0.80-L} kx\,dx$$

Expanding the first of the two integrals gives you

$$W = \int_{0.40-L}^{0.60-L} kx\,dx$$

$$25 = \left.\frac{kx^2}{2}\right|_{0.40-L}^{0.60-L}$$

$$25 = \frac{k}{2}\left[(0.60-L)^2 - (0.40-L)^2\right]$$

$$50 = k\left[0.2 - 0.4L\right]$$

Likewise, expanding the second integral gives you

$$W = \int_{0.60-L}^{0.80-L} kx\,dx$$

$$40 = \left.\frac{kx^2}{2}\right|_{0.60-L}^{0.80-L}$$

$$40 = \frac{k}{2}\left[(0.80-L)^2 - (0.60-L)^2\right]$$

$$80 = k\left[0.28 - 0.4L\right]$$

Now solve for the spring constant k. By subtracting $50 = k[0.2 - 0.4L]$ from $80 = k[0.28 - 0.4L]$, you're left with $30 = k(0.08)$ so that $k = 375$.

Finally, find the natural length of the spring. Substituting the value of k into the equation $50 = k[0.2 - 0.4L]$ gives you $50 = 375[0.2 - 0.4L]$; solving for L yields $L = \frac{1}{6}$ meters, so the natural length of the spring is approximately 16.67 centimeters.

---

733.   615,752 J

Let n be the number of subintervals of length Δx and let x_i^* be a sample point in the ith subinterval.

Here's a view of the tank underground:

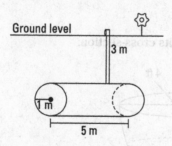

Ground level

3 m

1 m

5 m

And here's a cross-section of the tank:

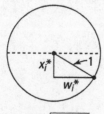

$$w_i^* = \sqrt{1 - (x_i^*)^2}$$

A horizontal slice of water that is Δx meters thick and is at a distance of x_i^* meters down from the middle of the tank has a volume of $5\left(2\sqrt{1-\left(x_i^*\right)^2}\right)\Delta x = \left(10\sqrt{1-\left(x_i^*\right)^2}\Delta x\right)$ cubic meters. To find the weight of the water, multiply the acceleration due to gravity ($g = 9.8$ m/s^2) by the *mass* of the water, which equals the density of water (1,000 kg/m^3) times the volume:

$$\left(9.8 \text{ m/s}^2\right)\left(1,000 \text{ kg/m}^3\right)\left(10\sqrt{1-\left(x_i^*\right)^2}\Delta x \text{ m}^3\right)$$

The slice of water must travel a distance of $\left(4+x_i^*\right)$ meters to reach ground level. (Note that negative x values correspond to slices of water that are above the middle of the tank.) Therefore, the total work required is

$$\lim_{n\to\infty}\sum_{i=1}^{n}(98,000)\left(4+x_i^*\right)\sqrt{1-\left(x_i^*\right)^2}\Delta x$$

$$= \int_{-1}^{1}98,000(4+x)\sqrt{1-x^2}\,dx$$

$$= 98,000\left[\int_{-1}^{1}4\sqrt{1-x^2}\,dx + \int_{-1}^{1}x\sqrt{1-x^2}\,dx\right]$$

Notice that second integral is zero because $g(x) = x\sqrt{1-x^2}$ is an odd function being integrated over an interval that's symmetric about the origin. To evaluate $98,000\left[4\int_{-1}^{1}\sqrt{1-x^2}\,dx\right]$, notice that the integral represents the area of a semicircle with a radius of 1, so the work equals

$$98,000\left[4\int_{-1}^{1}\sqrt{1-x^2}\,dx\right] = 392,000\left(\frac{\pi(1)^2}{2}\right)$$

$$= 196,000\pi \text{ J}$$

$$\approx 615,752 \text{ J}$$

734.   20,944 ft·lb

Here's a view of the tank and its cross-section:

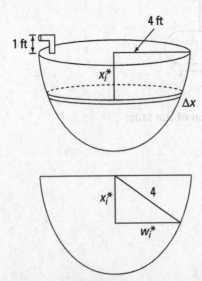

Let n be the number of subintervals of length Δx and let x_i^* be a sample point in the ith subinterval. Consider a horizontal slice of water that is Δx feet thick and is at a distance of x_i^* feet from the top of the tank; let w_i^* be the radius of the slice of water. Because the tank has a shape of a hemisphere of radius 4, you have the following relationship:

$$\left(x_i^*\right)^2 + \left(w_i^*\right)^2 = 16$$

Therefore, the slice has a volume of

$$V_i = \pi\left(w_i^*\right)^2 \Delta x = \pi\left(16 - \left(x_i^*\right)^2\right)\Delta x \text{ ft}^3$$

And the force on this slice is

$$F_i = (62.5)\pi\left(16 - \left(x_i^*\right)^2\right)\Delta x \text{ lb}$$

The slice of water must travel a distance of $\left(1 + x_i^*\right)$ feet to exit the tank. Therefore, the total work required is

$$W = \lim_{n \to \infty} \sum_{i=1}^{n} (62.5)\pi\left(1 + x_i^*\right)\left(16 - \left(x_i^*\right)^2\right)\Delta x$$

$$= 62.5\pi \int_0^4 (1 + x)\left(16 - x^2\right) dx$$

$$= 62.5\pi \int_0^4 \left(16 + 16x - x^2 - x^3\right) dx$$

$$= 62.5\pi \left(16x + 8x^2 - \frac{x^3}{3} - \frac{x^4}{4}\right)\Bigg|_0^4$$

$$= 62.5\pi \left(16(4) + 8(4)^2 - \frac{(4)^3}{3} - \frac{(4)^4}{4}\right)$$

$$= 62.5\pi \left(\frac{320}{3}\right) = \frac{20,000\pi}{3} \text{ ft}\cdot\text{lb}$$

$$\approx 20,944 \text{ ft}\cdot\text{lb}$$

735. 14,726 ft·lb

Here's a view of the tank and its cross-section:

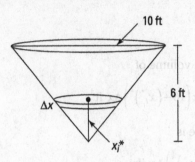

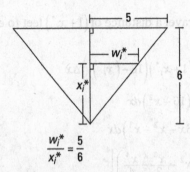

$$\frac{w_i^*}{x_i^*} = \frac{5}{6}$$

Let n be the number of subintervals of length Δx and let x_i^* be a sample point in the ith subinterval. Consider a horizontal slice of water that is Δx feet thick and is x_i^* feet from the bottom of the tank with a width of w_i^*. This slice has a volume of

$$V_i = \pi \left(w_i^* \right)^2 \Delta x \text{ ft}^3$$

By similar triangles, you have $\dfrac{w_i^*}{x_i^*} = \dfrac{5}{6}$, so $w_i^* = \dfrac{5}{6} x_i^*$; therefore, the volume becomes

$$V_i = \pi \left(w_i^* \right)^2 \Delta x$$

$$= \pi \left(\frac{5}{6} x_i^* \right)^2 \Delta x$$

$$= \frac{25\pi}{36} \left(x_i^* \right)^2 \Delta x \text{ ft}^3$$

The force on this slice of water is

$$F_i = 62.5 \left(\frac{25\pi}{36} \left(x_i^* \right)^2 \right) \Delta x \text{ lb}$$

The distance this slice must travel to exit the tank is $\left(6 - x_i^* \right)$ feet. Therefore, the total work required is

$$W = \lim_{n \to \infty} \sum_{i=1}^{n} 62.5\left(\left(6 - x_i^*\right)\right)\left(\frac{25\pi}{36}\left(x_i^*\right)^2\right)\Delta x$$

$$= (62.5)\frac{25\pi}{36}\int_0^6 (6 - x)\left(x^2\right)dx$$

$$= (62.5)\frac{25\pi}{36}\int_0^6 \left(6x^2 - x^3\right)dx$$

$$= (62.5)\frac{25\pi}{36}\left(2x^3 - \frac{x^4}{4}\right)\Big|_0^6$$

$$= (62.5)\frac{25\pi}{36}\left(2(6)^3 - \frac{(6)^4}{4}\right)$$

$$= \frac{9,375}{2}\pi \text{ ft} \cdot \text{lb}$$

$$\approx 14,726 \text{ ft} \cdot \text{lb}$$

---

736. $\dfrac{5}{4}$

Using the average value formula with $a = -1$, $b = 2$, and $f(x) = x^3$ gives you

$$f_{avg} = \frac{1}{2 - (-1)}\int_{-1}^{2} x^3 dx$$

$$= \frac{1}{3}\left(\frac{x^4}{4}\right)\Big|_{-1}^{2}$$

$$= \frac{1}{12}\left(2^4 - (-1)^4\right)$$

$$= \frac{5}{4}$$

---

737. $\dfrac{2}{3\pi}$

Using the average value formula with $a = 0$, $b = \dfrac{3\pi}{2}$, and $f(x) = \sin x$ gives you

$$f_{avg} = \frac{1}{3\pi/2 - 0}\int_0^{3\pi/2} \sin x\, dx$$

$$= \frac{2}{3\pi}\left(-\cos x\right)\Big|_0^{3\pi/2}$$

$$= \frac{-2}{3\pi}\left(\cos\frac{3\pi}{2} - \cos 0\right)$$

$$= \frac{-2}{3\pi}(0 - 1)$$

$$= \frac{2}{3\pi}$$

738. $\dfrac{1}{2\pi}$

Using the average value formula with $a = 0$, $b = \dfrac{\pi}{2}$, and $f(x) = (\sin^3 x)(\cos x)$ gives you

$$f_{avg} = \frac{1}{\pi/2 - 0} \int_0^{\pi/2} (\sin^3 x)(\cos x)\,dx$$

$$= \frac{2}{\pi} \int_0^{\pi/2} (\sin^3 x)(\cos x)\,dx$$

Next, use the substitution $u = \sin x$ so that $du = \cos x\,dx$. Find the new limits of integration by noting that if $x = \dfrac{\pi}{2}$, then $u = \sin\dfrac{\pi}{2} = 1$, and if $x = 0$, then $u = \sin 0 = 0$. With these new values, you get the following:

$$\frac{2}{\pi} \int_0^{\pi/2} (\sin^3 x)(\cos x)\,dx = \frac{2}{\pi} \int_0^1 u^3\,du$$

$$= \frac{2}{\pi}\left(\frac{u^4}{4}\right)\Bigg|_0^1$$

$$= \frac{2}{\pi}\left(\frac{1}{4} - 0\right)$$

$$= \frac{1}{2\pi}$$

739. $\dfrac{26}{9}$

Using the average value formula with $a = 0$, $b = 2$, and $f(x) = x^2\sqrt{1+x^3}$ gives you

$$f_{avg} = \frac{1}{2-0} \int_0^2 x^2\sqrt{1+x^3}\,dx$$

Begin by using the substitution $u = 1 + x^3$ so that $du = 3x^2\,dx$, or $\dfrac{1}{3}du = x^2\,dx$. You can also find the new limits of integration by noting that if $x = 2$, then $u = 1 + 2^3 = 9$, and if $x = 0$, then $u = 1 + 0^3 = 1$. With these values, you have the following:

$$\frac{1}{2-0} \int_0^2 x^2\sqrt{1+x^3}\,dx = \frac{1}{2} \int_1^9 \frac{1}{3}\sqrt{u}\,du$$

$$= \frac{1}{6} \int_1^9 u^{1/2}\,du$$

$$= \frac{1}{6}\left(\frac{2u^{3/2}}{3}\right)\Bigg|_1^9$$

$$= \frac{1}{9}\left(9^{3/2} - 1^{3/2}\right)$$

$$= \frac{1}{9}(27 - 1)$$

$$= \frac{26}{9}$$

740. $\dfrac{8}{9\ln 3}$

Using the average value formula with $a = 0$, $b = \ln 3$, and $f(x) = \sinh x \cosh x$, gives you

$$f_{avg} = \frac{1}{\ln 3 - 0} \int_0^{\ln 3} \sinh \cosh x \, dx$$

Use the substitution $u = \cosh x$ so that $du = \sinh x$. You can find the new limits of integration by noting that if $x = \ln 3$, then $u = \cosh(\ln 3) = \dfrac{e^{\ln 3} + e^{-\ln 3}}{2} = \dfrac{3 + \frac{1}{3}}{2} = \dfrac{5}{3}$, and if $x = 0$, then $u = \cosh(0) = \dfrac{e^0 + e^{-0}}{2} = 1$. With these new values, you have

$$\frac{1}{\ln 3} \int_1^{5/3} u \, du = \frac{1}{\ln 3} \frac{u^2}{2} \Big|_1^{5/3}$$

$$= \frac{1}{2\ln 3}\left[\left(\frac{5}{3}\right)^2 - 1\right]$$

$$= \frac{8}{9\ln 3}$$

741. $\dfrac{1}{2}$

Using the average value formula with $a = 1$, $b = 4$, and $f(r) = \dfrac{5}{(1+r)^2}$ gives you

$$f_{avg} = \frac{1}{4-1}\int_1^4 \frac{5}{(1+r)^2}\, dr$$

$$= \frac{1}{3}\int_1^4 \frac{5}{(1+r)^2}\, dr$$

Next, use the substitution $u = 1 + r$ so that $du = dr$. Find the new limits of integration by noting that if $r = 4$, then $u = 1 + 4 = 5$, and if $r = 1$, then $u = 1 + 1 = 2$. With these new values, you find that

$$\frac{1}{3}\int_2^5 \frac{5}{u^2}\, du = \frac{5}{3}\int_2^5 u^{-2} du$$

$$= \frac{5}{3}\left(-\frac{1}{u}\right)\Big|_2^5$$

$$= \frac{5}{3}\left(-\frac{1}{5} - \left(-\frac{1}{2}\right)\right)$$

$$= \frac{5}{3}\left(-\frac{1}{5} + \frac{1}{2}\right)$$

$$= \frac{5}{3}\left(\frac{3}{10}\right)$$

$$= \frac{1}{2}$$

742. $\frac{7}{2}\left(\sqrt{10}-\sqrt{2}\right)$

Using the average value formula with $a = 0$, $b = 8$, and $f(x) = \dfrac{14}{\sqrt{x+2}}$ gives you

$$f_{avg} = \frac{1}{8-0}\int_0^8 \frac{14}{\sqrt{x+2}}\,dx$$

$$= \frac{1}{8}\int_0^8 \frac{14}{\sqrt{x+2}}\,dx$$

Next, use the substitution $u = x + 2$ so that $du = dx$. You can find the new limits of integration by noting that if $x = 8$, then $u = 8 + 2 = 10$, and if $x = 0$, then $u = 0 + 2 = 2$. With these new values, you find that

$$\frac{1}{8}\int_2^{10} 14u^{-1/2}\,du = \frac{7}{4}\left(2u^{1/2}\right)\Big|_2^{10}$$

$$= \frac{7}{2}\left(\sqrt{10}-\sqrt{2}\right)$$

743. $d = 1$

Using the average value formula with $a = 0$, $b = d$, and $f(x) = 2 + 4x - 3x^2$ gives you

$$f_{avg} = \frac{1}{d-0}\int_0^d \left(2+4x-3x^2\right)dx$$

$$= \frac{1}{d}\left(2x+2x^2-x^3\right)\Big|_0^d$$

$$= \frac{1}{d}\left(2d+2d^2-d^3\right)$$

$$= 2+2d-d^2$$

Next, set this expression equal to 3 and solve for d:

$$2+2d-d^2 = 3$$

$$0 = d^2 - 2d + 1$$

$$0 = (d-1)(d-1)$$

$$d = 1$$

744. $d = 4$

Using the average value formula with $a = 0$, $b = d$, and $f(x) = 3 + 6x - 9x^2$ gives you

$$f_{avg} = \frac{1}{d-0}\int_0^d \left(3+6x-9x^2\right)dx$$

$$= \frac{1}{d}\left(3x+3x^2-3x^3\right)\Big|_0^d$$

$$= \frac{1}{d}\left(3d+3d^2-3d^3\right)$$

$$= 3+3d-3d^2$$

Next, set this expression equal to -33 and solve for d:

$$3 + 3d - 3d^2 = -33$$
$$1 + d - d^2 = -11$$
$$0 = d^2 - d - 12$$
$$0 = (d - 4)(d + 3)$$
$$d = 4, -3$$

Because the interval is of the form $[0, d]$, the only solution is $d = 4$.

---

745. $c = \sqrt{3}$

Begin by finding the average value of the function on the given interval by using the average value formula with $a = 1$, $b = 3$, and $f(x) = \dfrac{4(x^2 + 1)}{x^2}$:

$$f_{avg} = \frac{1}{3 - 1} \int_1^3 \frac{4(x^2 + 1)}{x^2} \, dx$$
$$= \frac{4}{2} \int_1^3 \left(\frac{x^2}{x^2} + \frac{1}{x^2} \right) dx$$
$$= 2 \int_1^3 \left(1 + x^{-2} \right) dx$$
$$= 2 \left(x - \frac{1}{x} \right) \Big|_1^3$$
$$= 2 \left(3 - \frac{1}{3} - (1 - 1) \right)$$
$$= \frac{16}{3}$$

Next, set the original function equal to the average value and solve for c:

$$\frac{4(c^2 + 1)}{c^2} = \frac{16}{3}$$
$$\frac{c^2 + 1}{c^2} = \frac{4}{3}$$
$$3c^2 + 3 = 4c^2$$
$$3 = c^2$$
$$\pm\sqrt{3} = c$$

Because $-\sqrt{3} = c$ isn't in the specified interval, the solution is $c = \sqrt{3}$.

746. $c = \dfrac{25}{4}$

Begin by finding the average value of the function on the given interval by using the average value formula with $a = 4$, $b = 9$, and $f(x) = \dfrac{1}{\sqrt{x}}$:

$$f_{avg} = \frac{1}{9-4} \int_4^9 \frac{1}{\sqrt{x}}\,dx$$

$$= \frac{1}{5} \int_4^9 x^{-1/2}\,dx$$

$$= \frac{1}{5}\left(2x^{1/2}\right)\Big|_4^9$$

$$= \frac{2}{5}\left(\sqrt{9}-\sqrt{4}\right)$$

$$= \frac{2}{5}$$

Next, set the original function equal to the average value and solve for c:

$$\frac{1}{\sqrt{c}} = \frac{2}{5}$$

$$5 = 2\sqrt{c}$$

$$\frac{5}{2} = \sqrt{c}$$

$$\frac{25}{4} = c$$

Because $c = \dfrac{25}{4}$ is in the given interval, you've found the solution.

747. $c = \pm\dfrac{2\sqrt{3}}{3}$

Begin by finding the average value of the function on the given interval by using the average value formula with $a = -2$, $b = 2$, and $f(x) = 5 - 3x^2$:

$$f_{avg} = \frac{1}{2-(-2)} \int_{-2}^2 \left(5-3x^2\right)dx$$

$$= \frac{1}{4}\left(5x - x^3\right)\Big|_{-2}^2$$

$$= \frac{1}{4}\left(5(2)-(2)^3 - \left(5(-2)-(-2)^3\right)\right)$$

$$= 1$$

Next, set the original function equal to the average value and solve for c:

$$5 - 3c^2 = 1$$

$$4 = 3c^2$$

$$\frac{4}{3} = c^2$$

$$\pm\sqrt{\frac{4}{3}} = c$$

$$\pm\frac{2}{\sqrt{3}} = c$$

$$\pm\frac{2\sqrt{3}}{3} = c$$

Both values fall in the given interval, so both are solutions.

---

748. $\dfrac{2}{\pi}$

Using the average value formula with $a = 0$, $b = \frac{\pi}{2}$, and $f(x) = x \sin x$ gives you

$$f_{\text{avg}} = \frac{1}{\pi/2 - 0} \int_0^{\pi/2} x \sin x \, dx$$

$$= \frac{2}{\pi} \int_0^{\pi/2} x \sin x \, dx$$

To integrate this function, you can use integration by parts. Notice that for the indefinite integral $\int x \sin x \, dx$, you can let $u = x$ so that $du = dx$ and let $dv = \sin x \, dx$ so that $v = -\cos x \, dx$. Using integration by parts formula gives you

$$\int x \sin x \, dx = -x \cos x + \int \cos x \, dx$$

$$= -x \cos x + \sin x + C$$

Therefore, to evaluate the definite integral $\dfrac{2}{\pi} \displaystyle\int_0^{\pi/2} x \sin x \, dx$, you have

$$\frac{2}{\pi} \int_0^{\pi/2} x \sin x \, dx = \frac{2}{\pi}(-x \cos x + \sin x)\Big|_0^{\pi/2}$$

$$= \frac{2}{\pi}\left(-\frac{\pi}{2}\cos\left(\frac{\pi}{2}\right) + \sin\left(\frac{\pi}{2}\right) - (0 + \sin(0))\right)$$

$$= \frac{2}{\pi}(1)$$

$$= \frac{2}{\pi}$$

749.

$$\frac{4}{\pi}\left[\frac{1}{2}\ln\left|\frac{\pi^2}{16}+1\right|+\tan^{-1}\frac{\pi}{4}\right]$$

Using the average value formula with $a = 0$, $b = \frac{\pi}{4}$, and $f(x) = \frac{x+1}{x^2+1}$ gives you

$$f_{avg} = \frac{1}{\pi/4 - 0}\int_0^{\pi/4}\frac{x+1}{x^2+1}dx$$

To evaluate this integral, begin by splitting up the fraction:

$$\frac{4}{\pi}\left[\int_0^{\pi/4}\frac{x}{x^2+1}dx + \int_0^{\pi/4}\frac{1}{x^2+1}dx\right]$$

For the first integral, use a substitution, letting $u = x^2 + 1$ so that $du = 2x\,dx$, or $\frac{1}{2}du = x\,dx$. You can also find the new limits of integration for the first integral by noting that if $x = \frac{\pi}{4}$, then $u = \left(\frac{\pi}{4}\right)^2 + 1 = \frac{\pi^2}{16} + 1$, and if $x = 0$, then $u = 0^2 + 1 = 1$. With these new values, you have

$$\frac{4}{\pi}\left[\int_0^{\pi/4}\frac{x}{x^2+1}dx + \int_0^{\pi/4}\frac{1}{x^2+1}dx\right]$$

$$= \frac{4}{\pi}\left[\int_1^{\frac{\pi^2}{16}+1}\frac{1}{2}\left(\frac{1}{u}\right)du + \int_0^{\pi/4}\frac{1}{x^2+1}dx\right]$$

$$= \frac{4}{\pi}\left[\frac{1}{2}\ln|u|\Big|_1^{\frac{\pi^2}{16}+1} + \tan^{-1}x\Big|_0^{\pi/4}\right]$$

$$= \frac{4}{\pi}\left[\frac{1}{2}\ln\left|\frac{\pi^2}{16}+1\right|+\tan^{-1}\frac{\pi}{4}\right]$$

750. $\dfrac{2}{\sqrt{2x-x^2}}$

To find the derivative of $y = 2\sin^{-1}(x-1)$, use the derivative formula for $\sin^{-1} x$ and the chain rule:

$$y' = 2\frac{1}{\sqrt{1-(x-1)^2}}$$

$$= \frac{2}{\sqrt{1-(x^2-2x+1)}}$$

$$= \frac{2}{\sqrt{2x-x^2}}$$

751. $\dfrac{-12x^3-3}{\sqrt{-x^8-2x^5-x^2+1}}$

To find the derivative of $y = 3\cos^{-1}(x^4+x)$, you need to know the derivative formula for the inverse cosine function and also use the chain rule:

$$y' = 3\left(-\frac{1}{\sqrt{1-(x^4+x)^2}}\right)(4x^3+1)$$

$$= \frac{-3(4x^3+1)}{\sqrt{1-(x^8+2x^5+x^2)}}$$

$$= \frac{-12x^3-3}{\sqrt{-x^8-2x^5-x^2+1}}$$

752. $\dfrac{1}{2\sqrt{\tan^{-1} x}\,(1+x^2)}$

First rewrite the radical using an exponent:

$$y = \sqrt{\tan^{-1} x} = \left(\tan^{-1} x\right)^{1/2}$$

Then use the derivative formula for $\tan^{-1} x$ and the chain rule to find the derivative:

$$y' = \frac{1}{2}\left(\tan^{-1} x\right)^{-1/2}\left(\frac{1}{1+x^2}\right)$$

$$= \frac{1}{2\sqrt{\tan^{-1} x}\,(1+x^2)}$$

753. $1 - \dfrac{x\sin^{-1} x}{\sqrt{1-x^2}}$

First rewrite the radical using an exponent:

$$y = \sqrt{1-x^2}\,\sin^{-1} x = \left(1-x^2\right)^{1/2}\sin^{-1} x$$

Find the derivative using the product rule and the derivative formula for $\sin^{-1} x$:

$$y' = \frac{1}{2}\left(1-x^2\right)^{-1/2}\left(-2x\right)\sin^{-1} x + \left(1-x^2\right)^{1/2}\frac{1}{\sqrt{1-x^2}}$$

$$= \frac{-x\sin^{-1} x}{\sqrt{1-x^2}} + \frac{\sqrt{1-x^2}}{\sqrt{1-x^2}}$$

$$= 1 - \frac{x\sin^{-1} x}{\sqrt{1-x^2}}$$

754. $\dfrac{-\sin x}{1+\cos^2 x}$

To find the derivative of $y = \tan^{-1}(\cos x)$, use the derivative formula for $\tan^{-1} x$ along with the chain rule:

$$y' = \frac{1}{1+(\cos x)^2}(-\sin x) = \frac{-\sin x}{1+\cos^2 x}$$

755. $\dfrac{e^{\sec^{-1} t}}{t\sqrt{t^2-1}}$

Find the derivative of $y = e^{\sec^{-1} t}$ using the chain rule and the derivative formula for $\sec^{-1} t$:

$$y' = e^{\sec^{-1} t}\frac{1}{t\sqrt{t^2-1}}$$

$$= \frac{e^{\sec^{-1} t}}{t\sqrt{t^2-1}}$$

756. $\dfrac{-2}{\sqrt{e^{4x}-1}}$

To find the derivative of $y = \csc^{-1} e^{2x}$, use the derivative formula for $\csc^{-1} x$ along with the chain rule:

$$y' = \frac{-1}{e^{2x}\sqrt{\left(e^{2x}\right)^2-1}}\left(e^{2x}(2)\right)$$

$$= \frac{-2}{\sqrt{e^{4x}-1}}$$

757. $e^{x\sec^{-1} x}\left(\sec^{-1} x + \dfrac{1}{\sqrt{x^2-1}}\right)$

To find the derivative of $y = e^{x\sec^{-1} x}$, use the derivative formula for e^x along with the chain rule and product rule:

$$y' = e^{x\sec^{-1} x}\left(1\sec^{-1} x + x\frac{1}{x\sqrt{x^2-1}}\right)$$

$$= e^{x\sec^{-1} x}\left(\sec^{-1} x + \frac{1}{\sqrt{x^2-1}}\right)$$

758. $\dfrac{-4}{\sqrt{9-x^2}}$

Rewrite the arccosine as the inverse cosine:

$$y = 4\arccos\frac{x}{3} = 4\cos^{-1}\frac{x}{3}$$

Now use the chain rule along with the derivative formula for $\cos^{-1}x$:

$$y' = 4\left(\frac{-\frac{1}{3}}{\sqrt{1-\left(\frac{x}{3}\right)^2}}\right)$$

Next, simplify the term underneath the square root. Begin by getting common denominators under the radical:

$$y' = \frac{-4}{3}\left(\frac{1}{\sqrt{\frac{9-x^2}{9}}}\right)$$

Then split up the square root and simplify:

$$y' = \frac{-4}{3}\left(\frac{1}{\sqrt{\frac{9-x^2}{9}}}\right)$$

$$= \frac{-4}{3}\left(\frac{1}{\frac{\sqrt{9-x^2}}{\sqrt{9}}}\right)$$

$$= \frac{-4}{3}\left(\frac{3}{\sqrt{9-x^2}}\right)$$

$$= \frac{-4}{\sqrt{9-x^2}}$$

759. $\sin^{-1}x$

First, rewrite the radical using an exponent:

$$y = x\sin^{-1}x + \sqrt{1-x^2}$$

$$= x\sin^{-1}x + \left(1-x^2\right)^{1/2}$$

Use the product rule on the first term and the chain rule on the second term to get the derivative:

$$y' = 1\sin^{-1}x + x\frac{1}{\sqrt{1-x^2}} + \frac{1}{2}\left(1-x^2\right)^{-1/2}(-2x)$$

$$= \sin^{-1}x + \frac{x}{\sqrt{1-x^2}} - \frac{x}{\sqrt{1-x^2}}$$

$$= \sin^{-1}x$$

760. 0

Rewrite $\frac{1}{x}$ in the given function using a negative exponent:

$$y = \cot^{-1} x + \cot^{-1} \frac{1}{x}$$

$$= \cot^{-1} x + \cot^{-1}\left(x^{-1}\right)$$

Use the derivative formula for $\cot^{-1} x$ for the first term, and use the same derivative formula and the chain rule for the second term. Here are the calculations:

$$y' = \frac{-1}{1+x^2} + \frac{-1}{1+\left(x^{-1}\right)^2}\left(-x^{-2}\right)$$

$$= \frac{-1}{1+x^2} + \frac{1}{1+\left(\frac{1}{x}\right)^2}\left(\frac{1}{x^2}\right)$$

$$= \frac{-1}{1+x^2} + \frac{1}{x^2\left(1+\frac{1}{x^2}\right)}$$

$$= \frac{-1}{1+x^2} + \frac{1}{x^2+1}$$

$$= 0$$

761. $\dfrac{2}{\left(1+x^2\right)^2}$

Here's the given function:

$$y = \tan^{-1} x + \frac{x}{1+x^2}$$

Begin by using the derivative formula for $\tan^{-1} x$ on the first term and using the quotient rule on the second term:

$$y' = \frac{1}{1+x^2} + \frac{\left(1+x^2\right)(1) - x(2x)}{\left(1+x^2\right)^2}$$

$$= \frac{1}{1+x^2} + \frac{1-x^2}{\left(1+x^2\right)^2}$$

Then get common denominators and simplify:

$$y' = \frac{1+x^2}{\left(1+x^2\right)^2} + \frac{1-x^2}{\left(1+x^2\right)^2} = \frac{2}{\left(1+x^2\right)^2}$$

762.
$$\frac{1}{2(1+x^2)}$$

Rewrite the radical in the given function using an exponent:

$$y = \tan^{-1}\left(x - \sqrt{1+x^2}\right)$$

$$= \tan^{-1}\left(x - \left(1+x^2\right)^{1/2}\right)$$

Use the derivative formula for $\tan^{-1}x$ along with the chain rule:

$$y' = \frac{1}{1 + \left(x - \sqrt{1+x^2}\right)^2}\left(1 - \frac{1}{2}\left(1+x^2\right)^{-1/2}(2x)\right)$$

$$= \left(\frac{1}{1 + x^2 - 2x\sqrt{1+x^2} + 1 + x^2}\right)\left(1 - \frac{x}{\sqrt{1+x^2}}\right)$$

Now the fun algebra simplification begins:

$$y' = \left(\frac{1}{1 + x^2 - 2x\sqrt{1+x^2} + 1 + x^2}\right)\left(1 - \frac{x}{\sqrt{1+x^2}}\right)$$

$$= \left(\frac{1}{2 + 2x^2 - 2x\sqrt{1+x^2}}\right)\left(\frac{\sqrt{1+x^2} - x}{\sqrt{1+x^2}}\right)$$

$$= \frac{\sqrt{1+x^2} - x}{2\sqrt{1+x^2}\left(1 + x^2 - x\sqrt{1+x^2}\right)}$$

$$= \frac{\sqrt{1+x^2} - x}{2\left(\sqrt{1+x^2}\left(1+x^2\right) - x\left(1+x^2\right)\right)}$$

$$= \frac{\sqrt{1+x^2} - x}{2\left(1+x^2\right)\left(\sqrt{1+x^2} - x\right)}$$

$$= \frac{1}{2(1+x^2)}$$

763.
$$\frac{\pi}{2}$$

Here's the initial problem:

$$\int_0^1 \frac{2}{x^2+1}\,dx$$

Because you have the antiderivative formula $\int \frac{1}{x^2+a^2}\,dx = \frac{1}{a}\tan^{-1}\left(\frac{x}{a}\right) + C$, you can simply apply that here to find the solution:

$$\int_0^1 \frac{2}{x^2+1}\,dx = 2\tan^{-1}x\Big|_0^1$$
$$= 2\left(\tan^{-1}1 - \tan^{-1}0\right)$$
$$= 2\left(\frac{\pi}{4} - 0\right)$$
$$= \frac{\pi}{2}$$

764. $\quad \dfrac{5\pi}{6}$

Here's the given problem:

$$\int_{1/2}^{\sqrt{3}/2} \frac{5}{\sqrt{1-x^2}}\,dx$$

Because you have the antiderivative formula $\int \dfrac{1}{\sqrt{1-x^2}}\,dx = \sin^{-1}x + C$, you can simply apply that here to find the solution:

$$\int_{1/2}^{\sqrt{3}/2} \frac{5}{\sqrt{1-x^2}}\,dx = 5\left(\sin^{-1}x\right)\Big|_{1/2}^{\sqrt{3}/2}$$
$$= 5\left(\sin^{-1}\frac{\sqrt{3}}{2} - \sin^{-1}\frac{1}{2}\right)$$
$$= 5\left(\frac{\pi}{3} - \frac{\pi}{6}\right)$$
$$= \frac{5\pi}{6}$$

765. $\quad \dfrac{1}{3}\sin^{-1}(3x) + C$

You may want to write the second term underneath the radical as a quantity squared so you can more easily see which substitution to use:

$$\int \frac{dx}{\sqrt{1-9x^2}} = \int \frac{dx}{\sqrt{1-(3x)^2}}$$

Now begin by using the substitution $u = 3x$ so that $du = 3\,dx$, or $\dfrac{1}{3}du = dx$. This gives you the following:

$$\int \frac{dx}{\sqrt{1-(3x)^2}} = \frac{1}{3}\int \frac{1}{\sqrt{1-u^2}}\,du$$
$$= \frac{1}{3}\sin^{-1}(u) + C$$
$$= \frac{1}{3}\sin^{-1}(3x) + C$$

766. $\dfrac{\pi^2}{72}$

Here's the given problem:

$$\int_0^{1/2} \frac{\sin^{-1} x}{\sqrt{1-x^2}}\, dx$$

Begin by using the substitution $u = \sin^{-1} x$ so that $du = \dfrac{1}{\sqrt{1-x^2}}\, dx$. Find the new limits of integration by noting that if $x = \dfrac{1}{2}$, then $u = \sin^{-1}\dfrac{1}{2} = \dfrac{\pi}{6}$, and if $x = 0$, then $u = \sin^{-1} 0 = 0$. With these new values, you get the following answer:

$$\int_0^{1/2} \frac{\sin^{-1} x}{\sqrt{1-x^2}}\, dx = \int_0^{\pi/6} u\, du$$

$$= \frac{u^2}{2}\Big|_0^{\pi/6}$$

$$= \frac{1}{2}\left(\left(\frac{\pi}{6}\right)^2 - (0)^2\right)$$

$$= \frac{\pi^2}{72}$$

767. $\dfrac{\pi}{4}$

Here's the given problem:

$$\int_0^{\pi/2} \frac{\cos x}{1+\sin^2 x}\, dx$$

Begin with the substitution $u = \sin x$ so that $du = \cos x\, dx$. Notice that if $x = \dfrac{\pi}{2}$, then $u = \sin\dfrac{\pi}{2} = 1$, and that if $x = 0$, then $u = \sin 0 = 0$. With these new values, you get the following:

$$\int_0^{\pi/2} \frac{\cos x}{1+\sin^2 x}\, dx = \int_0^1 \frac{1}{1+u^2}\, du$$

$$= \tan^{-1} u\Big|_0^1$$

$$= \tan^{-1} 1 - \tan^{-1} 0$$

$$= \frac{\pi}{4} - 0$$

$$= \frac{\pi}{4}$$

768. $\tan^{-1}(\ln x) + C$

Start with the given problem:

$$\int \frac{1}{x\left(1+(\ln x)^2\right)}\, dx$$

Use the substitution $u = \ln x$ so that $du = \frac{1}{x}\,dx$:

$$\int \frac{1}{x\left(1+\left(\ln x\right)^2\right)}\,dx = \int \frac{1}{1+u^2}\,du$$

$$= \tan^{-1} u + C$$

$$= \tan^{-1}\left(\ln x\right) + C$$

769. $2\tan^{-1}\sqrt{x} + C$

Here's the given problem:

$$\int \frac{dx}{\sqrt{x}\left(1+x\right)}$$

Begin with the substitution $u = \sqrt{x}$ so that $u^2 = x$ and $2u\,du = dx$. Using these substitutions gives you the following:

$$\int \frac{dx}{\sqrt{x}\left(1+x\right)} = \int \frac{2u}{u\left(1+u^2\right)}\,du$$

$$= 2\int \frac{1}{1+u^2}\,du$$

$$= 2\tan^{-1} u + C$$

$$= 2\tan^{-1}\sqrt{x} + C$$

770. $\frac{1}{5}\sec^{-1}\left|\frac{x}{5}\right| + C$

Here's the given problem:

$$\int \frac{1}{x\sqrt{x^2-25}}\,dx$$

Begin by doing a bit of algebra to manipulate the denominator of the integrand:

$$\int \frac{1}{x\sqrt{x^2-25}}\,dx = \int \frac{1}{x\sqrt{25\left(\frac{x^2}{25}-1\right)}}\,dx$$

$$= \int \frac{1}{5x\sqrt{\left(\frac{x}{5}\right)^2-1}}\,dx$$

Now use the substitution $u = \frac{x}{5}$ so that $du = \frac{1}{5}\,dx$, or $5\,du = dx$:

$$\int \frac{1}{5x\sqrt{\left(\frac{x}{5}\right)^2-1}}\,dx = \int \frac{5\,du}{5\left(5u\right)\sqrt{u^2-1}}$$

$$= \frac{1}{5}\int \frac{1}{u\sqrt{u^2-1}}\,du$$

$$= \frac{1}{5}\sec^{-1}|u| + C$$

$$= \frac{1}{5}\sec^{-1}\left|\frac{x}{5}\right| + C$$

771. $\sec^{-1}|x-1|+C$

Here's the given problem:

$$\int \frac{1}{(x-1)\sqrt{x^2-2x}}\,dx$$

Begin by completing the square on the expression underneath the square root.

$$x^2-2x=\left(x^2-2x+1\right)-1$$
$$=(x-1)^2-1$$

This gives you

$$\int \frac{1}{(x-1)\sqrt{x^2-2x}}\,dx = \int \frac{1}{(x-1)\sqrt{(x-1)^2-1}}\,dx$$

Now use the substitution $u = x - 1$ so that $du = dx$:

$$\int \frac{1}{(x-1)\sqrt{(x-1)^2-1}}\,dx = \int \frac{du}{u\sqrt{u^2-1}}$$
$$=\sec^{-1}|u|+C$$
$$=\sec^{-1}|x-1|+C$$

772. $\frac{1}{3}\sin^{-1}\left(e^{3x}\right)+C$

The given problem is

$$\int \frac{e^{3x}}{\sqrt{1-e^{6x}}}\,dx$$

You'd like to have "1 – (a quantity squared)" under the radical so that you can use the formula $\int \frac{1}{\sqrt{1-x^2}}\,dx = \sin^{-1}x+C$, so you may want to write the second term under the radical as a quantity squared and try a substitution to see whether it all works out. (Of course, you can use the substitution without rewriting, but rewriting may help you see which substitution to use.)

$$\int \frac{e^{3x}}{\sqrt{1-e^{6x}}}\,dx = \int \frac{e^{3x}}{\sqrt{1-\left(e^{3x}\right)^2}}\,dx$$

Now you can use the substitution $u = e^{3x}$ so that $du = 3e^{3x}\,dx$, or $\frac{1}{3}du = e^{3x}dx$:

$$\int \frac{e^{3x}}{\sqrt{1-\left(e^{3x}\right)^2}}\,dx = \frac{1}{3}\int \frac{1}{\sqrt{1-u^2}}\,du$$
$$=\frac{1}{3}\sin^{-1}u+C$$
$$=\frac{1}{3}\sin^{-1}(e^{3x})+C$$

773.

$$\frac{1}{2}\ln\left|x^2+4\right|+2\tan^{-1}\left(\frac{x}{2}\right)+C$$

Begin by splitting the integral:

$$\int\frac{x+4}{x^2+4}\,dx=\int\frac{x}{x^2+4}\,dx+\int\frac{4}{x^2+4}\,dx$$

For the first integral, you can use the substitution $u=x^2+4$ so that $du=2x\,dx$, or $\frac{1}{2}du=x\,dx$. For the second integral, simply use the $\tan^{-1}x$ formula for integration, which gives you the following:

$$\frac{1}{2}\int\frac{1}{u}\,du+4\int\frac{1}{x^2+2^2}\,dx$$

$$=\frac{1}{2}\ln|u|+4\left(\frac{1}{2}\tan^{-1}\frac{x}{2}\right)+C$$

$$=\frac{1}{2}\ln\left|x^2+4\right|+2\tan^{-1}\left(\frac{x}{2}\right)+C$$

774.

$$4-2\sqrt{3}+\frac{\pi}{6}$$

Here's the given problem:

$$\int_2^3\frac{2x-3}{\sqrt{4x-x^2}}\,dx$$

Begin by completing the square on the expression under the radical:

$$4x-x^2=-x^2+4x$$

$$=-\left(x^2-4x\right)$$

$$=-\left(x^2-4x+4\right)+4$$

$$=-\left(x-2\right)^2+4$$

$$=4-\left(x-2\right)^2$$

This gives you

$$\int_2^3\frac{2x-3}{\sqrt{4x-x^2}}\,dx=\int_2^3\frac{2x-3}{\sqrt{4-\left(x-2\right)^2}}\,dx$$

Now use the substitution $u=x-2$ so that $u+2=x$ and $du=dx$. Notice you can find the new limits of integration by noting that if $x=3$, then $u=3-2=1$, and if $x=2$, then $u=2-2=0$.

$$\int_2^3\frac{2x-3}{\sqrt{4-\left(x-2\right)^2}}\,dx=\int_0^1\frac{2\left(u+2\right)-3}{\sqrt{4-u^2}}\,du$$

Next, use some algebra to manipulate the expression under the radical:

$$\int_0^1 \frac{2(u+2)-3}{\sqrt{4-u^2}}\,du = \int_0^1 \frac{2u+1}{\sqrt{4\left(1-\dfrac{u^2}{4}\right)}}\,du$$

$$= \int_0^1 \frac{2u+1}{2\sqrt{1-\left(\dfrac{u}{2}\right)^2}}\,du$$

Now you can use the substitution $w = \frac{u}{2}$ so that $2w = u$ and $2\,dw = du$. You can again find the new upper limit of integration by using the substitution $w = \frac{u}{2}$ and noting that when $u = 1$, you have $w = \frac{1}{2}$. Likewise, you can find the new lower limit of integration by noting that when $u = 0$, you have $w = 0$.

$$\int_0^1 \frac{2u+1}{2\sqrt{1-\left(\dfrac{u}{2}\right)^2}}\,du = 2\int_0^{1/2} \frac{2(2w)+1}{2\sqrt{1-w^2}}\,dw$$

$$= \int_0^{1/2} \frac{4w+1}{\sqrt{1-w^2}}\,dw$$

Now split the integral:

$$\int_0^{1/2} \frac{4w+1}{\sqrt{1-w^2}}\,dw$$

$$= \int_0^{1/2} \frac{4w}{\sqrt{1-w^2}}\,dw + \int_0^{1/2} \frac{1}{\sqrt{1-w^2}}\,dw$$

In the first integral, use the substitution $v = 1 - w^2$ so that $dv = -2w\,dw$ and $-2\,dv = 4w\,dw$. You can again update the limits of integration by using the substitution $v = 1 - w^2$. To find the new upper limit of integration, note that when $w = \frac{1}{2}$, you have $v = 1 - \left(\frac{1}{2}\right)^2 = \frac{3}{4}$. To find the new lower limit of integration, note that when $w = 0$, you have $v = 1 - 0^2 = 1$. Again, use this substitution only for the first integral:

$$\int_0^{1/2} \frac{4w}{\sqrt{1-w^2}}\,dw + \int_0^{1/2} \frac{1}{\sqrt{1-w^2}}\,dw$$

$$= \int_1^{3/4} \frac{-2}{\sqrt{v}}\,dv + \int_0^{1/2} \frac{1}{\sqrt{1-w^2}}\,dw$$

Reverse the bounds in the first integral and change the sign. Then simplify:

$$= 2\int_{3/4}^1 v^{-1/2}\,dv + \int_0^{1/2} \frac{1}{\sqrt{1-w^2}}\,dw$$

$$= 2\left(2v^{1/2}\right)\Big|_{3/4}^1 + \sin^{-1} w\Big|_0^{1/2}$$

$$= 4\sqrt{1} - 4\sqrt{\frac{3}{4}} + \sin^{-1}\frac{1}{2} - \sin^{-1}0$$

$$= 4 - 2\sqrt{3} + \frac{\pi}{6}$$

Whew!

775.   0

Using the definition of hyperbolic sine gives you the following:

$$\sinh 0 = \frac{e^0 - e^{-0}}{2} = \frac{1-1}{2} = 0$$

776. $\frac{5}{4}$

Using the definition of hyperbolic cosine gives you the following:

$$\cosh(\ln 2) = \frac{e^{\ln 2} + e^{-\ln 2}}{2}$$

$$= \frac{e^{\ln 2} + \left(e^{\ln 2}\right)^{-1}}{2}$$

$$= \frac{2 + \frac{1}{2}}{2} = \frac{4+1}{4} = \frac{5}{4}$$

777. $\frac{37}{35}$

Using the definition of hyperbolic cotangent gives you the following:

$$\coth(\ln 6) = \frac{\cosh(\ln 6)}{\sinh(\ln 6)}$$

$$= \frac{\dfrac{e^{\ln 6} + e^{-\ln 6}}{2}}{\dfrac{e^{\ln 6} - e^{-\ln 6}}{2}}$$

$$= \frac{e^{\ln 6} + \left(e^{\ln 6}\right)^{-1}}{e^{\ln 6} - \left(e^{\ln 6}\right)^{-1}}$$

$$= \frac{6 + 6^{-1}}{6 - 6^{-1}} = \frac{6 + \frac{1}{6}}{6 - \frac{1}{6}}$$

$$= \frac{36+1}{36-1} = \frac{37}{35}$$

778. $\dfrac{e^2 - 1}{e^2 + 1}$

Using the definition of hyperbolic tangent gives you the following:

$$\tanh 1 = \frac{\sinh 1}{\cosh 1}$$

$$= \frac{\dfrac{e^1 - e^{-1}}{2}}{\dfrac{e^1 + e^{-1}}{2}} = \frac{e - \dfrac{1}{e}}{e + \dfrac{1}{e}} = \frac{e^2 - 1}{e^2 + 1}$$

779. $\dfrac{2\sqrt{e}}{e+1}$

Using the definition of hyperbolic cosine gives you the following:

$$\operatorname{sech}\frac{1}{2} = \frac{1}{\cosh\frac{1}{2}}$$

$$= \frac{2}{e^{1/2} + e^{-1/2}}$$

$$= \frac{2e^{1/2}}{e^{1} + e^{0}}$$

$$= \frac{2\sqrt{e}}{e+1}$$

780. $2\cosh x \sinh x$

The given function is $y = \cosh^2 x$, which equals $(\cosh x)^2$. Using the chain rule gives you the derivative $y' = 2\cosh x \sinh x$.

781. $2x\cosh\left(x^2\right)$

The given function is $y = \sinh\left(x^2\right)$. Using the chain rule gives you

$$y' = \left(\cosh\left(x^2\right)\right)(2x)$$

$$= 2x\cosh\left(x^2\right)$$

782. $-\dfrac{1}{3}\operatorname{csch}(2x)\coth(2x)$

Here's the given function:

$$y = \frac{1}{6}\operatorname{csch}(2x)$$

Use the chain rule to find the derivative:

$$y' = \frac{1}{6}\left(-\operatorname{csch}(2x)\coth(2x)\right)(2)$$

$$= -\frac{1}{3}\operatorname{csch}(2x)\coth(2x)$$

783. $e^x \operatorname{sech}^2\left(e^x\right)$

Here's the given function:

$$y = \tanh\left(e^x\right)$$

Use the chain rule to find the derivative:

$$y' = \left[\operatorname{sech}^2\left(e^x\right)\right]e^x$$

$$= e^x \operatorname{sech}^2\left(e^x\right)$$

784. $\cosh x \operatorname{sech}^2(\sinh x)$

The given function is

$$y = \tanh(\sinh x)$$

Use the chain rule to get the derivative:

$$y' = \left(\operatorname{sech}^2(\sinh x)\right)\cosh x$$
$$= \cosh x \operatorname{sech}^2(\sinh x)$$

785. $5\cosh(5x)e^{\sinh(5x)}$

Here's the given function:

$$y = e^{\sinh(5x)}$$

Use the chain rule to find the derivative:

$$y' = e^{\sinh(5x)}\left(\cosh(5x)\right)(5)$$
$$= 5\cosh(5x)e^{\sinh(5x)}$$

786. $-40\operatorname{sech}^4(10x)\tanh(10x)$

Rewrite the given function:

$$y = \operatorname{sech}^4(10x) = \left[\operatorname{sech}(10x)\right]^4$$

Then use the chain rule to find the derivative:

$$y' = 4\left[\operatorname{sech}(10x)\right]^3\left(-\operatorname{sech}(10x)\tanh(10x)\right)(10)$$
$$= -40\operatorname{sech}^4(10x)\tanh(10x)$$

787. $\dfrac{2t^3 \operatorname{sech}^2\sqrt{1+t^4}}{\sqrt{1+t^4}}$

Here's the given function:

$$y = \tanh\left(\sqrt{1+t^4}\right)$$

Use the chain rule to find the derivative:

$$y' = \left(\operatorname{sech}^2\left(1+t^4\right)^{1/2}\right)\left(\tfrac{1}{2}\left(1+t^4\right)^{-1/2}\right)\left(4t^3\right)$$
$$= \frac{2t^3 \operatorname{sech}^2\sqrt{1+t^4}}{\sqrt{1+t^4}}$$

788.
$2x^3 - x$

Before taking the derivative, simplify the function:

$$y = x^3 \sinh(\ln x)$$

$$= x^3 \left(\frac{e^{\ln x} - e^{-\ln x}}{2} \right)$$

$$= x^3 \left(\frac{e^{\ln x} - \left(e^{\ln x}\right)^{-1}}{2} \right)$$

$$= x^3 \left(\frac{x - x^{-1}}{2} \right)$$

$$= \frac{1}{2}x^4 - \frac{1}{2}x^2$$

Now simply take the derivative using the power rule:

$$y' = \frac{1}{2}\left(4x^3\right) - \frac{1}{2}(2x)$$

$$= 2x^3 - x$$

789.
$\frac{1}{3}\operatorname{csch}\frac{x}{3}\operatorname{sech}\frac{x}{3}$

The given function is

$$y = \ln\left(\tanh\left(\frac{x}{3}\right) \right)$$

Use the chain rule to find the derivative:

$$y' = \frac{1}{\tanh\left(\frac{x}{3}\right)}\left(\operatorname{sech}^2 \frac{x}{3} \right)\left(\frac{1}{3}\right)$$

Rewrite and simplify the equation to get the answer:

$$y' = \frac{1}{3}\left(\frac{\cosh\frac{x}{3}}{\sinh\frac{x}{3}} \right)\frac{1}{\cosh^2\frac{x}{3}} = \frac{1}{3}\left(\frac{1}{\sinh\frac{x}{3}} \right)\frac{1}{\cosh\frac{x}{3}}$$

$$= \frac{1}{3}\operatorname{csch}\frac{x}{3}\operatorname{sech}\frac{x}{3}$$

790.
$-\frac{1}{3}\cosh(1-3x)+C$

Here's the given problem:

$$\int \sinh(1-3x)\,dx$$

To find the antiderivative, use the substitution $u = 1 - 3x$ so that $du = -3\,dx$, or $-\frac{1}{3}du = dx$:

$$\int \sinh(1-3x)\,dx = -\frac{1}{3}\int \sinh(u)\,du$$

$$= -\frac{1}{3}\cosh u + C$$

$$= -\frac{1}{3}\cosh(1-3x) + C$$

791. $\frac{1}{3}\cosh^3(x-3)+C$

Here's the given problem:

$$\int \cosh^2(x-3)\sinh(x-3)\,dx$$

To find the antiderivative, begin by using the substitution $u = \cosh(x-3)$ so that $du = \sinh(x-3)\,dx$. With this substitution, you get the following:

$$\int \cosh^2(x-3)\sinh(x-3)\,dx$$

$$= \int u^2\,du$$

$$= \frac{u^3}{3}+C$$

$$= \frac{1}{3}\cosh^3(x-3)+C$$

792. $\ln|\sinh x|+C$

Rewrite the problem using the definition of hyperbolic cotangent:

$$\int \coth x\,dx = \int \frac{\cosh x}{\sinh x}\,dx$$

Begin finding the antiderivative by using the substitution $u = \sinh x$ so that $du = \cosh x\,dx$. With these substitutions, you get the following:

$$\int \frac{1}{u}\,du = \ln|u|+C = \ln|\sinh x|+C$$

793. $\frac{1}{3}\tanh(3x-2)+C$

Here's the given problem:

$$\int \operatorname{sech}^2(3x-2)\,dx$$

To find the antiderivative, use the substitution $u = 3x - 2$ to get $du = 3\,dx$ so that $\frac{1}{3}du = dx$. With these substitutions, you get the following:

$$\frac{1}{3}\int \operatorname{sech}^2(u)\,du = \frac{1}{3}\tanh(u)+C$$

$$= \frac{1}{3}\tanh(3x-2)+C$$

---

794. $\ln|2+\tanh x|+C$

The given problem is

$$\int \frac{\operatorname{sech}^2 x}{2+\tanh x}dx$$

Begin by using the substitution $u = 2 + \tanh x$ so that $du = \operatorname{sech}^2 x\,dx$:

$$\int \frac{1}{u}du = \ln|u|+C$$

$$= \ln|2+\tanh x|+C$$

---

795. $\frac{1}{6}x\sinh(6x)-\frac{1}{36}\cosh(6x)+C$

Here's the given problem:

$$\int x\cosh(6x)dx$$

To find the antiderivative, use integration by parts with $u = x$ so that $du = dx$, and let $dv = \cosh(6x)\,dx$ so that $v = \frac{1}{6}\sinh(6x)$. This gives you the following:

$$\int x\cosh(6x)dx$$

$$= x\left(\frac{1}{6}\sinh(6x)\right)-\int \frac{1}{6}\sinh(6x)dx$$

$$= \frac{1}{6}x\sinh(6x)-\frac{1}{36}\cosh(6x)+C$$

---

796. $-\frac{5}{2}\coth\left(\frac{x^2}{5}\right)+C$

Here's the given problem:

$$\int x\operatorname{csch}^2\left(\frac{x^2}{5}\right)dx$$

To find the antiderivative, begin by using the substitution $u = \frac{x^2}{5}$ so that $du = \frac{2}{5}x\,dx$, or $\frac{5}{2}du = x\,dx$. With these substitutions, you get the following:

$$\frac{5}{2}\int \operatorname{csch}^2(u)\,du = \frac{5}{2}(-\coth(u))+C$$

$$= -\frac{5}{2}\coth\left(\frac{x^2}{5}\right)+C$$

---

797. $\frac{1}{2}\operatorname{csch}\left(\frac{1}{x^2}\right)+C$

Start with the given problem:

$$\int \frac{\operatorname{csch}\left(\frac{1}{x^2}\right)\coth\left(\frac{1}{x^2}\right)}{x^3}dx$$

To find the antiderivative, begin by using the substitution $u = \frac{1}{x^2}$ so that $du = -2x^{-3}$, or $-\frac{1}{2}du = \frac{1}{x^3}dx$. With these substitutions, you get the following:

$$-\frac{1}{2}\int \operatorname{csch}(u)\coth(u)\,du$$

$$= -\frac{1}{2}(-\operatorname{csch}(u)) + C$$

$$= \frac{1}{2}\operatorname{csch}\left(\frac{1}{x^2}\right) + C$$

---

798. $\dfrac{7}{12}$

The given problem is

$$\int_{\ln 2}^{\ln 3}\frac{\cosh x}{\cosh^2 x - 1}\,dx$$

To find the antiderivative, begin by using the hyperbolic identity $\cosh^2 x - 1 = \sinh^2 x$:

$$\int_{\ln 2}^{\ln 3}\frac{\cosh x}{\cosh^2 x - 1}\,dx$$

$$= \int_{\ln 2}^{\ln 3}\frac{\cosh x}{\sinh^2 x}\,dx$$

$$= \int_{\ln 2}^{\ln 3}\coth x \operatorname{csch} x\,dx$$

$$= -\operatorname{csch} x\Big|_{\ln 2}^{\ln 3}$$

$$= -\left(\operatorname{csch}(\ln 3) - \operatorname{csch}(\ln 2)\right)$$

$$= -\left(\frac{2}{e^{\ln 3} - e^{-\ln 3}} - \frac{2}{e^{\ln 2} - e^{-\ln 2}}\right)$$

$$= -\left(\frac{2}{3 - \frac{1}{3}} - \frac{2}{2 - \frac{1}{2}}\right)$$

$$= -\left(\frac{6}{9-1} - \frac{4}{4-1}\right) = -\left(\frac{3}{4} - \frac{4}{3}\right) = \frac{7}{12}$$

---

799. $\sin^{-1}\left(\dfrac{1}{4}\right)$

Here's the given problem:

$$\int_0^{\ln 2}\frac{\cosh x}{\sqrt{9 - \sinh^2 x}}\,dx$$

To find the antiderivative, begin by using the substitution $u = \sinh x$ so that $du = \cosh x\,dx$. You can find the new limits of integration by noting that if $x = \ln 2$, then

$$u = \sinh(\ln 2) = \frac{e^{\ln 2} - e^{-\ln 2}}{2} = \frac{2 - \frac{1}{2}}{2} = \frac{3}{4}, \text{ and if } x = 0, \text{ then } u = \sinh(0) = \frac{e^0 - e^0}{2} = 0. \text{ With}$$

these substitutions, you produce the following:

$$\int_0^{3/4} \frac{1}{\sqrt{9-u^2}}\,du = \int_0^{3/4} \frac{1}{\sqrt{9\left(1-\dfrac{u^2}{9}\right)}}\,du$$

$$= \int_0^{3/4} \frac{1}{3\sqrt{1-\left(\dfrac{u}{3}\right)^2}}\,du$$

Now use a new substitution $w = \dfrac{u}{3}$ so that $dw = \dfrac{1}{3}\,du$, or $3\,dw = du$. You can again find the new limits of integration by noting that if $u = \dfrac{3}{4}$, then $w = \dfrac{3/4}{3} = \dfrac{1}{4}$, and if $u = 0$, then $w = \dfrac{0}{3} = 0$. With these substitutions, you get the following:

$$\int_0^{1/4} \frac{3\,dw}{3\sqrt{1-w^2}} = \sin^{-1} w \Big|_0^{1/4}$$

$$= \sin^{-1}\left(\frac{1}{4}\right) - \sin^{-1} 0$$

$$= \sin^{-1}\left(\frac{1}{4}\right)$$

---

800. 3

The given problem is

$$\lim_{x \to -1} \frac{x^3+1}{x+1}$$

Notice that if you substitute in the value $x = -1$, you get the indeterminate form $\dfrac{0}{0}$. Now use L'Hôpital's rule to find the limit:

$$\lim_{x \to -1} \frac{x^3+1}{x+1} = \lim_{x \to -1} \frac{3x^2}{1} = 3(-1)^2 = 3$$

---

801. $\dfrac{4}{3}$

Here's the given problem:

$$\lim_{x \to 1} \frac{x^4-1}{x^3-1}$$

Notice that if you substitute in the value $x = 1$, you get the indeterminate form $\dfrac{0}{0}$. Using L'Hôpital's rule gives you the limit:

$$\lim_{x \to 1} \frac{x^4-1}{x^3-1} = \lim_{x \to 1} \frac{4x^3}{3x^2} = \lim_{x \to 1} \frac{4x}{3} = \frac{4(1)}{3} = \frac{4}{3}$$

---

802. $\dfrac{1}{5}$

Start with the given problem:

$$\lim_{x \to 2} \frac{x-2}{x^2+x-6}$$

Notice that if you substitute in the value $x = 2$, you get the indeterminate form $\frac{0}{0}$. Use L'Hôpital's rule to find the limit:

$$\lim_{x \to 2} \frac{x-2}{x^2+x-6} = \lim_{x \to 2} \frac{1}{2x+1} = \frac{1}{2(2)+1} = \frac{1}{5}$$

803. ∞

The given problem is

$$\lim_{x \to \left(\frac{\pi}{2}\right)^-} \frac{\cos x}{1 - \sin x}$$

Notice that if you substitute in the value $x = \frac{\pi}{2}$, you get the indeterminate form $\frac{0}{0}$. Using L'Hôpital's rule gives you the limit:

$$\lim_{x \to \left(\frac{\pi}{2}\right)^-} \frac{\cos x}{1 - \sin x} = \lim_{x \to \left(\frac{\pi}{2}\right)^-} \frac{-\sin x}{-\cos x} = \lim_{x \to \left(\frac{\pi}{2}\right)^-} \tan x = \infty$$

804. 0

Start with the given problem:

$$\lim_{x \to 0} \frac{1 - \cos x}{\tan x}$$

Notice that if you substitute in the value $x = 0$, you get the indeterminate form $\frac{0}{0}$. Use L'Hôpital's rule to find the limit:

$$\lim_{x \to 0} \frac{1 - \cos x}{\tan x} = \lim_{x \to 0} \frac{\sin x}{\sec^2 x} = \frac{\sin 0}{\sec^2 0} = \frac{0}{1} = 0$$

805. 0

Here's the given problem:

$$\lim_{x \to \infty} \frac{\ln x}{x^2}$$

Notice that if you take the limit, you get the indeterminate form $\frac{\infty}{\infty}$. Using L'Hôpital's rule gives you the limit as follows:

$$\lim_{x \to \infty} \frac{1/x}{2x} = \lim_{x \to \infty} \frac{1}{2x^2} = 0$$

806. $-\frac{2}{\pi}$

Here's the given problem:

$$\lim_{x \to 1} \frac{\ln x}{\cos\left(\frac{\pi}{2} x\right)}$$

Notice that if you substitute in the value $x = 1$, you get the indeterminate form $\frac{0}{0}$. Using L'Hôpital's rule gives you

$$\lim_{x \to 1} \frac{\ln x}{\cos\left(\frac{\pi}{2}x\right)} = \lim_{x \to 1} \frac{\frac{1}{x}}{\left(-\sin\left(\frac{\pi}{2}x\right)\right)\frac{\pi}{2}} = \frac{\frac{1}{1}}{-\frac{\pi}{2}\sin\left(\frac{\pi}{2}\right)} = -\frac{2}{\pi}$$

807. 1

The given problem is

$$\lim_{x \to 0} \frac{\tan^{-1}x}{x}$$

Notice that if you substitute in the value $x = 0$, you get the indeterminate form $\frac{0}{0}$. Using L'Hôpital's rule gives you the limit:

$$\lim_{x \to 0} \frac{\tan^{-1}x}{x} = \lim_{x \to 0} \frac{\frac{1}{1+x^2}}{1} = \frac{1}{1+0^2} = 1$$

808. $-\frac{3}{2}$

Here's the given problem:

$$\lim_{x \to 2^+}\left(\frac{8}{x^2-4} - \frac{x}{x-2}\right)$$

Applying the limit gives you the indeterminate form $\infty - \infty$. Write the expression as a single fraction by getting common denominators and then apply L'Hôpital's rule:

$$\lim_{x \to 2^+}\left(\frac{8}{x^2-4} - \frac{x}{x-2}\right) = \lim_{x \to 2^+} \frac{8 - x(x+2)}{x^2-4}$$

$$= \lim_{x \to 2^+} \frac{-x^2-2x+8}{x^2-4}$$

$$= \lim_{x \to 2^+} \frac{-2x-2}{2x}$$

$$= \frac{-2(2)-2}{2(2)} = -\frac{3}{2}$$

809. 2

The given problem is

$$\lim_{x \to 0} \frac{\sin(4x)}{2x}$$

Notice that if you substitute in the value $x = 0$, you get the indeterminate form $\frac{0}{0}$. Apply L'Hôpital's rule to find the limit:

$$\lim_{x \to 0} \frac{\sin(4x)}{2x} = \lim_{x \to 0} \frac{4\cos(4x)}{2}$$
$$= 2\cos(0)$$
$$= 2$$

810.   0

Here's the given problem:

$$\lim_{x \to 0^+} x \ln x$$

Applying the limit gives you the indeterminate form $0(-\infty)$. Rewrite the product as a quotient and use L'Hôpital's rule to find the limit:

$$\lim_{x \to 0^+} \frac{\ln x}{x^{-1}} = \lim_{x \to 0^+} \frac{\frac{1}{x}}{-x^{-2}}$$
$$= \lim_{x \to 0^+} -\frac{1}{x}\left(x^2\right)$$
$$= \lim_{x \to 0^+} (-x)$$
$$= 0$$

811.   0

Here's the given problem:

$$\lim_{x \to \infty} \frac{\ln x^4}{x^3}$$

Applying the limit gives you the indeterminate form $\frac{\infty}{\infty}$. Use properties of logarithms to rewrite the expression and then use L'Hôpital's rule to find the limit:

$$\lim_{x \to \infty} \frac{4\ln x}{x^3} = \lim_{x \to \infty} \frac{\frac{4}{x}}{3x^2} = \lim_{x \to \infty} \frac{4}{3x^3} = 0$$

812.   1

The given problem is

$$\lim_{x \to 0} \frac{\sin^{-1} x}{x}$$

Substituting the value $x = 0$ gives you the indeterminate form $\frac{0}{0}$. Apply L'Hôpital's rule to find the limit:

$$\lim_{x \to 0} \frac{\frac{1}{\sqrt{1-x^2}}}{1} = \frac{1}{\sqrt{1-0^2}} = 1$$

813. 0

Here's the initial problem:

$$\lim_{x \to \infty} \frac{\tan^{-1} x - \frac{\pi}{4}}{x - 1}$$

Make sure you have an indeterminate form before applying L'Hospital's rule, or you can get incorrect results! Notice that as $x \to \infty$ in the numerator, $\left(\tan^{-1}(x) - \frac{\pi}{4}\right) \to \frac{\pi}{2} - \frac{\pi}{4} = \frac{\pi}{4}$, and as $x \to \infty$ in the denominator, $(x - 1) \to \infty$. Therefore, the solution becomes

$$\lim_{x \to \infty} \frac{\tan^{-1}(x) - \frac{\pi}{4}}{x - 1} = 0$$

814. $-\infty$

Here's the given problem:

$$\lim_{x \to \infty} \frac{\tan^{-1} x - \frac{\pi}{2}}{\frac{1}{1 + x^3}}$$

Applying the limit gives you the indeterminate form $\frac{0}{0}$. Applying L'Hôpital's rule gives you the following:

$$\lim_{x \to \infty} \frac{\tan^{-1} x - \frac{\pi}{2}}{\frac{1}{1 + x^3}} = \lim_{x \to \infty} \frac{\frac{1}{1 + x^2}}{\frac{-3x^2}{\left(1 + x^3\right)^2}}$$

$$= \lim_{x \to \infty} \frac{\left(1 + x^3\right)^2}{-3x^2 \left(1 + x^2\right)}$$

$$= -\lim_{x \to \infty} \frac{1 + 2x^3 + x^6}{3x^2 + 3x^4}$$

This limit is still an indeterminate form, and you can continue using L'Hôpital's rule four more times to arrive at the solution. However, because the degree of the numerator is larger than the degree of the denominator, you can find the limit by noting that

$$-\lim_{x \to \infty} \frac{1 + 2x^3 + x^6}{3x^2 + 3x^4} = -\lim_{x \to \infty} \frac{x^6}{3x^4} = -\lim_{x \to \infty} \frac{x^2}{3} = -\infty$$

815. 0

The given problem is

$$\lim_{x \to \frac{\pi}{2}^-} (\sec x - \tan x)$$

Applying the limit gives you the indeterminate form $\infty - \infty$. Rewrite the difference as a quotient and use L'Hôpital's rule:

$$\lim_{x \to \frac{\pi}{2}^-} (\sec x - \tan x) = \lim_{x \to \frac{\pi}{2}^-} \left(\frac{1}{\cos x} - \frac{\sin x}{\cos x} \right)$$

$$= \lim_{x \to \frac{\pi}{2}^-} \left(\frac{1 - \sin x}{\cos x} \right)$$

$$= \lim_{x \to \frac{\pi}{2}^-} \left(\frac{-\cos x}{-\sin x} \right)$$

$$= \frac{\cos \frac{\pi}{2}}{\sin \frac{\pi}{2}} = \frac{0}{1} = 0$$

816.   1

Start with the given problem:

$$\lim_{x \to \infty} x \sin\left(\frac{1}{x} \right)$$

Applying the limit gives you the indeterminate form $(\infty)(0)$. Rewrite the product as a quotient and use L'Hôpital's rule as follows:

$$\lim_{x \to \infty} \frac{\sin\left(\frac{1}{x} \right)}{x^{-1}} = \lim_{x \to \infty} \frac{\cos\left(\frac{1}{x} \right) \cdot \left(-\frac{1}{x^2} \right)}{-\frac{1}{x^2}}$$

$$= \lim_{x \to \infty} \cos\left(\frac{1}{x} \right)$$

$$= \cos(0)$$

$$= 1$$

817.   0

Here's the initial problem:

$$\lim_{x \to -\infty} x^3 e^x$$

Taking the limit gives you the indeterminate form $(-\infty)(0)$. Rewrite the product as a quotient and use L'Hôpital's rule:

$$\lim_{x \to -\infty} x^3 e^x = \lim_{x \to -\infty} \frac{x^3}{e^{-x}} = \lim_{x \to -\infty} \frac{3x^2}{-e^{-x}}$$

Applying the limit again gives you the indeterminate form $-\frac{\infty}{\infty}$. Using L'Hôpital's rule again gives you the following:

$$\lim_{x \to -\infty} \frac{3x^2}{-e^{-x}} = \lim_{x \to -\infty} \frac{6x}{e^{-x}}$$

If you again take the limit, you get the indeterminate form $-\frac{\infty}{\infty}$. Use L'Hôpital's rule one more time to get the final answer:

$$\lim_{x \to -\infty} \frac{6x}{e^{-x}} = \lim_{x \to -\infty} \frac{6}{-e^{-x}} = 0$$

818.   1

The given problem is

$$\lim_{x \to \infty} \left(xe^{1/x} - x \right)$$

Applying the limit gives you the indeterminate form $\infty - \infty$. Begin by factoring out x; then rewrite the expression as a fraction and use L'Hôpital's rule:

$$\lim_{x \to \infty} x \left(e^{1/x} - 1 \right) = \lim_{x \to \infty} \frac{e^{1/x} - 1}{x^{-1}}$$

$$= \lim_{x \to \infty} \frac{\left(e^{1/x} \right) \left(-\dfrac{1}{x^2} \right)}{-\dfrac{1}{x^2}}$$

$$= \lim_{x \to \infty} e^{1/x} = e^0 = 1$$

819. ∞

The given problem is

$$\lim_{x \to \infty} \frac{e^{3x+1}}{x^2}$$

Substituting in the value $x = \infty$ gives you the indeterminate form $\frac{\infty}{\infty}$, so use L'Hôpital's rule:

$$\lim_{x \to \infty} \frac{e^{3x+1}}{x^2} = \lim_{x \to \infty} \frac{\left(e^{3x+1} \right)(3)}{2x}$$

If you again substitute in the value $x = \infty$, you still get the indeterminate form $\frac{\infty}{\infty}$. Applying L'Hôpital's rule one more time gives you the limit:

$$\lim_{x \to \infty} \frac{\left(e^{3x+1} \right)(3)}{2x} = \lim_{x \to \infty} \frac{3 \left(e^{3x+1} \right)(3)}{2} = \infty$$

820. $-\dfrac{1}{\pi^2}$

Here's the initial problem:

$$\lim_{x \to 1} \frac{1 - x + \ln x}{1 + \cos \pi x}$$

Substituting in the value $x = 1$ gives you the indeterminate form $\frac{0}{0}$, so use L'Hôpital's rule:

$$\lim_{x \to 1} \frac{1 - x + \ln x}{1 + \cos \pi x} = \lim_{x \to 1} \frac{-1 + \frac{1}{x}}{(-\sin(\pi x))(\pi)}$$

If you again substitute in the value $x = 1$, you still get the indeterminate form $\frac{0}{0}$. Using L'Hôpital's rule again gives you the limit:

$$\lim_{x \to 1} \frac{-1 + \frac{1}{x}}{(-\sin(\pi x))(\pi)} = \lim_{x \to 1} \frac{-\frac{1}{x^2}}{-\pi(\cos(\pi x))(\pi)}$$

$$= \frac{\frac{1}{1^2}}{\pi^2 \cos(\pi)}$$

$$= -\frac{1}{\pi^2}$$

821. e^{-5}

Here's the given problem:

$$\lim_{x \to 0} (1 - 5x)^{1/x}$$

Substituting in the value $x = 0$ gives you the indeterminate form 1^{∞}. Create a new limit by taking the natural logarithm of the original expression, $y = (1 - 5x)^{1/x}$, so that $\ln y = \ln(1 - 5x)^{1/x}$, or $\ln y = \frac{\ln(1 - 5x)}{x}$. Take the limit of the new expression:

$$\lim_{x \to 0} \ln y = \lim_{x \to 0} \frac{\ln(1 - 5x)}{x}$$

Then apply L'Hôpital's rule:

$$\lim_{x \to 0} \frac{\ln(1 - 5x)}{x} = \lim_{x \to 0} \frac{\frac{-5}{1 - 5x}}{1}$$

$$= \frac{-5}{1 - 5(0)}$$

$$= -5$$

Because $(1 - 5x)^{1/x} = y$ and $y = e^{\ln y}$, it follows that

$$\lim_{x \to 0} (1 - 5x)^{1/x} = \lim_{x \to 0} y = \lim_{x \to 0} e^{\ln y} = e^{\lim_{x \to 0} \ln y} = e^{-5}$$

822.   1

Here's the initial problem:

$$\lim_{x \to \infty} x^{1/x^2}$$

Applying the limit gives you the indeterminate form ∞^0. Create a new limit by taking the natural logarithm of the original expression, $y = x^{1/x^2}$, so that $\ln y = \ln\left(x^{1/x^2}\right)$, or $\ln y = \dfrac{\ln x}{x^2}$. Then take the limit of this new expression:

$$\lim_{x \to \infty} \ln y = \lim_{x \to \infty} \frac{\ln x}{x^2}$$

Apply L'Hôpital's rule:

$$\lim_{x \to \infty} \frac{\ln x}{x^2} = \lim_{x \to \infty} \frac{1/x}{2x} = \lim_{x \to \infty} \frac{1}{2x^2} = 0$$

Because $x^{1/x^2} = y$ and $y = e^{\ln y}$, it follows that

$$\lim_{x \to \infty} x^{1/x^2} = \lim_{x \to \infty} y = \lim_{x \to \infty} e^{\ln y} = e^{\lim\limits_{x \to \infty} \ln y} = e^0 = 1$$

---

823. 1

Here's the given problem:

$$\lim_{x \to 0^+} (\cos x)^{2/x}$$

Applying the limit gives you the indeterminate form 1^∞. Create a new limit by taking the natural logarithm of the original expression, $y = (\cos x)^{2/x}$, so that $\ln y = \ln(\cos x)^{2/x}$, or $\ln y = \dfrac{2\ln(\cos x)}{x}$. Then take the limit of this new expression:

$$\lim_{x \to 0^+} \ln y = \lim_{x \to 0^+} \frac{2\ln(\cos x)}{x}$$

Apply L'Hôpital's rule:

$$\begin{aligned}\lim_{x \to 0^+} \frac{2\ln(\cos x)}{x} &= \lim_{x \to 0^+} \frac{2\left(\dfrac{1}{\cos x}\right)(-\sin x)}{1}\\ &= -2\tan(0)\\ &= 0\end{aligned}$$

Because $(\cos x)^{2/x} = y$ and $y = e^{\ln y}$, it follows that

$$\lim_{x \to 0^+} (\cos x)^{2/x} = \lim_{x \to 0^+} y = \lim_{x \to 0^+} e^{\ln y} = e^{\lim\limits_{x \to 0^+} \ln y} = e^0 = 1$$

---

824. $e^{-15/9}$

The given problem is

$$\lim_{x \to \infty} \left(\frac{3x-1}{3x+4}\right)^{x-1}$$

Applying the limit gives you the indeterminate form 1^∞. Create a new limit by taking the natural logarithm of the original expression, $y = \left(\dfrac{3x-1}{3x+4}\right)^{x-1}$, so that $\ln y = \ln\left(\dfrac{3x-1}{3x+4}\right)^{x-1}$, or $\ln y = (x-1)\left(\ln\dfrac{3x-1}{3x+4}\right)$. Next, rewrite the expression as a fraction and use properties of logarithms to expand the natural logarithm (this step will make finding the derivative a bit easier):

$$\lim_{x\to\infty}\ln y = \lim_{x\to\infty}\frac{\left(\ln\dfrac{3x-1}{3x+4}\right)}{(x-1)^{-1}} = \lim_{x\to\infty}\frac{\ln(3x-1)-\ln(3x+4)}{(x-1)^{-1}}$$

Apply L'Hôpital's rule:

$$\lim_{x\to\infty}\frac{\ln(3x-1)-\ln(3x+4)}{(x-1)^{-1}}$$

$$=\lim_{x\to\infty}\frac{\dfrac{3}{3x-1}-\dfrac{3}{3x+4}}{-(x-1)^{-2}}$$

$$=-\lim_{x\to\infty}(x-1)^2\left(\frac{15}{(3x-1)(3x+4)}\right)$$

$$=-\lim_{x\to\infty}\frac{15x^2-30x+15}{9x^2+9x-4}$$

$$=-\frac{15}{9}$$

Because $\left(\dfrac{3x-1}{3x+4}\right)^{x-1}=y$ and $y=e^{\ln y}$, it follows that

$$\lim_{x\to\infty}\left(\frac{3x-1}{3x+4}\right)^{x-1}=\lim_{x\to\infty}y=\lim_{x\to\infty}e^{\ln y}=e^{\lim\limits_{x\to\infty}\ln y}=e^{-15/9}$$

825.   1

Here's the given problem:

$$\lim_{x\to 0^+}(2x)^{x^2}$$

Substituting in the value $x = 0$ gives you the indeterminate form 0^0. Create a new limit by taking the natural logarithm of the original expression, $y=(2x)^{x^2}$, so that $\ln y = \ln\left((2x)^{x^2}\right)$, or $\ln y = x^2(\ln(2x))$. Then take the limit of the new expression:

$$\lim_{x\to 0^+}\ln y=\lim_{x\to 0^+}x^2\ln(2x)=\lim_{x\to 0^+}\frac{\ln(2x)}{x^{-2}}$$

Apply L'Hôpital's rule:

$$\lim_{x\to 0^+}\frac{\ln(2x)}{x^{-2}}=\lim_{x\to 0^+}\frac{\dfrac{1}{2x}(2)}{-2x^{-3}}=\lim_{x\to 0^+}\frac{x^2}{-2}=0$$

Because $(2x)^{x^2}=y$ and $y=e^{\ln y}$, it follows that

$$y=\lim_{x\to 0^+}(2x)^{x^2}=\lim_{x\to 0^+}y=\lim_{x\to 0^+}e^{\ln y}=e^{\lim\limits_{x\to 0^+}\ln y}=e^0=1$$

826.

1

Here's the given problem:

$$\lim_{x \to 0^+} (\tan 3x)^x$$

Substituting in the value $x = 0$ gives you the indeterminate form 0^0. Create a new limit by taking the natural logarithm of the original expression, $y = (\tan 3x)^x$, so that $\ln y = \ln(\tan 3x)^x$, or $\ln y = x \ln(\tan 3x)$. Next, take the limit of the new expression:

$$\lim_{x \to 0^+} \ln y = \lim_{x \to 0^+} x \ln(\tan 3x) = \lim_{x \to 0^+} \frac{\ln(\tan 3x)}{x^{-1}}$$

Apply L'Hôpital's rule:

$$\lim_{x \to 0^+} \frac{\ln(\tan 3x)}{x^{-1}} = \lim_{x \to 0^+} \frac{\dfrac{3 \sec^2(3x)}{\tan 3x}}{-x^{-2}}$$

Rewrite the expression and simplify:

$$\lim_{x \to 0^+} \frac{\dfrac{3 \sec^2(3x)}{\tan 3x}}{-x^{-2}} = -\lim_{x \to 0^+} (x^2) \left(\frac{3}{\cos^2(3x)} \right) \left(\frac{\cos(3x)}{\sin(3x)} \right)$$

$$= -\lim_{x \to 0^+} \left(\frac{1}{\cos(3x)} \right) \left(\frac{3x}{\sin(3x)} \right)(x)$$

$$= -(1)(1)(0) = 0$$

Because $(\tan 3x)^x = y$ and $y = e^{\ln y}$, it follows that

$$\lim_{x \to 0^+} (\tan 3x)^x = \lim_{x \to 0^+} y = \lim_{x \to 0^+} e^{\ln y} = e^{\lim_{x \to 0^+} \ln y} = e^0 = 1$$

827.

0

The given problem is

$$\lim_{x \to 0^+} \tan x \ln x$$

Applying the limit gives you the indeterminate form $0(-\infty)$. Rewrite the product as a quotient and use L'Hôpital's rule to get the limit:

$$\lim_{x \to 0^+} \tan x \ln x = \lim_{x \to 0^+} \frac{\ln x}{\cot x}$$

$$= \lim_{x \to 0^+} \frac{1/x}{-\csc^2 x}$$

$$= \lim_{x \to 0^+} \left(\frac{-\sin x}{x} \sin x \right)$$

$$= -\lim_{x \to 0^+} \left(\frac{\sin x}{x} \right) \cdot \lim_{x \to 0^+} (\sin x)$$

$$= -1 \cdot 0$$

$$= 0$$

828. e

Here's the initial problem:

$$\lim_{x \to \infty} \left(1 + \frac{1}{x}\right)^x$$

Applying the limit gives you the indeterminate form 1^∞. Create a new limit by taking the natural logarithm of the original expression, $y = \left(1 + \frac{1}{x}\right)^x$, so that $\ln y = \ln\left(1 + \frac{1}{x}\right)^x$, or $\ln y = x \ln\left(1 + \frac{1}{x}\right)$. Noting that $\lim_{x \to \infty} x \ln\left(1 + \frac{1}{x}\right)$ is an indeterminate product, you can rewrite this limit as

$$\lim_{x \to \infty} x \ln\left(1 + \frac{1}{x}\right) = \lim_{x \to \infty} \frac{\ln\left(1 + \frac{1}{x}\right)}{x^{-1}} = \lim_{x \to \infty} \frac{\ln\left(1 + x^{-1}\right)}{x^{-1}}$$

Next, apply L'Hôpital's rule:

$$\lim_{x \to \infty} \frac{\ln\left(1 + x^{-1}\right)}{x^{-1}} = \lim_{x \to \infty} \frac{\left(\frac{1}{1 + x^{-1}}\right)\left(-x^{-2}\right)}{-x^{-2}}$$

$$= \lim_{x \to \infty} \frac{1}{1 + x^{-1}}$$

$$= \lim_{x \to \infty} \frac{1}{1 + \frac{1}{x}}$$

$$= \frac{1}{1 + 0} = 1$$

Because $\left(1 + \frac{1}{x}\right)^x = y$ and $y = e^{\ln y}$, it follows that

$$\lim_{x \to \infty} \left(1 + \frac{1}{x}\right)^x = \lim_{x \to \infty} y = \lim_{x \to \infty} e^{\ln y} = e^{\lim_{x \to \infty} \ln y} e^1 = e$$

829. e^4

The given problem is

$$\lim_{x \to 0^+} \left(e^x + x\right)^{2/x}$$

Applying the limit gives you the indeterminate form 1^∞. Create a new limit by taking the natural logarithm of the original expression, $y = \left(e^x + x\right)^{2/x}$, so that $\ln y = \ln\left(e^x + x\right)^{2/x}$, or $\ln y = \frac{2\ln\left(e^x + x\right)}{x}$. Then use L'Hôpital's rule on the new limit:

$$\lim_{x \to 0^+} \ln y = \lim_{x \to 0^+} \frac{2\ln(e^x + x)}{x}$$

$$= \lim_{x \to 0^+} \frac{\frac{2}{(e^x + x)}(e^x + 1)}{1}$$

$$= \lim_{x \to 0^+} \frac{2(e^x + 1)}{(e^x + x)}$$

$$= \frac{2(e^0 + 1)}{e^0 + 0}$$

$$= \frac{2(2)}{1} = 4$$

Because $(e^x + x)^{2/x} = y$ and $y = e^{\ln y}$, it follows that

$$\lim_{x \to 0^+} (e^x + x)^{2/x} = \lim_{x \to 0^+} y = \lim_{x \to 0^+} e^{\ln y} = e^{\lim\limits_{x \to 0^+} \ln y} = e^4$$

830.

$-\infty$

Here's the given problem:

$$\lim_{x \to 1^+} \left(\frac{x^2}{x^2 - 1} - \frac{1}{\ln x} \right)$$

Applying the limit gives you the indeterminate form $\infty - \infty$, so write the expression as a single fraction by getting common denominators and then apply L'Hôpital's rule:

$$\lim_{x \to 1^+} \frac{x^2(\ln x) - x^2 + 1}{(\ln x)(x^2 - 1)} = \lim_{x \to 1^+} \frac{2x(\ln x) + x^2\left(\frac{1}{x}\right) - 2x}{\frac{1}{x}(x^2 - 1) + (\ln x)(2x)}$$

Multiply the numerator and denominator by x:

$$\lim_{x \to 1^+} \frac{2x^2(\ln x) - x^2}{x^2 - 1 + 2x^2 \ln x}$$

Notice that as $x \to 1^+$ in the numerator, $2x^2(\ln x) - x^2 \to -1$, and as $x \to 1^+$ in the denominator, $x^2 - 1 + 2x^2 \ln x \to 0^+$. Therefore, you get the following answer:

$$\lim_{x \to 1^+} \frac{2x^2 \ln x - x^2}{x^2 - 1 + 2x^2 \ln x} = -\infty$$

831.

$\frac{1}{2}$

Here's the given problem:

$$\lim_{x \to 1^+} \left(\frac{1}{\ln x} - \frac{1}{x - 1} \right)$$

Evaluating the limit gives you the indeterminate form $\infty - \infty$, so write the expression as a single fraction by getting a common denominator and then use L'Hôpital's rule:

$$\lim_{x \to 1^+}\left(\frac{x-1-\ln x}{(\ln x)(x-1)}\right) = \lim_{x \to 1^+}\left(\frac{1-\frac{1}{x}}{\frac{1}{x}(x-1)+\ln x(1)}\right)$$

Multiply the numerator and denominator by x:

$$\lim_{x \to 1^+}\frac{x-1}{x-1+x\ln x}$$

Substituting in the value $x = 1$ gives you the indeterminate form $\frac{0}{0}$, so you can again use L'Hôpital's rule:

$$\lim_{x \to 1^+}\frac{x-1}{x-1+x\ln x} = \lim_{x \to 1^+}\frac{1}{1+(1)\ln x+x\left(\frac{1}{x}\right)}$$

$$= \lim_{x \to 1^+}\frac{1}{1+\ln x+1}$$

$$= \frac{1}{1+\ln(1)+1} = \frac{1}{2}$$

832. $-\frac{1}{5}\cos(5x)+C$

Here's the given integral:

$$\int \sin(5x)\,dx$$

If you let $u = 5x$, then $du = 5dx$, or $\frac{1}{5}du = dx$. Substituting into the original integral, you get

$$\int \sin(u)\frac{1}{5}\,du = \frac{1}{5}\int \sin(u)\,du$$

$$= \frac{1}{5}\big(-\cos(u)\big)+C$$

$$= -\frac{1}{5}\cos(5x)+C$$

833. $\frac{1}{101}(x+4)^{101}+C$

Here's the given integral:

$$\int (x+4)^{100}\,dx$$

If you let $u = x + 4$, then $du = dx$. Substituting into the original integral gives you the following:

$$\int u^{100}\,du = \frac{u^{101}}{101}+C$$

$$= \frac{1}{101}(x+4)^{101}+C$$

834. $\frac{2}{3}(x^3+1)^{3/2}+C$

The given integral is

$$\int 3x^2\sqrt{x^3+1}\,dx$$

If you let $u = x^3 + 1$, then $du = 3x^2\,dx$. Substitute into the original integral and solve:

$$\int \sqrt{u}\,du = \int u^{1/2}\,du$$

$$= \frac{u^{3/2}}{3/2}+C$$

$$= \frac{2}{3}u^{3/2}+C$$

$$= \frac{2}{3}(x^3+1)^{3/2}+C$$

835. $2\tan(\sqrt{x})+C$

The given integral is

$$\int \frac{\sec^2\sqrt{x}}{\sqrt{x}}\,dx = \int \sec^2\sqrt{x}\,\frac{1}{\sqrt{x}}\,dx$$

If you let $u = \sqrt{x} = x^{1/2}$, then you get $du = \frac{1}{2}x^{-1/2}dx = \frac{1}{2\sqrt{x}}\,dx$, or equivalently,

$2du = \frac{1}{\sqrt{x}}\,dx$. Substitute into the original integral:

$$2\int \sec^2(u)\,du = 2\tan(u)+C$$

$$= 2\tan(\sqrt{x})+C$$

836. $-\dfrac{6}{35(4+5x)^7}+C$

Here's the given integral:

$$\int \frac{6}{(4+5x)^8}\,dx$$

If you let $u = 4 + 5x$, then you get $du = 5\,dx$, or $\frac{1}{5}du = dx$. Substitute into the original integral:

$$\int \frac{6}{u^8}\frac{1}{5}\,du = \frac{6}{5}\int u^{-8}\,du$$

$$= \frac{6}{5}\left(\frac{u^{-7}}{-7}\right)+C$$

$$= -\frac{6}{35u^7}+C$$

$$= -\frac{6}{35(4+5x)^7}+C$$

837. $\frac{1}{8}\sin^8\theta + C$

Start by rewriting the given integral:

$$\int \sin^7\theta\cos\theta\,d\theta = \int (\sin\theta)^7\cos\theta\,d\theta$$

If you let $u = \sin\theta$, then $du = \cos\theta\,d\theta$. Substitute into the original integral to find the answer:

$$\int u^7\,du = \frac{u^8}{8} + C$$

$$= \frac{1}{8}(\sin\theta)^8 + C$$

$$= \frac{1}{8}\sin^8\theta + C$$

838. 0

Recall that $\tan\theta$ is an odd function; that is, $\tan(-\theta) = -\tan\theta$. Likewise, $\tan^3(-\theta) = -\tan^3\theta$ so that $\tan^3(\theta)$ is also an odd function. Because this function is symmetric about the origin and you're integrating over an interval of the form $[-a, a]$, you get $\int_{-\pi/3}^{\pi/3}\tan^3\theta\,d\theta = 0$.

Here's an alternate approach: Begin by saving a factor of the tangent and use an identity to allow you to split up the integral:

$$\int_{-\pi/3}^{\pi/3}\tan^3\theta\,d\theta = \int_{-\pi/3}^{\pi/3}\tan^2\theta\tan\theta\,d\theta$$

$$= \int_{-\pi/3}^{\pi/3}(1+\sec^2\theta)\tan\theta\,d\theta$$

$$= \int_{-\pi/3}^{\pi/3}(\tan\theta + \tan\theta\sec^2\theta)\,d\theta$$

$$= \int_{-\pi/3}^{\pi/3}\tan\theta\,d\theta + \int_{-\pi/3}^{\pi/3}\tan\theta\sec^2\theta\,d\theta$$

To evaluate the first integral, rewrite the tangent:

$$\int_{-\pi/3}^{\pi/3}\tan\theta\,d\theta = \int_{-\pi/3}^{\pi/3}\frac{\sin\theta}{\cos\theta}\,d\theta$$

Then use the substitution $u = \cos\theta$ so that $du = -\sin\theta\,d\theta$, or $-du = \sin\theta\,d\theta$. You can find the new limits of integration by noting that if $\theta = \frac{\pi}{3}$ or if $\theta = -\frac{\pi}{3}$, you have $\cos\left(\frac{\pi}{3}\right) = \cos\left(-\frac{\pi}{3}\right) = \frac{1}{2}$:

$$\int_{-\pi/3}^{\pi/3}\frac{\sin\theta}{\cos\theta}\,d\theta = \int_{1/2}^{1/2}-\frac{1}{u}\,du$$

Because the upper and lower limits of integration are the same, you have the following for the first integral:

$$\int_{1/2}^{1/2}-\frac{1}{u}\,du = 0$$

The second integral is

$$\int_{-\pi/3}^{\pi/3} \tan\theta \sec^2\theta \, d\theta$$

To evaluate the second integral, use the substitution $u = \tan\theta$ so that $du = \sec^2\theta \, d\theta$. Notice that you can find the new limits of integration by noting that if $\theta = \frac{\pi}{3}$, then $\tan\left(\frac{\pi}{3}\right) = \sqrt{3}$, and if $\theta = -\frac{\pi}{3}$, then $\tan\left(-\frac{\pi}{3}\right) = \sqrt{3}$. Therefore, the second integral becomes

$$\int_{-\pi/3}^{\pi/3} \tan\theta \sec^2\theta \, d\theta = \int_{\sqrt{3}}^{\sqrt{3}} u \, du = 0$$

Combining the values of the first and second integrals gives you the answer:

$$\int_{-\pi/3}^{\pi/3} \tan^3\theta \, d\theta = 0$$

839.   4

The given integral is

$$\int_1^{e^2} \frac{(\ln x)^3}{x} \, dx$$

Start by using the substitution $u = \ln x$ so that $du = \frac{1}{x} \, dx$. You can find the new limits of integration by noting that if $x = e^2$, then $u = \ln e^2 = 2$, and if $x = 1$, then $u = \ln 1 = 0$. Therefore, the integral becomes

$$\int_1^{e^2} \frac{(\ln x)^3}{x} \, dx = \int_0^2 u^3 \, du$$

$$= \frac{u^4}{4}\bigg|_0^2$$

$$= \frac{2^4}{4} - 0$$

$$= 4$$

840. $-\frac{1}{3}\ln|4-3x| + C$

The given integral is

$$\int \frac{dx}{4-3x}$$

Start by letting $u = 4 - 3x$ so that $du = -3dx$, or $-\frac{1}{3}du = dx$. Substitute this back into the original integral:

$$\int -\frac{1}{3}\frac{1}{u} \, du = -\frac{1}{3}\ln|u| + C$$

$$= -\frac{1}{3}\ln|4-3x| + C$$

841. $\frac{1}{2}\left(\tan^{-1}x\right)^{2}+C$

Here's the given integral:

$$\int \frac{\tan^{-1}x}{1+x^2}\,dx$$

Let $u = \tan^{-1}x$ so that $du = \frac{1}{1+x^2}\,dx$. Substituting these values into the original integral, you get

$$\int u\,du = \frac{u^2}{2}+C$$

$$= \frac{1}{2}\left(\tan^{-1}x\right)^{2}+C$$

842. $\ln|\sec x|+C$

Start by rewriting the integral:

$$\int \tan x\,dx = \int \frac{\sin x}{\cos x}\,dx$$

Now you can perform a u-substitution with $u = \cos x$ so that $du = -\sin x\,dx$, or $-du = \sin x\,dx$. Substituting this expression into the original integral gives you the following:

$$\int \frac{-1}{u}\,du = -\ln|u|+C$$

$$= -\ln|\cos x|+C$$

You can now use the power property of logarithms to rewrite the integral:

$$-\ln|\cos x|+C = \ln\left|(\cos x)^{-1}\right|+C$$

$$= \ln\left|\frac{1}{\cos x}\right|+C$$

$$= \ln|\sec x|+C$$

843. $e - e^{1/2}$

The given integral is

$$\int_{0}^{\pi/3} e^{\cos x}\sin x\,dx$$

Begin by using the substitution $u = \cos x$ so that $du = -\sin x\,dx$, or $-du = \sin x\,dx$. Notice you can find the new limits of integration by noting that if $x = \frac{\pi}{3}$, then $u = \cos\frac{\pi}{3} = \frac{1}{2}$, and if $x = 0$, then $u = \cos 0 = 1$. With these values, you have the following:

$$-\int_{1}^{1/2} e^{u}\,du = \int_{1/2}^{1} e^{u}\,du$$

$$= e^{u}\Big|_{1/2}^{1}$$

$$= e^{1} - e^{1/2}$$

$$= e - e^{1/2}$$

844. $\frac{2}{3}(\sin 2 - \sin 1)$

Rewrite the given integral:

$$\int_0^1 \sqrt{x}\cos\left(1+x^{3/2}\right)dx = \int_0^1 x^{1/2}\cos\left(1+x^{3/2}\right)dx$$

Begin by using the substitution $u = 1 + x^{3/2}$ so that $du = \frac{3}{2}x^{1/2}dx$, or $\frac{2}{3}du = \sqrt{x}\,dx$. Notice you can find the new limits of integration by noting that if $x = 1$, then $u = 1 + 1^{3/2} = 2$, and if $x = 0$, then $u = 1 + 0^{3/2} = 1$. With these new values, you find that

$$\int_1^2 \frac{2}{3}\cos u\,du = \frac{2}{3}\sin u\Big|_1^2$$

$$= \frac{2}{3}(\sin 2 - \sin 1)$$

845. $\frac{7}{8}(\tan x)^{8/7} + C$

Rewrite the given integral:

$$\int \sqrt[7]{\tan x}\,\sec^2 x\,dx = \int (\tan x)^{1/7}\sec^2 x\,dx$$

Then use the substitution $u = \tan x$ so that $du = \sec^2 x\,dx$:

$$\int u^{1/7}du = \frac{u^{8/7}}{8/7} + C$$

$$= \frac{7}{8}(\tan x)^{8/7} + C$$

846. $\frac{1}{\pi}\cos\frac{\pi}{x} + C$

Here's the given integral:

$$\int \frac{\sin\left(\frac{\pi}{x}\right)}{x^2}dx$$

Use the substitution $u = \frac{\pi}{x} = \pi x^{-1}$ so that $du = -\pi x^{-2}dx = -\frac{\pi}{x^2}dx$, or $-\frac{1}{\pi}du = \frac{1}{x^2}dx$:

$$\int \frac{\sin\left(\frac{\pi}{x}\right)}{x^2}dx = -\frac{1}{\pi}\int \sin u\,du$$

$$= \frac{1}{\pi}\cos u + C$$

$$= \frac{1}{\pi}\cos\frac{\pi}{x} + C$$

847. $4\sqrt{1+2x+3x^2}+C$

Here's the given integral:

$$\int \frac{4+12x}{\sqrt{1+2x+3x^2}}\,dx$$

If you let $u = 1 + 2x + 3x^2$, then you get $du = (2 + 6x)dx$, or $2du = (4 + 12x)dx$. Substitute this value into the original integral:

$$\int \frac{2}{\sqrt{u}}\,du = \int 2u^{-1/2}\,du$$

$$= 2\frac{u^{1/2}}{1/2} + C$$

$$= 4\sqrt{1+2x+3x^2} + C$$

848. $-\frac{3}{4}\cot^{4/3}x + C$

Here's the given integral:

$$\int \sqrt[3]{\cot x}\,\csc^2 x\,dx$$

If you let $u = \cot x$, then you get $du = -\csc^2 x\,dx$, or $-du = \csc^2 x\,dx$. Substitute this value into the original integral:

$$-\int \sqrt[3]{u}\,du = -\int u^{1/3}\,du$$

$$= -\frac{u^{4/3}}{4/3} + C$$

$$= -\frac{3}{4}(\cot x)^{4/3} + C$$

$$= -\frac{3}{4}\cot^{4/3}x + C$$

849. $\frac{-\sqrt{2}+2}{4}$

Here's the given integral:

$$\int_0^{\sqrt{\pi}/2} x\sin\left(x^2\right)dx$$

If you let $u = x^2$, then $du = 2x\,dx$, or $\frac{1}{2}du = x\,dx$. You can also find the new limits of integration by using the original limits of integration along with the u-substitution. When $x = \frac{\sqrt{\pi}}{2}$, $u = \left(\frac{\sqrt{\pi}}{2}\right)^2 = \frac{\pi}{4}$. Likewise, if $x = 0$, then $u = (0)^2 = 0$.

Now substitute these values into the original integral:

$$\int_0^{\pi/4} \sin(u)\frac{1}{2}\,du = -\frac{1}{2}\cos(u)\Big|_0^{\pi/4}$$

$$= -\frac{1}{2}\cos\left(\frac{\pi}{4}\right) + \frac{1}{2}\cos 0$$

$$= -\frac{1}{2}\left(\frac{\sqrt{2}}{2}\right) + \frac{1}{2}$$

$$= \frac{-\sqrt{2}+2}{4}$$

850. $\quad 3\tan^{-1}x + \frac{1}{2}\ln(1+x^2) + C$

The given integral is

$$\int \frac{3+x}{1+x^2}\,dx$$

Begin by breaking the integral into two separate integrals:

$$\int\left(\frac{3}{1+x^2} + \frac{x}{1+x^2}\right)dx = \int\left(\frac{3}{1+x^2}\right)dx + \int\left(\frac{x}{1+x^2}\right)dx$$

Notice that $\int\left(\frac{3}{1+x^2}\right)dx = 3\tan^{-1}x + C$.

All that's left is to integrate the second integral; you can do this with a u-substitution. For $\int\left(\frac{x}{1+x^2}\right)dx$, let $u = 1 + x^2$ so that $du = 2x\,dx$, or $\frac{1}{2}du = x\,dx$. Substituting these values into the second integral gives you the following:

$$\frac{1}{2}\int\frac{1}{u}\,du = \frac{1}{2}\ln|u| + C$$

$$= \frac{1}{2}\ln|1+x^2| + C$$

The expression $1 + x^2$ is always positive, so you can drop the absolute value sign:

$$= \frac{1}{2}\ln(1+x^2) + C$$

Combining the solutions gives you

$$\int\left(\frac{3}{1+x^2} + \frac{x}{1+x^2}\right)dx$$

$$= \int\left(\frac{3}{1+x^2}\right)dx + \int\left(\frac{x}{1+x^2}\right)dx$$

$$= 3\tan^{-1}x + \frac{1}{2}\ln(1+x^2) + C$$

851. $2\sqrt{2}-2$

The given integral is

$$\int_e^{e^2} \frac{dx}{x\sqrt{\ln x}}$$

Begin with the u-substitution $u = \ln x$ so that $du = \frac{1}{x}dx$. You can also compute the new limits of integration by noting that if $x = e$, then $u = \ln e = 1$, and if $x = e^2$, then $u = \ln e^2 = 2$. Now you can integrate:

$$\int_1^2 \frac{1}{\sqrt{u}}\,du = \int_1^2 u^{-1/2}\,du$$

$$= \frac{u^{1/2}}{1/2}\bigg|_1^2$$

$$= 2\sqrt{u}\,\bigg|_1^2$$

$$= 2\sqrt{2}-2\sqrt{1}$$

$$= 2\sqrt{2}-2$$

852. $-\frac{1}{2}\left(\ln(\cos\theta)\right)^2 + C$

Here's the given integral:

$$\int \tan\theta \ln(\cos\theta)\,d\theta$$

Start with the substitution $u = \ln(\cos\theta)$ so that $du = \frac{1}{\cos\theta}(-\sin\theta)\,d\theta = -\tan\theta\,d\theta$. Substituting these values into the original integral, you get the following:

$$-\int u\,du = -\frac{u^2}{2} + C$$

$$= -\frac{1}{2}\left(\ln(\cos\theta)\right)^2 + C$$

853. $-\frac{1}{3}e^{-x^3} + C$

The given integral is

$$\int x^2 e^{-x^3}\,dx$$

Begin by letting $u = -x^3$ so that $du = -3x^2\,dx$, or $-\frac{1}{3}\,du = x^2\,dx$. Substitute these values into the original integral:

$$\int\left(-\frac{1}{3}\right)e^u\,du = -\frac{1}{3}e^u + C$$

$$= -\frac{1}{3}e^{-x^3} + C$$

854. $\dfrac{116}{15}$

Here's the given integral:

$$\int_0^3 x\sqrt{x+1}\,dx$$

Begin by using the substitution $u = x + 1$ so that $u - 1 = x$ and $du = dx$. Notice you can find the new limits of integration by noting that if $x = 3$, then $u = 3 + 1 = 4$, and if $x = 0$, then $u = 0 + 1 = 1$. With these new values, you have the following:

$$\int_1^4 (u-1)\sqrt{u}\,du = \int_1^4 (u-1)u^{1/2}du$$

$$= \int_1^4 \left(u^{3/2} - u^{1/2}\right)du$$

$$= \left(\frac{2}{5}u^{5/2} - \frac{2}{3}u^{3/2}\right)\Big|_1^4$$

$$= \left(\left(\frac{2}{5}4^{5/2} - \frac{2}{3}4^{3/2}\right) - \left(\frac{2}{5}1^{5/2} - \frac{2}{3}1^{3/2}\right)\right)$$

$$= \frac{64}{5} - \frac{16}{3} - \frac{2}{5} + \frac{2}{3}$$

$$= \frac{116}{15}$$

855. $\dfrac{\left(2^{7/3}\right)-1}{7} + \dfrac{1-\left(2^{4/3}\right)}{4}$

At first, you may have trouble seeing how a substitution will work for this problem. But note that you can split x^5 into $x^3 x^2$:

$$\int_0^1 x^5 \sqrt[3]{x^3+1}\,dx = \int_0^1 x^3 \sqrt[3]{x^3+1}\,x^2 dx$$

Now you can use the substitution $u = x^3 + 1$ so that $u - 1 = x^3$ and $du = 3x^2\,dx$, or $\frac{1}{3}du = x^2 dx$. Notice that you can find the new limits of integration by noting that if $x = 1$, then $u = 1^3 + 1 = 2$, and if $x = 0$, then $u = 0^3 + 1 = 1$. With these values, you get the following integral:

$$\frac{1}{3}\int_1^2 (u-1)u^{1/3}du = \frac{1}{3}\int_1^2 \left(u^{4/3} - u^{1/3}\right)du$$

$$= \frac{1}{3}\left(\frac{3}{7}u^{7/3} - \frac{3}{4}u^{4/3}\right)\Big|_1^2$$

$$= \frac{1}{3}\left(\left(\frac{3}{7}2^{7/3} - \frac{3}{4}2^{4/3}\right) - \left(\frac{3}{7}1^{7/3} - \frac{3}{4}1^{4/3}\right)\right)$$

$$= \frac{1}{3}\left(\frac{3}{7}2^{7/3} - \frac{3}{4}2^{4/3} - \frac{3}{7} + \frac{3}{4}\right)$$

$$= \frac{\left(2^{7/3}\right)-1}{7} + \frac{1-\left(2^{4/3}\right)}{4}$$

856. $\frac{6}{11}(x+3)^{11/6} - \frac{18}{5}(x+3)^{5/6} + C$

Here's the given integral:

$$\int \frac{x}{\sqrt[6]{x+3}}\,dx$$

If you let $u = x + 3$, then $du = dx$ and $u - 3 = x$. Substituting into the original integral, you get the following:

$$\int \frac{(u-3)}{\sqrt[6]{u}}\,du = \int \frac{(u-3)}{u^{1/6}}\,du$$

$$= \int \left(\frac{u}{u^{1/6}} - \frac{3}{u^{1/6}} \right)du$$

$$= \int \left(u^{5/6} - 3u^{-1/6} \right)du$$

$$= \frac{u^{11/6}}{11/6} - 3\left(\frac{u^{5/6}}{5/6} \right) + C$$

$$= \frac{6}{11}(x+3)^{11/6} - \frac{18}{5}(x+3)^{5/6} + C$$

857. $\frac{1}{10}\left(x^4+1\right)^{5/2} - \frac{1}{6}\left(x^4+1\right)^{3/2} + C$

At first glance, substitution may not seem to work for this problem. However, you can rewrite the integral as follows:

$$\int x^7 \sqrt{x^4+1}\,dx = \int x^4 x^3 \sqrt{x^4+1}\,dx$$

If you let $u = x^4 + 1$, then you get $du = 4x^3\,dx$, or $\frac{1}{4}du = x^3\,dx$. Notice also from the u-substitution that $u - 1 = x^4$. Substitute into the original integral and simplify:

$$\int (u-1)\sqrt{u}\left(\frac{1}{4} \right)du = \frac{1}{4}\int \left(u^{3/2} - u^{1/2} \right)du$$

$$= \frac{1}{4}\left(\left(\frac{u^{5/2}}{5/2} \right) - \left(\frac{u^{3/2}}{3/2} \right) \right) + C$$

$$= \frac{1}{4}\left(\frac{2}{5}\left(x^4+1\right)^{5/2} - \frac{2}{3}\left(x^4+1\right)^{3/2} \right) + C$$

$$= \frac{1}{10}\left(x^4+1\right)^{5/2} - \frac{1}{6}\left(x^4+1\right)^{3/2} + C$$

858. $\frac{1}{4}x\sin(4x) + \frac{1}{16}\cos(4x) + C$

The given integral is

$$\int x\cos(4x)\,dx$$

Use integration by parts with $u = x$ so that $du = dx$ and let $dv = \cos(4x)\,dx$ so that $v = \frac{1}{4}\sin(4x)$. Using the integration by parts formula gives you the following:

$$\int x \cos(4x)dx = x\left(\frac{1}{4}\sin(4x)\right) - \int \frac{1}{4}\sin(4x)dx$$

$$= \frac{1}{4}x\sin(4x) - \int \frac{1}{4}\sin(4x)dx$$

$$= \frac{1}{4}x\sin(4x) + \frac{1}{16}\cos(4x) + C$$

859. $xe^x - e^x + C$

Here's the given integral:

$$\int xe^x dx$$

Use integration by parts with $u = x$ so that $du = dx$ and let $dv = e^x\,dx$ so that $v = e^x$. Using the integration by parts formula gives you the following:

$$\int xe^x dx = xe^x - \int e^x dx$$

$$= xe^x - e^x + C$$

860. $\frac{1}{2}x\cosh(2x) - \frac{1}{4}\sinh(2x) + C$

Here's the given integral:

$$\int x\sinh(2x)dx$$

Use integration by parts with $u = x$ so that $du = dx$ and let $dv = \sinh(2x)\,dx$ so that $v = \frac{1}{2}\cosh(2x)$. Using the integration by parts formula gives you the following:

$$\int x\sinh(2x)dx = x\left(\frac{1}{2}\cosh(2x)\right) - \int \frac{1}{2}\cosh(2x)dx$$

$$= \frac{1}{2}x\cosh(2x) - \frac{1}{4}\sinh(2x) + C$$

861. $\frac{(3)6^3}{\ln 6} + \frac{1-6^3}{(\ln 6)^2}$

The given integral is

$$\int_0^3 x6^x dx$$

Use integration by parts with $u = x$ so that $du = dx$ and let $dv = 6^x\,dx$ so that $v = \frac{6^x}{\ln 6}$. Using the integration by parts formula gives you the following:

$$\int_0^3 x6^x dx = \frac{x6^x}{\ln 6}\Big|_0^3 - \int_0^3 \frac{1}{\ln 6}6^x dx$$

$$= \left(\frac{x6^x}{\ln 6} - \frac{6^x}{(\ln 6)^2}\right)\Big|_0^3$$

$$= \left(\frac{(3)6^3}{\ln 6} - \frac{6^3}{(\ln 6)^2} - \left(0 - \frac{1}{(\ln 6)^2}\right)\right)$$

$$= \frac{(3)6^3}{\ln 6} + \frac{1-6^3}{(\ln 6)^2}$$

862. $x \sec x - \ln|\sec x + \tan x| + C$

The given integral is

$$\int x \sec x \tan x \, dx$$

Use integration by parts with $u = x$ so that $du = dx$ and let $dv = \sec x \tan x dx$ so that $v = \sec x$:

$$\int x \sec x \tan x \, dx = x \sec x - \int \sec x \, dx$$
$$= x \sec x - \ln|\sec x + \tan x| + C$$

863. $-\dfrac{x \cos(5x)}{5} + \dfrac{1}{25} \sin(5x) + C$

The given integral is

$$\int x \sin(5x) dx$$

Begin by using integration by parts with $u = x$ so that $du = dx$ and let $dv = \sin(5x)\,dx$ so that $v = -\dfrac{\cos(5x)}{5}$:

$$\int x \sin(5x) dx = -\frac{1}{5} x \cos(5x) + \frac{1}{5} \int \cos(5x) dx$$
$$= -\frac{1}{5} x \cos(5x) + \frac{1}{25} \sin(5x) + C$$

864. $-x \cot x + \ln|\sin x| + C$

The given integral is

$$\int x \csc^2 x \, dx$$

Begin by using integration by parts with $u = x$ so that $du = dx$ and let $dv = \csc^2 x \, dx$ so that $v = -\cot x$:

$$\int x \csc^2 x dx = -x \cot x + \int \cot x dx$$
$$= -x \cot x + \ln|\sin x| + C$$

865. $-x \cos x + \sin x + C$

The given integral is

$$\int x \sin x \, dx$$

Use integration by parts with $u = x$ so that $du = dx$ and let $dv = \sin x\, dx$ so that $v = -\cos x$:

$$\int x \sin x\, dx = -x\cos x + \int \cos x\, dx$$
$$= -x\cos x + \sin x + C$$

866. $-x\csc x + \ln|\csc x - \cot x| + C$

Here's the given integral:

$$\int x\csc x \cot x\, dx$$

Use integration by parts with $u = x$ so that $du = dx$ and let $dv = \csc x \cot x\, dx$ so that $v = -\csc x$:

$$\int x\csc x \cot x\, dx = -x\csc x + \int \csc x\, dx$$
$$= -x\csc x + \ln|\csc x - \cot x| + C$$

867. $x\ln(3x+1) - x + \frac{1}{3}\ln|3x+1| + C$

The given integral is

$$\int \ln(3x+1)dx$$

Use integration by parts with $u = \ln(3x+1)$ so that $du = \dfrac{3}{3x+1}dx$ and let $dv = dx$ so that $v = x$. Using the integration by parts formula gives you

$$\int \ln(3x+1)dx = \left(\ln(3x+1)\right)x - \int \frac{3x}{3x+1}dx$$

To evaluate the integral $\int \dfrac{3x}{3x+1}dx$, you can use long division to simplify, or you can add 1 and subtract 1 from the numerator and split up the fraction:

$$\left(\ln(3x+1)\right)x - \int \frac{3x}{3x+1}dx = x\ln(3x+1) - \int \frac{3x+1-1}{3x+1}dx$$
$$= x\ln(3x+1) - \int \left(\frac{3x+1}{3x+1} - \frac{1}{3x+1}\right)dx$$
$$= x\ln(3x+1) - \int \left(1 - \frac{1}{3x+1}\right)dx$$
$$= x\ln(3x+1) - x + \frac{1}{3}\ln|3x+1| + C$$

868. $x\tan^{-1}x - \frac{1}{2}\ln\left|1+x^2\right| + C$

The given integral is

$$\int \tan^{-1}x\,dx$$

Use integration by parts with $u = \tan^{-1}x$ so that $du = \dfrac{1}{1+x^2}\,dx$ and let $dv = dx$ so that $v = x$. Using the integration by parts formula gives you

$$\int \tan^{-1}x\,dx = \left(\tan^{-1}x\right)x - \int \frac{x}{1+x^2}\,dx$$

To evaluate $\displaystyle\int \frac{x}{1+x^2}\,dx$, use the substitution $u_1 = 1 + x^2$ so that $du_1 = 2x\,dx$, or $\frac{1}{2}du_1 = x\,dx$. With these substitutions, you get the following:

$$\left(\tan^{-1}x\right)x - \int \frac{x}{1+x^2}\,dx = \left(\tan^{-1}x\right)x - \frac{1}{2}\int \frac{1}{u_1}\,du_1$$

$$= x\tan^{-1}x - \frac{1}{2}\ln|u_1| + C$$

$$= x\tan^{-1}x - \frac{1}{2}\ln\left|1+x^2\right| + C$$

869. $-\frac{1}{4}\ln 4 + \frac{3}{4}$

The given integral is

$$\int_1^4 \frac{\ln x}{x^2}\,dx$$

Use integration by parts with $u = \ln x$ so that $du = \dfrac{1}{x}\,dx$ and let $dv = x^{-2}\,dx$ so that $v = -x^{-1} = -\dfrac{1}{x}$. Using the integration by parts formula gives you the following:

$$\int_1^4 x^{-2}\ln x\,dx = \left(\ln x\left(-\frac{1}{x}\right)\right)\bigg|_1^4 - \int_1^4 \left(-\frac{1}{x}\cdot\frac{1}{x}\right)dx$$

$$= \left(-\frac{1}{x}\ln x\right)\bigg|_1^4 + \int_1^4 \left(x^{-2}\right)dx$$

$$= \left(-\frac{1}{x}\ln x - \frac{1}{x}\right)\bigg|_1^4$$

$$= \left(\left(-\frac{1}{4}\ln 4 - \frac{1}{4}\right) - \left(-\frac{1}{1}\ln 1 - \frac{1}{1}\right)\right)$$

$$= -\frac{1}{4}\ln 4 + \frac{3}{4}$$

870. $4e^3$

Here's the given integral:

$$\int_1^9 e^{\sqrt{x}}\,dx$$

Begin by using the substitution $w = \sqrt{x}$ so that $w^2 = x$ and $2w\,dw = dx$. You can find the new limits of integration by noting that if $x = 9$, then $w = \sqrt{9} = 3$, and if $x = 1$, then $w = \sqrt{1} = 1$. These substitutions give you the following:

$$\int_1^3 2we^w\,dw$$

Use integration by parts with $u = 2w$ so that $du = 2\,dw$ and let $dv = e^w\,dw$ so that $v = e^w$. Using the integration by parts formula gives you the following:

$$\int_1^3 2we^w\,dw = 2we^w\Big|_1^3 - \int_1^3 2e^w\,dw$$

$$= \left(2we^w - 2e^w\right)\Big|_1^3$$

$$= \left(6e^3 - 2e^3\right) - \left(2e^1 - 2e^1\right)$$

$$= 4e^3$$

871. $\dfrac{486}{5}\ln 9 - \dfrac{968}{25}$

The given integral is

$$\int_1^9 x^{3/2}\ln x\,dx$$

Use integration by parts with $u = \ln x$ so that $du = \dfrac{1}{x}\,dx$ and let $dv = x^{3/2}\,dx$ so that $v = \dfrac{2}{5}x^{5/2}$ to get the following:

$$\int_1^9 x^{3/2}\ln x\,dx = \frac{2}{5}x^{5/2}\ln x\Big|_1^9 - \int_1^9 \frac{2}{5}x^{5/2}\left(\frac{1}{x}\right)dx$$

$$= \frac{2}{5}x^{5/2}\ln x\Big|_1^9 - \frac{2}{5}\int_1^9 x^{3/2}\,dx$$

$$= \left(\frac{2}{5}x^{5/2}\ln x - \frac{2}{5}\left(\frac{2}{5}x^{5/2}\right)\right)\Big|_1^9$$

$$= \left(\frac{2}{5}9^{5/2}\ln 9 - \frac{4}{25}9^{5/2}\right) - \left(\frac{2}{5}1^{5/2}\ln 1 - \frac{4}{25}1^{5/2}\right)$$

$$= \frac{486}{5}\ln 9 - \frac{972}{25} + \frac{4}{25}$$

$$= \frac{486}{5}\ln 9 - \frac{968}{25}$$

872. $x\cos^{-1}x - \sqrt{1-x^2} + C$

Here's the given integral:

$$\int \cos^{-1}x\, dx$$

Use integration by parts with $u = \cos^{-1}x$ so that $du = -\dfrac{1}{\sqrt{1-x^2}}\,dx$ and let $dv = 1\,dx$ so that $v = x$:

$$\int \cos^{-1}x\, dx = x\cos^{-1}x + \int \frac{x}{\sqrt{1-x^2}}\,dx$$

Next, use the substitution $w = 1-x^2$ so that $dw = -2x\,dx$, or $-\dfrac{1}{2}\,dw = x\,dx$:

$$x\cos^{-1}x + \int \frac{x}{\sqrt{1-x^2}}\,dx = x\cos^{-1}x - \frac{1}{2}\int \frac{1}{\sqrt{w}}\,dw$$

$$= x\cos^{-1}x - \frac{1}{2}\int w^{-1/2}\,dw$$

$$= x\cos^{-1}x - \frac{1}{2}\left(2w^{1/2}\right) + C$$

$$= x\cos^{-1}x - \sqrt{1-x^2} + C$$

873. $x\sin^{-1}(5x) + \dfrac{1}{5}\sqrt{1-25x^2} + C$

The given integral is

$$\int \sin^{-1}(5x)\,dx$$

Use integration by parts with $u = \sin^{-1}(5x)$ so that $du = \dfrac{1}{\sqrt{1-(5x)^2}}5\,dx$ and let $dv = 1\,dx$ so that $v = x$:

$$\int \sin^{-1}(5x)\,dx = x\sin^{-1}(5x) - \int \frac{5x}{\sqrt{1-25x^2}}\,dx$$

Next, use the substitution $w = 1-25x^2$ so that $dw = -50x\,dx$, or $-\dfrac{1}{10}\,dw = 5x\,dx$:

$$x\sin^{-1}(5x) - \int \frac{5x}{\sqrt{1-25x^2}}\,dx$$

$$= x\sin^{-1}(5x) + \frac{1}{10}\int \frac{1}{\sqrt{w}}\,dw$$

$$= x\sin^{-1}(5x) + \frac{1}{10}\int w^{-1/2}\,dw$$

$$= x\sin^{-1}(5x) + \frac{1}{10}\left(2w^{1/2}\right) + C$$

$$= x\sin^{-1}(5x) + \frac{1}{5}\sqrt{1-25x^2} + C$$

874. $\quad -\dfrac{5}{4}e^{-2}+\dfrac{1}{4}$

Here's the given integral:

$$\int_0^1 \dfrac{x^2}{e^{2x}}\,dx$$

Begin by using integration by parts with $u = x^2$ so that $du = 2x\,dx$ and let $dv = e^{-2x}$ so that $v = -\dfrac{1}{2}e^{-2x}$:

$$\int_0^1 x^2 e^{-2x}\,dx = -\dfrac{1}{2}x^2 e^{-2x}\Big|_0^1 + \int_0^1 x e^{-2x}\,dx$$

Use integration by parts again with $u_1 = x$ so that $du_1 = dx$ and let $dv_1 = e^{-2x}$ so that $v_1 = -\dfrac{1}{2}e^{-2x}$:

$$-\dfrac{1}{2}x^2 e^{-2x}\Big|_0^1 + \int_0^1 x e^{-2x}\,dx$$

$$= -\dfrac{1}{2}x^2 e^{-2x}\Big|_0^1 + \left(-\dfrac{1}{2}x e^{-2x}\Big|_0^1 - \int_0^1 -\dfrac{1}{2}e^{-2x}\,dx\right)$$

$$= \left(-\dfrac{1}{2}x^2 e^{-2x} - \dfrac{1}{2}x e^{-2x} - \dfrac{1}{4}e^{-2x}\right)\Big|_0^1$$

$$= \left(-\dfrac{1}{2}1^2 e^{-2} - \dfrac{1}{2}e^{-2} - \dfrac{1}{4}e^{-2}\right) - \left(0 - 0 - \dfrac{1}{4}\right)$$

$$= -\dfrac{1}{2}1^2 e^{-2} - \dfrac{1}{2}e^{-2} - \dfrac{1}{4}e^{-2} + \dfrac{1}{4}$$

$$= -e^{-2} - \dfrac{1}{4}e^{-2} + \dfrac{1}{4}$$

$$= -\dfrac{5}{4}e^{-2} + \dfrac{1}{4}$$

875. $\quad -6e^{-1} + 3$

The given integral is

$$\int_0^1 \left(x^2 + 1\right)e^{-x}\,dx$$

Begin by using integration by parts with $u = x^2 + 1$ so that $du = 2x\,dx$ and let $dv = e^{-x}\,dx$ so that $v = -e^{-x}$:

$$\int_0^1 \left(x^2 + 1\right)e^{-x}\,dx = -\left(x^2 + 1\right)e^{-x}\Big|_0^1 + \int_0^1 2x e^{-x}\,dx$$

Now use integration by parts again with $u_1 = 2x$ so that $du_1 = 2\,dx$ and let $dv_1 = e^{-x}\,dx$ so that $v_1 = -e^{-x}$:

$$-\left(x^2 + 1\right)e^{-x}\Big|_0^1 + \left(-2x e^{-x}\right)\Big|_0^1 + \int_0^1 2e^{-x}\,dx$$

$$= \left(-\left(x^2 + 1\right)e^{-x} - 2x e^{-x} - 2e^{-x}\right)\Big|_0^1$$

$$= \left(-\left(1^2 + 1\right)e^{-1} - 2e^{-1} - 2e^{-1}\right) - \left(-(0 + 1)e^{-0} - 0 - 2e^0\right)$$

$$= -2e^{-1} - 2e^{-1} - 2e^{-1} + 1 + 2$$

$$= -6e^{-1} + 3$$

876.
$$\frac{1}{2}x^2 \sin x^2 + \frac{1}{2}\cos x^2 + C$$

Here's the given integral:

$$\int x^3 \cos(x^2) dx = \int x^2 \cos(x^2) x\, dx$$

Begin by using the substitution $w = x^2$ so that $dw = 2x\, dx$, or $\frac{1}{2}dw = x\, dx$, to get the following:

$$\int x^2 \cos(x^2) x\, dx = \frac{1}{2}\int w \cos w\, dw$$

Next, use integration by parts with $u = w$ so that $du = dw$ and let $dv = \cos w\, dw$ so that $v = \sin w$:

$$\frac{1}{2}\int w \cos w\, dw = \frac{1}{2}\left(w \sin w - \int \sin w\, dw\right)$$
$$= \frac{1}{2}w \sin w + \frac{1}{2}\cos w + C$$
$$= \frac{1}{2}x^2 \sin x^2 + \frac{1}{2}\cos x^2 + C$$

877.
$$-\frac{1}{m}x^2 \cos(mx) + \frac{2}{m^2}x \sin(mx) + \frac{2}{m^3}\cos(mx) + C$$

The given integral is

$$\int x^2 \sin(mx) dx, \quad \text{where } m \neq 0$$

Use integration by parts with $u = x^2$ so that $du = 2x\, dx$ and let $dv = \sin(mx) dx$ so that $v = -\frac{1}{m}\cos(mx)$. Using the integration by parts formula gives you the following:

$$\int x^2 \sin(mx)dx$$
$$= x^2\left(-\frac{1}{m}\cos(mx)\right) - \int\left(-\frac{1}{m}\cos(mx)\right)2x\, dx$$
$$= -\frac{1}{m}x^2 \cos(mx) + \frac{2}{m}\int x \cos(mx)dx$$

To evaluate $\int x \cos(mx)dx$, use integration by parts again with $u_1 = x$ so that $du_1 = dx$ and let $dv_1 = \cos(mx)$ so that $v_1 = \frac{1}{m}\sin(mx)$. This gives you the following:

$$\int x \cos(mx)dx = x\left(\frac{1}{m}\sin(mx)\right) - \int \frac{1}{m}\sin(mx)dx$$
$$= \frac{1}{m}x \sin(mx) + \frac{1}{m^2}\cos(mx) + C$$

Therefore,

$$-\frac{1}{m}x^2 \cos(mx) + \frac{2}{m}\int x \cos(mx)dx$$
$$= -\frac{1}{m}x^2 \cos(mx) + \frac{2}{m}\left(\frac{1}{m}x \sin(mx) + \frac{1}{m^2}\cos(mx)\right) + C$$
$$= -\frac{1}{m}x^2 \cos(mx) + \frac{2}{m^2}x \sin(mx) + \frac{2}{m^3}\cos(mx) + C$$

878.

$$\frac{3^6}{6}(\ln 3)^2 - \frac{3^6}{18}\ln 3 + \frac{3^6}{108} - \frac{1}{108}$$

Here's the given integral:

$$\int_1^3 x^5 (\ln x)^2 dx$$

Begin by using integration by parts with $u = (\ln x)^2$ so that $du = 2(\ln x)\frac{1}{x}dx$ and let $dv = x^5 dx$ so that $v = \frac{x^6}{6}$. This gives you the following:

$$\int_1^3 x^5 (\ln x)^2 dx = \frac{x^6}{6}(\ln x)^2\Big|_1^3 - \int_1^3 \frac{x^6}{6}\left(2(\ln x)\frac{1}{x}\right)dx$$

$$= \frac{x^6}{6}(\ln x)^2\Big|_1^3 - \frac{1}{3}\int_1^3 x^5 \ln x\, dx$$

To evaluate the new integral, use integration by parts with $u_1 = \ln x$ so that $du_1 = \frac{1}{x}dx$ and let $dv_1 = x^5 dx$ so that $v_1 = \frac{x^6}{6}$. This gives you the following:

$$\frac{x^6}{6}(\ln x)^2\Big|_1^3 - \frac{1}{3}\int_1^3 x^5 \ln x\, dx$$

$$= \frac{x^6}{6}(\ln x)^2\Big|_1^3 - \frac{1}{3}\left(\frac{x^6}{6}\ln x\Big|_1^3 - \int_1^3 \frac{x^6}{6}\frac{1}{x}dx\right)$$

$$= \frac{x^6}{6}(\ln x)^2\Big|_1^3 - \frac{x^6}{18}\ln x\Big|_1^3 + \frac{1}{3}\int_1^3 \frac{1}{6}x^5 dx$$

$$= \left(\frac{x^6}{6}(\ln x)^2 - \frac{x^6}{18}\ln x + \frac{1}{108}x^6\right)\Big|_1^3$$

$$= \frac{3^6}{6}(\ln 3)^2 - \frac{3^6}{18}\ln 3 + \frac{3^6}{108} - \left(0 - 0 + \frac{1}{108}\right)$$

$$= \frac{3^6}{6}(\ln 3)^2 - \frac{3^6}{18}\ln 3 + \frac{3^6}{108} - \frac{1}{108}$$

879.

$$\frac{1}{3}\left(e^x \cos(2x) + e^x \sin(2x)\right) + C$$

The given integral is

$$\int e^x \cos(2x)dx$$

Begin by using integration by parts with $u = \cos(2x)$ so that $du = -2\sin(2x)dx$ and let $dv = e^x dx$ so that $v = e^x$. Using the integration by parts formula gives you the following:

$$\int e^x \cos(2x)dx$$

$$= e^x \cos(2x) - \int e^x (-2\sin(2x))dx$$

$$= e^x \cos(2x) + \int 2e^x \sin(2x)dx$$

Use integration by parts again with $u_1 = \sin(2x)$ so that $du_1 = 2\cos(2x)$ and let $dv_1 = e^x\,dx$ so that $v_1 = e^x$. Now you have the following:

$$e^x\cos(2x)+\int 2e^x\sin(2x)\,dx$$

$$= e^x\cos(2x)+e^x\sin(2x)-\int e^x\bigl(2\cos(2x)\bigr)\,dx$$

$$= e^x\cos(2x)+e^x\sin(2x)-2\int e^x\cos(2x)\,dx$$

Set the expression equal to the original integral:

$$\int e^x\cos(2x)\,dx = e^x\cos(2x)+e^x\sin(2x)-2\int e^x\cos(2x)\,dx$$

Add $2\int e^x\cos(2x)\,dx$ to both sides of the equation (note the addition of C_1 on the right):

$$3\int e^x\cos(2x)\,dx = e^x\cos(2x)+e^x\sin(2x)+C_1$$

Dividing both sides of the equation by 3 yields the solution:

$$\int e^x\cos(2x)\,dx = \frac{1}{3}\bigl(e^x\cos(2x)+e^x\sin(2x)\bigr)+C$$

Note: Don't be thrown off by the constant of integration that switches from C_1 to C; they're both arbitrary constants, but simply using C in both equations would be poor notation.

---

880. $-\cos x\ln(\cos x)+\cos x+C$

Here's the given integral:

$$\int \sin x\ln(\cos x)\,dx$$

Begin by using the substitution $w = \cos x$ so that $dw = -\sin x\,dx$, or $-dw = \sin x\,dx$. Using these substitutions gives you the following:

$$\int \sin x\,\ln(\cos x)\,dx = \int -1\ln w\,dw$$

Now use integration by parts with $u = \ln w$ so that $du = \frac{1}{w}\,dw$ and let $dv = -1\,dw$ so that $v = -w$. Using the integration by parts formula now gives you

$$\int -1\ln w\,dw = -w\ln w-\int -w\frac{1}{w}\,dw$$

$$= -w\ln w+\int 1\,dw$$

$$= -w\ln w+w+C$$

Replacing w with the original substitution, $\cos x$, gives you the solution:

$$-w\ln w+w+C$$

$$= -\cos x\ln(\cos x)+\cos x+C$$

881. $\quad 2\sqrt{x}\sin\sqrt{x}+2\cos\sqrt{x}+C$

The given integral is

$$\int\cos\sqrt{x}\,dx$$

Begin by using the substitution $w=\sqrt{x}$ so that $w^2=x$ and $2w\,dw=dx$. This gives you the following:

$$\int\cos\sqrt{x}\,dx=\int 2w\cos w\,dw$$

Now use integration by parts with $u=2w$ so that $du=2\,dw$ and let $dv=\cos w\,dw$ so that $v=\sin w$:

$$\int 2w\cos w\,dw=2w\sin w-\int 2\sin w\,dw$$

$$=2w\sin w+2\cos w+C$$

Using the original substitution, $w=\sqrt{x}$, gives you the solution:

$$2w\sin w+2\cos w+C$$

$$=2\sqrt{x}\sin\sqrt{x}+2\cos\sqrt{x}+C$$

882. $\quad \dfrac{x^5}{5}(\ln x)^2-\dfrac{2}{25}x^5\ln x+\dfrac{2}{125}x^5+C$

Here's the given integral:

$$\int x^4(\ln x)^2\,dx$$

Begin by using integration by parts with $u=(\ln x)^2$ so that $du=2(\ln x)\dfrac{1}{x}\,dx$ and let $dv=x^4\,dx$ so that $v=\dfrac{x^5}{5}$:

$$\int x^4(\ln x)^2\,dx=\frac{x^5}{5}(\ln x)^2-\frac{2}{5}\int x^5(\ln x)\frac{1}{x}\,dx$$

$$=\frac{x^5}{5}(\ln x)^2-\frac{2}{5}\int x^4(\ln x)\,dx$$

To evaluate the new integral, use integration by parts again with $u_1=\ln x$ so that $du_1=\dfrac{1}{x}\,dx$ and let $dv=x^4\,dx$ so that $v=\dfrac{x^5}{5}$:

$$\frac{x^5}{5}(\ln x)^2-\frac{2}{5}\left(\frac{x^5}{5}\ln x-\int\frac{x^5}{5}\frac{1}{x}\,dx\right)$$

$$=\frac{x^5}{5}(\ln x)^2-\frac{2}{25}x^5\ln x+\frac{2}{25}\int x^4\,dx$$

$$=\frac{x^5}{5}(\ln x)^2-\frac{2}{25}x^5\ln x+\frac{2}{25}\left(\frac{x^5}{5}\right)+C$$

$$=\frac{x^5}{5}(\ln x)^2-\frac{2}{25}x^5\ln x+\frac{2}{125}x^5+C$$

883. $\dfrac{x^2}{2}\tan^{-1}x - \dfrac{1}{2}x + \dfrac{1}{2}\tan^{-1}x + C$

Here's the given integral:

$$\int x\tan^{-1}x\,dx$$

Use integration by parts with $u = \tan^{-1}x$ so that $du = \dfrac{1}{1+x^2}\,dx$ and let $dv = x\,dx$ so that $v = \dfrac{x^2}{2}$:

$$\int x\tan^{-1}x\,dx = \frac{x^2}{2}\tan^{-1}x - \frac{1}{2}\int\frac{x^2}{1+x^2}\,dx$$

$$= \frac{x^2}{2}\tan^{-1}x - \frac{1}{2}\int\frac{x^2+1-1}{1+x^2}\,dx$$

$$= \frac{x^2}{2}\tan^{-1}x - \frac{1}{2}\int\left(1 - \frac{1}{1+x^2}\right)dx$$

$$= \frac{x^2}{2}\tan^{-1}x - \frac{1}{2}x + \frac{1}{2}\tan^{-1}x + C$$

Note: The numerator and denominator are almost the same for the expression $\dfrac{x^2}{1+x^2}$, allowing you to add 1 and subtract 1 (in the second line) and then break it up into two fractions (in the third line); this approach lets you avoid long division. Another way to simplify $\dfrac{x^2}{1+x^2}$ is to use polynomial long division to arrive at $\dfrac{x^2}{1+x^2} = 1 - \dfrac{1}{1+x^2}$.

884. $\dfrac{3\pi}{4}$

Here's the given integral:

$$\int_0^{3\pi/2}\sin^2(2\theta)\,d\theta$$

Recall that $\sin^2(\theta) = \dfrac{1}{2}\big(1-\cos(2\theta)\big)$ so that $\sin^2(2\theta) = \dfrac{1}{2}\big(1-\cos(4\theta)\big)$. Using this identity in the integral gives you

$$\frac{1}{2}\int_0^{3\pi/2}\big(1-\cos(4\theta)\big)\,d\theta$$

$$= \frac{1}{2}\left(\theta - \frac{1}{4}\sin(4\theta)\right)\Big|_0^{3\pi/2}$$

$$= \frac{1}{2}\left(\frac{3\pi}{2} - \frac{1}{4}\sin(6\pi) - \left(0 - \frac{1}{4}\sin(0)\right)\right)$$

$$= \frac{3\pi}{4}$$

885. $\dfrac{\pi}{8} + \dfrac{1}{4}$

The given integral is

$$\int_0^{\pi/4}\cos^2(\theta)\,d\theta$$

Recall that $\cos^2(\theta) = \dfrac{1}{2}\big(1+\cos(2\theta)\big)$. Using this identity gives you the following:

$$\frac{1}{2}\int_0^{\pi/4}\left(1+\cos(2\theta)\right)d\theta$$

$$=\frac{1}{2}\left(\theta+\frac{1}{2}\sin(2\theta)\right)\Big|_0^{\pi/4}$$

$$=\frac{1}{2}\left(\frac{\pi}{4}+\frac{1}{2}\sin\left(\frac{\pi}{2}\right)-\left(0+\frac{1}{2}\sin(0)\right)\right)$$

$$=\frac{\pi}{8}+\frac{1}{4}$$

886. $\tan x - x + C$

Here's the given integral:

$$\int \tan^2 x\,dx$$

As long as you remember your fundamental trigonometric identities, this problem is an easy one!

$$\int \tan^2 x\,dx = \int\left(\sec^2 x - 1\right)dx$$

$$= \tan x - x + C$$

887. $\tan t + \frac{1}{3}\tan^3 t + C$

The given integral is

$$\int \sec^4 t\,dt$$

Begin by factoring out $\sec^2 t$ and using a trigonometric identity on one $\sec^2 t$ to write it in terms of $\tan t$:

$$\int \sec^4 t\,dt = \int \sec^2 t\,\sec^2 t\,dt$$

$$= \int\left(1+\tan^2 t\right)\sec^2 t\,dt$$

Now use the substitution $u = \tan t$ so that $du = \sec^2 t\,dt$. Using these values in the integral gives you

$$\int\left(1+\tan^2 t\right)\sec^2 t\,dt = \int\left(1+u^2\right)du$$

$$= u + \frac{1}{3}u^3 + C$$

$$= \tan t + \frac{1}{3}\tan^3 t + C$$

888. $\frac{1}{2}\sin x - \frac{1}{10}\sin(5x) + C$

The given integral is

$$\int \sin(3x)\sin(2x)\,dx$$

Begin by applying the identity $\sin(A)\sin(B) = \frac{1}{2}\big(\cos(A-B) - \cos(A+B)\big)$ to the integrand to produce the following integral:

$$\int \sin(3x)\sin(2x)\,dx$$

$$= \int \frac{1}{2}\big(\cos(3x - 2x) - \cos(3x + 2x)\big)\,dx$$

$$= \frac{1}{2}\int \big(\cos x - \cos(5x)\big)\,dx$$

The first term has an elementary antiderivative. To integrate the second term, you can use the substitution $u = 5x$:

$$\frac{1}{2}\int \big(\cos x - \cos(5x)\big)\,dx$$

$$= \frac{1}{2}\left(\sin x - \frac{1}{5}\sin(5x)\right) + C$$

$$= \frac{1}{2}\sin x - \frac{1}{10}\sin(5x) + C$$

889. $\frac{1}{6}\sin(3x) + \frac{1}{14}\sin(7x) + C$

The given integral is

$$\int \cos(5x)\cos(2x)\,dx$$

Begin by using the identity $\cos(A)\cos(B) = \frac{1}{2}\big(\cos(A-B) + \cos(A+B)\big)$ on the integrand to produce the following integral:

$$\int \cos(5x)\cos(2x)\,dx$$

$$= \int \frac{1}{2}\big(\cos(5x - 2x) + \cos(5x + 2x)\big)\,dx$$

$$= \frac{1}{2}\int \big(\cos(3x) + \cos(7x)\big)\,dx$$

To integrate, use the substitution $u = 3x$ on the first term and the substitution $u = 7x$ on the second term:

$$\frac{1}{2}\int \big(\cos(3x) + \cos(7x)\big)\,dx$$

$$= \frac{1}{2}\left(\frac{1}{3}\sin(3x) + \frac{1}{7}\sin(7x)\right)$$

$$= \frac{1}{6}\sin(3x) + \frac{1}{14}\sin(7x) + C$$

890. $\frac{1}{5}(\sin x)^5 + C$

The given integral is

$$\int \sin^4 x \cos x \, dx$$

Use the substitution $u = \sin x$ so that $du = \cos x \, dx$:

$$\int u^4 \, du = \frac{u^5}{5} + C = \frac{1}{5}(\sin x)^5 + C$$

891. $\frac{1}{3}(\tan x)^3 + C$

The given integral is

$$\int \tan^2 x \sec^2 x \, dx$$

Begin by using the substitution $u = \tan x$ so that $du = \sec^2 x \, dx$:

$$\int u^2 \, du = \frac{u^3}{3} + C = \frac{1}{3}(\tan x)^3 + C$$

892. $-\frac{1}{2}\cos x - \frac{1}{18}\cos(9x) + C$

Here's the given integral:

$$\int \sin(5x)\cos(4x)\,dx$$

Begin by applying the identity $\sin(A)\cos(B) = \frac{1}{2}\left(\sin(A-B) + \sin(A+B)\right)$ to the integrand to produce the integral $\int \frac{1}{2}\left(\sin(5x-4x) + \sin(5x+4x)\right)dx$. Now simplify and integrate:

$$\int \frac{1}{2}\left(\sin(5x-4x) + \sin(5x+4x)\right)dx$$
$$= \frac{1}{2}\int \left(\sin x + \sin(9x)\right)dx$$
$$= \frac{1}{2}\left(-\cos x - \frac{1}{9}\cos(9x)\right) + C$$
$$= -\frac{1}{2}\cos x - \frac{1}{18}\cos(9x) + C$$

893. $\sec x + C$

Here's the given integral:

$$\int \sec x \tan x \, dx$$

There's a simple solution for this function:

$$\int \sec x \tan x \, dx = \sec x + C$$

894. $-\frac{2}{7}(\csc x)^{7/2}+C$

Start by rewriting the radical:

$$\int \sqrt{\csc^5 x}\, \csc x \cot x\, dx$$

$$=\int (\csc x)^{5/2}\csc x \cot x\, dx$$

Use the substitution $u = \csc x$ so that $du = -\csc x \cot x\, dx$, or $-du = \csc x \cot x\, dx$:

$$\int -(u)^{5/2}\, du = -\frac{u^{7/2}}{7/2}+C$$

$$=-\frac{2}{7}(\csc x)^{7/2}+C$$

895. $\frac{1}{3}\sin^3 x - \frac{1}{5}\sin^5 x + C$

Here's the given integral:

$$\int \cos^3 x \sin^2 x\, dx$$

Begin by factoring out $\cos x$ and using a trigonometric identity to write $\cos^2 x$ in terms of $\sin x$:

$$\int \cos^2 x \sin^2 x \cos x\, dx$$

$$=\int \left(1-\sin^2 x\right)\sin^2 x \cos x\, dx$$

Now use the substitution $u = \sin x$, which gives you $du = \cos x\, dx$. Putting these values into the integral gives you the following:

$$\int \left(1-\sin^2 x\right)\sin^2 x \cos x\, dx$$

$$=\int \left(1-u^2\right)u^2 du$$

$$=\int \left(u^2 - u^4\right)du$$

$$=\frac{1}{3}u^3 - \frac{1}{5}u^5 + C$$

$$=\frac{1}{3}\sin^3 x - \frac{1}{5}\sin^5 x + C$$

896. $\frac{27}{4}$

The given integral is

$$\int_0^{\pi/3} \tan^3 x \sec^4 x\, dx$$

Begin by factoring out $\sec^2 x$ and using a trigonometric identity on one factor of $\sec^2 x$ to write it in terms of $\tan x$:

$$\int_0^{\pi/3} \tan^3 x \sec^4 x\, dx$$

$$=\int_0^{\pi/3} \tan^3 x \sec^2 x \sec^2 x\, dx$$

$$=\int_0^{\pi/3} \tan^3 x \left(1+\tan^2 x\right)\sec^2 x\, dx$$

Now use the substitution $u = \tan x$ so that $du = \sec^2 x \, dx$. You can also change the limits of integration by noting that if $x = \frac{\pi}{3}$, then $u = \tan\frac{\pi}{3} = \sqrt{3}$, and if $x = 0$, then $u = \tan 0 = 0$.

With these new values, the integral becomes

$$\int_0^{\sqrt{3}} u^3 \left(1+u^2\right) du = \int_0^{\sqrt{3}} \left(u^3 + u^5\right) du$$

$$= \left(\frac{u^4}{4} + \frac{u^6}{6}\right)\Bigg|_0^{\sqrt{3}}$$

$$= \left(\frac{\left(\sqrt{3}\right)^4}{4} + \frac{\left(\sqrt{3}\right)^6}{6}\right) - \left(\frac{0^4}{4} + \frac{0^6}{6}\right)$$

$$= \frac{9}{4} + \frac{27}{6}$$

$$= \frac{27}{4}$$

897. $\quad -\frac{1}{6}(\cot x)^6 - \frac{1}{4}(\cot x)^4 + C$

The given integral is

$$\int \frac{\cot^3 x}{\sin^4 x} dx$$

Begin by rewriting the integral and factoring out $\csc^2 x$:

$$\int \frac{\cot^3 x}{\sin^4 x} dx = \int \cot^3 x \csc^4 x \, dx$$

$$= \int \cot^3 x \csc^2 x \csc^2 x \, dx$$

Now use an identity to get the following:

$$\int \cot^3 x \left(\cot^2 x + 1\right)\csc^2 x \, dx$$

Next, use the substitution $u = \cot x$ to get $du = -\csc^2 x \, dx$ so that

$$\int -u^3 \left(u^2 + 1\right) du = \int \left(-u^5 - u^3\right) du$$

$$= \frac{-u^6}{6} - \frac{u^4}{4} + C$$

$$= -\frac{1}{6}(\cot x)^6 - \frac{1}{4}(\cot x)^4 + C$$

898. $\quad \frac{2}{3}$

Here's the given integral:

$$\int_0^{\pi/2} \cos^3 x \, dx$$

Begin by factoring out $\cos x$ and using a trigonometric identity to write $\cos^2 x$ in terms of $\sin x$:

$$\int_0^{\pi/2} \cos^2 x \cos x \, dx = \int_0^{\pi/2} \left(1 - \sin^2 x\right)\cos x \, dx$$

Now use the substitution $u = \sin x$, which gives you $du = \cos x\,dx$. You can also find the new limits of integration. If $x = \frac{\pi}{2}$, then $u = \sin\frac{\pi}{2} = 1$, and if $x = 0$, then $u = \sin 0 = 0$. Using these values in the integral gives you

$$\int_0^1 (1-u^2)\,du = \left(u - \frac{u^3}{3}\right)\Big|_0^1$$

$$= \left(1 - \frac{1}{3} - (0-0)\right)$$

$$= \frac{2}{3}$$

---

899. $\ln|\csc x - \cot x| - \ln|\sin x| + C$

The given integral is

$$\int \frac{1-\cos x}{\sin x}\,dx$$

Start by rewriting the integrand as two terms:

$$\int\left(\frac{1}{\sin x} - \frac{\cos x}{\sin x}\right)dx$$

$$= \int \frac{1}{\sin x}\,dx - \int \frac{\cos x}{\sin x}\,dx$$

$$= \int \csc x\,dx - \int \frac{\cos x}{\sin x}\,dx$$

Recall that $\int \csc x\,dx = \ln|\csc x - \cot x| + C$. To evaluate $\int \frac{\cos x}{\sin x}\,dx$, use the substitution $u = \sin x$ so that $du = \cos x\,dx$:

$$\int \frac{\cos x}{\sin x}\,dx = \int \frac{1}{u}\,du$$

$$= \ln|u| + C$$

$$= \ln|\sin x| + C$$

Combining the antiderivatives of each indefinite integral gives you the solution:

$$\int \csc x\,dx - \int \frac{\cos x}{\sin x}\,dx$$

$$= \ln|\csc x - \cot x| - \ln|\sin x| + C$$

---

900. $\frac{3}{2}\theta - 2\cos\theta - \frac{1}{4}\sin(2\theta) + C$

Begin by expanding the integrand:

$$\int (1+\sin\theta)^2\,d\theta = \int\left(1 + 2\sin\theta + \sin^2\theta\right)d\theta$$

Next, use the identity $\sin^2(\theta) = \frac{1}{2}(1 - \cos(2\theta))$ to produce the following integral:

$$\int\left(1 + 2\sin\theta + \frac{1}{2}(1-\cos(2\theta))\right)d\theta$$

$$= \int\left(\frac{3}{2} + 2\sin\theta - \frac{1}{2}\cos(2\theta)\right)d\theta$$

$$= \frac{3}{2}\theta - 2\cos\theta - \frac{1}{4}\sin(2\theta) + C$$

901. $\frac{1}{2}\left(\ln|\csc x - \cot x| - \ln|\sec x + \tan x|\right) + C$

Here's the given integral:

$$\int \frac{\cos x - \sin x}{\sin(2x)} dx$$

Begin by applying the identity $\sin(2x) = 2\sin x \cos x$ to the denominator of the integrand. Then split up the fraction and simplify:

$$\int \frac{\cos x - \sin x}{\sin(2x)} dx = \int \frac{\cos x - \sin x}{2\sin x \cos x} dx$$

$$= \frac{1}{2} \int \left(\frac{\cos x}{\sin x \cos x} - \frac{\sin x}{\sin x \cos x} \right) dx$$

$$= \frac{1}{2} \int \left(\frac{1}{\sin x} - \frac{1}{\cos x} \right) dx$$

$$= \frac{1}{2} \int (\csc x - \sec x) dx$$

You can now apply basic antiderivative formulas to arrive at the solution:

$$\frac{1}{2} \int (\csc x - \sec x) dx$$

$$= \frac{1}{2} \left(\ln|\csc x - \cot x| - \ln|\sec x + \tan x| \right) + C$$

902. $\frac{1}{4}(\tan x)^4 + C$

Here's the given integral:

$$\int \frac{\tan^3 x}{\cos^2 x} dx$$

Begin by rewriting the integral:

$$\int \frac{\tan^3 x}{\cos^2 x} dx = \int \tan^3 x \sec^2 x$$

Then use the substitution $u = \tan x$ so that $du = \sec^2 x \, dx$:

$$\int u^3 du = \frac{u^4}{4} + C$$

$$= \frac{1}{4}(\tan x)^4 + C$$

903. $\frac{1}{4}x^2 - \frac{1}{4}x\sin(2x) - \frac{1}{8}\cos(2x) + C$

The given integral is

$$\int x \sin^2 x \, dx$$

To integrate this function, use integration by parts: $\int u\,dv = uv - \int v\,du$. Let $u = x$ so that $du = dx$ and let $dv = \sin^2 x$. Note that to find v by integrating dv, you have to use a trigonometric identity:

$$\int \sin^2 x \, dx = \int \frac{1}{2}\big(1-\cos(2x)\big) \, dx$$

$$= \frac{1}{2}\Big(x - \frac{1}{2}\sin(2x)\Big) + C$$

$$= \frac{1}{2}x - \frac{1}{4}\sin(2x) + C$$

Therefore, you get the following:

$$\int x \sin^2 x \, dx = x\Big(\frac{1}{2}x - \frac{1}{4}\sin(2x)\Big) - \int\Big(\frac{1}{2}x - \frac{1}{4}\sin(2x)\Big) dx$$

$$= \frac{1}{2}x^2 - \frac{1}{4}x\sin(2x) - \frac{1}{4}x^2 - \frac{1}{8}\cos(2x) + C$$

$$= \frac{1}{4}x^2 - \frac{1}{4}x\sin(2x) - \frac{1}{8}\cos(2x) + C$$

904. $\quad \frac{1}{4}(\sin x)^4 - \frac{1}{6}(\sin x)^6 + C$

Here's the given integral:

$$\int \sin^3 x \cos^3 x \, dx$$

For this problem, you can factor out either $\sin x$ or $\cos x$ and then use an identity. For example, you can factor out $\cos x$ and then use the identity $\cos^2 x = 1 - \sin^2 x$:

$$\int \sin^3 x \cos^3 x \, dx = \int \sin^3 x \cos^2 x \cos x \, dx$$

$$= \int \sin^3 x \big(1 - \sin^2 x\big)\cos x \, dx$$

Then use the substitution $u = \sin x$ so that $du = \cos x \, dx$ to get

$$\int u^3 \big(1 - u^2\big) \, du = \int \big(u^3 - u^5\big) \, du$$

$$= \frac{u^4}{4} - \frac{u^6}{6} + C$$

$$= \frac{1}{4}(\sin x)^4 - \frac{1}{6}(\sin x)^6 + C$$

905. $\quad \frac{1}{3}\sec^3 x - \sec x + C$

The given integral is

$$\int \tan^3 x \sec x \, dx$$

Begin by factoring out $\tan x$ and using an identity:

$$\int \tan^3 x \sec x \, dx$$

$$= \int \tan^2 x \sec x \tan x \, dx$$

$$= \int \big(\sec^2 x - 1\big)\sec x \tan x \, dx$$

Now use the substitution $u = \sec x$ so that $du = \sec x \tan x\, dx$. This gives you

$$\int \left(u^2 - 1 \right) du = \frac{u^3}{3} - u + C$$
$$= \frac{1}{3} \sec^3 x - \sec x + C$$

906. $-\frac{1}{3}(\csc x)^3 + \csc x + C$

Begin by factoring out $\cot x$ and using an identity:

$$\int \cot^3 x \csc x\, dx = \int \cot^2 x \csc x \cot x\, dx$$
$$= \int \left(\csc^2 x - 1 \right) \csc x \cot x\, dx$$

Now use the substitution $u = \csc x$ so that $du = -\csc x \cot x\, dx$, or $-du = \csc x \cot x\, dx$. This gives you the following:

$$\int -\left(u^2 - 1 \right) du = -\frac{u^3}{3} + u + C$$
$$= -\frac{1}{3}(\csc x)^3 + \csc x + C$$

907. $\frac{2}{3}(\sin x)^{3/2} - \frac{2}{7}(\sin x)^{7/2} + C$

Here's the given integral:

$$\int \cos^3 x \sqrt{\sin x}\, dx$$

Begin by factoring out $\cos x$ and using a trigonometric identity to write $\cos^2 x$ in terms of $\sin x$:

$$\int \cos^3 x \sqrt{\sin x}\, dx$$
$$= \int \cos^2 x \sqrt{\sin x} \cos x\, dx$$
$$= \int \left(1 - \sin^2 x \right) \sqrt{\sin x} \cos x\, dx$$

Now use u-substitution: $u = \sin x$ so that $du = \cos x$. Using these values in the integral gives you

$$\int \left(1 - u^2 \right) \sqrt{u}\, du = \int \left(1 - u^2 \right) u^{1/2} du$$
$$= \int \left(u^{1/2} - u^{5/2} \right) du$$
$$= \frac{2}{3} u^{3/2} - \frac{2}{7} u^{7/2} + C$$
$$= \frac{2}{3}(\sin x)^{3/2} - \frac{2}{7}(\sin x)^{7/2} + C$$

908. $-\frac{1}{3}(\cot x)^3 - \frac{1}{5}(\cot x)^5 + C$

Begin by factoring out $\csc^2 x$ and using an identity to get the following:

$$\int \cot^2 x \csc^4 x \, dx = \int \cot^2 x \csc^2 x \csc^2 x \, dx$$

$$= \int \cot^2 x \left(1 + \cot^2 x\right) \csc^2 x \, dx$$

Now use the substitution $u = \cot x$ so that $du = -\csc^2 x \, dx$, or $-du = \csc^2 x \, dx$:

$$\int -u^2 \left(1 + u^2\right) du = \int -u^2 - u^4 du$$

$$= -\frac{u^3}{3} - \frac{u^5}{5} + C$$

$$= -\frac{1}{3}(\cot x)^3 - \frac{1}{5}(\cot x)^5 + C$$

909. $\cos(\cos\theta) - \frac{2}{3}\cos^3(\cos\theta) + \frac{1}{5}\cos^5(\cos\theta) + C$

Here's the given integral:

$$\int \sin\theta \sin^5(\cos\theta) d\theta$$

Begin by performing the substitution $u = \cos\theta$ so that $du = -\sin\theta \, d\theta$, or $-du = \sin\theta \, d\theta$. Using these values in the integral gives you $-\int \sin^5 u \, du$. Now factor out $\sin u$ and use some algebra and a trigonometric identity to write $\sin^2 u$ in terms of $\cos u$:

$$-\int \sin^5 u \, du = -\int \sin^4 u \sin u \, du$$

$$= -\int \left(\sin^2 u\right)^2 \sin u \, du$$

$$= -\int \left(1 - \cos^2 u\right)^2 \sin u \, du$$

Again use a substitution to integrate: Let $w = \cos u$ so that $dw = -\sin u \, du$, or $-dw = \sin u \, du$. Using these values yields the integral $\int \left(1 - w^2\right)^2 dw$. Now simply expand the integrand and integrate:

$$\int \left(1 - w^2\right)^2 dw = \int \left(1 - 2w^2 + w^4\right) dw$$

$$= w - \frac{2}{3}w^3 + \frac{1}{5}w^5 + C$$

Then undo all the substitutions:

$$w - \frac{2}{3}w^3 + \frac{1}{5}w^5 + C$$

$$= \cos u - \frac{2}{3}\cos^3 u + \frac{1}{5}\cos^5 u + C$$

$$= \cos(\cos\theta) - \frac{2}{3}\cos^3(\cos\theta) + \frac{1}{5}\cos^5(\cos\theta) + C$$

910. $-\frac{1}{5}\cos^5 x + \frac{2}{7}\cos^7 x - \frac{1}{9}\cos^9 x + C$

The given integral is

$$\int \sin^5 x \cos^4 x\, dx$$

Begin by factoring out $\sin x$ and using a trigonometric identity to write $\sin^2 x$ in terms of $\cos x$:

$$\int \sin^4 x \cos^4 x \sin x\, dx$$

$$= \int \left(\sin^2 x\right)^2 \cos^4 x \sin x\, dx$$

$$= \int \left(1 - \cos^2 x\right)^2 \cos^4 x \sin x\, dx$$

Now use the substitution $u = \cos x$, which gives you $du = -\sin x\, dx$, or $-du = \sin x\, dx$. Using these values in the integral gives you

$$-\int \left(1 - u^2\right)^2 u^4\, du$$

$$= -\int \left(1 - 2u^2 + u^4\right)u^4\, du$$

$$= -\int \left(u^4 - 2u^6 + u^8\right)du$$

$$= -\left(\frac{u^5}{5} - \frac{2u^7}{7} + \frac{u^9}{9}\right) + C$$

$$= -\frac{1}{5}\cos^5 x + \frac{2}{7}\cos^7 x - \frac{1}{9}\cos^9 x + C$$

911. $\frac{1}{2}\left(\sec x \tan x + \ln|\sec x + \tan x|\right) + C$

To evaluate this integral, you can use integration by parts. You may want to first rewrite $\sec^3 x$ as $\sec^2 x \sec x$ (rewriting isn't necessary, but it may help you see things a bit better!):

$$\int \sec^3 x\, dx = \int \sec^2 x \sec x\, dx$$

Now use integration by parts with $u = \sec x$ so that $du = \sec x \tan x\, dx$ and let $dv = \sec^2 x\, dx$ so that $v = \tan x$. This gives you the following:

$$\int \sec^2 x \sec x\, dx$$

$$= \sec x \tan x - \int \tan x(\sec x \tan x)dx$$

$$= \sec x \tan x - \int \tan^2 x \sec x\, dx$$

Now use an identity to get

$$\sec x \tan x - \int \tan^2 x \sec x\, dx$$

$$= \sec x \tan x - \int \left(\sec^2 x - 1\right)\sec x\, dx$$

$$= \sec x \tan x - \int \sec^3 x\, dx + \int \sec x\, dx$$

Set this expression equal to the given integral:

$$\int \sec^3 x\, dx = \sec x \tan x - \int \sec^3 x\, dx + \int \sec x\, dx$$

After evaluating $\int \sec x\, dx$ on the right side of the equation, you get

$$\int \sec^3 x\, dx = \sec x \tan x - \int \sec^3 x\, dx + \ln|\sec x + \tan x|$$

Add $\int \sec^3 x\, dx$ to both sides of the equation and then divide by 2 to get the solution:

$$\int \sec^3 x\, dx = \frac{1}{2}\left(\sec x \tan x + \ln|\sec x + \tan x|\right) + C$$

---

912. $-\sec x - \tan x + C$

The given integral is

$$\int \frac{dx}{\sin x - 1}$$

Begin by multiplying the numerator and denominator of the integrand by the conjugate of the denominator. Recall that the conjugate of $(a - b)$ is $(a + b)$. In this case, you multiply the numerator and denominator by $(\sin x + 1)$:

$$\int \frac{1}{\sin x - 1}\, dx = \int \left(\frac{1}{(\sin x - 1)}\frac{(\sin x + 1)}{(\sin x + 1)}\right) dx$$

$$= \int \frac{\sin x + 1}{\sin^2 x - 1}\, dx$$

Now use the identity $\sin^2 x - 1 = -\cos^2 x$ and split apart the fraction to further simplify the integrand:

$$\int \frac{\sin x + 1}{\sin^2 x - 1}\, dx = \int \frac{\sin x + 1}{-\cos^2 x}\, dx$$

$$= -\int \left(\frac{\sin x}{\cos^2 x} + \frac{1}{\cos^2 x}\right) dx$$

$$= -\int \left(\frac{\sin x}{\cos x}\frac{1}{\cos x} + \sec^2 x\right) dx$$

$$= -\int \left(\tan x \sec x + \sec^2 x\right) dx$$

$$= -(\sec x + \tan x) + C$$

$$= -\sec x - \tan x + C$$

913. $\frac{1}{28}\sec^7(4x)-\frac{1}{20}\sec^5(4x)+C$

Here's the given integral:

$$\int \tan^3(4x)\sec^5(4x)dx$$

Begin by factoring out $\sec(4x)\tan(4x)$:

$$\int \tan^2(4x)\sec^4(4x)\big(\sec(4x)\tan(4x)\big)dx$$

Then use a trigonometric identity to arrive at

$$\int \big(\sec^2(4x)-1\big)\sec^4(4x)\big(\sec(4x)\tan(4x)\big)dx$$

Now use the substitution $u = \sec(4x)$ so that $du = 4\sec(4x)\tan(4x)\,dx$, or $\frac{1}{4}du = \sec(4x)\tan(4x)dx$. Using these substitutions, you now have

$$\frac{1}{4}\int\big(u^2-1\big)u^4 du$$

$$=\frac{1}{4}\int\big(u^6-u^4\big)du$$

$$=\frac{1}{4}\left(\frac{u^7}{7}-\frac{u^5}{5}\right)$$

$$=\frac{1}{28}\sec^7(4x)-\frac{1}{20}\sec^5(4x)+C$$

914. $\ln\left|\dfrac{\sqrt{x^2+4}+x}{2}\right|+C$

The given integral is

$$\int\frac{dx}{\sqrt{x^2+4}}$$

Start by using the substitution $x=2\tan\theta$ so that $dx=2\sec^2\theta\,d\theta$. Using these substitutions in the integral, you get the following:

$$\int\frac{2\sec^2\theta}{\sqrt{(2\tan\theta)^2+4}}\,d\theta$$

$$=\int\frac{2\sec^2\theta}{\sqrt{4\big(\tan^2\theta+1\big)}}\,d\theta$$

$$=\int\frac{2\sec^2\theta}{\sqrt{4\sec^2\theta}}\,d\theta$$

$$=\int\frac{2\sec^2\theta}{2\sec\theta}\,d\theta$$

$$=\int\sec\theta\,d\theta$$

$$=\ln\left|\sec\theta+\tan\theta\right|+C$$

From the substitution $x = 2\tan\theta$, you get $\frac{x}{2} = \tan\theta$, from which you can deduce that

$\frac{\sqrt{x^2+4}}{2} = \sec\theta$. (To deduce that $\frac{\sqrt{x^2+4}}{2} = \sec\theta$, you can label a right triangle using

$\frac{x}{2} = \tan\theta$ and find the missing side using the Pythagorean theorem.) Using these

values, you get

$$\ln|\sec\theta + \tan\theta| + C = \ln\left|\frac{\sqrt{x^2+4}}{2} + \frac{x}{2}\right| + C$$

$$= \ln\left|\frac{\sqrt{x^2+4}+x}{2}\right| + C$$

915. $\quad \frac{\pi}{8} - \frac{1}{4}$

Here's the given integral:

$$\int_0^{\sqrt{2}/2} \frac{x^2}{\sqrt{1-x^2}}\,dx$$

Begin by using the substitution $x = \sin\theta$ so that $dx = \cos\theta\,d\theta$. You can find the new

limits of integration by noting that if $x = \frac{\sqrt{2}}{2}$, then $\frac{\sqrt{2}}{2} = \sin\theta$ so that $\frac{\pi}{4} = \theta$, and if $x = 0$,

then $0 = \sin\theta$ so that $\theta = 0$:

$$\int_0^{\pi/4} \frac{\sin^2\theta\cos\theta\,d\theta}{\sqrt{1-\sin^2\theta}} = \int_0^{\pi/4} \sin^2\theta\,d\theta$$

$$= \int_0^{\pi/4} \frac{1}{2}\left(1-\cos(2\theta)\right)d\theta$$

$$= \frac{1}{2}\left(\theta - \frac{\sin(2\theta)}{2}\right)\Bigg|_0^{\pi/4}$$

$$= \frac{1}{2}\left(\frac{\pi}{4} - \frac{\sin\left(\frac{\pi}{2}\right)}{2} - (0-0)\right)$$

$$= \frac{\pi}{8} - \frac{1}{4}$$

916. $\quad \sqrt{x^2-1} + C$

Here's the given integral:

$$\int \frac{x}{\sqrt{x^2-1}}\,dx$$

Use the trigonometric substitution $x = \sec\theta$ so that $dx = \sec\theta\tan\theta\,d\theta$:

$$\int \frac{\sec\theta}{\sqrt{\sec^2\theta-1}}\sec\theta\tan\theta\,d\theta$$

$$= \int \frac{\sec^2\theta\tan\theta\,d\theta}{\sqrt{\tan^2\theta}}$$

$$= \int \sec^2\theta\,d\theta$$

$$= \tan\theta + C$$

From $x = \sec\theta$, you can deduce that $\tan\theta = \sqrt{x^2 - 1}$. (To deduce that $\tan\theta = \sqrt{x^2 - 1}$, you can label a right triangle using $\frac{x}{1} = \sec\theta$ and then use the Pythagorean theorem to find the missing side.) Therefore, the solution becomes

$$\tan\theta + C = \sqrt{x^2 - 1} + C$$

Note that you can also solve the problem with the substitution $u = x^2 - 1$.

---

917. $-\dfrac{\sqrt{1+x^2}}{x} + C$

The given integral is

$$\int \frac{1}{x^2 \sqrt{1+x^2}}\, dx$$

Begin with the trigonometric substitution $x = \tan\theta$ to get $dx = \sec^2\theta\, d\theta$, which gives you the following:

$$\int \frac{\sec^2\theta\, d\theta}{\tan^2\theta\sqrt{1+\tan^2\theta}} = \int \frac{\sec^2\theta}{\tan^2\theta\sqrt{\sec^2\theta}}\, d\theta$$

$$= \int \frac{\sec\theta}{\tan^2\theta}\, d\theta$$

$$= \int \frac{1}{\cos\theta}\frac{\cos^2\theta}{\sin^2\theta}\, d\theta$$

$$= \int \frac{\cos\theta}{\sin\theta}\frac{1}{\sin\theta}\, d\theta$$

$$= \int \cot\theta\csc\theta\, d\theta$$

$$= -\csc\theta + C$$

From $x = \tan\theta$, you can deduce that $\csc\theta = \dfrac{\sqrt{1+x^2}}{x}$. (To deduce that $\csc\theta = \dfrac{\sqrt{1+x^2}}{x}$, you can label a right triangle using $\frac{x}{1} = \tan\theta$ and then find the missing side using the Pythagorean theorem.) Therefore, you get the following solution:

$$-\csc\theta + C = -\frac{\sqrt{1+x^2}}{x} + C$$

---

918. $\dfrac{1}{16}\dfrac{\sqrt{x^2 - 16}}{x} + C$

The given integral is

$$\int \frac{1}{x^2 \sqrt{x^2 - 16}}\, dx$$

Begin by using the substitution $x = 4\sec\theta$ to get $dx = 4\sec\theta\tan\theta\, d\theta$. Substituting these values into the original integral, you have

$$\int \frac{4\sec\theta\tan\theta}{(4\sec\theta)^2\sqrt{(4\sec\theta)^2-16}}\,d\theta$$

$$=\int \frac{4\sec\theta\tan\theta}{(4\sec\theta)^2\sqrt{16\sec^2\theta-16}}\,d\theta$$

$$=\int \frac{4\sec\theta\tan\theta}{(4\sec\theta)^2\sqrt{16(\sec^2\theta-1)}}\,d\theta$$

$$=\int \frac{4\sec\theta\tan\theta}{4^2\sec^2\theta\sqrt{16(\tan^2\theta)}}\,d\theta$$

$$=\int \frac{4\sec\theta\tan\theta}{4^2\sec^2\theta(4\tan\theta)}\,d\theta$$

$$=\frac{1}{4^2}\int \frac{1}{\sec\theta}\,d\theta$$

$$=\frac{1}{16}\int\cos\theta\,d\theta$$

$$=\frac{1}{16}\sin\theta+C$$

From your original substitution, you have $\frac{x}{4}=\sec\theta$. From this, you can deduce that

$\sin\theta=\frac{\sqrt{x^2-16}}{x}$. (To deduce that $\sin\theta=\frac{\sqrt{x^2-16}}{x}$, you can label a right triangle using $\frac{x}{4}=\sec\theta$ and then use the Pythagorean theorem to find the missing side.) Therefore, the solution becomes

$$\frac{1}{16}\sin\theta+C=\frac{1}{16}\frac{\sqrt{x^2-16}}{x}+C$$

919.

$$\frac{9}{2}\left(\sin^{-1}\frac{x}{3}+\frac{x\sqrt{9-x^2}}{9}\right)+C$$

Here's the given integral:

$$\int\sqrt{9-x^2}\,dx$$

Begin by using the trigonometric identity $x=3\sin\theta$ so that $dx=3\cos\theta\,d\theta$:

$$\int\left(\sqrt{9-9\sin^2\theta}\right)3\cos\theta\,d\theta$$

$$=9\int\sqrt{1-\sin^2\theta}\,\cos\theta\,d\theta$$

$$=9\int\cos^2\theta\,d\theta$$

Now use an identity to get

$$9\int\frac{1}{2}(1+\cos(2\theta))\,d\theta$$

$$=\frac{9}{2}\left(\theta+\frac{\sin 2\theta}{2}\right)+C$$

$$=\frac{9}{2}\left(\theta+\frac{2\sin\theta\cos\theta}{2}\right)+C$$

$$=\frac{9}{2}(\theta+\sin\theta\cos\theta)+C$$

From the substitution $x = 3\sin\theta$, you get $\frac{x}{3} = \sin\theta$, so you can deduce that $\sin^{-1}\frac{x}{3} = \theta$ and that $\cos\theta = \frac{\sqrt{9-x^2}}{3}$. (To deduce that $\cos\theta = \frac{\sqrt{9-x^2}}{3}$, you can label a right triangle using $\frac{x}{3} = \sin\theta$ and then find the missing side with the Pythagorean theorem.) Therefore, the solution becomes

$$\frac{9}{2}(\theta + \sin\theta\cos\theta) + C$$

$$= \frac{9}{2}\left(\sin^{-1}\frac{x}{3} + \frac{x}{3}\left(\frac{\sqrt{9-x^2}}{3}\right)\right) + C$$

$$= \frac{9}{2}\left(\sin^{-1}\frac{x}{3} + \frac{x\sqrt{9-x^2}}{9}\right) + C$$

---

920. $\frac{1}{10}\left(\sin^{-1}(5x) + 5x\sqrt{1-25x^2}\right) + C$

Start by rewriting the given integral:

$$\int\sqrt{1-25x^2}\,dx = \int\sqrt{1-(5x)^2}\,dx$$

Now you can use the substitution $5x = \sin\theta$ to get $x = \frac{1}{5}\sin\theta$ and $dx = \frac{1}{5}\cos\theta\,d\theta$. Substituting these values into the integral, you get

$$\int\sqrt{1-\sin^2\theta}\,\frac{1}{5}\cos\theta\,d\theta$$

$$= \frac{1}{5}\int\sqrt{\cos^2\theta}\,\cos\theta\,d\theta$$

$$= \frac{1}{5}\int\cos^2\theta\,d\theta$$

Then you can use the identity $\cos^2\theta = \frac{1}{2}(1+\cos(2\theta))$ to simplify the integral:

$$\frac{1}{5}\int\frac{1}{2}(1+\cos(2\theta))\,d\theta$$

$$= \frac{1}{10}\int(1+\cos(2\theta))\,d\theta$$

$$= \frac{1}{10}\left(\theta + \frac{1}{2}\sin(2\theta)\right) + C$$

$$= \frac{1}{10}\left(\theta + \frac{1}{2}2\sin\theta\cos\theta\right) + C$$

$$= \frac{1}{10}(\theta + \sin\theta\cos\theta) + C$$

From the substitution $5x = \sin\theta$, you can deduce that $\sin^{-1}(5x) = \theta$ and also that $\cos\theta = \sqrt{1-25x^2}$. (To deduce that $\cos\theta = \sqrt{1-25x^2}$, you can label a right triangle using $\frac{5x}{1} = \sin\theta$ and find the missing side of the triangle with the Pythagorean theorem.) Using these values, you get the solution:

$$\frac{1}{10}(\theta + \sin\theta\cos\theta) + C$$

$$= \frac{1}{10}\left(\sin^{-1}(5x) + 5x\sqrt{1-25x^2}\right) + C$$

921. $\dfrac{1}{\sqrt{7}}\ln\left|\dfrac{\sqrt{7}-\sqrt{7-x^2}}{x}\right|+C$

The given integral is

$$\int \frac{dx}{x\sqrt{7-x^2}}$$

Begin by using the substitutions $x=\sqrt{7}\sin\theta$ and $dx=\sqrt{7}\cos\theta\,d\theta$:

$$\int\frac{\sqrt{7}\cos\theta\,d\theta}{\sqrt{7}\sin\theta\sqrt{7-7\sin^2\theta}}$$

$$=\int\frac{\cos\theta}{\sin\theta\sqrt{7\left(1-\sin^2\theta\right)}}\,d\theta$$

$$=\int\frac{\cos\theta}{\sin\theta\sqrt{7\cos^2\theta}}\,d\theta$$

$$=\int\frac{\cos\theta}{\sin\theta\sqrt{7}\cos\theta}\,d\theta$$

$$=\frac{1}{\sqrt{7}}\int\frac{1}{\sin\theta}\,d\theta\frac{1}{\sqrt{7}}\int\csc\theta\,d\theta$$

$$=\frac{1}{\sqrt{7}}\ln|\csc\theta-\cot\theta|+C$$

From the substitution $x=\sqrt{7}\sin\theta$, you get $\dfrac{x}{\sqrt{7}}=\sin\theta$, from which you can deduce that $\dfrac{\sqrt{7}}{x}=\csc\theta$ and also that $\cot\theta=\dfrac{\sqrt{7-x^2}}{x}$. (To deduce that $\cot\theta=\dfrac{\sqrt{7-x^2}}{x}$, you can label a right triangle using $\dfrac{x}{\sqrt{7}}=\sin\theta$ and find the missing side of the triangle with the

Pythagorean theorem.) Using these values, you get the following solution:

$$\frac{1}{\sqrt{7}}\ln\left|\frac{\sqrt{7}-\sqrt{7-x^2}}{x}\right|+C$$

922. $\dfrac{1}{2\sqrt{5}}\left(\sin^{-1}\left(x\sqrt{5}\right)+x\sqrt{5-25x^2}\right)+C$

Rewrite the integral with a radical sign:

$$\int\sqrt{1-5x^2}\,dx=\int\sqrt{1-\left(x\sqrt{5}\right)^2}\,dx$$

Now begin by using the substitution $x\sqrt{5}=\sin\theta$, or $x=\dfrac{1}{\sqrt{5}}\sin\theta$, so that $dx=\dfrac{1}{\sqrt{5}}\cos\theta\,d\theta$:

$$\int \sqrt{1-\sin^2\theta}\left(\frac{1}{\sqrt{5}}\cos\theta\,d\theta\right)$$

$$=\frac{1}{\sqrt{5}}\int\cos^2\theta\,d\theta$$

$$=\frac{1}{\sqrt{5}}\int\frac{1}{2}(1+\cos(2\theta))\,d\theta$$

$$=\frac{1}{2\sqrt{5}}\left(\theta+\frac{\sin 2\theta}{2}\right)+C$$

$$=\frac{1}{2\sqrt{5}}\left(\theta+\frac{2\sin\theta\cos\theta}{2}\right)+C$$

$$=\frac{1}{2\sqrt{5}}\left(\theta+\sin\theta\cos\theta\right)+C$$

From the substitution $x\sqrt{5}=\sin\theta$, you can deduce that $\sin^{-1}\left(x\sqrt{5}\right)=\theta$ and that $\cos\theta=\sqrt{1-5x^2}$. (To deduce that $\cos\theta=\sqrt{1-5x^2}$, you can label a right triangle using $\frac{x\sqrt{5}}{1}=\sin\theta$ and find the missing side of the triangle with the Pythagorean theorem.) Therefore, the solution becomes

$$\frac{1}{2\sqrt{5}}\left(\sin^{-1}\left(x\sqrt{5}\right)+x\sqrt{5}\sqrt{1-5x^2}\right)+C$$

$$=\frac{1}{2\sqrt{5}}\left(\sin^{-1}\left(x\sqrt{5}\right)+x\sqrt{5-25x^2}\right)+C$$

---

923. $\quad\frac{1}{3}\ln\left|\frac{3}{x}-\frac{\sqrt{9-x^2}}{x}\right|+C$

The given integral is

$$\int\frac{dx}{x\sqrt{9-x^2}}$$

Begin by using the trigonometric substitution $x=3\sin\theta$ so that $dx=3\cos\theta\,d\theta$:

$$\int\frac{3\cos\theta}{3\sin\theta\sqrt{9-9\sin^2\theta}}\,d\theta$$

$$=\int\frac{\cos\theta\,d\theta}{\sin\theta\sqrt{9\left(1-\sin^2\theta\right)}}\,d\theta$$

$$=\int\frac{\cos\theta\,d\theta}{3\sin\theta\sqrt{\cos^2\theta}}\,d\theta$$

$$=\frac{1}{3}\int\csc\theta\,d\theta$$

$$=\frac{1}{3}\ln\left|\csc\theta-\cot\theta\right|+C$$

From the substitution $x=3\sin\theta$, you can deduce that $\csc\theta=\frac{3}{x}$ and that $\cot\theta=\frac{\sqrt{9-x^2}}{x}$. (To deduce that $\cot\theta=\frac{\sqrt{9-x^2}}{x}$, you can label a right triangle using $\frac{x}{3}=\sin\theta$ and find the missing side of the triangle with the Pythagorean theorem.) Therefore, the solution becomes

$$\frac{1}{3}\ln\left|\frac{3}{x}-\frac{\sqrt{9-x^2}}{x}\right|+C$$

924. $\frac{1}{4}\left(\sec^{-1}\frac{x}{2} - \frac{2\sqrt{x^2-4}}{x^2}\right) + C$

The given integral is

$$\int \frac{\sqrt{x^2-4}}{x^3}\,dx$$

Begin with the trigonometric substitution $x = 2\sec\theta$ so that $dx = 2\sec\theta\tan\theta\,d\theta$:

$$\int \frac{\left(\sqrt{4\sec^2\theta-4}\right)2\sec\theta\tan\theta\,d\theta}{(2\sec\theta)^3}$$

$$= \int \frac{\left(2\sqrt{\tan^2\theta}\right)2\sec\theta\tan\theta\,d\theta}{8\sec^3\theta}$$

$$= \frac{1}{2}\int \frac{\tan^2\theta}{\sec^2\theta}\,d\theta$$

$$= \frac{1}{2}\int \frac{\sin^2\theta}{\cos^2\theta}\left(\cos^2\theta\right)d\theta$$

$$= \frac{1}{2}\int \sin^2\theta\,d\theta$$

Now use an identity to get

$$\frac{1}{2}\int \frac{1}{2}(1-\cos(2\theta))\,d\theta$$

$$= \frac{1}{4}\left(\theta - \frac{\sin 2\theta}{2}\right) + C$$

$$= \frac{1}{4}\left(\theta - \frac{2\sin\theta\cos\theta}{2}\right) + C$$

$$= \frac{1}{4}(\theta - \sin\theta\cos\theta) + C$$

From the substitution $x = 2\sec\theta$, you can deduce that $\sec^{-1}\frac{x}{2} = \theta$, $\sin\theta = \frac{\sqrt{x^2-4}}{x}$, and $\cos\theta = \frac{2}{x}$. (To find the values of $\sin\theta$ and $\cos\theta$, you can label a right triangle using $\frac{x}{2} = \sec\theta$ and find the missing side of the triangle with the Pythagorean theorem.) Therefore, the solution becomes

$$\frac{1}{4}\left(\sec^{-1}\frac{x}{2} - \left(\frac{\sqrt{x^2-4}}{x}\right)\left(\frac{2}{x}\right)\right) + C$$

$$= \frac{1}{4}\left(\sec^{-1}\frac{x}{2} - \frac{2\sqrt{x^2-4}}{x^2}\right) + C$$

925. $\frac{1}{4}\left(\sin^{-1}\left(x^2\right) + x^2\sqrt{1-x^4}\right) + C$

The given integral is

$$\int x\sqrt{1-x^4}\,dx = \int x\sqrt{1-(x^2)^2}\,dx$$

Begin with the substitution $u = x^2$ so that $du = 2x\,dx$, or $\frac{1}{2}du = x\,dx$:

$$\int \frac{1}{2}\sqrt{1-u^2}\,du$$

Now use the substitution $u = \sin\theta$ so that $du = \cos\theta\,d\theta$:

$$\frac{1}{2}\int \sqrt{1-\sin^2\theta}\,\cos\theta\,d\theta$$

$$= \frac{1}{2}\int \cos^2\theta\,d\theta$$

$$= \frac{1}{2}\int \frac{1}{2}(1+\cos 2\theta)\,d\theta$$

$$= \frac{1}{4}\left(\theta + \frac{\sin 2\theta}{2}\right) + C$$

$$= \frac{1}{4}\left(\theta + \frac{2\sin\theta\cos\theta}{2}\right) + C$$

$$= \frac{1}{4}(\theta + \sin\theta\cos\theta) + C$$

From the substitution $u = \sin\theta$, you can deduce that $\sin^{-1}u = \theta$ and that $\cos\theta = \sqrt{1-u^2}$. Therefore, the answer is

$$\frac{1}{4}(\theta + \sin\theta\cos\theta) + C$$

$$= \frac{1}{4}\left(\sin^{-1}u + u\sqrt{1-u^2}\right) + C$$

$$= \frac{1}{4}\left(\sin^{-1}\left(x^2\right) + x^2\sqrt{1-x^4}\right) + C$$

---

926. $\ln\left|\sqrt{2}+1\right|$

Here's the given integral:

$$\int_0^{\pi/2} \frac{\sin t}{\sqrt{1+\cos^2 t}}\,dt$$

First begin with the substitution $u = \cos t$, which gives you $du = -\sin t\,dt$, or $-du = \sin t\,dt$. You can also compute the new limits of integration: If $t = \frac{\pi}{2}$, then $u = \cos\left(\frac{\pi}{2}\right) = 0$, and if $t = 0$, then $u = \cos 0 = 1$. Substituting these values into the integral gives you

$$-\int_1^0 \frac{1}{\sqrt{1+u^2}}\,du = \int_0^1 \frac{1}{\sqrt{1+u^2}}\,du$$

Next, use the trigonometric substitution $u = \tan\theta$, which gives you $du = \sec^2\theta\,d\theta$. You can again find the new limits of integration: If $u = 1$, then $1 = \tan\theta$ so that $\frac{\pi}{4} = \theta$, and if $u = 0$, then $0 = \tan\theta$ so that $0 = \theta$. Using these values, you produce the integral

$$\int_0^{\pi/4} \frac{\sec^2\theta}{\sqrt{1+\tan^2\theta}}\,d\theta$$

Now use a trigonometric identity and simplify:

$$\int_0^{\pi/4} \frac{\sec^2\theta}{\sqrt{1+\tan^2\theta}}\, d\theta$$

$$= \int_0^{\pi/4} \frac{\sec^2\theta}{\sqrt{\sec^2\theta}}\, d\theta$$

$$= \int_0^{\pi/4} \sec\theta\, d\theta$$

$$= \left(\ln|\sec\theta + \tan\theta|\right)\Big|_0^{\pi/4}$$

$$= \left(\ln\left|\sec\frac{\pi}{4} + \tan\frac{\pi}{4}\right| - \ln|\sec 0 + \tan 0|\right)$$

$$= \ln\left|\sqrt{2}+1\right| - \ln|1+0|$$

$$= \ln\left|\sqrt{2}+1\right|$$

---

927. $\sqrt{x^2-1} - \sec^{-1}x + C$

Here's the given integral:

$$\int \frac{\sqrt{x^2-1}}{x}\, dx$$

Begin with the substitution $x = \sec\theta$ so that $dx = \sec\theta\tan\theta\, d\theta$:

$$\int \frac{\sqrt{\sec^2\theta - 1}}{\sec\theta} \sec\theta\tan\theta\, d\theta$$

$$= \int \sqrt{\tan^2\theta}\, \tan\theta\, d\theta$$

$$= \int \tan^2\theta\, d\theta$$

Next, use an identity and integrate to get

$$\int \left(\sec^2\theta - 1\right)d\theta = \tan\theta - \theta + C$$

From the substitution $x = \sec\theta$, you can deduce that $\sec^{-1}x = \theta$ and that $\tan\theta = \sqrt{x^2-1}$ to get the solution:

$$\sqrt{x^2-1} - \sec^{-1}x + C$$

---

928. $\dfrac{\sqrt{3}}{2}$

The given integral is

$$\int_1^2 \frac{1}{x^2\sqrt{x^2-1}}\, dx$$

Use the trigonometric substitution $x = \sec\theta$ so that $dx = \sec\theta\tan\theta\, d\theta$. You can find the new limits of integration by noting that if $x = 1$, then $1 = \sec\theta$ and $\theta = 0$ and that if $x = 2$, then $2 = \sec\theta$ and $\theta = \frac{\pi}{3}$. With these new values, you get

$$\int_0^{\pi/3} \frac{\sec\theta\tan\theta}{\sec^2\theta\sqrt{\sec^2\theta-1}}\,d\theta = \int_0^{\pi/3} \frac{\tan\theta}{\sec\theta\sqrt{\tan^2\theta}}\,d\theta$$

$$= \int_0^{\pi/3} \frac{1}{\sec\theta}\,d\theta$$

$$= \int_0^{\pi/3} \cos\theta\,d\theta$$

$$= (\sin\theta)\big|_0^{\pi/3}$$

$$= \sin\frac{\pi}{3}-\sin 0$$

$$= \frac{\sqrt{3}}{2}$$

929. $\quad \dfrac{x}{\sqrt{4-x^2}}-\sin^{-1}\left(\dfrac{x}{2}\right)+C$

Start by rewriting the denominator with a radical:

$$\int \frac{x^2}{\left(4-x^2\right)^{3/2}}\,dx = \int \frac{x^2}{\left(\sqrt{4-x^2}\right)^3}\,dx$$

Use the trigonometric substitution $x=2\sin\theta$ to get $dx=2\cos\theta\,d\theta$. These substitutions give you the following:

$$\int \frac{(2\sin\theta)^2 2\cos\theta\,d\theta}{\left(\sqrt{4-4\sin^2\theta}\right)^3} = \int \frac{8\sin^2\theta\cos\theta\,d\theta}{\left(2\sqrt{\cos^2\theta}\right)^3}$$

$$= \int \frac{\sin^2\theta\cos\theta}{\cos^3\theta}\,d\theta$$

$$= \int \tan^2\theta\,d\theta$$

$$= \int \left(\sec^2\theta-1\right)d\theta$$

$$= \tan\theta-\theta+C$$

From the substitution $x=2\sin\theta$, or $\frac{x}{2}=\sin\theta$, you can deduce that $\sin^{-1}\left(\frac{x}{2}\right)=\theta$ and that $\tan\theta=\dfrac{x}{\sqrt{4-x^2}}$. (To deduce that $\tan\theta=\dfrac{x}{\sqrt{4-x^2}}$, you can label a right triangle using $\frac{x}{2}=\sin\theta$ and then find the missing side with the Pythagorean theorem.) Therefore, the solution becomes

$$\tan\theta-\theta+C = \frac{x}{\sqrt{4-x^2}}-\sin^{-1}\left(\frac{x}{2}\right)+C$$

930. $\quad \dfrac{1}{8}\left(\dfrac{\sqrt{x^2-4}}{x}\right)+C$

Start by rewriting the integrand:

$$\int \frac{1}{x^2\sqrt{4x^2-16}}\,dx = \int \frac{1}{x^2\sqrt{(2x)^2-4^2}}\,dx$$

Use the substitution $2x = 4\sec\theta$, or $x = 2\sec\theta$, so that $dx = 2\sec\theta\tan\theta\,d\theta$:

$$\int \frac{2\sec\theta\tan\theta\,d\theta}{(2\sec\theta)^2\sqrt{(4\sec\theta)^2 - 4^2}}$$

$$= \int \frac{2\tan\theta\,d\theta}{4\sec\theta\sqrt{16(\sec^2\theta - 1)}}$$

$$= \int \frac{2\tan\theta\,d\theta}{4\sec\theta\sqrt{16(\tan^2\theta)}}$$

$$= \frac{1}{8}\int \cos\theta\,d\theta$$

$$= \frac{1}{8}\sin\theta + C$$

From $x = 2\sec\theta$, or $\frac{x}{2} = \sec\theta$, you can deduce that $\sin\theta = \frac{\sqrt{x^2-4}}{x}$. (To deduce that $\sin\theta = \frac{\sqrt{x^2-4}}{x}$, you can label a right triangle using $\frac{x}{2} = \sec\theta$ and then find the missing side of the triangle with the Pythagorean theorem.) Therefore, the solution becomes

$$\frac{1}{8}\sin\theta + C = \frac{1}{8}\left(\frac{\sqrt{x^2-4}}{x}\right) + C$$

931. $\dfrac{(16+x^2)^{3/2}}{3} - 16\sqrt{16+x^2} + C$

Here's the given integral:

$$\int \frac{x^3}{\sqrt{x^2+16}}\,dx$$

Begin with the substitution $x = 4\tan\theta$ to get $dx = 4\sec^2\theta\,d\theta$. Substituting this value into the original integral gives you the following:

$$\int \frac{(4\tan\theta)^3}{\sqrt{(4\tan\theta)^2+16}}4\sec^2\theta\,d\theta$$

$$= 4^4\int \frac{\tan^3\theta\sec^2\theta}{\sqrt{16(\tan^2\theta+1)}}\,d\theta$$

$$= 4^4\int \frac{\tan^3\theta\sec^2\theta}{\sqrt{16(\sec^2\theta)}}\,d\theta$$

$$= 4^4\int \frac{\tan^3\theta\sec^2\theta}{4\sec\theta}\,d\theta$$

$$= 4^3\int \tan^3\theta\sec\theta\,d\theta$$

Now factor out $\tan\theta$ and use a trigonometric identity along with a u-substitution:

$$4^3\int \tan^2\theta\tan\theta\sec\theta\,d\theta$$

$$= 4^3\int (\sec^2\theta - 1)\sec\theta\tan\theta\,d\theta$$

Let $u = \sec\theta$ to get $du = \sec\theta\tan\theta\,d\theta$. Using these values in the integral, you get

$$4^3\int(u^2-1)\,du = 4^3\left(\frac{u^3}{3}-u\right)+C$$

$$= 4^3\left(\frac{1}{3}(\sec\theta)^3-\sec\theta\right)+C$$

From the first substitution, you get $\frac{x}{4} = \tan\theta$, from which you can deduce that $\frac{\sqrt{16+x^2}}{4} = \sec\theta$. (You can find the value $\frac{\sqrt{16+x^2}}{4} = \sec\theta$ by using a trigonometric identity for tangent and secant.) Inserting this value into the preceding equation and simplifying a bit gives you the answer:

$$\frac{4^3}{3}\left(\frac{\sqrt{16+x^2}}{4}\right)^3 - 4^3\left(\frac{\sqrt{16+x^2}}{4}\right)+C$$

$$= \frac{\left(\sqrt{16+x^2}\right)^3}{3} - 16\sqrt{16+x^2}+C$$

$$= \frac{\left(16+x^2\right)^{3/2}}{3} - 16\sqrt{16+x^2}+C$$

932. $\quad \frac{1}{5}\left(1+x^2\right)^{5/2} - \frac{2}{3}\left(1+x^2\right)^{3/2} + \sqrt{1+x^2}+C$

Here's the given integral:

$$\int\frac{x^5}{\sqrt{x^2+1}}\,dx$$

Begin with the substitution $x = \tan\theta$ so that $dx = \sec^2\theta\,d\theta$:

$$\int\frac{(\tan\theta)^5\sec^2\theta\,d\theta}{\sqrt{\tan^2\theta+1}} = \int\tan^5\theta\sec\theta\,d\theta$$

$$= \int\tan^4\theta\sec\theta\tan\theta\,d\theta$$

Next, use an identity to get

$$\int\left(\tan^2\theta\right)^2\sec\theta\tan\theta\,d\theta$$

$$= \int\left(\sec^2\theta-1\right)^2\sec\theta\tan\theta\,d\theta$$

Now use the substitution $u = \sec\theta$ so that $du = \sec\theta\tan\theta\,d\theta$:

$$\int\left(u^2-1\right)^2\,du$$

$$= \int\left(u^4-2u^2+1\right)du$$

$$= \frac{u^5}{5} - \frac{2u^3}{3} + u + C$$

$$= \frac{1}{5}(\sec\theta)^5 - \frac{2}{3}(\sec\theta)^3 + \sec\theta + C$$

From the substitution $x = \tan\theta$, you can deduce that $\sec\theta = \sqrt{1+x^2}$ to get the solution:

$$\frac{1}{5}\left(\sqrt{1+x^2}\right)^5 - \frac{2}{3}\left(\sqrt{1+x^2}\right)^3 + \sqrt{1+x^2} + C$$

$$= \frac{1}{5}\left(1+x^2\right)^{5/2} - \frac{2}{3}\left(1+x^2\right)^{3/2} + \sqrt{1+x^2} + C$$

Note that you can also do this problem starting with the substitution $u = \sqrt{x^2+1}$ and a bit of algebra.

---

933. $\quad 3^5\left(\dfrac{4\sqrt{2}-1}{5} - \dfrac{2\sqrt{2}-1}{3}\right)$

The given expression is

$$\int_0^3 x^3\sqrt{x^2+9}\,dx$$

Let $x = 3\tan\theta$ to get $dx = 3\sec^2\theta\,d\theta$. You can also find the new limits of integration by using the original limits along with the substitution $x = 3\tan\theta$. If $x = 3$, you get $3 = 3\tan\theta$, or $1 = \tan\theta$, so that $\frac{\pi}{4} = \theta$. Likewise, if $x = 0$, you get $0 = 3\tan\theta$, or $0 = \tan\theta$, so that $0 = \theta$. Using these substitutions and the new limits of integration gives you the following:

$$\int_0^{\pi/4} (3\tan\theta)^3 \sqrt{(3\tan\theta)^2+9}\left(3\sec^2\theta\right)d\theta$$

$$= \int_0^{\pi/4} (3\tan\theta)^3 \sqrt{9\left(\tan^2\theta+1\right)}\left(3\sec^2\theta\right)d\theta$$

$$= \int_0^{\pi/4} (3\tan\theta)^3 3\sqrt{\sec^2\theta}\left(3\sec^2\theta\right)d\theta$$

$$= \int_0^{\pi/4} 3^5 \tan^3\theta \sec^3\theta\,d\theta$$

Now factor out $\sec\theta\tan\theta$ and use a trigonometric identity:

$$\int_0^{\pi/4} 3^5 \tan^3\theta \sec^3\theta\,d\theta$$

$$= \int_0^{\pi/4} 3^5 \tan^2\theta \sec^2\theta \sec\theta\tan\theta\,d\theta$$

$$= \int_0^{\pi/4} 3^5 \left(\sec^2\theta - 1\right)\sec^2\theta \sec\theta\tan\theta\,d\theta$$

Use the substitution $u = \sec\theta$ so that $du = \sec\theta\tan\theta\,d\theta$. You can again find the new limits of integration; with $\theta = \frac{\pi}{4}$, you get $u = \sec\frac{\pi}{4} = \frac{2}{\sqrt{2}} = \sqrt{2}$, and with $\theta = 0$, you get $u = \sec 0 = 1$. Using the new limits and the substitution, you can get the answer as follows:

$$3^5 \int_1^{\sqrt{2}} \left(u^2-1\right)u^2\,du$$

$$= 3^5 \int_1^{\sqrt{2}} \left(u^4-u^2\right)du$$

$$= 3^5 \left(\frac{u^5}{5} - \frac{u^3}{3}\right)\Bigg|_1^{\sqrt{2}}$$

$$= 3^5 \left(\frac{\sqrt{2}^5}{5} - \frac{\sqrt{2}^3}{3} - \left(\frac{1}{5} - \frac{1}{3}\right)\right)$$

$$= 3^5 \left(\frac{4\sqrt{2}-1}{5} - \frac{2\sqrt{2}-1}{3}\right)$$

934. $\frac{1}{6}\left(\sin^{-1}(x^3)+x^3\sqrt{1-x^6}\right)+C$

Start by rewriting the integrand:

$$\int x^2\sqrt{1-x^6}\,dx = \int x^2\sqrt{1-(x^3)^2}\,dx$$

Use a substitution, letting $w = x^3$, to get $dw = 3x^2\,dx$, or $\frac{1}{3}\,dw = x^2\,dx$. Putting these values into the integral gives you

$$\int x^2\sqrt{1-(x^3)^2}\,dx = \int \frac{1}{3}\sqrt{1-w^2}\,dw$$

Now use the trigonometric substitution $w = 1\sin\theta$, which gives you $dw = \cos\theta\,d\theta$. Using these values in the integral gives you the following:

$$\int \frac{1}{3}\sqrt{1-\sin^2\theta}\,\cos\theta\,d\theta$$
$$=\frac{1}{3}\int\sqrt{\cos^2\theta}\,\cos\theta\,d\theta$$
$$=\frac{1}{3}\int\cos^2\theta\,d\theta$$

Now you can use the identity $\cos^2\theta = \frac{1}{2}(1+\cos(2\theta))$ to rewrite the integral:

$$\frac{1}{3}\int\frac{1}{2}(1+\cos(2\theta))\,d\theta$$
$$=\frac{1}{6}\int(1+\cos(2\theta))\,d\theta$$
$$=\frac{1}{6}\left(\theta+\frac{1}{2}\sin(2\theta)\right)+C$$
$$=\frac{1}{6}\left(\theta+\frac{1}{2}2\sin\theta\cos\theta\right)+C$$
$$=\frac{1}{6}(\theta+\sin\theta\cos\theta)+C$$

From $w = \sin\theta$, you can deduce that $\sin^{-1}w = \theta$ and that $\cos\theta = \sqrt{1-w^2}$. (Note that you can find the value for $\cos\theta$ by using an identity for sine and cosine.) Substituting these values into the integral gives you

$$\frac{1}{6}\left(\sin^{-1}w+w\sqrt{1-w^2}\right)+C$$

Replacing w with x^3 gives you the solution:

$$\frac{1}{6}\left(\sin^{-1}(x^3)+x^3\sqrt{1-x^6}\right)+C$$

935. $\frac{1}{16}\left(\dfrac{x+3}{\sqrt{7-6x-x^2}}\right)+C$

Rewrite the denominator as a radical:

$$\int\frac{dx}{\left(7-6x-x^2\right)^{3/2}} = \int\frac{dx}{\left(\sqrt{7-6x-x^2}\right)^3}$$

Now you complete the square on the quadratic expression:

$$7 - 6x - x^2 = -x^2 - 6x + 7$$
$$= -(x^2 + 6x) + 7$$
$$= -(x^2 + 6x + 9) + 7 + 9$$
$$= -(x + 3)^2 + 16$$
$$= 16 - (x + 3)^2$$

So the integral becomes

$$\int \frac{dx}{\left(\sqrt{16 - (x+3)^2}\right)^3}$$

Now use the substitution $(x + 3) = 4\sin\theta$, which gives you $dx = 4\cos\theta\, d\theta$. Substituting these values into the integral gives you

$$\int \frac{4\cos\theta\, d\theta}{\left(\sqrt{16 - (4\sin\theta)^2}\right)^3} = \int \frac{4\cos\theta\, d\theta}{\left(\sqrt{16(1 - \sin^2\theta)}\right)^3}$$

$$= \int \frac{4\cos\theta\, d\theta}{\left(4\sqrt{\cos^2\theta}\right)^3}$$

$$= \int \frac{4\cos\theta\, d\theta}{4^3\cos^3\theta}$$

$$= \frac{1}{4^2}\int \frac{1}{\cos^2\theta}\, d\theta$$

$$= \frac{1}{16}\int \sec^2\theta\, d\theta$$

$$= \frac{1}{16}\tan\theta + C$$

From the substitution $(x + 3) = 4\sin\theta$, or $\frac{x+3}{4} = \sin\theta$, you can deduce that

$\tan\theta = \dfrac{x+3}{\sqrt{7 - 6x - x^2}}$. (To deduce that $\tan\theta = \dfrac{x+3}{\sqrt{7 - 6x - x^2}}$, you can label a right triangle

using $\frac{x+3}{4} = \sin\theta$ and then find the missing side with the Pythagorean theorem.)

Therefore, the solution is

$$\frac{1}{16}\left(\frac{x+3}{\sqrt{7 - 6x - x^2}}\right) + C$$

936. $\dfrac{\sqrt{2}}{2}\ln\left|\dfrac{x + 2 - \sqrt{2}}{\sqrt{x^2 + 4x + 2}}\right| + C$

Here's the given integral:

$$\int \frac{dx}{x^2 + 4x + 2}$$

This may not look like a trigonometric substitution problem at first glance because so many of those problems involve some type of radical. However, many integrals that contain quadratic expressions can still be solved with this method.

First complete the square on the quadratic expression:

$$x^2 + 4x + 2 = (x^2 + 4x) + 2$$
$$= (x^2 + 4x + 4) + 2 - 4$$
$$= (x+2)^2 - 2$$

So the integral becomes

$$\int \frac{dx}{(x^2 + 4x + 2)} = \int \frac{dx}{(x+2)^2 - 2}$$

Use the substitution $(x+2) = \sqrt{2}\sec\theta$, which gives you $dx = \sqrt{2}\sec\theta\tan\theta\,d\theta$. Putting this value into your integral yields

$$\int \frac{dx}{((x+2)^2 - 2)} = \int \frac{\sqrt{2}\sec\theta\tan\theta\,d\theta}{((\sqrt{2}\sec\theta)^2 - 2)}$$
$$= \int \frac{\sqrt{2}\sec\theta\tan\theta\,d\theta}{(2\sec^2\theta - 2)}$$
$$= \int \frac{\sqrt{2}\sec\theta\tan\theta\,d\theta}{(2(\sec^2\theta - 1))}$$
$$= \int \frac{\sqrt{2}\sec\theta\tan\theta\,d\theta}{(2(\tan^2\theta))}$$
$$= \frac{\sqrt{2}}{2}\int \frac{\sec\theta\,d\theta}{\tan\theta}$$

Because $\frac{\sec\theta}{\tan\theta} = \frac{1}{\cos\theta}\frac{\cos\theta}{\sin\theta} = \csc\theta$, you have

$$\frac{\sqrt{2}}{2}\int \frac{\sec\theta\,d\theta}{\tan\theta} = \frac{\sqrt{2}}{2}\int \csc\theta\,d\theta$$
$$= \frac{\sqrt{2}}{2}\ln|\csc\theta - \cot\theta| + C$$

From the substitution $(x+2) = \sqrt{2}\sec\theta$, or $\frac{(x+2)}{\sqrt{2}} = \sec\theta$, you can deduce that $\csc\theta = \frac{x+2}{\sqrt{x^2 + 4x + 2}}$ and that $\cot\theta = \frac{\sqrt{2}}{\sqrt{x^2 + 4x + 2}}$. (To deduce these values for $\csc\theta$ and $\cot\theta$, you can label a right triangle using $\frac{(x+2)}{\sqrt{2}} = \sec\theta$ and then find the missing side using the Pythagorean theorem.) Putting these values into the antiderivative gives you the solution:

$$\frac{\sqrt{2}}{2}\ln\left|\frac{x+2}{\sqrt{x^2+4x+2}} - \frac{\sqrt{2}}{\sqrt{x^2+4x+2}}\right| + C$$
$$= \frac{\sqrt{2}}{2}\ln\left|\frac{x+2-\sqrt{2}}{\sqrt{x^2+4x+2}}\right| + C$$

937. $\dfrac{81}{40}$

You can begin by rewriting the expression underneath of the radical to make the required substitution clearer:

$$\int_0^{3/2} x^3 \sqrt{9-4x^2}\,dx = \int_0^{3/2} x^3 \sqrt{3^2 - (2x)^2}\,dx$$

Use the substitution $2x = 3\sin\theta$ to get $x = \dfrac{3}{2}\sin\theta$ so that $dx = \dfrac{3}{2}\cos\theta\,d\theta$. You can find the new limits of integration by using the original limits of integration and the substitution. If $x = \dfrac{3}{2}$, then $\dfrac{3}{2} = \dfrac{3}{2}\sin\theta$, or $1 = \sin\theta$, so that $\dfrac{\pi}{2} = \theta$. Likewise, if $x = 0$, then $0 = \dfrac{3}{2}\sin\theta$ so that $0 = \sin\theta$, or $0 = \theta$. Using this information, you can produce the following integral:

$$\int_0^{\pi/2} \left(\frac{3}{2}\sin\theta\right)^3 \sqrt{9 - (3\sin\theta)^2}\,\frac{3}{2}\cos\theta\,d\theta$$

$$= \int_0^{\pi/2} \left(\frac{3}{2}\right)^4 \sin^3\theta \sqrt{9\left(1 - \sin^2\theta\right)}\cos\theta\,d\theta$$

$$= \left(\frac{3}{2}\right)^4 \int_0^{\pi/2} \sin^3\theta \sqrt{9\cos^2\theta}\,\cos\theta\,d\theta$$

$$= \frac{3^5}{2^4} \int_0^{\pi/2} \sin^3\theta \cos^2\theta\,d\theta$$

$$= \frac{3^5}{2^4} \int_0^{\pi/2} \sin^2\theta \cos^2\theta \sin\theta\,d\theta$$

$$= \frac{3^5}{2^4} \int_0^{\pi/2} \left(1 - \cos^2\theta\right)\cos^2\theta \sin\theta\,d\theta$$

Now use the substitution $u = \cos\theta$ so that $du = -\sin\theta\,d\theta$, or $-du = \sin\theta\,d\theta$. You can again find the new limits of integration: If $\theta = \dfrac{\pi}{2}$, then $u = \cos\dfrac{\pi}{2} = 0$. Likewise, if $\theta = 0$, then $u = \cos 0 = 1$. Using the substitution along with the new limits of integration, you get the following (recall that when you switch the limits of integration, the sign on the integral changes):

$$-\frac{3^5}{2^4} \int_1^0 \left(1 - u^2\right)u^2\,du$$

$$= \frac{3^5}{2^4} \int_0^1 \left(1 - u^2\right)u^2\,du$$

$$= \frac{3^5}{2^4} \int_0^1 \left(u^2 - u^4\right)du$$

Now you have the following:

$$\frac{3^5}{2^4} \int_0^1 \left(u^2 - u^4\right)du = \frac{3^5}{2^4}\left(\frac{u^3}{3} - \frac{u^5}{5}\right)\Bigg|_0^1$$

$$= \frac{3^5}{2^4}\left(\frac{1}{3} - \frac{1}{5}\right)$$

$$= \frac{3^5}{2^4}\left(\frac{2}{15}\right)$$

$$= \frac{81}{40}$$

938.

$$8\sin^{-1}\left(\frac{x-3}{4}\right)+\frac{(x-3)\sqrt{7+6x-x^2}}{2}+C$$

Here's the given integral:

$$\int\sqrt{7+6x-x^2}\,dx$$

First complete the square on the quadratic expression underneath the square root:

$$\begin{aligned} 7+6x-x^2 &= -x^2+6x+7 \\ &= -\left(x^2-6x\right)+7 \\ &= -\left(x^2-6x+9\right)+7+9 \\ &= -(x-3)^2+16 \\ &= 16-(x-3)^2 \end{aligned}$$

Therefore, you have the integral

$$\int\sqrt{7+6x-x^2}\,dx = \int\sqrt{16-(x-3)^2}\,dx$$

Now you can use the substitution $(x-3)=4\sin\theta$, which gives you $dx = 4\cos\theta\,d\theta$. Substituting these values into the integral gives you

$$\begin{aligned} &\int\sqrt{16-(x-3)^2}\,dx \\ &= \int\sqrt{16-(4\sin\theta)^2}\,4\cos\theta\,d\theta \\ &= \int\sqrt{16\left(1-\sin^2\theta\right)}\,4\cos\theta\,d\theta \\ &= \int\sqrt{16\cos^2\theta}\,4\cos\theta\,d\theta \\ &= 16\int\cos^2\theta\,d\theta \end{aligned}$$

Then use the identity $\cos^2\theta = \frac{1}{2}(1+\cos(2\theta))$ to simplify the integral:

$$\begin{aligned} 16\int\cos^2\theta\,d\theta &= 16\int\frac{1}{2}(1+\cos(2\theta))\,d\theta \\ &= 8\int(1+\cos(2\theta))\,d\theta \\ &= 8\left(\theta+\frac{1}{2}\sin(2\theta)\right)+C \\ &= 8\left(\theta+\frac{1}{2}2\sin\theta\cos\theta\right)+C \\ &= 8(\theta+\sin\theta\cos\theta)+C \end{aligned}$$

From the substitution $(x-3)=4\sin\theta$, you get $\frac{x-3}{4}=\sin\theta$ and $\sin^{-1}\left(\frac{x-3}{4}\right)=\theta$. You can deduce that $\cos\theta = \dfrac{\sqrt{7+6x-x^2}}{4}$ by labeling a right triangle using $\frac{x-3}{4}=\sin\theta$ and then finding the missing side of the triangle with the Pythagorean theorem. Putting these values into the antiderivative gives you the solution:

$$8\left(\sin^{-1}\left(\frac{x-3}{4}\right)+\left(\frac{x-3}{4}\right)\left(\frac{\sqrt{7+6x-x^2}}{4}\right)\right)+C$$

$$=8\sin^{-1}\left(\frac{x-3}{4}\right)+\frac{(x-3)\sqrt{7+6x-x^2}}{2}+C$$

939. $-\dfrac{16}{3}\left(16-x^2\right)^{3/2}+\dfrac{1}{5}\left(16-x^2\right)^{5/2}+C$

Here's the given integral:

$$\int x^3\sqrt{16-x^2}\,dx$$

Start with the substitution $x=4\sin\theta$ so that $dx=4\cos\theta\,d\theta$. Substituting these values into the original integral, you get the following:

$$\int (4\sin\theta)^3\sqrt{16-(4\sin\theta)^2}\,4\cos\theta\,d\theta$$
$$=\int (4\sin\theta)^3\sqrt{16-16\sin^2\theta}\,4\cos\theta\,d\theta$$
$$=\int (4\sin\theta)^3\sqrt{16\left(1-\sin^2\theta\right)}\,4\cos\theta\,d\theta$$
$$=\int (4\sin\theta)^3\sqrt{16\cos^2\theta}\,4\cos\theta\,d\theta$$
$$=\int 4^3\sin^3\theta(4\cos\theta)4\cos\theta\,d\theta$$
$$=4^5\int \sin^3\theta\cos^2\theta\,d\theta$$

Now factor out $\sin\theta$ and use a trigonometric identity along with a u-substitution:

$$4^5\int \sin^2\theta\cos^2\theta\sin\theta\,d\theta$$
$$=4^5\int\left(1-\cos^2\theta\right)\cos^2\theta\sin\theta\,d\theta$$

With $u=\cos\theta$, you get $du=-\sin\theta\,d\theta$, or $-du=\sin\theta\,d\theta$. Substituting these values into the integral gives you

$$-4^5\int\left(1-u^2\right)u^2du$$
$$=-4^5\int\left(u^2-u^4\right)du$$
$$=-4^5\left(\dfrac{u^3}{3}-\dfrac{u^5}{5}\right)+C$$
$$=-4^5\left(\dfrac{(\cos\theta)^3}{3}-\dfrac{(\cos\theta)^5}{5}\right)+C$$

From the original substitution, you have $\dfrac{x}{4}=\sin\theta$, from which you can deduce that

$\dfrac{\sqrt{16-x^2}}{4}=\cos\theta$. (To deduce that $\dfrac{\sqrt{16-x^2}}{4}=\cos\theta$, you can label a right triangle using

$\dfrac{x}{4}=\sin\theta$ and then find the missing side with the Pythagorean theorem.) With these values, you arrive at the following solution:

$$-\dfrac{4^5}{3}\left(\dfrac{\sqrt{16-x^2}}{4}\right)^3+\dfrac{4^5}{5}\left(\dfrac{\sqrt{16-x^2}}{4}\right)^5+C$$
$$=-\dfrac{4^5}{3}\dfrac{\left(16-x^2\right)^{3/2}}{4^3}+\dfrac{4^5}{5}\dfrac{\left(16-x^2\right)^{5/2}}{4^5}+C$$
$$=-\dfrac{16}{3}\left(16-x^2\right)^{3/2}+\dfrac{1}{5}\left(16-x^2\right)^{5/2}+C$$

940. $\dfrac{A}{x}+\dfrac{B}{x^2}+\dfrac{C}{x^3}+\dfrac{D}{(x+1)}+\dfrac{E}{(x+1)^2}$

The given expression is

$$\frac{4x+1}{x^3(x+1)^2}$$

Notice that you have the linear factor x raised to the third power and the linear factor $(x+1)$ raised to the second power. Therefore, in the decomposition, all the numerators will be constants:

$$\frac{4x+1}{x^3(x+1)^2}=\frac{A}{x}+\frac{B}{x^2}+\frac{C}{x^3}+\frac{D}{(x+1)}+\frac{E}{(x+1)^2}$$

941. $\dfrac{A}{x-1}+\dfrac{B}{x+1}+\dfrac{Cx+D}{x^2+1}$

The given expression is

$$\frac{2x}{x^4-1}$$

Begin by factoring the denominator completely:

$$\frac{2x}{x^4-1}=\frac{2x}{\left(x^2-1\right)\left(x^2+1\right)}=\frac{2x}{(x-1)(x+1)\left(x^2+1\right)}$$

In this expression, you have two distinct linear factors and one irreducible quadratic factor. Therefore, the fraction decomposition becomes

$$\frac{2x}{(x-1)(x+1)\left(x^2+1\right)}=\frac{A}{x-1}+\frac{B}{x+1}+\frac{Cx+D}{x^2+1}$$

942. $\dfrac{A}{x+1}+\dfrac{B}{\left(x+1\right)^2}+\dfrac{Cx+D}{x^2+5}+\dfrac{Ex+F}{\left(x^2+5\right)^2}+\dfrac{Gx+H}{\left(x^2+5\right)^3}$

Here's the given expression:

$$\frac{5x^2+x-4}{(x+1)^2\left(x^2+5\right)^3}$$

In this expression, a linear factor is being squared, and an irreducible quadratic factor is raised to the third power. Therefore, the fraction decomposition becomes

$$\frac{5x^2+x-4}{(x+1)^2\left(x^2+5\right)^3}$$

$$=\frac{A}{x+1}+\frac{B}{\left(x+1\right)^2}+\frac{Cx+D}{x^2+5}+\frac{Ex+F}{\left(x^2+5\right)^2}+\frac{Gx+H}{\left(x^2+5\right)^3}$$

943. $\dfrac{A}{x}+\dfrac{B}{x^2}+\dfrac{C}{\left(x-1\right)}+\dfrac{D}{\left(x-1\right)^2}+\dfrac{E}{\left(x-1\right)^3}+\dfrac{Fx+G}{x^2+17}$

The given expression is

$$\frac{4x^3+19}{x^2(x-1)^3\left(x^2+17\right)}$$

In this example, you have the linear factor x that's being squared, the linear factor $(x - 1)$ that's being cubed, and the irreducible quadratic factor $(x^2 + 17)$. Therefore, the fraction decomposition becomes

$$\frac{4x^3 + 19}{x^2(x-1)^3(x^2+17)}$$

$$= \frac{A}{x} + \frac{B}{x^2} + \frac{C}{(x-1)} + \frac{D}{(x-1)^2} + \frac{E}{(x-1)^3} + \frac{Fx+G}{x^2+17}$$

944. $\dfrac{A}{x-3} + \dfrac{B}{x+3} + \dfrac{Cx+D}{x^2+1} + \dfrac{Ex+F}{\left(x^2+1\right)^2}$

Begin by factoring the denominator of the expression:

$$\frac{3x^2+4}{\left(x^2-9\right)\left(x^4+2x^2+1\right)} = \frac{3x^2+4}{(x-3)(x+3)\left(x^2+1\right)^2}$$

In the expression on the right side of the equation, you have the distinct linear factors $(x - 3)$ and $(x + 3)$ and the irreducible quadratic factor $(x^2 + 1)$ that's being squared. Therefore, the fraction decomposition becomes

$$\frac{3x^2+4}{(x-3)(x+3)\left(x^2+1\right)^2} = \frac{A}{x-3} + \frac{B}{x+3} + \frac{Cx+D}{x^2+1} + \frac{Ex+F}{\left(x^2+1\right)^2}$$

945. $\dfrac{-\frac{1}{3}}{x+2} + \dfrac{\frac{1}{3}}{x-1}$

Start by performing the decomposition:

$$\frac{1}{(x+2)(x-1)} = \frac{A}{x+2} + \frac{B}{x-1}$$

Multiply both sides of the equation by $(x + 2)(x - 1)$:

$$(x+2)(x-1)\frac{1}{(x+2)(x-1)} = (x+2)(x-1)\left(\frac{A}{x+2} + \frac{B}{x-1}\right)$$

$$1 = A(x-1) + B(x+2)$$

Now let $x = 1$ to get $1 = A(1 - 1) + B(1 + 2)$ so that $1 = 3B$, or $\frac{1}{3} = B$. Also let $x = -2$ to get $1 = A(-2 - 1) + B(-2 + 2)$ so that $1 = -3A$, or $-\frac{1}{3} = A$. Therefore, you arrive at

$$\frac{A}{x+2} + \frac{B}{x-1} = \frac{-\frac{1}{3}}{x+2} + \frac{\frac{1}{3}}{x-1}$$

946. $\dfrac{2}{x} + \dfrac{-2x+1}{x^2+1}$

Start by factoring the denominator:

$$\frac{x+2}{x^3+x} = \frac{x+2}{x\left(x^2+1\right)}$$

Then perform the decomposition:

$$\frac{x+2}{x\left(x^2+1\right)} = \frac{A}{x} + \frac{Bx+C}{x^2+1}$$

Multiply both sides of the equation by $x(x^2 + 1)$:

$$x\left(x^2+1\right)\left(\frac{x+2}{x\left(x^2+1\right)}\right)=x\left(x^2+1\right)\left(\frac{A}{x}+\frac{Bx+C}{x^2+1}\right)$$

$$x+2=A\left(x^2+1\right)+(Bx+C)x$$

Note that if $x = 0$, then $0+2 = A\left(0^2+1\right)+(B(0)+C)(0)$ so that $2 = A$. Next, expand the right side of the equation and equate the coefficients to find the remaining coefficients:

$$x+2=Ax^2+A+Bx^2+Cx$$

After rearranging, you get

$$0x^2+1x+2=(A+B)x^2+Cx+A$$

Now equate coefficients to arrive at the equation $0 = A + B$; however, you know that $2 = A$, so $0 = 2 + B$, or $-2 = B$. Likewise, by equating coefficients, you immediately find that $C = 1$. Therefore, the partial fraction decomposition becomes

$$\frac{x+2}{x\left(x^2+1\right)}=\frac{2}{x}+\frac{-2x+1}{x^2+1}$$

---

947. $\dfrac{5}{x-3}+\dfrac{16}{(x-3)^2}$

Start by factoring the denominator:

$$\frac{5x+1}{x^2-6x+9}=\frac{5x+1}{(x-3)^2}$$

Then perform the fraction decomposition:

$$\frac{5x+1}{(x-3)^2}=\frac{A}{x-3}+\frac{B}{(x-3)^2}$$

Multiply both sides of the equation by $(x-3)^2$:

$$(x-3)^2\frac{5x+1}{(x-3)^2}=(x-3)^2\left(\frac{A}{x-3}+\frac{B}{(x-3)^2}\right)$$

$$5x+1=A(x-3)+B$$

By letting $x = 3$ in the last equation, you arrive at $5(3) + 1 = A(3 - 3) + B$, so $16 = B$. By expanding the right side of the equation, you get

$$5x+1=Ax-3A+B$$

So by equating coefficients, you immediately see that $5 = A$. Therefore, the fraction decomposition becomes

$$\frac{5x+1}{(x-3)^2}=\frac{5}{x-3}+\frac{16}{(x-3)^2}$$

---

948. $\dfrac{\frac{1}{2}}{\left(x^2+1\right)}+\dfrac{\frac{1}{2}}{\left(x^2+3\right)}$

Begin by factoring the denominator:

$$\frac{x^2+2}{x^4+4x^2+3}=\frac{x^2+2}{\left(x^2+1\right)\left(x^2+3\right)}$$

Then perform the decomposition:

$$\frac{x^2+2}{\left(x^2+1\right)\left(x^2+3\right)}=\frac{Ax+B}{\left(x^2+1\right)}+\frac{Cx+D}{\left(x^2+3\right)}$$

Multiply both sides of the equation by $(x^2 + 1)(x^2 + 3)$ to get

$$x^2+2=\left(Ax+B\right)\left(x^2+3\right)+\left(Cx+D\right)\left(x^2+1\right)$$

Expand the right side of the equation and collect like terms:

$$x^2+2=Ax^3+3Ax+Bx^2+3B+Cx^3+Cx+Dx^2+D$$

$$0x^3+1x^2+0x+2=(A+C)x^3+(B+D)x^2+(3A+C)x+(3B+D)$$

Equating coefficients gives you the equations $A + C = 0$, $B + D = 1$, $3A + C = 0$, and $3B + D = 2$. Because $A = -C$, you get $3(-C) + C = 0$, so $-2C = 0$, or $C = 0$; that means $A = 0$ as well. Likewise, because $B = 1 - D$, you get $3(1 - D) + D = 2$, so $3 - 2D = 2$, or $D = \frac{1}{2}$; you also get $B = 1 - \frac{1}{2} = \frac{1}{2}$. With these values, you get the following solution:

$$\frac{x^2+2}{x^4+4x^2+3}=\frac{x^2+2}{\left(x^2+1\right)\left(x^2+3\right)}$$

$$=\frac{\frac{1}{2}}{\left(x^2+1\right)}+\frac{\frac{1}{2}}{\left(x^2+3\right)}$$

949. $\dfrac{1}{x^2+5}-\dfrac{4}{\left(x^2+5\right)^2}$

Begin by performing the decomposition:

$$\frac{x^2+1}{\left(x^2+5\right)^2}=\frac{Ax+B}{x^2+5}+\frac{Cx+D}{\left(x^2+5\right)^2}$$

Multiplying both sides of the equation by $(x^2 + 5)^2$ gives you

$$\left(x^2+5\right)^2\left(\frac{x^2+1}{\left(x^2+5\right)^2}\right)=\left(x^2+5\right)^2\left(\frac{Ax+B}{x^2+5}+\frac{Cx+D}{\left(x^2+5\right)^2}\right)$$

$$x^2+1=\left(Ax+B\right)\left(x^2+5\right)+\left(Cx+D\right)$$

Expand the right side of the equation and regroup:

$$x^2+1=Ax^3+5Ax+Bx^2+5B+Cx+D$$

$$0x^3+1x^2+0x+1=Ax^3+Bx^2+(5A+C)x+(5B+D)$$

By equating coefficients, you find $A = 0$, $B = 1$, $5A + C = 0$, and $5B + D = 0$. Using $A = 0$ and $5A + C = 0$, you get $5(0) + C = 0$, or $C = 0$. Likewise, using $B = 1$ and $5B + D = 1$, you find that $5(1) + D = 1$ so that $D = -4$. Therefore, the fraction decomposition becomes

$$\frac{x^2+1}{\left(x^2+5\right)^2}=\frac{1}{x^2+5}+\frac{-4}{\left(x^2+5\right)^2}$$

$$=\frac{1}{x^2+5}-\frac{4}{\left(x^2+5\right)^2}$$

950. $x + 5\ln|x - 5| + C$

The given integral is

$$\int \frac{x}{x-5} dx$$

Notice that the degree of the numerator is equal to the degree of the denominator, so you must divide. You could use long division, but for simple expressions of this type, you can simply subtract 5 and add 5 to the numerator and then split the fraction:

$$\int \frac{x}{x-5} dx = \int \frac{(x-5)+5}{x-5} dx$$

$$= \int \left(\frac{x-5}{x-5} + \frac{5}{x-5} \right) dx$$

$$= \int \left(1 + \frac{5}{x-5} \right) dx$$

Now apply elementary antiderivative formulas to get the solution:

$$\int \left(1 + \frac{5}{x-5} \right) dx = x + 5\ln|x-5| + C$$

951. $\frac{x^2}{2} - 6x + 36\ln|x + 6| + C$

The given integral is

$$\int \frac{x^2}{x+6} dx$$

The degree of the numerator is greater than or equal to the degree of the denominator, so use polynomial long division to get the following:

$$\int \frac{x^2}{x+6} dx = \int \left(x - 6 + \frac{36}{x+6} \right) dx$$

Then apply basic antiderivative formulas:

$$\int \left(x - 6 + \frac{36}{x+6} \right) dx = \frac{x^2}{2} - 6x + 36\ln|x+6| + C$$

952. $\frac{7}{9}\ln|x + 4| + \frac{2}{9}\ln|x - 5| + C$

The given integral is

$$\int \frac{x-3}{(x+4)(x-5)} dx$$

First perform a fraction decomposition on the integrand:

$$\frac{x-3}{(x+4)(x-5)} = \frac{A}{x+4} + \frac{B}{x-5}$$

Multiplying both sides of the equation by $(x + 4)(x - 5)$ and simplifying gives you

$$(x+4)(x-5)\left(\frac{x-3}{(x+4)(x-5)}\right)=(x+4)(x-5)\left(\frac{A}{x+4}+\frac{B}{x-5}\right)$$
$$x-3 = A(x-5)+B(x+4)$$

Now you can find the values of the coefficients by picking appropriate values of x and solving the resulting equations. So if you let $x = 5$, you get $5 - 3 = A(5 - 5) + B(5 + 4)$, or $2 = 9B$, so that $\frac{2}{9} = B$. Likewise, if $x = -4$, then you get $-4 - 3 = A(-4 - 5) + B(-4 + 4)$, or $-7 = A(-9)$, so that $\frac{7}{9} = A$. This gives you

$$\int\frac{x-3}{(x+4)(x-5)}\,dx = \int\left(\frac{A}{x+4}+\frac{B}{x-5}\right)dx$$
$$= \int\left(\frac{\frac{7}{9}}{x+4}+\frac{\frac{2}{9}}{x-5}\right)dx$$

Applying elementary antiderivative formulas gives you the solution:

$$\frac{7}{9}\ln|x+4|+\frac{2}{9}\ln|x-5|+C$$

953. $\ln\left(\frac{\sqrt{10}}{3}\right)$

The given expression is

$$\int_4^5\frac{1}{x^2-1}\,dx$$

First factor the denominator of the integrand:

$$\frac{1}{x^2-1}=\frac{1}{(x+1)(x-1)}$$

Then perform the fraction decomposition:

$$\frac{1}{(x+1)(x-1)}=\frac{A}{x+1}+\frac{B}{x-1}$$

Multiplying both sides of the equation by $(x + 1)(x - 1)$ and simplifying gives you

$$(x+1)(x-1)\left(\frac{1}{(x+1)(x-1)}\right)=(x+1)(x-1)\left(\frac{A}{x+1}+\frac{B}{x-1}\right)$$
$$1 = A(x-1)+B(x+1)$$

Now you can solve for the coefficients by picking appropriate values of x. Notice that if $x = 1$, you get $1 = A(1 - 1) + B(1 + 1)$ so that $1 = 2B$, or $\frac{1}{2} = B$. Likewise, if $x = -1$, you get $1 = A(-1 - 1) + B(-1 + 1)$ so that $1 = -2A$, or $\frac{-1}{2} = A$. This gives you

$$\int_4^5\frac{1}{x^2-1}\,dx = \int_4^5\left(\frac{1}{(x+1)(x-1)}\right)dx$$
$$= \int_4^5\left(\frac{A}{x+1}+\frac{B}{x-1}\right)dx$$
$$= \int_4^5\left(\frac{-\frac{1}{2}}{x+1}+\frac{\frac{1}{2}}{x-1}\right)dx$$

Applying elementary antiderivatives gives you the following:

$$\int_4^5 \left(\frac{-\frac{1}{2}}{x+1} + \frac{\frac{1}{2}}{x-1} \right) dx = \left(-\frac{1}{2}\ln|x+1| + \frac{1}{2}\ln|x-1| \right)\Big|_4^5$$

$$= \frac{1}{2}\left(\ln|x-1| - \ln|x+1| \right)\Big|_4^5$$

$$= \frac{1}{2}\ln\left|\frac{x-1}{x+1}\right|\Big|_4^5$$

$$= \frac{1}{2}\left(\ln\left|\frac{5-1}{5+1}\right| - \ln\left|\frac{4-1}{4+1}\right| \right)$$

$$= \frac{1}{2}\left(\ln\left|\frac{2}{3}\right| - \ln\left|\frac{3}{5}\right| \right)$$

$$= \frac{1}{2}\left(\ln\left(\frac{10}{9}\right) \right)$$

$$= \ln\left(\frac{10}{9}\right)^{1/2}$$

$$= \ln\left(\frac{\sqrt{10}}{\sqrt{9}} \right) = \ln\left(\frac{\sqrt{10}}{3} \right)$$

954. $\quad 3\ln|x+1| - \dfrac{2}{x+1} + C$

The given integral is

$$\int \frac{3x+5}{x^2+2x+1} dx$$

Begin by factoring the denominator:

$$\frac{3x+5}{x^2+2x+1} = \frac{3x+5}{(x+1)^2}$$

And then perform a fraction decomposition:

$$\frac{3x+5}{(x+1)^2} = \frac{A}{x+1} + \frac{B}{(x+1)^2}$$

Multiplying both sides by $(x+1)^2$ gives you the following:

$$(x+1)^2 \frac{3x+5}{(x+1)^2} = (x+1)^2 \left(\frac{A}{x+1} + \frac{B}{(x+1)^2} \right)$$

$$3x+5 = A(x+1) + B$$

Expanding the right side and equating coefficients gives you $3x+5 = Ax + (A+B)$ so that $A = 3$ and $A + B = 5$; in turn, that gives you $3 + B = 5$ so that $B = 2$. Now you have

$$\int \frac{3x+5}{x^2+2x+1} dx = \int \left(\frac{A}{x+1} + \frac{B}{(x+1)^2} \right) dx$$

$$= \int \left(\frac{3}{x+1} + \frac{2}{(x+1)^2} \right) dx$$

For the first term in the integrand, simply use an elementary antiderivative:
$\int \frac{3}{x+1} dx = 3\ln|x+1| + C$. For the second term in the integrand, use a substitution on
$\int \frac{3}{(x+1)^2} dx$ where $u = x + 1$ so that $du = dx$. Using these values gives you

$$\int \frac{2}{u^2} du = 2\int u^{-2} du$$

$$= 2\frac{u^{-1}}{-1} + C$$

$$= \frac{-2}{u} + C$$

$$= -\frac{2}{x+1} + C$$

Therefore, the answer is

$$\int \left(\frac{3}{x+1} + \frac{2}{(x+1)^2} \right) dx = 3\ln|x+1| - \frac{2}{x+1} + C$$

Note: You can also evaluate $\int \frac{3x+5}{(x+1)^2} dx$ with the substitution $u = x + 1$ so that $du = dx$
and $u - 1 = x$.

955. $\quad \frac{x^2}{2} + x + \frac{14}{5}\ln|x-3| - \frac{9}{5}\ln|x+2| + C$

The given expression is

$$\int \frac{x^3 - 6x + 5}{x^2 - x - 6} dx$$

The degree of the numerator is greater than or equal to the degree of the denominator,
so use polynomial long division to get the following:

$$\int \frac{x^3 - 6x + 5}{x^2 - x - 6} dx = \int \left(x + 1 + \frac{x+11}{x^2 - x - 6} \right) dx$$

Next, perform a fraction decomposition:

$$\frac{x+11}{(x-3)(x+2)} = \frac{A}{x-3} + \frac{B}{x+2}$$

Multiply both sides of the equation by $(x - 3)(x + 2)$:

$$x + 11 = A(x+2) + B(x-3)$$

If you let $x = 2$, you get $9 = B(-2 - 3)$ so that $-\frac{9}{5} = B$. And if you let $x = 3$, you get
$14 = A(3 + 2)$ so that $\frac{14}{5} = A$. With these values, you produce the following integral:

$$\int \left(x + 1 + \frac{x+11}{x^2 - x - 6} \right) dx = \int \left(x + 1 + \frac{14/5}{x-3} + \frac{-9/5}{x+2} \right) dx$$

To integrate, use basic antiderivative formulas to get

$$\int \left(x + 1 + \frac{14/5}{x-3} + \frac{-9/5}{x+2} \right) dx = \frac{x^2}{2} + x + \frac{14}{5}\ln|x-3| - \frac{9}{5}\ln|x+2| + C$$

956. $8\ln|x| - 4\ln|x^2+1| + C$

The given expression is

$$\int \frac{8}{x^3+x}\,dx$$

First factor the denominator of the integrand:

$$\frac{8}{x^3+x} = \frac{8}{x(x^2+1)}$$

And then perform the fraction decomposition:

$$\frac{8}{x(x^2+1)} = \frac{A}{x} + \frac{Bx+C}{x^2+1}$$

Multiply both sides by $x(x^2+1)$ and simplify to get

$$x(x^2+1)\left(\frac{8}{x(x^2+1)}\right) = x(x^2+1)\left(\frac{A}{x} + \frac{Bx+C}{x^2+1}\right)$$

$$8 = A(x^2+1) + (Bx+C)(x)$$

Expanding the right side, rearranging terms, and rewriting the left side yields the following:

$$8 = Ax^2 + A + Bx^2 + Cx$$

$$8 = (A+B)x^2 + Cx + A$$

$$0x^2 + 0x + 8 = (A+B)x^2 + Cx + A$$

Equating coefficients gives you $0 = A + B$, $0 = C$, and $8 = A$. Using $8 = A$ and $0 = A + B$ gives you $0 = 8 + B$, or $-8 = B$. These values give you the following integral:

$$\int \frac{8}{x^3+x}\,dx = \int \left(\frac{A}{x} + \frac{Bx+C}{x^2+1}\right)dx$$

$$= \int \left(\frac{8}{x} + \frac{-8x}{x^2+1}\right)dx$$

To evaluate the first term in the integrand, use elementary antiderivatives: $\int \frac{8}{x}\,dx = 8\ln|x| + C$. To evaluate the second term in the integrand, use a substitution on $\int \frac{-8x}{x^2+1}\,dx$ where $u = x^2 + 1$ so that $du = 2x\,dx$. Multiplying both sides of last equation by -4 gives you $-4\,du = -8x\,dx$. Using these substitutions gives you the following:

$$\int \frac{-8x}{x^2+1}\,dx = \int \frac{-4}{u}\,du$$

$$= -4\ln|u| + C$$

$$= -4\ln|x^2+1| + C$$

Combining the two antiderivatives gives you the solution:

$$\int \left(\frac{8}{x} + \frac{-8x}{x^2+1}\right)dx = 8\ln|x| - 4\ln|x^2+1| + C$$

957. $\dfrac{19\sqrt{3}}{18}\tan^{-1}\left(\dfrac{x}{\sqrt{3}}\right)-\dfrac{17x}{6\left(x^2+3\right)}+C$

The given expression is

$$\int\frac{6x^2+1}{x^4+6x^2+9}\,dx$$

Begin by factoring the denominator of the integrand:

$$\frac{6x^2+1}{x^4+6x^2+9}=\frac{6x^2+1}{\left(x^2+3\right)^2}$$

And do a fraction decomposition:

$$\frac{6x^2+1}{\left(x^2+3\right)^2}=\frac{Ax+B}{x^2+3}+\frac{Cx+D}{\left(x^2+3\right)^2}$$

Multiply both sides by $(x^2+3)^2$ and simplify:

$$\left(x^2+3\right)^2\frac{6x^2+1}{(x^2+3)^2}=\left(x^2+3\right)^2\left(\frac{Ax+B}{x^2+3}+\frac{Cx+D}{\left(x^2+3\right)^2}\right)$$

$$6x^2+1=(Ax+B)\left(x^2+3\right)+(Cx+D)$$

Expanding the right side, rearranging and collecting like terms, and rewriting the left side gives you

$$6x^2+1=Ax^3+3Ax+Bx^2+3B+Cx+D$$

$$0x^3+6x^2+0x+1=Ax^3+Bx^2+(3A+C)x+(3B+D)$$

Equating coefficients gives you $A=0$, $B=6$, $3A+C=0$, and $3B+D=1$. Using $3B+D=1$ and $B=6$ gives you $3(6)+D=1$, or $D=-17$. From $A=0$ and $3A+C=0$, you can conclude that $C=0$. So the original integral becomes

$$\int\frac{6x^2+1}{x^4+6x^2+9}\,dx=\int\frac{6x^2+1}{\left(x^2+3\right)^2}\,dx$$

$$=\int\left(\frac{6}{x^2+3}+\frac{-17}{\left(x^2+3\right)^2}\right)dx$$

To integrate the first term in the integrand, simply use an elementary antiderivative formula:

$$\int\frac{6}{x^2+3}\,dx=\int\frac{6}{x^2+\sqrt{3}^2}\,dx$$

$$=\frac{6}{\sqrt{3}}\tan^{-1}\frac{x}{\sqrt{3}}+C$$

Integrating the second term of the integrand is a bit more difficult. Begin by using a trigonometric substitution: $\int \frac{-17}{\left(x^2+3\right)^2}\,dx$ with $x=\sqrt{3}\tan\theta$ so that $dx=\sqrt{3}\sec^2\theta\,d\theta$.

Using these substitutions gives you the following:

$$\int \frac{-17}{\left(x^2+3\right)^2}\,dx = \int \frac{-17\sqrt{3}\sec^2\theta}{\left(\left(\sqrt{3}\tan\theta\right)^2+3\right)^2}\,d\theta$$

$$= \int \frac{-17\sqrt{3}\sec^2\theta}{\left(3\left(\tan^2\theta+1\right)\right)^2}\,d\theta$$

$$= \int \frac{-17\sqrt{3}\sec^2\theta}{9\left(\sec^2\theta\right)^2}\,d\theta$$

$$= \frac{-17\sqrt{3}}{9}\int \frac{1}{\sec^2\theta}\,d\theta$$

$$= \frac{-17\sqrt{3}}{9}\int \cos^2\theta\,d\theta$$

$$= \frac{-17\sqrt{3}}{9}\int \frac{1}{2}\left(1+\cos(2\theta)\right)d\theta$$

$$= \frac{-17\sqrt{3}}{18}\int \left(1+\cos(2\theta)\right)d\theta$$

$$= -\frac{-17\sqrt{3}}{18}\left(\theta+\frac{1}{2}\sin(2\theta)\right)+C$$

$$= \frac{-17\sqrt{3}}{18}\left(\theta+\frac{1}{2}2\sin\theta\cos\theta\right)+C$$

$$= \frac{-17\sqrt{3}}{18}\left(\theta+\sin\theta\cos\theta\right)+C$$

Using the substitution $x=\sqrt{3}\tan\theta$ gives you $\frac{x}{\sqrt{3}}=\tan\theta$, from which you can deduce that $\tan^{-1}\left(\frac{x}{\sqrt{3}}\right)=\theta$, $\sin\theta=\frac{x}{\sqrt{x^2+3}}$, and $\cos\theta=\frac{\sqrt{3}}{\sqrt{x^2+3}}$.

So the antiderivative of the second term becomes

$$\frac{-17\sqrt{3}}{18}\left(\theta+\sin\theta\cos\theta\right)+C$$

$$= \frac{-17\sqrt{3}}{18}\left(\tan^{-1}\left(\frac{x}{\sqrt{3}}\right)+\left(\frac{x}{\sqrt{x^2+3}}\right)\left(\frac{\sqrt{3}}{\sqrt{x^2+3}}\right)\right)+C$$

$$= \frac{-17\sqrt{3}}{18}\left(\tan^{-1}\left(\frac{x}{\sqrt{3}}\right)+\frac{x\sqrt{3}}{x^2+3}\right)+C$$

Combining the two solutions gives you the answer:

$$\int\left(\frac{6}{x^2+3}+\frac{-17}{\left(x^2+3\right)^2}\right)dx$$

$$=\frac{6}{\sqrt{3}}\tan^{-1}\left(\frac{x}{\sqrt{3}}\right)-\frac{17\sqrt{3}}{18}\left(\tan^{-1}\left(\frac{x}{\sqrt{3}}\right)+\frac{x\sqrt{3}}{x^2+3}\right)+C$$

$$=\frac{6}{\sqrt{3}}\tan^{-1}\left(\frac{x}{\sqrt{3}}\right)-\frac{17\sqrt{3}}{18}\tan^{-1}\left(\frac{x}{\sqrt{3}}\right)-\frac{51x}{18\left(x^2+3\right)}+C$$

$$=\frac{6\left(6\sqrt{3}\right)}{\sqrt{3}\left(6\sqrt{3}\right)}\tan^{-1}\left(\frac{x}{\sqrt{3}}\right)-\frac{17\sqrt{3}}{18}\tan^{-1}\left(\frac{x}{\sqrt{3}}\right)-\frac{17x}{6\left(x^2+3\right)}+C$$

$$=\frac{36\sqrt{3}}{18}\tan^{-1}\left(\frac{x}{\sqrt{3}}\right)-\frac{17\sqrt{3}}{18}\tan^{-1}\left(\frac{x}{\sqrt{3}}\right)-\frac{17x}{6\left(x^2+3\right)}+C$$

$$=\frac{19\sqrt{3}}{18}\tan^{-1}\left(\frac{x}{\sqrt{3}}\right)-\frac{17x}{6\left(x^2+3\right)}+C$$

958. $\frac{1}{3}\ln|x+1|-\frac{1}{6}\ln|x^2-x+1|+\frac{\sqrt{3}}{3}\tan^{-1}\left(\frac{2x-1}{\sqrt{3}}\right)+C$

Begin by factoring the denominator:

$$\int\frac{1}{x^3+1}dx=\int\frac{1}{(x+1)\left(x^2-x+1\right)}dx$$

Then find the fraction decomposition:

$$\frac{1}{(x+1)\left(x^2-x+1\right)}=\frac{A}{x+1}+\frac{Bx+C}{x^2-x+1}$$

Multiply both sides by $(x + 1)(x^2 - x + 1)$, which yields

$$1=A\left(x^2-x+1\right)+(Bx+C)(x+1)$$

Expanding the right side and collecting like terms gives you

$$1=Ax^2-Ax+A+Bx^2+Bx+Cx+C$$

$$0x^2+0x+1=(A+B)x^2+(-A+B+C)+(A+C)$$

Equating coefficients gives you the equations $0 = A + B$, $0 = -A + B + C$, and $1 = A + C$. From the first of the three equations, you have $A = -B$, and from the third, you get $1 - A = C$ so that $1 + B = C$. Using the equation $0 = -A + B + C$ with $A = -B$ and with $1 + B = C$ gives you $0 = -(-B) + B + (1 + B)$ so that $-\frac{1}{3} = B$; therefore, $A = \frac{1}{3}$ and $\frac{2}{3} = C$.

With these coefficients, you now have the following integral:

$$\int\frac{1}{(x+1)\left(x^2-x+1\right)}dx=\int\left(\frac{\frac{1}{3}}{x+1}+\frac{-\frac{1}{3}x+\frac{2}{3}}{x^2-x+1}\right)dx$$

Next, complete the square on the expression $x^2 - x + 1$:

$$x^2 - x + 1 = \left(x^2 - x + \frac{1}{4}\right) + 1 - \frac{1}{4}$$

$$= \left(x - \frac{1}{2}\right)^2 + \frac{3}{4}$$

Splitting up the integral gives you

$$\int \left(\frac{\frac{1}{3}}{x+1} + \frac{-\frac{1}{3}x + \frac{2}{3}}{x^2 - x + 1}\right) dx = \int \frac{\frac{1}{3}}{x+1} dx + \int \frac{-\frac{1}{3}x + \frac{2}{3}}{\left(x - \frac{1}{2}\right)^2 + \frac{3}{4}} dx$$

Using the substitution $u = x - \frac{1}{2}$, you get $du = dx$ and $u + \frac{1}{2} = x$ so that $-\frac{1}{3}\left(u + \frac{1}{2}\right) = -\frac{1}{3}x$ and $-\frac{1}{3}u + \frac{1}{2} = -\frac{1}{3}x + \frac{2}{3}$. Using these values gives you

$$\int \frac{\frac{1}{3}}{x+1} dx + \int \frac{-\frac{1}{3}x + \frac{2}{3}}{\left(x - \frac{1}{2}\right)^2 + \frac{3}{4}} dx$$

$$= \int \frac{\frac{1}{3}}{x+1} dx + \int \frac{-\frac{1}{3}u + \frac{1}{2}}{(u)^2 + \left(\frac{\sqrt{3}}{2}\right)^2} dx$$

$$= \int \frac{\frac{1}{3}}{x+1} dx + \int \frac{-\frac{1}{3}u}{u^2 + \frac{3}{4}} du + \frac{1}{2}\int \frac{1}{(u)^2 + \left(\frac{\sqrt{3}}{2}\right)^2} du$$

To evaluate the second integral, $\frac{1}{3}\int \frac{-u}{u^2 + \frac{3}{4}} du$, use the substitution $w = u^2 + \frac{3}{4}$ so that $dw = 2u\,du$, or $-\frac{1}{2} dw = -u\,du$. This substitution gives you

$$-\frac{1}{6}\int \frac{1}{w} dw = -\frac{1}{6}\ln|w| + C$$

$$= -\frac{1}{6}\ln\left|u^2 + \frac{3}{4}\right| + C$$

$$= -\frac{1}{6}\ln\left|x^2 - x + 1\right| + C$$

Therefore, the solution is

$$\int \frac{\frac{1}{3}}{x+1} dx + \int \frac{-\frac{1}{3}u}{u^2 + \frac{3}{4}} du + \frac{1}{2}\int \frac{1}{(u)^2 + \left(\frac{\sqrt{3}}{2}\right)^2} du$$

$$= \frac{1}{3}\ln|x+1| - \frac{1}{6}\ln\left|x^2 - x + 1\right| + \frac{1}{2}\left(\frac{2}{\sqrt{3}}\tan^{-1}\left(\frac{2u}{\sqrt{3}}\right)\right) + C$$

$$= \frac{1}{3}\ln|x+1| - \frac{1}{6}\ln\left|x^2 - x + 1\right| + \frac{\sqrt{3}}{3}\tan^{-1}\left(\frac{2x-1}{\sqrt{3}}\right) + C$$

959. $\ln\left|\dfrac{\sqrt{x+1}-1}{\sqrt{x+1}+1}\right|+C$

Here's the given problem:

$$\int \frac{1}{x\sqrt{x+1}}\,dx$$

Begin by using the rationalizing substitution $u=\sqrt{x+1}$ so that $u^2 = x + 1$, or $u^2 - 1 = x$, and $2u\,du = dx$. This substitution gives you

$$\int \frac{2u\,du}{(u^2-1)u} = 2\int \frac{1}{(u-1)(u+1)}\,du$$

Next, find the fraction decomposition:

$$\frac{1}{(u-1)(u+1)} = \frac{A}{u-1} + \frac{B}{u+1}$$

Multiply both sides by $(u-1)(u+1)$:

$$1 = A(u+1) + B(u-1)$$

Notice that if $u = -1$, you get $1 = B(-2)$ so that $-\dfrac{1}{2} = B$, and if $u = 1$, you get $1 = A(2)$ so that $\dfrac{1}{2} = A$. With these values, you get the following:

$$2\int \frac{1}{(u-1)(u+1)}\,du = 2\int \left(\frac{\frac{1}{2}}{u-1} - \frac{\frac{1}{2}}{u+1}\right)du$$

$$= \ln|u-1| - \ln|u+1| + C$$

$$= \ln\left|\frac{u-1}{u+1}\right| + C$$

$$= \ln\left|\frac{\sqrt{x+1}-1}{\sqrt{x+1}+1}\right| + C$$

960. $2+\ln\dfrac{3}{2}$

Here's the given expression:

$$\int_4^9 \frac{\sqrt{x}}{x-1}\,dx$$

Begin by using the rationalizing substitution $u=\sqrt{x}$ so that $u^2 = x$ and $2u\,du = dx$.

Note that if $x = 9$, you have $u=\sqrt{9}=3$, and if $x = 4$, you have $u=\sqrt{4}=2$, producing the following integral:

$$\int_2^3 \frac{u(2u)}{u^2-1}\,du = 2\int_2^3 \frac{u^2-1+1}{u^2-1}\,du$$

$$= 2\int_2^3 (1)\,du + 2\int_2^3 \frac{1}{(u-1)(u+1)}\,du$$

Next, perform the fraction decomposition:

$$\frac{1}{(u-1)(u+1)} = \frac{A}{u-1} + \frac{B}{u+1}$$

And multiply both sides by $(u-1)(u+1)$ to get

$$1 = A(u+1) + B(u-1)$$

Notice that if $u = -1$, you get $1 = B(-2)$ so that $-\frac{1}{2} = B$, and if $u = 1$, you get $1 = A(2)$ so that $\frac{1}{2} = A$. With these values, you get the following:

$$2\int_2^3 (1)\,du + 2\int_2^3 \left(\frac{1/2}{u-1} - \frac{1/2}{u+1}\right) du$$

$$= 2\left(u + \frac{1}{2}\ln|u-1| - \frac{1}{2}\ln|u+1|\right)\Big|_2^3$$

$$= \left(2u + \ln\left|\frac{u-1}{u+1}\right|\right)\Big|_2^3$$

$$= \left(6 + \ln\frac{1}{2} - \left(4 + \ln\frac{1}{3}\right)\right)$$

$$= 2 + \ln\frac{1}{2} - \ln\frac{1}{3}$$

$$= 2 + \ln\frac{3}{2}$$

961.

$$12\left(\frac{x^{1/2}}{3} + \frac{x^{1/3}}{2} + x^{1/6} + \ln\left|x^{1/6} - 1\right|\right) + C$$

The given expression is

$$\int \frac{2}{\sqrt{x} - \sqrt[3]{x}}\,dx$$

Begin by using the substitution $u = \sqrt[6]{x} = x^{1/6}$ so that $u^6 = x$ and $6u^5\,du = dx$. Also notice that $u^2 = x^{1/3}$ and that $u^3 = x^{1/2}$. With these values, you produce the following integral:

$$2\int \frac{6u^5}{u^3 - u^2}\,du = 12\int \frac{u^5}{u^2(u-1)}\,du$$

$$= 12\int \frac{u^3}{(u-1)}\,du$$

Because the degree of the numerator is greater than or equal to the degree of the denominator, use polynomial long division to get

$$12\int \frac{u^3}{(u-1)}\,du = 12\int \left(u^2 + u + 1 + \frac{1}{u-1}\right) du$$

Now use elementary antiderivative formulas to get the solution:

$$12\int \left(u^2 + u + 1 + \frac{1}{u-1}\right) du$$

$$= 12\left(\frac{u^3}{3} + \frac{u^2}{2} + u + \ln|u-1|\right) + C$$

$$= 12\left(\frac{x^{1/2}}{3} + \frac{x^{1/3}}{2} + x^{1/6} + \ln\left|x^{1/6} - 1\right|\right) + C$$

962.

$$12\left(\frac{x^{2/3}}{8}+\frac{x^{7/12}}{7}+\frac{x^{1/2}}{6}+\frac{x^{5/12}}{5}+\frac{x^{1/3}}{4}+\frac{x^{1/4}}{3}+\frac{x^{1/6}}{2}+x^{1/12}+\ln\left|x^{1/12}-1\right|\right)+C$$

The given problem is

$$\int\frac{1}{\sqrt[3]{x}-\sqrt[4]{x}}\,dx$$

Begin by using the substitution $u=\sqrt[12]{x}=x^{1/12}$ so that $u^{12}=x$ and so that $12u^{11}\,du=dx$. Also notice that $u^4=x^{1/3}$ and that $u^3=x^{1/4}$. With these values, you produce the integral

$$\int\frac{12u^{11}}{u^4-u^3}\,du=12\int\frac{u^{11}}{u^3(u-1)}\,du$$

$$=12\int\frac{u^8}{(u-1)}\,du$$

Because the degree of the numerator is greater than or equal to the degree of the denominator, use polynomial long division:

$$12\int\frac{u^8}{(u-1)}\,du$$

$$=12\int\left(u^7+u^6+u^5+u^4+u^3+u^2+u+1+\frac{1}{u-1}\right)du$$

Then use elementary antiderivative formulas to get the answer:

$$12\int\left(u^7+u^6+u^5+u^4+u^3+u^2+u+1+\frac{1}{u-1}\right)du$$

$$=12\left(\frac{u^8}{8}+\frac{u^7}{7}+\frac{u^6}{6}+\frac{u^5}{5}+\frac{u^4}{4}+\frac{u^3}{3}+\frac{u^2}{2}+u+\ln|u-1|\right)+C$$

$$=12\left(\frac{x^{2/3}}{8}+\frac{x^{7/12}}{7}+\frac{x^{1/2}}{6}+\frac{x^{5/12}}{5}+\frac{x^{1/3}}{4}+\frac{x^{1/4}}{3}+\frac{x^{1/6}}{2}+x^{1/12}+\ln\left|x^{1/12}-1\right|\right)+C$$

963.

$$\ln\left|\frac{e^{2x}+4e^x+4}{e^x+1}\right|+C$$

The given expression is

$$\int\frac{e^{2x}}{e^{2x}+3e^x+2}\,dx$$

Begin with the substitution $u=e^x$ so that $du=e^x\,dx$:

$$\int\frac{u}{u^2+3u+2}\,du=\int\frac{u}{(u+1)(u+2)}\,du$$

Then perform the fraction decomposition:

$$\frac{u}{(u+1)(u+2)}=\frac{A}{u+1}+\frac{B}{u+2}$$

Multiplying both sides by $(u+1)(u+2)$ gives you $u=A(u+2)+B(u+1)$. If you let $u=-2$, you get $-2=B(-2+1)$ so that $2=B$, and if $u=-1$, you get $-1=A(-1+2)$ so that $-1=A$. With these values, you have

$$\int\left(\frac{-1}{u+1}+\frac{2}{u+2}\right)du$$

$$=-\ln|u+1|+2\ln|u+2|+C$$

$$=\ln\left|(u+2)^2\right|-\ln|u+1|+C$$

$$=\ln\left|\frac{u^2+4u+4}{u+1}\right|+C$$

$$=\ln\left|\frac{e^{2x}+4e^x+4}{e^x+1}\right|+C$$

---

964. convergent, $\frac{1}{2}$

Begin by rewriting the integral using a limit:

$$\int_1^\infty\frac{1}{(x+1)^2}\,dx=\lim_{a\to\infty}\int_1^a\frac{1}{(x+1)^2}\,dx$$

First evaluate $\int\frac{1}{(x+1)^2}\,dx$ using the substitution $u=x+1$ so that $du=dx$. Using these substitutions gives you

$$\int\frac{1}{u^2}\,du=\int u^{-2}du$$

$$=-u^{-1}+C$$

$$=-\frac{1}{x+1}+C$$

Because $\int\frac{1}{(x+1)^2}\,dx=-\frac{1}{x+1}+C$, you have the following after replacing the limit at infinity and the limits of integration:

$$\lim_{a\to\infty}\int_1^a\frac{1}{(x+1)^2}\,dx=\lim_{a\to\infty}\left(-\frac{1}{x+1}\Big|_1^a\right)$$

Now evaluate the limit:

$$\lim_{a\to\infty}\left(-\frac{1}{x+1}\Big|_1^a\right)$$

$$=\lim_{a\to\infty}\left(-\frac{1}{a+1}+\frac{1}{1+1}\right)$$

$$=\left(0+\frac{1}{2}\right)$$

$$=\frac{1}{2}$$

Because the value of the integral is finite, the improper integral is convergent.

965. divergent

Note that the function $f(x) = \dfrac{1}{x\sqrt{x}}$ has an infinite discontinuity when $x = 0$, which is included in the interval of integration, so you have an improper integral. Begin by rewriting the definite integral using limits; then integrate:

$$\int_0^5 \frac{1}{x\sqrt{x}}\,dx = \lim_{a\to 0^+} \int_a^5 \frac{1}{x^{3/2}}\,dx$$

$$= \lim_{a\to 0^+} \int_a^5 x^{-3/2}\,dx$$

$$= \lim_{a\to 0^+} \left(\left. \frac{x^{-1/2}}{-1/2} \right|_a^5 \right)$$

$$= -2\lim_{a\to 0^+} \left(\left. \frac{1}{\sqrt{x}} \right|_a^5 \right)$$

$$= -2\lim_{a\to 0^+} \left(\frac{1}{\sqrt{5}} - \frac{1}{\sqrt{a}} \right)$$

Because $\lim\limits_{a\to 0^+} \dfrac{1}{\sqrt{a}} = \infty$, you can conclude that $\int_0^5 \dfrac{1}{x\sqrt{x}}\,dx$ is divergent.

966. divergent

Note that the function $f(x) = \dfrac{1}{x}$ has an infinite discontinuity when $x = 0$, which is included in the interval of integration, so you have an improper integral. Begin by rewriting the definite integral using limits; then integrate:

$$\int_0^2 \frac{1}{x}\,dx = \lim_{a\to 0^+} \int_a^2 \frac{1}{x}\,dx$$

$$= \lim_{a\to 0^+} \left(\ln|x| \Big|_a^2 \right)$$

$$= \lim_{a\to 0^+} \left(\ln 2 - \ln a \right)$$

Because $\lim\limits_{a\to 0^+} \left(\ln a \right) = -\infty$, you know that $\int_0^2 \dfrac{1}{x}\,dx$ is divergent.

967. divergent

Begin by rewriting the integral using a limit:

$$\int_{-\infty}^1 e^{-4x}\,dx = \lim_{a\to -\infty} \int_a^1 e^{-4x}\,dx$$

Then use elementary antiderivatives to get the following:

$$\int_{-\infty}^1 e^{-4x}\,dx = \lim_{a\to -\infty} \int_a^1 e^{-4x}\,dx$$

$$= \lim_{a\to -\infty} \left(\left. -\frac{1}{4}e^{-4x} \right|_a^1 \right)$$

$$= \lim_{a\to -\infty} \left(-\frac{1}{4}e^{-4(1)} - \left(-\frac{1}{4}e^{-4a} \right) \right)$$

Because $\lim\limits_{a\to -\infty} e^{-4a} = \infty$, the integral is divergent.

968.   convergent, $\dfrac{32}{3}$

Notice that the function $f(x) = (x-1)^{-1/4}$ has an infinite discontinuity when $x = 1$, which is included in the interval of integration, so you have an improper integral. Begin by writing the integral using a limit:

$$\int_1^{17} (x-1)^{-1/4}\, dx = \lim_{a \to 1^+} \int_a^{17} (x-1)^{-1/4}\, dx$$

Notice that $\displaystyle\int (x-1)^{-1/4}\, dx = \frac{4(x-1)^{3/4}}{3} + C$ so that

$$\lim_{a \to 1^+} \int_a^{17} (x-1)^{-1/4}\, dx = \lim_{a \to 1^+} \left(\frac{4(x-1)^{3/4}}{3} \right)\Bigg|_a^{17}$$

Substituting in the limits of integration and evaluating the limit gives you

$$\lim_{a \to 1^+} \left(\frac{4(x-1)^{3/4}}{3} \right)\Bigg|_a^{17}$$

$$= \lim_{a \to 1^+} \left(\frac{4(17-1)^{3/4}}{3} - \frac{4(a-1)^{3/4}}{3} \right)$$

$$= \frac{32}{3}$$

The answer is finite, so the integral is convergent.

969.   divergent

Begin by rewriting the integral using a limit:

$$\int_{-\infty}^{-3} \frac{1}{x+1}\, dx = \lim_{a \to -\infty} \int_a^{-3} \frac{1}{x+1}\, dx$$

To evaluate this integral, use elementary antiderivatives:

$$\lim_{a \to -\infty} \int_a^{-3} \frac{1}{x+1}\, dx$$

$$= \lim_{a \to -\infty} \left(\ln|x+1|\Big|_a^{-3} \right)$$

$$= \lim_{a \to -\infty} \left(\ln|-3+1| - \ln|a+1| \right)$$

Because $\displaystyle\lim_{a \to -\infty} \ln|a+1| = \infty$, the integral is divergent.

970. divergent

Begin by rewriting the improper integral using a definite integral and a limit; then integrate:

$$\int_{-\infty}^{2}\left(x^2-5\right)dx$$

$$= \lim_{a\to-\infty}\int_{a}^{2}\left(x^2-5\right)dx$$

$$= \lim_{a\to-\infty}\left(\frac{x^3}{3}-5x\Bigg|_{a}^{2}\right)$$

$$= \lim_{a\to-\infty}\left(\frac{2^3}{3}-5(2)-\left(\frac{a^3}{3}-5a\right)\right)$$

Notice that $\lim\limits_{a\to-\infty}\left(\dfrac{a^3}{3}-5a\right)=\lim\limits_{a\to-\infty}\left(\dfrac{a^3-15a}{3}\right)$. Because the first term in the numerator is the dominant term, you have

$$\lim_{a\to-\infty}\left(\frac{a^3-15a}{3}\right)=\lim_{a\to-\infty}\left(\frac{a^3}{3}\right)=-\infty$$

Therefore, it follows that

$$\lim_{a\to-\infty}\left(\frac{2^3}{3}-5(2)-\left(\frac{a^3}{3}-5a\right)\right)=\frac{2^3}{3}-5(2)+\infty$$

and you can conclude the integral is divergent.

971. divergent

Begin by splitting up the integral into two separate integrals, using limits to rewrite each integral:

$$\int_{-\infty}^{\infty}\left(3-x^4\right)dx$$

$$= \lim_{a\to-\infty}\int_{a}^{0}\left(3-x^4\right)dx+\lim_{b\to\infty}\int_{0}^{b}\left(3-x^4\right)dx$$

Evaluate the first integral by integrating; then evaluate the resulting limit:

$$\lim_{a\to-\infty}\int_{a}^{0}\left(3-x^4\right)dx$$

$$= \lim_{a\to-\infty}\left(3x-\frac{x^5}{5}\Bigg|_{a}^{0}\right)$$

$$= \lim_{a\to-\infty}\left(3(0)-\frac{0^5}{5}-\left(3a-\frac{a^5}{5}\right)\right)$$

Notice that $\lim\limits_{a\to-\infty}\left(3a-\dfrac{a^5}{5}\right)=\lim\limits_{a\to-\infty}\left(\dfrac{15a-a^5}{5}\right)$. Because the second term in the numerator is the dominant term, you have

$$\lim_{a\to-\infty}\left(\frac{15a-a^5}{5}\right)=\lim_{a\to-\infty}\left(\frac{-a^5}{5}\right)=\infty$$

Therefore, it follows that

$$\lim_{a \to -\infty}\left(3(0)-\frac{0^5}{5}-\left(3a-\frac{a^5}{5}\right)\right)=0-0-\infty$$

and you can conclude that the integral is divergent. Because the first integral is divergent, you don't need to evaluate the second integral; $\int_{-\infty}^{\infty}\left(3-x^4\right)dx$ is divergent.

---

972. convergent, $3e^{-2/3}$

Begin by rewriting the integral using a limit:

$$\int_2^{\infty} e^{-x/3}dx = \lim_{a \to \infty}\int_2^a e^{-x/3}dx$$

To evaluate $\int e^{-x/3}dx$, let $u = -\frac{1}{3}x$ so that $du = -\frac{1}{3}dx$, or $-3\,du = dx$. This gives you

$$\int e^{-x/3}dx = -3\int e^u du$$
$$= -3e^u + C$$
$$= -3e^{-x/3} + C$$

Because $\int e^{-x/3}dx = -3e^{-x/3}+C$, it follows that

$$\lim_{a \to \infty}\int_2^a e^{-x/3}dx = \lim_{a \to \infty}\left(-3e^{-x/3}\,\Big|_2^a\right)$$

Now simply evaluate the limit:

$$\lim_{a \to \infty}\left(-3e^{-x/3}\,\Big|_2^a\right) = \lim_{a \to \infty}\left(-3e^{-a/3}+3e^{-2/3}\right)$$
$$= -3e^{-\infty/3}+3e^{-2/3}$$
$$= 0+3e^{-2/3}$$
$$= 3e^{-2/3}$$

The answer is finite, so the integral is convergent.

---

973. convergent, 1

Begin by writing the integral using a limit:

$$\int_e^{\infty}\frac{1}{x(\ln x)^2}\,dx = \lim_{a \to \infty}\int_e^a \frac{1}{x(\ln x)^2}\,dx$$

To evaluate $\int \frac{1}{x(\ln x)^2}\,dx$, use the substitution $u = \ln x$ so that $du = \frac{1}{x}dx$:

$$\int \frac{1}{u^2}\,du = \int u^{-2}du$$
$$= -u^{-1}+C$$
$$= -\frac{1}{u}+C$$
$$= -\frac{1}{\ln x}+C$$

Because $\int \frac{1}{u^2} du = -\frac{1}{\ln x} + C$, it follows that

$$\lim_{a \to \infty} \int_e^a \frac{1}{x(\ln x)^2} dx = \lim_{a \to \infty} \left(-\frac{1}{\ln x} \right) \Big|_e^a$$

Evaluating the limit gives you

$$\int_e^\infty \frac{1}{x(\ln x)^2} dx = \lim_{a \to \infty} \int_e^a \frac{1}{x(\ln x)^2} dx$$

$$= \lim_{a \to \infty} \left(-\frac{1}{\ln x} \right) \Big|_e^a$$

$$= \lim_{a \to \infty} \left(-\frac{1}{\ln a} + \frac{1}{\ln e} \right)$$

$$= \left(0 + \frac{1}{1} \right)$$

$$= 1$$

The answer is finite, so the integral is convergent.

974. convergent, $\frac{\pi}{3}$

Begin by splitting the integral into two integrals, using limits to rewrite each integral:

$$\int_{-\infty}^\infty \frac{x^2}{1 + x^6} dx$$

$$= \lim_{a \to -\infty} \int_a^0 \frac{x^2}{1 + x^6} dx + \lim_{b \to \infty} \int_0^b \frac{x^2}{1 + x^6} dx$$

To evaluate $\int \frac{x^2}{1 + x^6} dx$, rewrite the integral as $\int \frac{x^2}{1 + \left(x^3 \right)^2} dx$. Then use a substitution, letting $u = x^3$ so that $du = 3x^2 dx$, or $\frac{1}{3} du = x^2 dx$. Using these substitutions gives you

$$\int \frac{x^2}{1 + \left(x^3 \right)^2} dx = \frac{1}{3} \int \frac{1}{1 + u^2} du$$

$$= \frac{1}{3} \tan^{-1} u + C$$

$$= \frac{1}{3} \tan^{-1} \left(x^3 \right) + C$$

Therefore, as a approaches $-\infty$, you have

$$\lim_{a \to -\infty} \int_a^0 \frac{x^2}{1 + x^6} dx = \lim_{a \to -\infty} \left(\frac{1}{3} \tan^{-1} \left(x^3 \right) \Big|_a^0 \right)$$

$$= \frac{1}{3} \lim_{a \to -\infty} \left(\tan^{-1} \left(0^3 \right) - \tan^{-1} \left(a^3 \right) \right)$$

$$= \frac{1}{3} \left(0 - \left(-\frac{\pi}{2} \right) \right)$$

$$= \frac{\pi}{6}$$

And as b approaches ∞, you have

$$\lim_{b\to\infty}\int_0^b \frac{x^2}{1+x^6}dx = \lim_{b\to\infty}\left(\frac{1}{3}\tan^{-1}\left(x^3\right)\Big|_0^b\right)$$

$$= \lim_{b\to\infty}\left(\frac{1}{3}\tan^{-1}\left(b^3\right)-\frac{1}{3}\tan^{-1}(0)\right)$$

$$= \frac{1}{3}\left(\frac{\pi}{2}\right)-0$$

$$= \frac{\pi}{6}$$

Combining the two values gives you the answer:

$$\lim_{a\to-\infty}\int_a^0 \frac{x^2}{1+x^6}dx + \lim_{b\to\infty}\int_0^b \frac{x^2}{1+x^6}dx$$

$$= \frac{\pi}{6}+\frac{\pi}{6}$$

$$= \frac{2\pi}{6}$$

$$= \frac{\pi}{3}$$

The answer is finite, so the integral is convergent.

Note: If you noticed that the integrand is even, you could've simply computed one of the integrals and multiplied by 2 to arrive at the solution.

975. convergent, $\frac{1}{4}$

Begin by rewriting the integral using a limit:

$$\int_0^\infty xe^{-2x}dx = \lim_{a\to\infty}\int_0^a xe^{-2x}dx$$

To evaluate $\int xe^{-2x}dx$, use integration by parts. If $u = x$, then $du = dx$, and if $dv = e^{-2x}$, then $v = -\frac{1}{2}e^{-2x}$:

$$\int xe^{-2x}dx = x\left(-\frac{1}{2}e^{-2x}\right)-\int\left(-\frac{1}{2}e^{-2x}\right)dx$$

$$= -\frac{1}{2}xe^{-2x}-\frac{1}{4}e^{-2x}+C$$

Because $\int xe^{-2x}dx = -\frac{1}{2}xe^{-2x}-\frac{1}{4}e^{-2x}+C$, it follows that

$$\lim_{a\to\infty}\int_0^a xe^{-2x}dx = \lim_{a\to\infty}\left(-\frac{1}{2}xe^{-2x}-\frac{1}{4}e^{-2x}\Big|_0^a\right)$$

Next, evaluate the limit:

$$\lim_{a\to\infty}\left(-\frac{1}{2}xe^{-2x}-\frac{1}{4}e^{-2x}\Big|_0^a\right)$$

$$= \lim_{a\to\infty}\left(-\frac{1}{2}ae^{-2a}-\frac{1}{4}e^{-2a}-\left(-\frac{1}{2}(0)e^{-2(0)}-\frac{1}{4}e^{-2(0)}\right)\right)$$

Notice that $\lim\limits_{a \to \infty}\left(-\frac{1}{4}e^{-2a}\right)=0$ and that $\lim\limits_{a \to \infty}\left(-\frac{1}{2}ae^{-2a}\right)=(-\infty)(0)$, which is an indeterminate form. To evaluate $\lim\limits_{a \to \infty}\left(-\frac{1}{2}ae^{-2a}\right)$, use L'Hôpital's rule:

$$\lim_{a \to \infty}\left(-\frac{1}{2}ae^{-2a}\right)=-\frac{1}{2}\lim_{a \to \infty}\frac{a}{e^{2a}}$$

$$=-\frac{1}{2}\lim_{a \to \infty}\frac{1}{2e^{2a}}$$

$$=-\frac{1}{2}(0)$$

$$=0$$

Using these values gives you

$$\lim_{a \to \infty}\left(-\frac{1}{2}ae^{-2a}-\frac{1}{4}e^{-2a}-\left(-\frac{1}{2}(0)e^{-2(0)}-\frac{1}{4}e^{-2(0)}\right)\right)$$

$$=0+0+\frac{1}{2}(0)e^{-2(0)}+\frac{1}{4}e^{-2(0)}$$

$$=\frac{1}{4}$$

The answer is finite, so the integral is convergent.

976. divergent

Begin by splitting up the integral into two separate integrals, using limits to rewrite each integral:

$$\int_{-\infty}^{\infty}x^4e^{-x^5}\,dx=\lim_{a \to -\infty}\int_a^0 x^4e^{-x^5}\,dx+\lim_{b \to \infty}\int_0^b x^4e^{-x^5}\,dx$$

To evaluate $\int x^4e^{-x^5}\,dx$, use the substitution $u=-x^5$ so that $du=-5x^4\,dx$, or $-\frac{1}{5}\,du=x^4\,dx$:

$$\int x^4e^{-x^5}\,dx=-\frac{1}{5}\int e^u\,du$$

$$=-\frac{1}{5}e^u+C$$

$$=-\frac{1}{5}e^{-x^5}+C$$

Because $\int x^4e^{-x^5}\,dx=-\frac{1}{5}e^{-x^5}+C$, it follows that

$$\lim_{a \to -\infty}\int_a^0 x^4e^{-x^5}\,dx=\lim_{a \to -\infty}\left(-\frac{1}{5}e^{-x^5}\Big|_a^0\right)$$

Next, evaluate the limit:

$$\lim_{a \to -\infty}\left(-\frac{1}{5}e^{-x^5}\Big|_a^0\right)=\lim_{a \to -\infty}\left(-\frac{1}{5}e^{-(0)^5}-\left(-\frac{1}{5}e^{-a^5}\right)\right)$$

Notice that $\lim\limits_{a \to -\infty}\left(-\frac{1}{5}e^{-a^5}\right)=\infty$ so that $\lim\limits_{a \to -\infty}\int_a^0 x^4e^{-x^5}\,dx$ is divergent. Because this integral is divergent, you don't need to evaluate $\lim\limits_{b \to \infty}\int_0^b x^4e^{-x^5}\,dx$. You can conclude that $\int_{-\infty}^{\infty}x^4e^{-x^5}\,dx$ is divergent.

977.

convergent, $\frac{3}{2}\left(\tan^{-1}(16) - \frac{\pi}{2}\right)$

Begin by rewriting the integral using a limit:

$$\int_{-\infty}^{-4} \frac{3x}{1+x^4}\,dx = \lim_{a \to -\infty} \int_a^{-4} \frac{3x}{1+x^4}\,dx$$

To evaluate $\int \frac{3x}{1+x^4}\,dx$, rewrite the integral as $\int \frac{3x}{1+\left(x^2\right)^2}\,dx$. Then use a substitution where $u = x^2$ so that $du = 2x\,dx$, or $\frac{1}{2}du = x\,dx$:

$$\int \frac{3x}{1+\left(x^2\right)^2}\,dx = \frac{3}{2} \int \frac{1}{1+u^2}\,du$$

$$= \frac{3}{2}\tan^{-1}(u) + C$$

$$= \frac{3}{2}\tan^{-1}\left(x^2\right) + C$$

Because $\int \frac{3x}{1+\left(x^2\right)^2}\,dx = \frac{3}{2}\tan^{-1}\left(x^2\right) + C$, it follows that

$$\lim_{a \to -\infty} \int_a^{-4} \frac{3x}{1+x^4}\,dx = \lim_{a \to -\infty}\left(\frac{3}{2}\tan^{-1}\left(x^2\right)\Big|_a^{-4} \right)$$

Now evaluate the limit:

$$\lim_{a \to -\infty}\left(\frac{3}{2}\tan^{-1}\left(x^2\right)\Big|_a^{-4} \right)$$

$$= \frac{3}{2}\lim_{a \to -\infty}\left(\tan^{-1}(16) - \tan^{-1}\left(a^2\right) \right)$$

Notice that $\lim_{a \to -\infty} \tan^{-1}\left(a^2\right) = \frac{\pi}{2}$ so that

$$\frac{3}{2}\lim_{a \to -\infty}\left(\tan^{-1}(16) - \tan^{-1}\left(a^2\right) \right)$$

$$= \frac{3}{2}\left(\tan^{-1}(16) - \frac{\pi}{2} \right)$$

The answer is finite, so the integral is convergent.

978.

convergent, $\frac{\ln 2}{24} + \frac{1}{72}$

Begin by rewriting the integral using a limit:

$$\int_2^{\infty} \frac{\ln x}{x^4}\,dx = \lim_{a \to \infty} \int_2^a \frac{\ln x}{x^4}\,dx$$

To evaluate $\int \frac{\ln x}{x^4}\,dx = \int x^{-4}\ln x\,dx$, use integration by parts. If $u = \ln x$, then $du = \frac{1}{x}\,dx$, and if $dv = x^{-4}\,dx$, then $v = \frac{x^{-3}}{-3} = -\frac{1}{3x^3}$:

$$\int x^{-4}\ln x\, dx = -\frac{1}{3x^3}\ln x - \int\left(-\frac{1}{3x^3}\right)\frac{1}{x}\, dx$$

$$= -\frac{1}{3x^3}\ln x + \int \frac{1}{3x^4}\, dx$$

$$= -\frac{1}{3x^3}\ln x + \frac{1}{3}\int x^{-4}\, dx$$

$$= -\frac{1}{3x^3}\ln x + \frac{1}{3}\left(-\frac{1}{3x^3}\right) + C$$

$$= -\frac{1}{3x^3}\ln x - \frac{1}{9x^3} + C$$

Because $\int x^{-4}\ln x\, dx = -\frac{1}{3x^3}\ln x - \frac{1}{9x^3} + C$, it follows that

$$\lim_{a\to\infty}\int_2^a \frac{\ln x}{x^4}\, dx = \lim_{a\to\infty}\left(-\frac{1}{3x^3}\ln x - \frac{1}{9x^3}\Big|_2^a\right)$$

Evaluating the limit gives you

$$\lim_{a\to\infty}\left(-\frac{1}{3x^3}\ln x - \frac{1}{9x^3}\Big|_2^a\right)$$

$$= \lim_{a\to\infty}\left(-\frac{1}{3a^3}\ln a - \frac{1}{9a^3} - \left(-\frac{1}{3(2)^3}\ln 2 - \frac{1}{9(2)^3}\right)\right)$$

Notice that $\lim_{a\to\infty}\frac{\ln a}{3a^3}$ has the indeterminate form $\frac{\infty}{\infty}$, so you can use L'Hôpital's rule to evaluate it:

$$\lim_{a\to\infty}\frac{\ln a}{3a^3} = \lim_{a\to\infty}\frac{1/a}{9a^2} = \lim_{a\to\infty}\frac{1}{9a^3} = 0$$

Because $\lim_{a\to\infty}\frac{1}{9a^3} = 0$, the limit becomes

$$\lim_{a\to\infty}\left(\frac{1}{3a^3}\ln a - \frac{1}{9a^3} - \left(-\frac{1}{3(2)^3}\ln 2 - \frac{1}{9(2)^3}\right)\right)$$

$$= \frac{1}{3(2)^3}\ln 2 + \frac{1}{9(2)^3}$$

$$= \frac{\ln 2}{24} + \frac{1}{72}$$

The answer is finite, so the integral is convergent.

979.   divergent

Begin by rewriting the integral using a limit:

$$\int_0^\infty \frac{x}{x^2+4}\, dx = \lim_{a\to\infty}\int_0^a \frac{x}{x^2+4}\, dx$$

To find $\int \frac{x}{x^2+4}dx$, let $u = x^2 + 4$ so that $du = 2x\,dx$, or $\frac{1}{2}du = x\,dx$. This substitution gives you

$$\int \frac{x}{x^2+4}dx = \frac{1}{2}\int \frac{1}{u}du$$

$$= \frac{1}{2}\ln|u| + C$$

$$= \frac{1}{2}\ln|x^2+4| + C$$

Because $\int \frac{x}{x^2+4}dx = \frac{1}{2}\ln|x^2+4| + C$, you have

$$\lim_{a\to\infty}\int_0^a \frac{x}{x^2+4}dx = \lim_{a\to\infty}\frac{1}{2}\ln|x^2+4|\Big|_0^a$$

Evaluating the limit gives you

$$\lim_{a\to\infty}\frac{1}{2}\ln|x^2+4|\Big|_0^a = \frac{1}{2}\lim_{a\to\infty}\left(\ln|a^2+4| - \ln 4\right) = \infty$$

Therefore, the integral is divergent.

980. convergent, $-\frac{3}{2}(7)^{2/3}$

Begin by rewriting the integral using a limit:

$$\int_1^8 \frac{1}{\sqrt[3]{x-8}}dx = \lim_{a\to 8^-}\int_1^a \frac{1}{\sqrt[3]{x-8}}dx$$

To evaluate $\int \frac{1}{\sqrt[3]{x-8}}dx$, let $u = x - 8$ so that $du = dx$. This substitution gives you

$$\int \frac{1}{\sqrt[3]{u}}du = \int u^{-1/3}du$$

$$= \frac{u^{2/3}}{2/3} + C$$

$$= \frac{3}{2}(x-8)^{2/3} + C$$

Because $\int \frac{1}{\sqrt[3]{u}}du = \frac{3}{2}(x-8)^{2/3} + C$, you have

$$\lim_{a\to 8^-}\int_1^a \frac{1}{\sqrt[3]{x-8}}dx = \lim_{a\to 8^-}\left(\frac{3}{2}(x-8)^{2/3}\Big|_1^a\right)$$

Evaluating the limit gives you

$$\lim_{a\to 8^-}\left(\frac{3}{2}(x-8)^{2/3}\Big|_1^a\right)$$

$$= \lim_{a\to 8^-}\left(\frac{3}{2}(a-8)^{2/3} - \frac{3}{2}(1-8)^{2/3}\right)$$

Noting that $\lim\limits_{a \to 8^-} \frac{3}{2}(a-8)^{2/3} = 0$, you get the following:

$$\lim_{a \to 8^-}\left(\frac{3}{2}(a-8)^{2/3} - \frac{3}{2}(1-8)^{2/3}\right)$$
$$= 0 - \frac{3}{2}(1-8)^{2/3}$$
$$= -\frac{3}{2}(-7)^{2/3}$$
$$= -\frac{3}{2}(7)^{2/3}$$

The answer is finite, so the integral is convergent.

981. divergent

Begin by rewriting the integral using a limit:

$$\int_0^{\pi/2} \tan^2 x \, dx = \lim_{a \to \left(\frac{\pi}{2}\right)^-} \int_0^a \tan^2 x \, dx$$

To evaluate $\int \tan^2 x \, dx$, use a trigonometric identity:

$$\int \tan^2 x \, dx = \int \left(\sec^2 x - 1\right) dx$$
$$= \tan x - x + C$$

Because $\int \tan^2 x \, dx = \tan x - x + C$, you have

$$\lim_{a \to \left(\frac{\pi}{2}\right)^-} \int_0^a \tan^2 x \, dx = \lim_{a \to \left(\frac{\pi}{2}\right)^-} \left(\tan x - x \Big|_0^a\right)$$

Evaluating the limit gives you

$$\lim_{a \to \left(\frac{\pi}{2}\right)^-} \left(\tan x - x \Big|_0^a\right)$$
$$= \lim_{a \to \left(\frac{\pi}{2}\right)^-} \left(\tan a - a - (\tan 0 - 0)\right)$$

However, because $\lim\limits_{a \to \left(\frac{\pi}{2}\right)^-} \tan a = \infty$, you can conclude that $\int_0^{\pi/2} \tan^2 x \, dx$ is divergent.

982. divergent

Begin by rewriting the integral using a limit:

$$\int_0^\infty \sin^2 x \, dx = \lim_{a \to \infty} \int_0^a \sin^2 x \, dx$$

To evaluate $\int \sin^2 x \, dx$, use the trigonometric identity $\sin^2 x = \frac{1}{2}(1 - \cos(2x))$ so that

$$\int \sin^2 x \, dx = \int \frac{1}{2}(1 - \cos(2x)) dx$$
$$= \frac{1}{2}\left(x - \frac{1}{2}\sin(2x)\right) + C$$

Because $\int \sin^2 x\, dx = \frac{1}{2}\left(x - \frac{1}{2}\sin(2x)\right) + C$, it follows that

$$\lim_{a \to \infty} \int_0^a \sin^2 x\, dx = \lim_{a \to \infty} \frac{1}{2}\left(x - \frac{1}{2}\sin(2x)\right)\Big|_0^a$$

Evaluating the limit gives you

$$\lim_{a \to \infty} \frac{1}{2}\left(x - \frac{1}{2}\sin(2x)\right)\Big|_0^a$$

$$= \frac{1}{2}\lim_{a \to \infty}\left(a - \frac{1}{2}\sin(2a) - \left(0 - \frac{1}{2}\sin(2(0))\right)\right)$$

Because $\lim_{a \to \infty} a = \infty$ and because $\sin(2a)$ is always between -1 and 1, the integral is divergent.

983.

divergent

Notice that the function $\dfrac{e^x}{e^x - 1}$ has an infinite discontinuity at $x = 0$. Rewrite the integral using limits:

$$\int_{-1}^{1} \frac{e^x}{e^x - 1}\, dx = \lim_{a \to 0^-} \int_{-1}^{a} \frac{e^x}{e^x - 1}\, dx + \lim_{b \to 0^+} \int_{b}^{1} \frac{e^x}{e^x - 1}\, dx$$

To evaluate $\int \dfrac{e^x}{e^x - 1}\, dx$, use the substitution $u = e^x - 1$ so that $du = e^x\, dx$ to get

$$\int \frac{1}{u}\, du = \ln|u| + C$$

$$= \ln|e^x - 1| + C$$

Because $\int \dfrac{e^x}{e^x - 1}\, dx = \ln|e^x - 1| + C$, it follows that

$$\lim_{a \to 0^-} \int_{-1}^{a} \frac{e^x}{e^x - 1}\, dx = \lim_{a \to 0^-}\left(\ln|e^x - 1|\right)\Big|_{-1}^{a}$$

Evaluating the limit gives you

$$\lim_{a \to 0^-}\left(\ln|e^x - 1|\right)\Big|_{-1}^{a}$$

$$= \lim_{a \to 0^-}\left(\ln|e^a - 1| - \ln|e^{-1} - 1|\right)$$

However, because $\lim_{a \to 0^-} \ln|e^a - 1| = -\infty$, the integral $\lim_{a \to 0^-} \int_{-1}^{a} \dfrac{e^x}{e^x - 1}\, dx$ is divergent,

so you can conclude that $\int_{-1}^{1} \dfrac{e^x}{e^x - 1}\, dx$ is divergent.

984.

convergent, $\frac{1}{5}\ln\left(\frac{7}{2}\right)$

Begin by writing the integral using a limit:

$$\int_{4}^{\infty} \frac{1}{x^2 + x - 6}\, dx = \lim_{a \to \infty} \int_{4}^{a} \frac{1}{x^2 + x - 6}\, dx$$

To evaluate $\int \dfrac{1}{x^2 + x - 6}\, dx$, use a partial fraction decomposition:

$$\frac{1}{x^2 + x - 6} = \frac{1}{(x+3)(x-2)} = \frac{A}{x+3} + \frac{B}{x-2}$$

Multiply both sides of the equation $\frac{1}{(x+3)(x-2)} = \frac{A}{x+3} + \frac{B}{x-2}$ by $(x + 3)(x - 2)$ to get

$$1 = A(x - 2) + B(x + 3)$$

If you let $x = 2$, you have $1 = B(2 + 3)$ so that $\frac{1}{5} = B$. And if $x = -3$, then $1 = A(-3 - 2)$ so that $-\frac{1}{5} = A$. Entering these values, you get

$$\int \frac{1}{x^2 + x - 6} dx = \int \left(\frac{-\frac{1}{5}}{x+3} + \frac{\frac{1}{5}}{x-2} \right) dx$$

$$= -\frac{1}{5} \ln|x+3| + \frac{1}{5} \ln|x-2| + C$$

$$= \frac{1}{5} \ln \left| \frac{x-2}{x+3} \right| + C$$

Because $\int \frac{1}{x^2 + x - 6} dx = \frac{1}{5} \ln \left| \frac{x-2}{x+3} \right| + C$, it follows that

$$\lim_{a \to \infty} \int_4^a \frac{1}{x^2 + x - 6} dx = \lim_{a \to \infty} \frac{1}{5} \ln \left| \frac{x-2}{x+3} \right| \Big|_4^a$$

Evaluating the limit gives you

$$\lim_{a \to \infty} \frac{1}{5} \ln \left| \frac{x-2}{x+3} \right| \Big|_4^a$$

$$= \lim_{a \to \infty} \left(\frac{1}{5} \ln \left| \frac{a-2}{a+3} \right| - \frac{1}{5} \ln \left(\frac{2}{7} \right) \right)$$

Because $\lim_{a \to \infty} \frac{1}{5} \ln \left| \frac{a-2}{a+3} \right| = \frac{1}{5} \ln 1 = 0$, you get the following solution:

$$\lim_{a \to \infty} \frac{1}{5} \ln \left| \frac{x-2}{x+3} \right| \Big|_4^a = \lim_{a \to \infty} \left(\frac{1}{5} \ln \left| \frac{a-2}{a+3} \right| - \frac{1}{5} \ln \left(\frac{2}{7} \right) \right)$$

$$= -\frac{1}{5} \ln \left(\frac{2}{7} \right)$$

$$= \frac{1}{5} \ln \left(\frac{7}{2} \right)$$

The answer is finite, so the integral is convergent.

985. convergent, $\frac{\pi}{2}$

Begin by writing the integral using a limit:

$$\int_0^2 \frac{dx}{\sqrt{4 - x^2}} = \lim_{a \to 2^-} \int_0^a \frac{dx}{\sqrt{4 - x^2}}$$

To evaluate $\int \frac{dx}{\sqrt{4 - x^2}}$, use the trigonometric substitution $x = 2 \sin \theta$ so that $dx = 2 \cos \theta \, d\theta$. This gives you

$$\int \frac{2\cos\theta\,d\theta}{\sqrt{4-(2\sin\theta)^2}} = \int \frac{2\cos\theta\,d\theta}{\sqrt{4(1-\sin^2\theta)}}$$

$$= \int \frac{2\cos\theta\,d\theta}{\sqrt{4(\cos^2\theta)}}$$

$$= \int \frac{2\cos\theta\,d\theta}{2\cos\theta}$$

$$= \int 1\,d\theta$$

$$= \theta + C$$

From the substitution $x = 2\sin\theta$, or $\frac{x}{2} = \sin\theta$, you have $\sin^{-1}\left(\frac{x}{2}\right) = \theta$; therefore,

$$\theta + C = \sin^{-1}\left(\frac{x}{2}\right) + C$$

Because $\int \frac{dx}{\sqrt{4-x^2}} = \sin^{-1}\left(\frac{x}{2}\right) + C$, it follows that

$$\lim_{a\to 2^-} \int_0^a \frac{dx}{\sqrt{4-x^2}} = \lim_{a\to 2^-} \left(\sin^{-1}\left(\frac{x}{2}\right)\bigg|_0^a \right)$$

Evaluating the limit gives you

$$\lim_{a\to 2^-} \left(\sin^{-1}\left(\frac{x}{2}\right)\bigg|_0^a \right)$$

$$= \lim_{a\to 2^-} \left(\sin^{-1}\left(\frac{a}{2}\right) - \sin^{-1}0 \right)$$

$$= \sin^{-1}\left(\frac{2}{2}\right) - 0$$

$$= \sin^{-1}(1)$$

$$= \frac{\pi}{2}$$

The answer is finite, so the integral is convergent.

986. convergent, 2

Rewrite the integral using limits:

$$\int_{-\infty}^{\infty} e^{-|x|}\,dx = \lim_{a\to-\infty}\int_{a}^{0} e^{-|x|}\,dx + \lim_{b\to\infty}\int_{0}^{b} e^{-|x|}\,dx$$

On the interval $(-\infty, 0)$, you have $-|x| = x$, and on the interval $(0, \infty)$, you have $-|x| = -x$. Therefore, the limit is

$$\lim_{a\to-\infty}\int_{a}^{0} e^{-|x|}\,dx + \lim_{b\to\infty}\int_{0}^{b} e^{-|x|}\,dx$$

$$= \lim_{a\to-\infty}\int_{a}^{0} e^{x}\,dx + \lim_{b\to\infty}\int_{0}^{b} e^{-x}\,dx$$

$$= \lim_{a\to-\infty}\left(e^{x}\right)\Big|_{a}^{0} + \lim_{b\to\infty}\left(-e^{-x}\right)\Big|_{0}^{b}$$

$$= \lim_{a\to-\infty}\left(e^{0} - e^{a}\right) + \lim_{b\to\infty}\left(-e^{-b} - \left(-e^{0}\right)\right)$$

$$= (1-0) + (0+1)$$

$$= 2$$

The answer is finite, so the integral is convergent.

987. convergent, $\dfrac{-14\sqrt{2}}{3}$

Begin by rewriting the integral using a limit:

$$\int_{3}^{5} \frac{x}{\sqrt{x-3}}\,dx = \lim_{a\to 3^{+}}\int_{a}^{5} \frac{x}{\sqrt{x-3}}\,dx$$

To evaluate $\displaystyle\int \frac{x}{\sqrt{x-3}}\,dx$, let $u = x-3$ so that $u-3 = x$ and $du = dx$. This gives you

$$\int \frac{u-3}{u^{1/2}}\,du = \int\left(\frac{u}{u^{1/2}} - \frac{3}{u^{1/2}}\right)du$$

$$= \int\left(u^{1/2} - 3u^{-1/2}\right)du$$

$$= \frac{u^{3/2}}{3/2} - 3\frac{u^{1/2}}{1/2} + C$$

$$= \frac{2}{3}(x-3)^{3/2} - 6(x-3)^{1/2} + C$$

Because $\displaystyle\int \frac{x}{\sqrt{x-3}}\,dx = \frac{2}{3}(x-3)^{3/2} - 6(x-3)^{1/2} + C$, it follows that

$$\int_{3}^{5} \frac{x}{\sqrt{x-3}}\,dx = \lim_{a\to 3^{+}}\int_{a}^{5} \frac{x}{\sqrt{x-3}}\,dx$$

$$= \lim_{a\to 3^{+}}\left(\frac{2}{3}(x-3)^{3/2} - 6(x-3)^{1/2}\,\Big|_{a}^{5}\right)$$

Substitute in the limits of integration and evaluate the resulting limit:

$$\lim_{a \to 3^+} \left(\frac{2}{3}(x-3)^{3/2} - 6(x-3)^{1/2} \Big|_a^5 \right)$$

$$= \lim_{a \to 3^+} \left(\frac{2}{3}(5-3)^{3/2} - 6(5-3)^{1/2} - \left(\frac{2}{3}(a-3)^{3/2} - 6(a-3)^{1/2} \right) \right)$$

$$= \left(\frac{2}{3}(2)^{3/2} - 6(2)^{1/2} - (0-0) \right)$$

$$= \frac{2}{3}(2)^{3/2} - 6(2)^{1/2}$$

$$= 2^{1/2} \left(\frac{4}{3} - 6 \right)$$

$$= \frac{-14\sqrt{2}}{3}$$

The answer is finite, so the integral is convergent.

---

988. convergent, compare to $\int_1^\infty \frac{1}{x^2} dx$

Recall the comparison theorem: Suppose that f and g are continuous functions with $f(x) \geq g(x) \geq 0$ for $x \geq a$.

✔ If $\int_a^\infty f(x)dx$ is convergent, then $\int_a^\infty g(x)dx$ is convergent.

✔ If $\int_a^\infty g(x)dx$ is divergent, then $\int_a^\infty f(x)dx$ is divergent.

The given improper integral is $\int_1^\infty \frac{\sin^2 x}{1+x^2} dx$. Notice that on the interval $[1, \infty]$, the following inequalities are true: $-1 \leq \sin x \leq 1$; therefore, $0 \leq \sin^2 x \leq 1$. Because $1 + x^2 \geq 0$, you have $0 \leq \frac{\sin^2 x}{1+x^2} \leq \frac{1}{1+x^2}$. However, $\frac{1}{1+x^2} \leq \frac{1}{x^2}$ on $[1, \infty]$ so that $\int_1^\infty \frac{1}{1+x^2} dx \leq \int_1^\infty \frac{1}{x^2} dx$. An integral of the form $\int_1^\infty \frac{1}{x^p} dx$ converges if and only if $p > 1$, so $\int_1^\infty \frac{1}{x^2} dx$ is convergent (you can also show this directly by using a limit to write the integral and evaluating). Because $0 \leq \int_1^\infty \frac{\sin^2 x}{1+x^2} dx \leq \int_1^\infty \frac{1}{1+x^2} dx \leq \int_1^\infty \frac{1}{x^2} dx$, you can conclude that $\int_1^\infty \frac{\sin^2 x}{1+x^2} dx$ is convergent.

---

989. convergent, compare to $\int_1^\infty \frac{1}{x^4} dx$

The given improper integral is $\int_1^\infty \frac{dx}{x^4 + e^{3x}}$. Notice that $e^{3x} \geq 0$ on the interval $[1, \infty]$, so you have $x^4 + e^{3x} \geq x^4$. It follows that $\frac{1}{x^4 + e^{3x}} \leq \frac{1}{x^4}$ and that $0 \leq \int_1^\infty \frac{1}{x^4 + e^{3x}} dx \leq \int_1^\infty \frac{1}{x^4} dx$. An integral of the form $\int_1^\infty \frac{1}{x^p} dx$ converges if and only if $p > 1$, so $\int_1^\infty \frac{1}{x^4} dx$ is convergent (you can also show this directly by using a limit to write the integral and evaluating). Because $0 \leq \int_1^\infty \frac{1}{x^4 + e^{3x}} dx \leq \int_1^\infty \frac{1}{x^4} dx$, you know that $\int_1^\infty \frac{dx}{x^4 + e^{3x}}$ also converges.

990. divergent, compare to $\int_1^\infty \frac{1}{x}\,dx$

The given improper integral is $\int_1^\infty \frac{x+2}{\sqrt{x^4-1}}\,dx$. Notice that on the interval $[1, \infty]$, you have $0 \le x^4 - 1 \le x^4$, so it follows that $0 \le \sqrt{x^4-1} \le \sqrt{x^4}$ and that $\frac{1}{\sqrt{x^4-1}} \ge \frac{1}{\sqrt{x^4}}$. Also,

because $x + 2 > x$, you get $\frac{x+2}{\sqrt{x^4-1}} \ge \frac{x}{\sqrt{x^4}} = \frac{x}{x^2} = \frac{1}{x}$. Therefore, it follows that

$\int_1^\infty \frac{x+2}{\sqrt{x^4-1}}\,dx \ge \int_1^\infty \frac{x}{\sqrt{x^4}}\,dx = \int_1^\infty \frac{x}{x^2}\,dx = \int_1^\infty \frac{1}{x}\,dx \ge 0$. An integral of the form $\int_1^\infty \frac{1}{x^p}\,dx$

converges if and only if $p > 1$, so $\int_1^\infty \frac{1}{x}\,dx$ is divergent (you can also show this

directly by using a limit to write the integral and evaluating). Therefore, because

$\int_1^\infty \frac{x+2}{\sqrt{x^4-1}}\,dx \ge \int_1^\infty \frac{1}{x}\,dx \ge 0$, you can conclude that $\int_1^\infty \frac{x+2}{\sqrt{x^4-1}}\,dx$ is divergent.

991. convergent, compare to $\frac{\pi}{2}\int_1^\infty \frac{1}{x^5}\,dx$

The given improper integral is $\int_1^\infty \frac{\tan^{-1}x}{x^5}\,dx$. Notice that $0 \le \tan^{-1}x \le \frac{\pi}{2}$ on the interval

$[1, \infty)$, so $0 \le \frac{\tan^{-1}x}{x^5} \le \frac{\pi/2}{x^5}$. Note that you could actually bound $\tan^{-1}x$ below by $\frac{\pi}{4}$,

but zero also works and makes the inequalities cleaner.

It follows that $0 \le \int_1^\infty \frac{\tan^{-1}x}{x^5}\,dx \le \int_1^\infty \frac{\pi/2}{x^5} = \frac{\pi}{2}\int_1^\infty \frac{1}{x^5}\,dx$. An integral of the form $\int_1^\infty \frac{1}{x^p}\,dx$ con-

verges if and only if $p > 1$; because $\int_1^\infty \frac{1}{x^5}\,dx$ is convergent, $\frac{\pi}{2}\int_1^\infty \frac{1}{x^5}\,dx$ is also convergent

(you can also show this directly by using a limit to write the integral and evaluating).

Because $0 \le \int_1^\infty \frac{\tan^{-1}x}{x^5}\,dx \le \int_1^\infty \frac{\pi/2}{x^5} = \frac{\pi}{2}\int_1^\infty \frac{1}{x^5}\,dx$, you can conclude that $\int_1^\infty \frac{\tan^{-1}x}{x^5}\,dx$ is also

convergent.

992. divergent, compare to $\int_1^\infty \frac{1}{x}\,dx$

The given improper integral is $\int_2^\infty \frac{x^2}{\sqrt{x^6-1}}\,dx$. Notice that on the interval $[2, \infty)$, you have

$\frac{x^2}{\sqrt{x^6-1}} \ge \frac{x^2}{\sqrt{x^6}} = \frac{x^2}{x^3} = \frac{1}{x} \ge 0$. Therefore, you know that $\int_2^\infty \frac{x^2}{\sqrt{x^6-1}}\,dx \ge \int_2^\infty \frac{1}{x}\,dx$. The inte-

gral $\int_1^\infty \frac{1}{x^p}\,dx$ diverges if and only if $p \le 1$, so $\int_2^\infty \frac{1}{x}\,dx$ also diverges (note that the lower

limit of 2 instead of 1 does not affect the divergence). Therefore, because

$\int_2^\infty \frac{x^2}{\sqrt{x^6-1}}\,dx \ge \int_2^\infty \frac{1}{x}\,dx$, you can conclude that $\int_2^\infty \frac{x^2}{\sqrt{x^6-1}}\,dx$ also diverges.

993. divergent, compare to $\int_1^\infty \frac{1}{x}\,dx$

The given improper integral is $\int_1^\infty \frac{5+e^{-x}}{x}\,dx$. Notice that on the interval $[1, \infty)$, you have $e^{-x} > 0$ so that $5 + e^{-x} > 5$ and also $\frac{5+e^{-x}}{x} > \frac{5}{x} = 5\frac{1}{x}$. Therefore, it follows that $\int_1^\infty \frac{5+e^{-x}}{x}\,dx > 5\int_1^\infty \frac{1}{x}\,dx$. The integral $\int_1^\infty \frac{1}{x^p}\,dx$ diverges if and only if $p \leq 1$, so $5\int_1^\infty \frac{1}{x}\,dx$ diverges. Because $\int_1^\infty \frac{5+e^{-x}}{x}\,dx > 5\int_1^\infty \frac{1}{x}\,dx$, you know that $\int_1^\infty \frac{5+e^{-x}}{x}\,dx$ also diverges.

994. 18.915

Recall that the trapezoid rule states the following:

$$\int_a^b f(x) \approx \frac{\Delta x}{2}\left(f(x_0) + 2f(x_1) + 2f(x_2) + \ldots + 2f(x_{n-1}) + f(x_n)\right)$$

where $\Delta x = \frac{b-a}{n}$ and $x_i = a + i\Delta x$.

You want to approximate $\int_0^6 \sqrt[3]{1+x^3}\,dx$ with $n = 6$. Using the trapezoid rule, $\Delta x = \frac{b-a}{n} = \frac{6-0}{6} = 1$ so that

$$\int_0^6 \sqrt[3]{1+x^3}\,dx \approx \frac{1}{2}\left(f(0) + 2f(1) + 2f(2) + 2f(3) + 2f(4) + 2f(5) + f(6)\right)$$

$$= \frac{1}{2}\left(\sqrt[3]{1+0^3} + 2\sqrt[3]{1+1^3} + 2\sqrt[3]{1+2^3} + 2\sqrt[3]{1+3^3} + 2\sqrt[3]{1+4^3} + 2\sqrt[3]{1+5^3} + \sqrt[3]{1+6^3}\right)$$

$$\approx 18.915$$

995. 0.105

You want to approximate $\int_1^2 \frac{\ln x}{1+x^2}\,dx$ with $n = 4$. Using the trapezoid rule, you have $\Delta x = \frac{2-1}{4} = \frac{1}{4} = 0.25$ so that

$$\int_1^2 \frac{\ln x}{1+x^2}\,dx \approx \frac{0.25}{2}\left(f(1) + 2f(1.25) + 2f(1.5) + 2f(1.75) + f(2)\right)$$

$$= \frac{0.25}{2}\left(\frac{\ln 1}{1+1^2} + 2\frac{\ln(1.25)}{1+(1.25)^2} + 2\frac{\ln(1.5)}{1+(1.5)^2} + 2\frac{\ln(1.75)}{1+(1.75)^2} + \frac{\ln 2}{1+2^2}\right)$$

$$\approx 0.105$$

996. −0.210

You want to approximate $\int_1^3 \frac{\cos x}{x}\,dx$ with $n = 8$. Using the trapezoid rule, you have $\Delta x = \frac{b-a}{n} = \frac{3-1}{8} = 0.25$ so that

$$\int_1^3 \frac{\cos x}{x}\,dx \approx \frac{0.25}{2}\left(\begin{array}{l} f(1) + 2f(1.25) + 2f(1.5) + 2f(1.75) + 2f(2) \\ + 2f(2.25) + 2f(2.5) + 2f(2.75) + f(3)\end{array}\right)$$

$$= \frac{0.25}{2}\left(\begin{array}{l} \frac{\cos(1)}{1} + 2\frac{\cos(1.25)}{1.25} + 2\frac{\cos(1.5)}{1.5} + 2\frac{\cos(1.75)}{1.75} + 2\frac{\cos(2)}{2} \\ + 2\frac{\cos(2.25)}{2.25} + 2\frac{\cos(2.5)}{2.5} + 2\frac{\cos(2.75)}{2.75} + \frac{\cos(3)}{3}\end{array}\right)$$

$$\approx -0.210$$

997. 0.216

You want to approximate $\int_0^{1/2} \sin\sqrt{x}\, dx$ with $n = 4$. Using the trapezoid rule, you have $\Delta x = \dfrac{\frac{1}{2}-0}{4} = \dfrac{1}{8} = 0.125$ so that

$$\int_0^{1/2} \sin\sqrt{x}\, dx \approx \frac{0.125}{2}\big(f(0)+2f(0.125)+2f(0.25)+2f(0.375)+f(0.5)\big)$$

$$= \frac{0.125}{2}\Big(\sin\sqrt{0}+2\sin\sqrt{0.125}+2\sin\sqrt{0.25}+2\sin\sqrt{0.375}+\sin\sqrt{0.5}\Big)$$

$$\approx 0.216$$

998. 18.817

Recall that Simpson's rule states

$$\int_a^b f(x) \approx \frac{\Delta x}{3}\big(f(x_0)+4f(x_1)+2f(x_2)+4f(x_3)+\ldots+2f(x_{n-2})+4f(x_{n-1})+f(x_n)\big)$$

where n is even, $\Delta x = \dfrac{b-a}{n}$, and $x_i = a + i\Delta x$.

You want to approximate $\int_0^6 \sqrt[3]{1+x^3}\, dx$ with $n = 6$. Using Simpson's rule with $n = 6$ gives you $\Delta x = \dfrac{b-a}{n} = \dfrac{6-0}{6} = 1$ so that

$$\int_0^6 \sqrt[3]{1+x^3}\, dx \approx \frac{1}{3}\big(f(0)+4f(1)+2f(2)+4f(3)+2f(4)+4f(5)+f(6)\big)$$

$$= \frac{1}{3}\Big(\sqrt[3]{1+0^3}+4\sqrt[3]{1+1^3}+2\sqrt[3]{1+2^3}+4\sqrt[3]{1+3^3}+2\sqrt[3]{1+4^3}+4\sqrt[3]{1+5^3}+\sqrt[3]{1+6^3}\Big)$$

$$\approx 18.817$$

999. 0.107

You want to approximate $\int_1^2 \dfrac{\ln x}{1+x^2}\, dx$ with $n = 4$. Using Simpson's rule with $n = 4$ gives you $\Delta x = \dfrac{2-1}{4} = \dfrac{1}{4} = 0.25$ so that

$$\int_1^2 \frac{\ln x}{1+x^2}\, dx \approx \frac{0.25}{3}\big(f(1)+4f(1.25)+2f(1.5)+4f(1.75)+f(2)\big)$$

$$= \frac{0.25}{3}\left(\frac{\ln 1}{1+1^2}+4\frac{\ln(1.25)}{1+(1.25)^2}+2\frac{\ln(1.5)}{1+(1.5)^2}+4\frac{\ln(1.75)}{1+(1.75)^2}+\frac{\ln 2}{1+2^2}\right)$$

$$\approx 0.107$$

1,000. −0.218

You want to approximate $\int_1^3 \frac{\cos x}{x}\,dx$ with $n = 8$. Using Simpson's rule with $n = 8$ gives you $\Delta x = \frac{3-1}{8} = \frac{2}{8} = 0.25$ so that

$$\int_1^3 \frac{\cos x}{x}\,dx \approx \frac{0.25}{3}\begin{pmatrix} f(1) + 4f(1.25) + 2f(1.5) + 4f(1.75) + 2f(2) \\ + 4f(2.25) + 2f(2.5) + 4f(2.75) + f(3) \end{pmatrix}$$

$$= \frac{0.25}{3}\begin{pmatrix} \frac{\cos(1)}{1} + 4\frac{\cos(1.25)}{1.25} + 2\frac{\cos(1.5)}{1.5} + 4\frac{\cos(1.75)}{1.75} + 2\frac{\cos(2)}{2} \\ + 4\frac{\cos(2.25)}{2.25} + 2\frac{\cos(2.5)}{2.5} + 4\frac{\cos(2.75)}{2.75} + \frac{\cos(3)}{3} \end{pmatrix}$$

$$\approx -0.218$$

1,001. 0.221

You want to approximate $\int_0^{1/2} \sin\sqrt{x}\,dx$ with $n = 4$. Using Simpson's rule with $n = 4$ gives you $\Delta x = \frac{\frac{1}{2}-0}{4} = \frac{1}{8} = 0.125$ so that

$$\int_0^{1/2} \sin\sqrt{x}\,dx \approx \frac{0.125}{3}\left(f(0) + 4f(0.125) + 2f(0.25) + 4f(0.375) + f(0.5) \right)$$

$$= \frac{0.125}{3}\left(\sin\sqrt{0} + 4\sin\sqrt{0.125} + 2\sin\sqrt{0.25} + 4\sin\sqrt{0.375} + \sin\sqrt{0.5} \right)$$

$$\approx 0.221$$

Index

Notes

About the Author

Patrick Jones (also known as PatrickJMT) has a master's degree in Mathematics from the University of Louisville. Although he also did a bit of PhD work at Vanderbilt University, he decided that the academic life was not the correct path for him because he wanted to focus more on the teaching aspect instead of spending time on research. He has one mathematics paper published, for which he received Erdős number 2, and after that his research days came to an end.

In addition to teaching at the University of Louisville and Vanderbilt University, he has taught classes at Austin Community College.

While teaching, he also did some tutoring on the side and decided that he enjoyed the one-on-one aspect of tutoring over the stand-and-lecture mode of teaching. Because every student has questions outside of the classroom, he started posting short supplements for his students on YouTube under the name PatrickJMT. Although he thought only his students would watch the videos, positive feedback from all over the world came in, and he started posting more and more videos — so began the video-making career of PatrickJMT. Now Patrick primarily spends his time expanding his video library, although he does regular consulting work for many other projects as well.

In addition to being one very happy guy professionally, he is an avid cyclist and chess player, although he doesn't claim to be very good at either activity.

Author's Acknowledgments

I would like to thank Matt Wagner of Fresh Books, Inc., for approaching me with the offer of writing this book; Lindsay Lefevere, who was my initial contact at Wiley and got the ball rolling; Chrissy Guthrie at Wiley for being such an easy and accommodating person to work with on this project; and Mary Jane Sterling, who was my mentor for the project and gave much-appreciated and encouraging feedback early on. In addition, I would like to give an absolute booming shout out of gratitude to Jeane Wenzel, who typed up many, many pages of messy handwritten notes and did a fabulous job of proofing the work both mathematically and grammatically.

Publisher's Acknowledgments

Executive Editor: Lindsay Sandman Lefevere

Senior Project Editor: Christina Guthrie

Senior Copy Editor: Danielle Voirol

Technical Editors: Julie Dilday, Scott Parsell

Project Coordinator: Patrick Redmond

Project Managers: Laura Moss-Hollister, Jay Kern

Cover Image: ©iStockphoto.com/xiaoke ma

Math & Science

Algebra I For Dummies,
2nd Edition
978-0-470-55964-2

Anatomy and Physiology
For Dummies, 2nd Edition
978-0-470-92326-9

Astronomy For Dummies,
3rd Edition
978-1-118-37697-3

Biology For Dummies,
2nd Edition
978-0-470-59875-7

Chemistry For Dummies,
2nd Edition
978-1-118-00730-3

1001 Algebra II Practice
Problems For Dummies
978-1-118-44662-1

Microsoft Office

Excel 2013 For Dummies
978-1-118-51012-4

Office 2013 All-in-One
For Dummies
978-1-118-51636-2

PowerPoint 2013
For Dummies
978-1-118-50253-2

Word 2013 For Dummies
978-1-118-49123-2

Music

Blues Harmonica
For Dummies
978-1-118-25269-7

Guitar For Dummies,
3rd Edition
978-1-118-11554-1

iPod & iTunes For Dummies,
10th Edition
978-1-118-50864-0

Programming

Beginning Programming
with C For Dummies
978-1-118-73763-7

Excel VBA Programming
For Dummies, 3rd Edition
978-1-118-49037-2

Java For Dummies,
6th Edition
978-1-118-40780-6

Religion & Inspiration

The Bible For Dummies
978-0-7645-5296-0

Buddhism For Dummies,
2nd Edition
978-1-118-02379-2

Catholicism For Dummies,
2nd Edition
978-1-118-07778-8

Self-Help & Relationships

Beating Sugar Addiction
For Dummies
978-1-118-54645-1

Meditation For Dummies,
3rd Edition
978-1-118-29144-3

Seniors

Laptops For Seniors
For Dummies, 3rd Edition
978-1-118-71105-7

Computers For Seniors
For Dummies, 3rd Edition
978-1-118-11553-4

iPad For Seniors
For Dummies, 6th Edition
978-1-118-72826-0

Social Security For Dummies
978-1-118-20573-0

Smartphones & Tablets

Android Phones
For Dummies, 2nd Edition
978-1-118-72030-1

Nexus Tablets For Dummies
978-1-118-77243-0

Samsung Galaxy S 4
For Dummies
978-1-118-64222-1

Samsung Galaxy Tabs
For Dummies
978-1-118-77294-2

Test Prep

ACT For Dummies,
5th Edition
978-1-118-01259-8

ASVAB For Dummies,
3rd Edition
978-0-470-63760-9

GRE For Dummies,
7th Edition
978-0-470-88921-3

Officer Candidate Tests
For Dummies
978-0-470-59876-4

Physician's Assistant Exam
For Dummies
978-1-118-11556-5

Series 7 Exam For Dummies
978-0-470-09932-2

Windows 8

Windows 8.1 All-in-One
For Dummies
978-1-118-82087-2

Windows 8.1 For Dummies
978-1-118-82121-3

Windows 8.1 For Dummies,
Book + DVD Bundle
978-1-118-82107-7

Available in print and e-book formats.

Available wherever books are sold. **For more information or to order direct visit www.dummies.com**

Take Dummies with you everywhere you go!

Whether you are excited about e-books, want more from the web, must have your mobile apps, or are swept up in social media, Dummies makes everything easier.

Leverage the Power

For Dummies is the global leader in the reference category and one of the most trusted and highly regarded brands in the world. No longer just focused on books, customers now have access to the For Dummies content they need in the format they want. Let us help you develop a solution that will fit your brand and help you connect with your customers.

Advertising & Sponsorships

Connect with an engaged audience on a powerful multimedia site, and position your message alongside expert how-to content.

Targeted ads • Video • Email marketing • Microsites • Sweepstakes sponsorship